Springer Series in

MATERIALS SCIENCE 48

Springer
Berlin
Heidelberg
New York
Barcelona
Hong Kong
London
Milan
Paris
Singapore
Tokyo

Springer Series in
MATERIALS SCIENCE

The Springer Series in Materials Science covers the complete spectrum of materials physics, including fundamental principles, physical properties, materials theory and design. Recognizing the increasing importance of materials science in future device technologies, the book titles in this series reflect the state-of-the-art in understanding and controlling the structure and properties of all important classes of materials.

27 **Physics of New Materials**
Editor: F. E. Fujita 2nd Edition

28 **Laser Ablation**
Principles and Applications
Editor: J. C. Miller

29 **Elements of Rapid Solidification**
Fundamentals and Applications
Editor: M. A. Otooni

30 **Process Technology for Semiconductor Lasers**
Crystal Growth and Microprocesses
By K. Iga and S. Kinoshita

31 **Nanostructures and Quantum Effects**
By H. Sakaki and H. Noge

32 **Nitride Semiconductors and Devices**
By H. Morkoç

33 **Supercarbon**
Synthesis, Properties and Applications
Editors: S. Yoshimura and R. P. H. Chang

34 **Computational Materials Design**
Editor: T. Saito

35 **Macromolecular Science and Engineering**
New Aspects
Editor: Y. Tanabe

36 **Ceramics**
Mechanical Properties, Failure Behaviour, Materials Selection
By D. Munz and T. Fett

37 **Technology and Applications of Amorphous Silicon**
Editor: R. A. Street

38 **Fullerene Polymers and Fullerene Polymer Composites**
Editors: P. C. Eklund and A. M. Rao

39 **Semiconducting Silicides**
Editor: V.E. Borisenko

40 **Reference Materials in Analytical Chemistry**
A Guide for Selection and Use
Editor: A. Zschunke

41 **Organic Electronic Materials**
Conjugated Polymers and Low-Molecular-Weight Organic Solids
Editors: R. Farchioni and G. Grosso

42 **Raman Scattering in Materials Science**
Editors: W. H. Weber and R. Merlin

43 **The Atomistic Nature of Crystal Growth**
By B. Mutaftschiev

44 **Thermodynamic Basis of Crystal Growth**
P–T–X Phase Equilibrium and Nonstoichiometry
By J.H. Greenberg

45 **Principles of Thermoelectrics**
Basics and New Materials Developments
By G.S. Nolas, J. Sharp, and H.J. Goldsmid

46 **Fundamental Aspects of Silicon Oxidation**
Editor: Y. J. Chabal

47 **Disorder and Order in Strongly Non-Stoichiometric Compounds**
Transition Metal Carbides, Nitrides and Oxides
By A.I. Gusev, A.A. Rempel, and A.J. Magerl

48 **The Glass Transition**
Relaxation Dynamics in Liquids and Disordered Materials
By E. Donth

Series homepage – http://www.springer.de/phys/books/ssms/

Volumes 1–26 are listed at the end of the book.

E. Donth

The Glass Transition

Relaxation Dynamics in Liquids and Disordered Materials

With 56 Figures and 11 Tables

Springer

Prof. Ernst-Joachim Donth
University of Halle
Department of Physics
06099 Halle (Saale)
Germany

Series Editors:

Prof. Alex Zunger
NREL
National Renewable Energy Laboratory
1617 Cole Boulevard
Golden Colorado 80401-3393, USA

Prof. Robert Hull
University of Virginia
Dept. of Materials Science and Engineering
Thornton Hall
Charlottesville, VA 22903-2442, USA

Prof. R. M. Osgood, Jr.
Microelectronics Science Laboratory
Department of Electrical Engineering
Columbia University
Seeley W. Mudd Building
New York, NY 10027, USA

Prof. H. Sakaki
Institute of Industrial Science
University of Tokyo
7-22-1 Roppongi, Minato-ku
Tokyo 106, Japan

ISSN 0933-033x

Library of Congress Cataloging-in-Publication Data applied for.

Die Deutsche Bibliothek - CIP-Einheitsaufnahme
Donth, Ernst-Joachim:
The glass transition: relaxation dynamics in liquids and disordered materials / Ernst-Joachim Donth. -
Berlin; Heidelberg; New York; Barcelona; Hong Kong; London; Milan; Paris; Singapore; Tokyo: Springer, 2001
(Springer series in materials science; Vol. 48)
(Physics and astronomy online library)

http://www.springer.de

ISBN 978-3-642-07519-3 ISBN 978-3-662-04365-3 (eBook)
DOI 10.1007/978-3-662-04365-3

Originally published by Springer-Verlag Berlin Heidelberg in 2001.

Softcover reprint of the hardcover 1st edition 2001

Typesetting: Camera-ready copy produced by the author using a Springer TeX macro package
Cover concept: eStudio Calamar Steinen
Cover production: *design & production* GmbH, Heidelberg

Printed on acid-free paper SPIN: 10797900 57/3141/göh 5 4 3 2 1 0

Preface

The glass transition is well known to glass makers or from the common experience of drying a used chewing gum. A liquid melt or a rubber becomes a solid glass when its temperature is lowered or a solvent is extracted without crystallization. There are also dynamic effects. The viscosity of a liquid is small at high temperatures but increases dramatically as cooling proceeds down to the glass temperature T_g. The increase is continuous and amounts to about fifteen orders of magnitude! The technical importance of the glass transition cannot be overestimated. A few examples will be presented in the Introduction. Most practical knowledge of the glass transition needed for glass or plastic technologies and applications is now readily available.

Where then is the problem? In the last few years, glass transition research has enormously intensified. We now have several hundred papers a year in expensive, top scientific journals. In a 1995 Science magazine ranking, the glass transition belongs to the six major physical quests, along with broken charges, physical input for low-dimensional geometry, measurement philosophy in quantum mechanics, coherent X-ray radiation for materials research, and applications of superconductivity. On the other hand, interested people outside the glass transition community have difficulty seeing exactly what the glass transition problems are. In addition, even insiders split into groups over which question could be the most important for slow dynamics in cold liquids. There is much misunderstanding and sometimes intolerance, even within the glass transition community itself. Some theorists think it may be impossible to invent a general theory of liquid dynamics because liquid structures are too varied – compare a metallic glass with a polymer such as polyvinylacetate – and because no small parameter for a Taylor series expansion could be detected. In spite of the multifarious structural peculiarities in liquids and disordered materials, we do observe a general relaxation dynamics, i.e., a general scenario for the glass transition.

I would not expect anybody to find one single decisive experiment or idea that could solve the problem. A good counterexample from 20th century physics, in which this is not the case, is superconductivity. There were two basic experiments: that of Kamerlingh and Onnes which found zero conductivity in 1904, and that of Meissner and Ochsenfeld which found zero induction in 1933; and there was one decisive idea, the Cooper pairs of the elegant

BCS theory 1956. Research has never ceased since that time, of course, but the basic problem was solved.

The history of the glass transition is quite different. In 1970 everything seemed clear. Viscosity increases with the shortage of free volume, and molecular cooperativity must help to save fluidity at low temperatures. The essential problems evolved later because it was not only details that were left to be solved, but an unexpected and previously hidden complexity.

(i) There is a mysterious crossover region of dynamic glass transition at medium viscosity.
(ii) The two processes separated there are distinct and independent: a nontrivial high-temperature process above and a cooperative process below the crossover. The molecular dynamics differs in warm and cold liquids.
(iii) Two fast picosecond processes were discovered in the angstrom and ten-angstrom scales, together with an ultraslow megasecond process in the thousand angstrom scale. These may be connected with the dynamic glass transition.
(iv) Despite much progress over the last decade, long-running issues have survived from glass and polymer physics that may depend on essentials of the glass transition: tunneling systems at temperatures near 1 kelvin, the mixed alkali effect in silicate glasses, the flow transition of polymers, and others.

The glass transition problem acquires a 'biological' level of complexity, whereby we combine poorly understood elements into a better understanding of a less well understood whole. Will this perhaps be a feature of 21st century physics, without elegant breakthroughs?

The central question was mentioned above: How can we get a general liquid dynamics, with a clear architecture in the Arrhenius diagram, from the multifarious liquid structures of the different substance classes? Is there any common medium, such as the spatio-temporal pattern of some dynamic heterogeneity? The central concept used in this book is a dynamic approach to molecular cooperativity for the cold liquid. The field used to combine unknowns is the Arrhenius diagram with traces for the different processes and with one central point, the crossover region.

The method in the first half of this book is a verbal description of the phenomena, reflecting history to a certain extent. When we have the choice, thermodynamic or phenomenological arguments will be preferred. Most experimental methods for glass transition research (e.g., linear response and dynamic scattering) consist in detecting and evaluating subsystem fluctuations. To be precise, we start with a definite model: the Glarum–Levy defect diffusion model. Although we do not have unequivocal evidence for this model, we use it as a vehicle for expressing ideas and speculation.

The presentation is inductive, i.e., generalizing from examples. In the present state of such a complex subject as the glass transition, the truth of elements increases not only through decisive experiments (these are scarce), theory, and computer simulation, but also through their ability to fit into a consistent scenario. Put simply, I describe the tested pictures. For clarity, repetitions are not always avoided. I describe many aspects of few facts, rather than the opposite, and I will concentrate upon general features of glass formers with moderate complexity. The particularities of certain substances, especially exceptional glass formers such as amorphous water, SiO_2, BeF_2, and proteins, will not be considered, apart from a few examples. This book therefore differs from a review. The culmination of the descriptive part is a preliminary physical picture of the main transition. A deductive scheme for a possible explanation of the cooperative process in the cold liquid is indicated in Appendix B.

The second half of this book is a compilation of theoretical concepts adapted to the glass transition (Chap. 3) and an attempt to present a theoretical explanation for the slowing down of each of the following processes: the high-temperature process, Johari–Goldstein process, cooperative process, Fischer modes, and structural relaxation below the glass temperature (Chap. 4).

The explanation of the cooperative process and Fischer modes contains elements of nonconventional thermodynamics. I believe it is legitimate to raise the question of nonconventional aspects after such an intensive period of research since 1970. The risk in using such features is that they may provide merely ad hoc explanations which, in the end, only explain the particular phenomenon in question. Chapter 3 makes a careful analysis of what should be explained before conventional statistical mechanics can be applied. I consider representative thermodynamic subsystems with temperature fluctuations, here discussed within the framework of the von Laue approach to thermodynamics, and the fluctuation–dissipation theorem, here discussed from the standpoint of the quantum mechanical measuring process. I think that the dynamic glass transition, especially near the crossover, is sensitive to such nonconventional features. This differs from the example provided by superconductivity, where a conventional explanation has been successful.

According to the French mathematician and philosopher Poincaré, a limited number of experimental facts can be explained by an unlimited number of theoretical models. But the idea is not to wait until an unlimited number of experimental facts has been collected before seeking the true explanation through one theoretical model.

I would like to thank the Deutsche Forschungsgemeinschaft DFG and the Fonds Chemische Industrie FCI for financial support.

I had useful and mostly controversial discussions with the following colleagues: H. Rötger (1975), S. Kästner (1977), W. Holzmüller (1978), D.R. Uhlmann (1982, 1990), C.T. Moynihan (1982, 1997), W. Vogel (Jena) (1983), G.M. Barteniev (1985), O.V. Mazurin (1985), E.W. Fischer (1985 ff.), H. Sillescu (1986 ff.), G.H. Michler (1986 ff.), V.P. Privalko (1988, 1998 ff.),

J. Krüger (1988, 1999), J. Jäckle (1990, 1997), D. Richter (1990 ff.), U. Buchenau (1990 ff.), K. Binder (1990 ff.), J.M. O'Reilly (1990), I.M. Hodge (1990, 1996), C.A. Angell (1990, 1999), H.W. Spiess (1990 ff.), F. Kremer (1990 ff.), M. Schulz (1992 ff.), M.D. Ediger (1994 ff.), S.R. Nagel (1994), W. Pechhold (1995), G. Heinrich (1995 ff.), G.B. McKenna (1996 ff.), E.A. Di-Marzio (1996), D.J. Plazek (1996 ff.), K.L. Ngai (1996), W. Götze (1996 ff.), M. Fuchs (1996 ff.), R. Richert (1996 ff.), J.M. Hutchinson (1997 ff.), Y.H. Jeong (1997 ff.), E. Rössler (1997 ff.), S Hunklinger (1997 ff.), W. Petry (1998), J. Wuttke (1998), P. Maass (1998), G.P. Johari (1998), R. Böhmer (1999), B. Roling (1999), B. Wunderlich (2000), K. Funke (2000), D. Fioretto (2000), N.O. Birge (2000), and A. Heuer (2000). I also thank colleagues from our former group in Merseburg, mainly C. Schick (now in Rostock), A. Schönhals (now in Berlin), and K. Schneider (now in Dresden), and from our group in Halle, mainly K. Schröter, M. Beiner, F. Garwe, S. Kahle, and Elke Hempel for many discussions that have gone to form the picture presented in the book, and K. Schröter, M. Beiner, H. Huth, and Elke Hempel for their help with literature, files, and tables.

I also thank Springer-Verlag in Heidelberg, especially Angela M. Lahee for her pleasant collaboration, and Stephen Lyle for improving the English.

Katrin Herfurt typeset the manuscript, produced the figures with uncommon skill, and never complained about my endless additions and revisions. I thank her. My wife, Jutta Donth, provided the love, support, and understanding without which this book would not have been completed.

E. Donth   Halle (Saale), April 2001

Symbols and Names for Traces in Arrhenius Diagrams

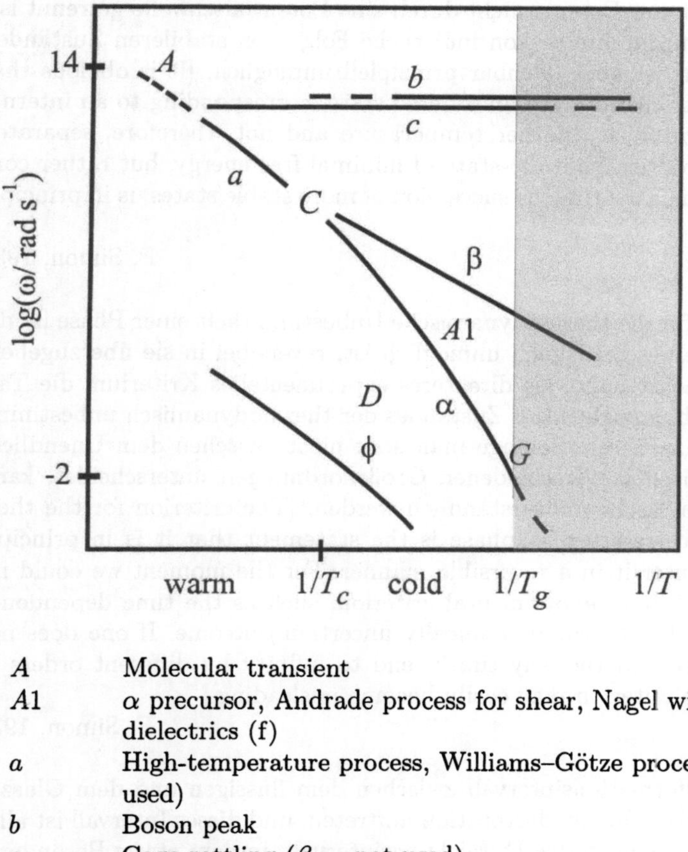

A	Molecular transient
$A1$	α precursor, Andrade process for shear, Nagel wing for dielectrics (f)
a	High-temperature process, Williams–Götze process (α or $\alpha\beta$ not used)
b	Boson peak
c	Cage rattling (β_{fast} not used)
C	Crossover region
D	Diffusion modes
G	Thermal glass transition, freezing-in
T_c	Crossover temperature
$T > T_c$	Warm liquid
$T < T_c$	Cold liquid
ϕ	Ultraslow modes (usm), Fischer modes
α	Cooperative process
$\{a,\alpha\}$	Dynamic glass transition, main transition
β	Local mode, Johari–Goldstein process (β_{slow} not used)

Gray shading represents the glass state and *white* the liquid state.

Quotations from the Classical Age of Glass Transition Research

Der isotherme Übergang in einen Ordnungszustand, der dem inneren thermischen Gleichgewicht einer anderen Temperatur entspricht und daher vom Zustand kleinster freier Energie nicht durch eine Potentialschwelle getrennt ist, sondern mit ihm durch eine kontinuierliche Folge von stabileren Zuständen zusammenhängt, ist aber offenbar prinzipiell unmöglich. [It is obvious that the isothermal transition into an ordered state, corresponding to an internal thermal equilibrium at another temperature and not, therefore, separated by a potential barrier from the state of minimal free energy, but rather connected with it via a continuous succession of more stable states, is in principle impossible.]

F. Simon, 1930

Das Kriterium für die thermodynamische Unbestimmtheit einer Phase ist die Feststellung, daß es prinzipiell unmöglich ist, reversibel in sie überzugehen. – Es läge zunächst nahe, als direkteres experimentelles Kriterium die Tatsache der Zeitabhängigkeit des Zustandes der thermodynamisch unbestimmten Gebilde einzuführen. Solange man aber nicht zwischen dem Unendlichwerden von Zeiten in verschiedenen Größenordnungen unterscheidet, kann dieses Kriterium leicht mißverständlich werden. [The criterion for the thermodynamic uncertainty of a phase is the statement that it is in principle impossible to enter it in a reversible manner. For the moment we could introduce a more direct experimental criterion, such as the time dependence of the state for the thermodynamically uncertain outcome. If one does not discriminate between the way times tend to infinity by different orders of magnitude, this criterion may easily become misleading.]

F. Simon, 1930

In einem Transformationsintervall zwischen dem flüssigen und dem Glaszustande muß also die Molekularrotation auftreten, und dieses Intervall ist aller Wahrscheinlichkeit nach das Erweichungsintervall, in dem etwas Besonderes vor sich geht, was allen Gläsern eigentümlich ist. [Molecular rotation must occur in a transformation interval between the liquid and the glass state, and this interval is probably the softening interval where something particular happens that is characteristic of all glasses.]

G. Tammann, 1933

Daher ist es wahrscheinlich, daß bei hinreichender Abkühlungsgeschwindigkeit alle oder doch die meisten Stoffe, wenn auch nur in geringen Mengen, in den Glaszustand übergeführt werden können. [It is therefore probable that, at the right cooling rate, all or at least most substances, even if in small amounts, can be transformed into the glass state.]

G. Tammann, 1933

Although the condition of an undercooled glass is not one of equilibrium at that temperature of that glass, it usually corresponds to one of equilibrium at some temperature within the annealing range .... This latter has been termed the equilibrium, or fictive, temperature of the glass ....

A.Q. Tool, 1946

... a glass is a liquid in which certain degrees of freedom characteristic of liquids are 'frozen in' and can no longer contribute to the specific heat and thermal expansion.

W. Kauzmann, 1948

Certainly it is unthinkable that the entropy of the liquid can ever be very much less than that of the solid.

W. Kauzmann, 1948

This treatment is quite independent of the nature of the relaxation spectrum and the time dependence of mechanical and electrical properties; it appears to be equally applicable to narrow and broad relaxation distributions.

M.L. Williams, R.F. Landel, J.D. Ferry, 1955

Стеклование жидкостей ... представляет собой релаксационный процесс и подчиняется кинетическим а не термодинамическим закономерностям ... при стекловании замораживается структура вещества соответствующая условиям при которых молекулярные перегруппировки становятся настолько медленнымие, что изменение структуры не успевает следовать за изменением внешних параметров .... [Vitrification of liquids ... turns out to be a relaxation process and obeys kinetic rather than thermodynamic laws .... At vitrification the structure of the substances freezes in under conditions where molecular rearrangements are so slow that a change of structure cannot keep up with changes in external parameters ....]

M.V. Vol'kenshtejn, O.B. Ptitsyn, 1956

М.В. Волькенштеин, О.Б. Птицын, 1956

For simplicity we shall suppose that when a defect reaches a dipole the latter relaxes completely and instantly.

S.H. Glarum, 1960

Le volume libre est défini par les divers auteurs de deux manières essentiellement différentes. [The free volume is defined in two essentially different ways by the various authors.]

A.J. Kovacs, 1963

We define a cooperatively rearranging region as a subsystem of the sample which, upon a sufficient fluctuation in energy (or, more correctly, enthalpy) can rearrange into another configuration independently of its environment.

G. Adam, J.H. Gibbs, 1965

It is appropriate to 'adopt a cautious view and assume only that the second-order transition temperature T_2 predicted by the quasi-lattice calculation is a convenient reference point on the bend of a curve of at present undetermined sharpness'.

J.H. Gibbs, E.A. DiMarzio, 1960, 1976

There are wide differences in the glassiness of the glassy state, both ... as a result of previous thermal history, and among polymers in general as a result of transitions lying below T_G.

R.F. Boyer, 1963

Thus, the ($\alpha\beta$) process was a distinct and separate process in this polymer ....

G. Williams, 1966

[On multiplicity of transitions in polymers] ... , there are more than just one glass transition, but rather a number of such glass-like transitions below and above $T_\mathbf{g}$.

M.C. Shen, A. Eisenberg, 1966

However, we will assume that a transition over a potential barrier in $\mathcal{U}$ space [potential energy as function of *all* the atomic coordinates] is in some sense 'local', in that in the rearrangement process leading from one minimum to a 'near-by' one, most atomic coordinates change very little, and only those in a small region of the substance change by appreciable amounts .... The potential barrier description of flow will have a range of validity at low temperatures, but gradually wash out as the temperature is raised and the liquid becomes very fluid.

M. Goldstein, 1969

The β-process can be envisaged as arising from *non-cooperative* rearrangement of some molecules engaged by large regions in which a stringent requirement of a cooperative motion has relatively fixed the orientation of a large number of molecules.

G.P. Johari, 1972

Glasses formed by different thermodynamic histories may have different thermodynamic properties at the same temperature and pressure. They thus may be regarded as different substances even though they have identical chemical composition.

J.E. McKinney, M. Goldstein, 1974

... the viscosity–temperature relations for several different substances, including weak and strong inorganic network liquids, ionic liquids, and two molecular liquids, are brought together by plotting viscosities on a reduced temperature scale, T_g/T, ...

C.A. Angell, W. Sichina, 1975

Jeder reale Glasübergang liegt in gewisser Weise zwischen dem idealen Glasübergang und einem Arrheniusmechanismus. [Any real glass transition lies, in a way, somewhere between the ideal glass transition and an Arrhenius mechanism.]

E.D. Schatzhauser, 1979

Contents

1. Introduction

This chapter describes examples illustrating the relation between the glass transition and other areas of science and technology.

Consider the use of the glass transition made by glass makers. At high temperatures the glass melt is a viscous liquid. The viscosity increases with falling temperatures and a 'forming interval' is passed where the glass makers can work. This interval can be characterized by a viscosity range or by a corresponding temperature range. Further lowering the temperature, the glass melt becomes a solid, viz., glass. This continuous transition from a liquid to a glass is called the *glass transition* and is the subject of our book.

The glass transition is different from solidification by crystallization. The latter is a phase transition with a well defined thermodynamic transition temperature, the crystallization temperature. From a dynamic point of view the matter is not so simple. Crystal formation needs time for crystal nuclei of the new phase to form and time for crystals to grow.

Crystallization is also relevant for the glass industry. The aim is usually to avoid crystallization in order to get a clear optical glass. Other industries, dealing with ceramics or enamels, for instance, are interested in a manageable interplay between the glass and phase transitions. The situation is usually expressed in a *t t t diagram* (time, temperature, transition), where a quantity such as the viscosity can be compared with a crystallization rate after a suitable reduction of the two quantities to comparable time scales (Fig. 1.1). For case A shown in the figure, the crystallization rate in the forming interval is slow so that it is easy to get a clear glass by sufficiently quick action. For case B, the reciprocal crystallization rate at the 'nose' is comparable with the viscosity time scale in the forming interval so that a reasonable interplay can be organized. This diagram depends on parameters such as sample composition, external pressure, or to some extent, sample preparation. The technological objectives are optimized temperature–time programs (cooling, annealing, tempering) which may also be represented by lines in the *t t t* diagram.

Such diagrams can, in principle, also be used for the food industry. The temperature axis may be replaced or completed by a solvent percentage axis. In a certain simplification, food materials can be considered as systems of water-plasticized natural polymers. A waffle, for instance, corresponds to a partly crystallized glass produced by water extraction and heating.

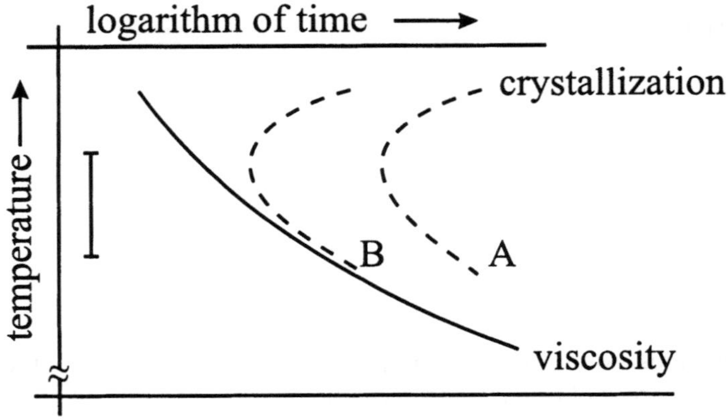

Fig. 1.1. The *ttt* diagram (time, temperature, transition) which compares times from viscosity with times from a suitably defined crystallization rate. The *bar* indicates the forming interval for viscosity

Consider further physicochemical or biological aspects of the glass transition, e.g., formation of a glassy matrix or freezing of biological materials at low temperatures. People involved in such activities used to find the glass transition by dynamic calorimetry with a given cooling or heating rate, $\dot{T} = \mathrm{d}T/\mathrm{d}t$, e.g., by DSC (Dynamic Scanning Calorimetry, see Fig. 1.2). The glass transition is indicated by a continuous step in heat capacity. This calorimetric glass transition is characterized by a typical temperature T_g, the *glass temperature*, by a step height ΔC_p, and by a surprisingly small transformation interval ΔT of order 10 K. T_g corresponds approximately to the solidification or freezing temperature that is defined via a standardized viscosity, e.g., $\eta = 10^{12.3}$ Pa s (pascal × seconds).

The glass state indicated by the thermogram is a nonequilibrium state. Waiting a day at a temperature five kelvins below T_g, we would find an equilibrated liquid with a 'static' heat capacity typical for the liquid state. Static means measured with very slow heating rates, smaller than a few kelvins per day. Correspondingly, the DSC thermograms for glass transitions are characterized by two peculiarities: the heating curve can be quite different from the cooling curve, and the entire phenomenon is shifted to higher temperatures when the temperature rates $\dot{T}$ are increased. For instance, an increase by one decade, e.g., from 1 K per minute to 10 K per minute, makes a positive T_g shift of 3–20 K. The actual shift depends on the substance. This dynamic aspect of the glass transition resembles the viscosity curve in the *ttt* diagram.

The viscosity curve in the *ttt* diagram is associated with the term 'dynamic glass transition'. The solidification at T_g is then called the 'thermal glass transition'. Above T_g the glass transition is a dynamic phenomenon in the equilibrium liquid. We learn from the *ttt* diagram that we have 'slow

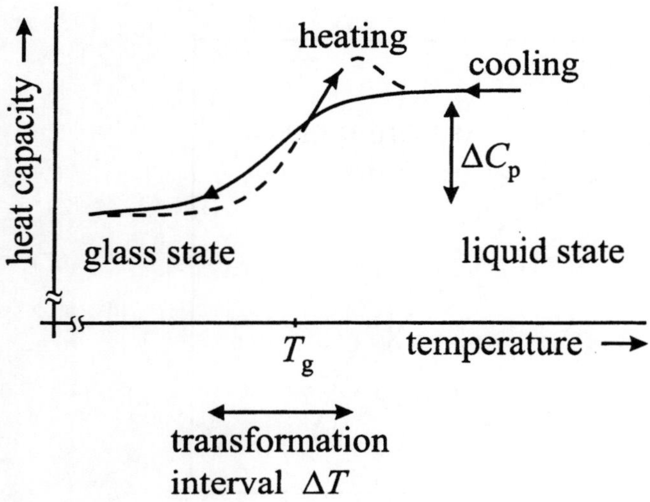

Fig. 1.2. Thermogram near the glass temperature T_g

dynamics' near T_g that becomes faster and faster at higher temperatures. The lower branch (mobility branch) of the nose is a spatiotemporal interplay of the crystallization (or another phase transition) curve with the dynamic glass transition. The upper branch (thermodynamic branch) of the nose is dominated by thermodynamic driving forces such as differences in chemical potentials. The dynamic glass transition is here reflected by global values of transport coefficients at the higher temperatures.

The dynamic glass transition may be of interest in itself, not only as a high-temperature precursor for the thermal glass transition. For example, it is useful in understanding why we need different car tires for summer and winter (Fig. 1.3). The transition can be represented by a typical relaxation time, on a logarithmic scale, as a function of reciprocal temperature. This gives a trace in an Arrhenius diagram, $\log t$ vs. $1/T$, or $\log \nu$ vs. $1/T$, where ν stands for frequency, a reciprocal time. The trace is fixed for a given rubber. On the other hand, we have three interesting regions in this diagram that are more or less defined by the use of the car. R stands for rolling resistance due to the periodic deformation of the tire. Its frequency is given by car velocity divided by wheel diameter. S stands for skid resistance. Its frequency is given by the car velocity devided by a typical length associated with road surface roughness (between micro and millimeter). Typical frequencies are $\nu_R = 100$ hertz for R and $\nu_S = 10^5$ hertz for S, i.e., $\nu_S > \nu_R$. The relevant temperatures depend on the place where heat is produced: near the tire surface for S and in the bulk for R, i.e., $T_R < T_S$. The third region T will be considered later.

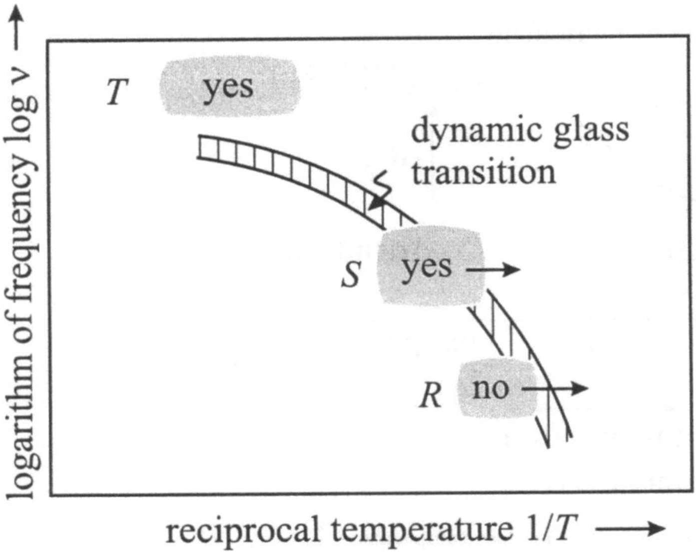

Fig. 1.3. Summer and winter tires. The trace is a stripe of large mechanical loss for the dynamic glass transition of a tire rubber in the Arrhenius diagram (schematic). R, S, T are regions of interest for the materials scientist in tire development. *Horizontal arrows* indicate the shift from summer to winter. The shift of the S region should be followed by the trace. This means changing the tires. *Yes* and *no* indicate whether the trace should cross the region. See text for details

The temperatures of both the R and S regions depend on the season. For lower tire temperatures, both regions shift to the right in the Arrhenius diagram. The problem arises from the fact that, for a given tire, the position of the trace does not depend on the season. It is a material property which does not shift: decreasing the temperature, we move along the given trace to smaller molecular mobility, $\log \nu$.

Materials science tries to find optimal rubbers for tires, in this case, with an optimal position for the trace of the dynamic glass transition in the Arrhenius diagram. We simplify the discussion by the assumption that only one sort of rubber is used for all components of a given tire. The effects of large strain amplitudes and of rubber-filler compounds are also ignored. The trace of the dynamic glass transition is accompanied by a stripe of high loss, i.e., transition from work to heat. This is desirable for S – yes, it prevents the tires from skidding – but is not desirable for R – no, periodic deformation leads to a loss that must be paid at the filling station. This means that the trace of the dynamic glass transition in Fig. 1.3 must be brought into a suitable relation with the S and R regions. Exactly which relation is influenced by considerations such as ecology and petrol prices. The conflict between yes for S and no for R arises because the trace cannot be arbitrarily deformed.

The dynamic glass transition has its own physical laws. The possibilities for distributing the loss intensity across the stripe are also limited.

As the S region shifts to the right in the winter the trace should follow. Since this is not possible for a given rubber, we need another rubber for the winter. Winter tires have a newly optimized trace in relation to the winter positions of R and S. The simplest requirement is that the glass temperature T_g should be lower by about twenty kelvins.

The position of the tire trace in the high-frequency region (e.g., $\nu = 10^{10}$ or 10^{11} Hz) is also of immediate interest. The T region of Fig. 1.3 refers to tear strength or tear resistance. Its position in the Arrhenius diagram is given by molecular parameters for frequency and an effective (high) temperature at the crack tip. It is desirable to have the trace of the dynamic glass transition near or inside the T region because the enhanced loss can transform work to heat, thereby heating rather than cracking the rubber. Enhanced loss can then help to avoid tearing. For a good tire, the trace is at high frequency at the T-region temperature. This can be achieved by using an optimal isomer composition for the rubber, for example.

For rubber and certain other materials, the state above the glass temperature T_g is not a true liquid, i.e., not a liquid which could really flow. This is prevented by a polymer network, i.e., chemical or physical crosslinking between macromolecular chains. Such materials are sometimes said to be soft. Increasing crosslink density shifts the glass temperature to higher values. An instructive example is the curing of an epoxy. Before, with zero or low crosslink density, the epoxy is a soft material. After curing, i.e., after the glass temperature goes above room temperature, the epoxy is a solid material, a glass.

The material below T_g is not completely frozen in. Examples attesting to some remaining molecular mobility are secondary relaxations at low temperatures that may mediate a certain compliance of the material, structural relaxation that may change the material properties with time, or ionic conductivity that may be essentially higher than expected from viscosities extrapolated to low temperatures $T < T_g$. As a rule, a glass below T_g is not a dead material, and not always a brittle material. Materials scientists know many methods (e.g., partial crystallization, stretching, mixing, compounding, and so on) for tailoring desired material properties below and above T_g.

An example of how the dynamic glass transition can contribute to material toughness below T_g is crazing of polymers. A craze is a precrack with highly stretched material fibrils within. This stretching shows that a certain flow of the material is also possible below T_g (Fig. 1.4). Consider an external impact on the sample. Characterize the three-axial stress state at the craze tip by a negative pressure and assume that the ability to flow is connected with crossing the glass transition curve. We are then interested in the pressure (P) dependence of the glass temperature, $T_g = T_g(P)$, not only for positive ($P > 0$), but also for negative ($P < 0$) pressures. Glass transition experience and

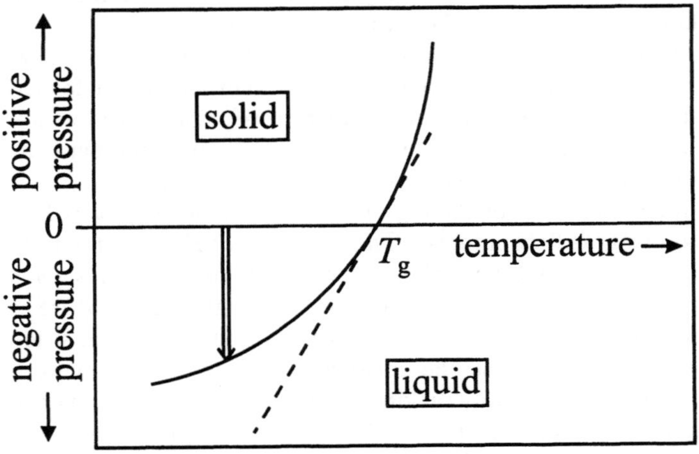

Fig. 1.4. Pressure dependence of glass temperature $T_g(P)$. The *double arrow* maps an external impact into this graph and can explain schematically why, in crazing materials, some toughness is also obtained far below the normal glass temperature T_g. Toughness will be connected with the material ability to flow and to dissipate the impact energy, both properties of a material near the dynamic glass transition

theory indicate a hyperbolic form for the $T_g(P)$ graph (Fig. 1.4). The absolute stress values at low temperature are much smaller than expected from a linear extrapolation of the dT_g/dP slope near normal pressure, i.e., near the standard glass temperature T_g. This means that far below T_g surprisingly small values of stress are required to reach the glass transition. The softening associated with the transition may initiate a craze or, more generally, may initiate a local toughness of the material.

The pressure and composition dependence of the glass transition is also important in geology. It may be decisive, for instance, in determining the kind of volcanic eruption that will take place, depending on whether the magma in the chimney can flow (i.e., it is well above the glass temperature of a given magma composition for the corresponding pressure distribution) or cannot flow (i.e., it is below the glass transition).

The above examples show how external and internal parameters affect the glass transition. Increasing the frequency or shortening the time intervals always yields a larger dynamic glass temperature. Usually, larger pressure, densification, extraction of solvent (deplasticizing), and chemical crosslinking (as for polymers) also raise T_g. As a physicochemical rule of thumb, greater molecular order (i.e., less disorder, as for folded lamellas in polyethylene, for example) and larger intermolecular energy (cohesion energy, as for chlorination of polyethylene, for example) raise the glass temperature.

The interrelation between the dynamic glass transition and protein folding is a more speculative example. Protein folding has several facets, including biomedical, biochemical, biophysical, and biomathematical aspects. Let us assume that essential features of protein folding can be grasped by polymer physics. We start from a model *ttt* diagram similar to Fig. 1.1, as for polyethylene crystallization. The macromolecular chains are folded into lamellas with a thickness of order ten nanometers. The folding is more irregular for thin lamellas, and more regular for thick lamellas. Apart from side groups and branching, polyethylene is homogeneous along the chain, so that chain flexibility is constant.

Proteins, however, are linear chains consisting of sequences of different amino acids. Flexibility is not constant along the chain. For example, it is particularly small near a proline unit. We know from synthetic polymers that chain flexibility provides a measure of the mobility of monomeric units. In polybutadienes, for instance, the molecular mobility of the cis isomer is ten decades higher than that of the trans isomer, i.e., the frequency ratio is 10^{10}. To avoid misunderstanding, both mobilities are related to the bulk material at the same temperature T_g of the low-mobility isomer. Considering also the various energetic interactions between different amino acids, with preferences for some specific (e.g., helical) local order, additionally modified by water molecules, helper proteins, chaperons, or enzymes in the neighborhood, we arrive at the complex and nanoheterogeneous folding pictures for proteins in the textbooks. In physical terms again, we get a (possibly multiple) nanophase separation for the folded state. Nevertheless, we will assume that, in principle, the folded or native state corresponds to the lamellas of polyethylene, and the unfolded or denatured state corresponds to the interpenetrating molecular coils of amorphous polyethylene.

Now for the differences. The glass transition has a characteristic length of a few nanometers. It can therefore serve as a nanoprobe for the nanometer heterogeneity of energy and mobility in the folded protein. Consider first the upper branch of the *ttt* diagram nose (Fig. 1.5). This branch is dominated by thermodynamic preferences. Molecular mobility is high and operates only by means of general transport properties. Diffusion is then no problem so that specific environments can energetically 'condense' to local helices or other structural units. The number of effective degrees of freedom in this energetically driven prestructure is strongly reduced by the larger units. Near the *ttt* nose, the stability of units may be further increased by 'immobile' proline segments, and so on.

Now consider the lower branch of the *ttt* diagram. Folding polyethylene is only confronted with a homogeneous increasing viscosity on the melt side so that the homogeneous lamella structure will not be changed. After folding, we have low mobility with just one high T_g. A homogeneous increase in viscosity is also expected for the denatured protein state. We expect only a small stripe for its dynamic glass transition in the *ttt* diagram. In prestructured proteins, however, with environments of different energy and disorder, mobility is locally regulated by their different dynamic glass transitions. Put

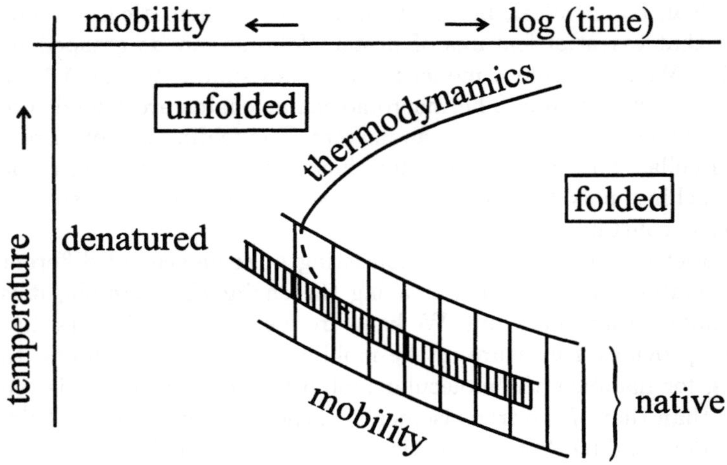

Fig. 1.5. Spatiotemporal interplay of the glass transition and folding nanoheterogeneities (nanophase separation) in the mobility branch of the *ttt* diagram for protein folding (schematic)

simply, the heterogeneity of energy and disorder is reflected in a heterogeneity of T_gs, i.e., a broad distribution of glass temperatures. (This T_g heterogeneity should not be confused with the dynamic heterogeneity of a uniform glass transition, with just one T_g, which plays an important role in other parts of this book. Nanophase separation is a simplified picture: each nanophase is assumed to have its own glass temperature, although a nanophase can only be understood by a picture combining inter/intramolecular interactions and thermodynamics.) During folding, the protein is confronted with the enormous heterogeneity in molecular mobility, expressed by the large variation in T_g. Since the spatial scales of protein heterogeneity and of the glass transition lie in the same range, the later stages of folding involve a long-term interplay between disorder and mobility heterogeneity. Small spatial scales lead to glass transitions in confining geometries, with confining effects that change the mobility, and surface effects that change the order. These effects generate an additional cooperativity between the subsystems and further complicate the process.

Let us consider one last example. The absence of a glass transition can also have technological consequences. The ceramic silicon boron nitride $Si_3B_3N_7$, for instance, has a thermal degradation temperature of about $T = 1700°C$. The absence of crystallization below this temperature indicates the absence of sufficiently extended atom mobility, which is usually initiated by the glass transition. We therefore expect a glass temperature T_g well above 1700°C. Technology gives stable amorphous materials. Only local relaxations can be used in the material, involving precursor fragments such as local BN_3 and

SiN_4 crystal fragments. We will call such a ceramic an amorphous solid, rather than a glass, because it is not produced via a glass transition.

All these examples are intended to show that the glass transition is of general interest in natural science, materials science, and the life sciences. Analysis of such complex situations requires a reliable knowledge of the glass transition. The aim of our book is a physical description of the phenomenon, including attempts to achieve a certain molecular understanding. The main problem is that no generally accepted microscopic theory is so far available. Hence all pictures remain vague to a certain degree, and we must use general principles for the description.

Firstly, our description is based on a view from the liquid side, in order to embrace the dynamic glass transition far above T_g. Apart from some picosecond/terahertz processes, the material is in a state that cannot be understood from the solid standpoint. Moreover, the phenomena that will be described are typical for all liquids and disordered materials. I believe that the molecular dynamics in liquids is synonymous with the dynamic glass transition.

Secondly, we shall collect the somewhat scarce information about typical ('characteristic') microscopic lengths for the glass transition, and try to combine it with the large body of information concerning various activities in the time or frequency domain. We shall attempt to construct a closed spatiotemporal picture for the main processes of the glass transition.

Essential problems arise from the fact that the characteristic length of the glass transition is small. As mentioned above, it is in the range of a few nanometers for the dynamic glass transition near T_g and decreases for higher temperatures. On the one hand, the phenomenology of slow dynamics is similar in different substances. On the other hand, everyone expects the molecular individuality of such varied substances as a silicate glass, a polymer, a metallic glass, or a protein to become apparent for any phenomenon at such small lengths. I try to concentrate on the similarities, and to indicate where individuality shines through. Although interesting, a discussion of the individuality of different substances is outside the scope of this book.

One task would be to find a bridge between the general dynamics and the multifarious structures, to construct a dense enough medium to overcome excessive individualism down to the scale of a few nanometers. Two suggestions follow this line. Firstly, to apply the concept of dynamic heterogeneity, e.g., constructing a fluctuating spatiotemporal mobility field $\log \omega(\boldsymbol{r}, t)$. Secondly, to use the generalizing power of thermodynamics, e.g., inventing suitable order parameters as spatiotemporal fields $\chi(\boldsymbol{r}, t)$ or as hidden entities, e.g., in a replica technique.

I hope to contribute to lowering the language barriers between people of different disciplines when they debate about the glass transition. This concerns also the barriers between different schools of professional glass-transition research carried out by physicists, physico-chemists, and engineers. A general communication problems is related to time. Two and, to some extent, even three different time concepts must be used simultaneously: a logarithmic time measure is equivalent to state changes, and as a result of the continuous slowing down from micro to kiloseconds below the crossover region, there is no 'typical time' for the dynamic glass transition. From experience, it seems rather difficult to agree over basic definitions because we are used to thinking and feeling in only one linear time, counted by well defined units.

More or less complete representations of the salient facts of the glass transition can be found in several books [1–10] or in chapters of more general books [11–22] often restricted to special substance classes, such as silicate glasses or polymers. There are also accessible review papers [23–55] and reports on international conferences devoted to the glass transition [56–63,704].

2. Description of the Phenomenon

The glass transition is a complex phenomenon burdened with many substantial misunderstandings. The description in Chap. 2 consists of experimental facts, selected concepts, and terminology, and is divided into five sections. The classical physicochemical picture achieved in about 1970 (Sect. 2.1) seemed neither complicated nor problematic, but the following decades have produced many serious problems and puzzles (Sect. 2.2). The high-temperature process of the dynamic glass transition, however, was still thought to be simple. Surprisingly, both experiment and theory also proved the contrary in the late 1980s (Sect. 2.3). The present state of affairs may be characterized by some long-running issues (Sect. 2.4) that are still awaiting a complete solution. Chapter 2 concludes with a suggestion for a seemingly simple molecular picture for All That (Sect. 2.5).

2.1 The Classical Picture. No Serious Problems

2.1.1 Viscosity

In all substances with a certain molecular disorder the velocity of thermal fluctuations dramatically slows down with falling temperatures. Assuming that no phase transition intervenes that would kill or change the disorder, we then have, for the present, no typical times. Instead the fluctuation time τ increases continuously from 10^{-12} s (picoseconds) to 10^3 s (about 20 minutes), i.e., fifteen orders of magnitude. This variation is out of all proportion with concomitant changes in density or structure. If the fluctuation time τ arrives at a typical experimental time, e.g., 20 minutes, the substance vitrifies when cooling is continued. A disordered material formed in this way is called a *glass*.

Well-known examples are glass-forming liquids or melts. A glass is then an amorphous solid formed from the melt by cooling to rigidity without crystallization. Rigidity means zero or weak steady response to a permanent shear stress. In the liquid we observe such a response: the viscosity η. To a certain approximation, it is connected with the fluctuation time τ by the mechanical Maxwell equation,

$$\eta \approx G_g \tau , \tag{2.1}$$

where G_g is a glass modulus of order 10^9 to 10^{12} Pa. The increase in fluctuation time is reflected by an increase in viscosity (Fig. 2.1a).

Without crystallization, the solidification of glasses is gradual, with no discontinuity. For ordinary glasses, we pass a pouring interval (10^1–10^3 Pa s, where Pa s is pascal second, 0.1 Pa s = 1 dPa s = 1 Poise), a forming or working interval, important for glass makers (10^5–10^8 Pa s), and other intervals [17, 19, 22]. Vitrification is characterized by a transition point (normalized at $\eta = 10^{12.3}$ Pa s, for example) whose temperature is called T_g, the *glass temperature*. A detailed discussion of T_g is postponed until Sect. 2.1.4.

The mystery of glasses is thus hidden in the liquid. What happens in the liquid to make thermal fluctuations slow down dramatically by 15 orders of magnitude, and what happens to the fluctuations at and below their freezing-in at T_g?

The alternative to our 'view from the liquid' is a 'view from the solid'. The latter is outside the scope of our book. One might try to understand an amorphous solid from a heavy distortion of a crystal, resulting, for example, in a 'paracrystal' [64]; from too high a density of disclinations or dislocations [65]; from a structural mismatch originating from a map of regular structures on small-sized compact, topologically nontrivial spaces on the common space [66, 67]. One might try to understand the picosecond vibrations and relaxations in amorphous materials from a modification of solid state dynamics, e.g., soft modes [68]. An energetic view of slow dynamics is useless as can be seen from an estimate via the equation $E = \hbar\omega = k_B T$, giving the orders $E = 10^{-15}$ eV and $T = 10^{-11}$ K for $\omega = 1$ rad/s.

Let us continue with liquids. Important for glass makers is the temperature interval ΔT corresponding to the working interval of viscosity shown in Fig. 2.1b. (This ΔT should not be confused with the transformation interval in Fig. 1.2.) For a given cooling rate $\dot{T} = \mathrm{d}T/\mathrm{d}t$, they have a lot of time for glasses with large ΔT. These are the long glasses, and the others the short glasses.

Having a three-parameter equation for the $\log \eta$–T relation, including both the working interval and T_g, our property can be simply characterized by one parameter [69] that may then be determined from properties near T_g. This parameter was called *fragility* by Angell [70, 71]. There is much misunderstanding outside the glass transition community because this concept has, in general, nothing to do with fragility of glasses in the common sense, i.e., breaking easily, or being brittle. Instead, Angell's pristine interpretation was related to molecular pictures for dynamics or thermodynamics in the liquid. Briefly [72, 73], the fragility of the α process is considered as a measure for generation of configurational entropy in the liquid [see (2.29) in Sect. 2.1.3].

A widely used three-parameter equation is the VFT equation (Vogel–Fulcher–Tammann) [74–76],

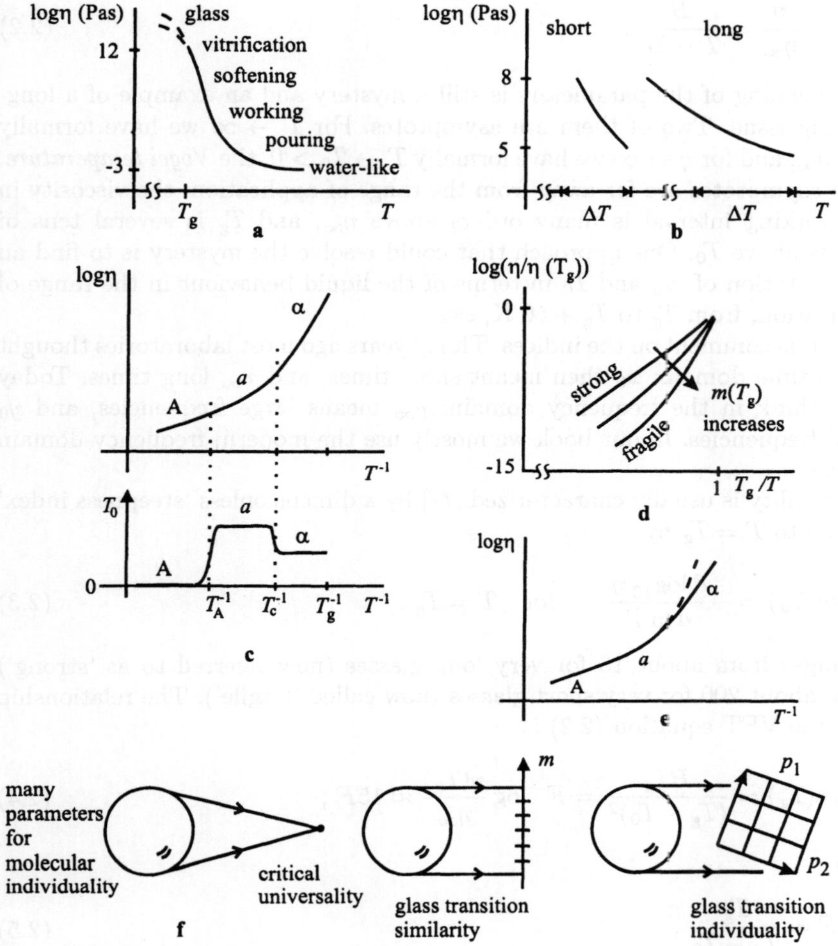

Fig. 2.1. Temperature dependence of viscosity η. (**a**) Continuity of glass solidification. The properties of a glass below vitrification depend on the cooling rate near T_g. (**b**) Short and long glasses. (**c**) *Top*: partition of the continuity in an Arrhenius plot. A is the molecular transient, a the high temperature process, and α the cooperative process. *Bottom*: Vogel temperature T_0 for the three parts. (**d**) Angell plot. Both η and T are reduced to 1 at the glass temperature T_g. (**e**) Arrhenius-like behavior of the α process, $T_0(\alpha) = 0$, is sometimes observed. The prefactor η_∞ in the VFT equation (2.2) is then much smaller than for any real liquid. (**f**) Homomorphism from the many parameters of molecular individuality to no parameter (universality), to one parameter m (similarity), or to two parameters $\{p_1, p_2\}$ (individuality of glass transition behavior)

$$\log \frac{\eta}{\eta_\infty} = \frac{B}{T - T_0} \,. \tag{2.2}$$

The meaning of the parameters is still a mystery and an example of a long-running issue. Two of them are asymptotes. For $T \to \infty$ we have formally $\eta \to \eta_\infty$, and for $\eta \to \infty$ we have formally $T \to T_0 > 0$, the *Vogel temperature*. Both asymptotes are far away from the range of application: the viscosity in the working interval is many orders above η_∞, and T_g is several tens of kelvins above T_0. One approach that could resolve the mystery is to find an interpretation of η_∞ and T_0 in terms of the liquid behaviour in the range of application, from T_g to $T_g + 50$ K, say.

Let us comment on the indices. Thirty years ago most laboratories thought in the time domain: x_0 then meant short times, and x_∞ long times. Today most think in the frequency domain: y_∞ means large frequencies, and y_0 small frequencies. In our book we mostly use the modern, frequency-domain indices.

Fragility is usually characterized [77] by a dimensionless 'steepness index' related to $T = T_g$ by

$$m(T_g) = -\frac{\mathrm{d}\log_{10}\eta}{\mathrm{d}\ln T} \quad \text{for} \quad T = T_g \,. \tag{2.3}$$

It ranges from about 15 for very long glasses (now referred to as 'strong') up to about 200 for very short glasses (now called 'fragile'). The relationship with the VFT equation (2.2) is

$$m(T_g) = \frac{BT_g}{(T_g - T_0)^2} = F \cdot \log \frac{\eta(T_g)}{\eta_\infty} \approx 15F \,, \tag{2.4}$$

where

$$F = \frac{T_g}{T_g - T_0} \,, \tag{2.5}$$

another measure of fragility. Relating m to a current temperature T instead of T_g, we would have

$$m(T) = m(T_g) \frac{T}{T_g} \frac{(T_g - T_0)^2}{(T - T_0)^2} \,, \tag{2.6}$$

decreasing with increasing temperature. The actual steepness index $m(T)$ in the working interval is much smaller than $m(T_g)$. For general properties of VFT-type equations, we refer to Sect. 4.3, (4.70)–(4.89).

A partition of the viscosity–temperature behavior was observed (e.g., by Uhlmann et al.) irrespectively of the continuity of the $\log \eta$–$(1/T)$ graph and its first derivative (Fig. 2.1c) [4, 78–81]. We find, probably typically, three regions: an Arrhenius region at high temperatures corresponding to low viscosities, A with $T_0(A) = 0$, and two successive, more or less true

VFT regions with different parameters at low temperatures. The underlying fluctuations or relaxations of the former will be called A process, or 'molecular transient', and of the latter two 'high-temperature a process' and 'cooperative α process', respectively. As a rule,

$$T_0(A) = 0 < T_0(\alpha) < T_0(a) , \tag{2.7}$$

i.e., the Vogel temperatures are not monotonic with temperature (lower part of Fig. 2.1c). The Vogel temperatures for the α process are sometimes, i.e., for certain substances, surprisingly small, viz., $T_0(\alpha)/T_0(a) < 0.5$. In a few cases, the α process even tends to a low-temperature Arrhenius process [80, 82], $T_0(\alpha) \approx 0$. The dividing temperatures are T_A, Barlow's Arrhenius temperature, and T_c, in modern terms, the *crossover temperature*. As a rule, T_c corresponds to a viscosity somewhere in the range 10 to 10^6 Pa s, and the corresponding fluctuation times are around the microsecond range (see Table 2.10 in Sect. 2.5.4). The partition of the trace was later confirmed by careful linear-response methods [83–87]. The crossover is sometimes reflected by the equilibrium curve of heat capacity in the liquid state, $C_p(T)$ (Fig. 2.30 in Sect. 2.5), e.g., as a weak bend at T_c for the scenario I (Sect. 2.2.1). The liquid can thus be partitioned into a *warm liquid* ($T > T_c$) and a *cold liquid* ($T < T_c$). The crossover plays a central role in current discussions of the glass transition (Sects. 2.2.1, 2.3.2, 2.5.1–3).

The fluctuation time at the crossover temperature defines some kind of a typical time for the dynamic glass transition. However, the general problem survives. Below T_c, as well as above T_c, the fluctuation time changes continuously by many orders of magnitude, so that we cannot truly speak about a typical time, either for the α process or for the a process.

We thus have, in a reasonable or even good approximation, one VFT equation for the α process, and a different VFT equation for the a process. This implies the possibility of constructing one-parameter diagrams for the logarithm of reduced viscosity against reduced reciprocal temperature, for each of α and a. The one-parameter diagram for the α process [79, 88, 89] is now called the *Angell plot*, and the parameter is the fragility (Fig. 2.1d). If the steepness index is included in the reduction, collapsed plots are obtained (one line for different substances [90]). Of course, the regularity of such diagrams is violated outside the range of application, i.e., when $T < T_g$ and $T > T_c$ for the α process, and when $T < T_c$ and $T > T_A$ for the a process.

If we are not interested in the T_c partition, the whole $\{a, \alpha\}$ trace, or even the $\{A, a, \alpha\}$ trace, is called the *main transition* or dynamic glass transition in the wider sense.

This section will be closed with a general remark about three concepts: universality, similarity, and individuality. Let us first recall the geometry of plane triangles. Congruent triangles cannot be distinguished by any parameter, similar triangles can be distinguished by one parameter, a length, and all other triangles need two for equilateral ones or, in general, three parameters to describe their individuality.

A critical state is universal. A given universality class has a given set of critical exponents and, inside the class, there is no free parameter for them. This corresponds to the congruent triangles in geometry. The parametrization of the glass transition by fragility m would correspond to *similarity*. The glass transitions in different substances would be similar if all their properties could be parametrized by m. We shall see, however, that not all properties depend only on m. More than one parameter is sometimes needed. The question is: when is a glass transition in a given substance similar to that in others, and when do individualities shine through.

Usually, the molecules of glass formers are individual, so that several or many parameters are needed for their identification and their interaction. Thermodynamic equations of state have a tendency to reduce the number of relevant parameters; a similar tendency is expected for phenomenological equations of glass-transition dynamics. The map of molecular individuality to the universality of critical state with a given set of critical indices reduces the number of free parameters to zero (Fig. 2.1f). The map to the VFT equation for the viscosity of the α process reduces the number to one, and we have similarity with respect to the fragility m. The map to individual properties of the glass transition reduces the number to two $\{p_1, p_2\}$ or more. A map with such a reduction of dimensionality is called a homomorphism. We will sometimes use the term 'general' if we have two or three parameters of individuality, or if the number of parameters for glass-transition individuality is small compared with the number of parameters for molecular individuality in a larger assortment of glass-forming substances.

A further example of such a reduction is the fluctuation–dissipation theorem (FDT) of linear response. Three-, four-, or many-time correlations of mechanical molecular motions are reduced to two-time correlation functions of thermodynamic responses.

2.1.2 Linear Response

Consider an equilibrium liquid at a temperature T well above the glass temperature T_g so that the fluctuation time τ is several decades shorter than 1000 seconds, e.g., $\tau = 0.1$ s, frequency $\nu \approx 1/\tau = 10$ Hz, viscosity of order 10^8 Pa s. Consider a perturbation by a thermodynamic variable so small that the response is proportional to the perturbation amplitude (that is, a linear response), e.g., perturbing by $\Delta p(t)$ = pressure–time program gives an answer $\Delta V(t)$ = volume–time response. There are two basic experiments, namely the response after a step perturbation program which is discussed in the time domain, and the stationary response during a periodic experiment which is discussed in the frequency domain. The former is called a relaxation experiment, the latter a dynamic experiment. We mostly use the angular frequency $\omega = 2\pi\nu$, measured in radian per second [rad/s].

The linear response is completely determined by properties of the equilibrium liquid if thermal fluctuations are included in the equilibrium concept:

there is no response, no internal equilibrium, and no 'quasistatic' change of thermodynamic variables without the relevant thermal fluctuations. The fluctuation–dissipation theorem (FDT) of statistical physics implies the important statement that it is exclusively thermal fluctuations which determine the linear response. This statement is not trivial with respect to two points.

- Any linear response experiment uses an external disturbance and is therefore, irrespective of the smallness of its amplitude, connected with entropy production. The Second Law of thermodynamics implies irreversibility of the experiment. The FDT also holds for the variables temperature and entropy, $\Delta T(t)$ and $\Delta S(t)$ [91–93] (Sects. 3.2–3.4). On the other hand, thermal fluctuations are reversible.
- According to the FDT [94,95], the linear response is completely determined by the correlation function (time domain) or by the spectral density (frequency domain) of the relevant thermal fluctuation. The coupling between response and fluctuation is made by the general Boltzmann constant, and no specific coupling constant is needed. Related to a sample of millimole or mole order, the relative fluctuation amplitude is small, of order 10^{-10}, whereas the perturbation and response amplitudes of linear experiments are large, usually of order 10^{-2}, i.e., the experimental linear region is in the per cent range.

A necessary condition for linearity seems to be that the perturbation amplitude is smaller than, or at most of the same order as, the fluctuation amplitude of the smallest representative subsystems of the sample. The relative fluctuation amplitude decreases as the square root of the subsystem size. The $10^{-10}/10^{-2}$ ratio of the preceding paragraph can only be understood when the size of these representative subsystems is in the nanometer range.

Thermodynamically intensive and extensive variables have different linear responses (Fig. 2.2a, b). Intensive variables are temperature, pressure, and so on. They will be denoted by $\Delta f = \Delta T$, Δp, and so on. Extensive variables such as entropy, volume, etc., will be denoted by $\Delta x = \Delta S$, ΔV, and so on. In the time domain, an extensive response after an intensive step perturbation is characterized by a compliance $J(t)$ (details in Sects. 3.3 and 3.4). This is an increasing function of time. In the frequency domain, we have a dynamic compliance $J^* = J' - \mathrm{i}J''$, with a continuous downward step in J' as a function of $\log\omega$, $J'(\log\omega)$, and a peak in $J''(\log\omega)$, both at the dynamic glass transition. For the intensive response, we have a decreasing modulus $G(t)$ and a dynamic modulus $G^* = G' + \mathrm{i}G''$, with an upward step in $G'(\log\omega)$ and a peak in $G''(\log\omega)$. The relation

$$G^* \cdot J^* = 1 \tag{2.8}$$

is an implication from the equivalence of Figs. 2.2a and b. From the different time monotonicity, it follows that the G'' maximum is at higher frequencies than the J'' maximum.

The bottom graphs of Figs. 2.2a and b describe a situation with a relaxation contribution at much shorter times than within the apparatus window. The response observed in the time window is then smaller than the equilibrium change of variables, e.g., $\Delta f(\text{observed}) = \Delta f(\text{equilibrium}) - \Delta f(\text{instantaneous})$. The difference comes from the 'instantaneous response' at the shorter times. This picture will become important, e.g., for the vault effect (Sects. 2.2.6, 2.1.5) or dynamic neutron scattering (Sect. 2.3.1).

If the distinction between extensive and intensive variables is important, the response of extensive variables, e.g., $\Delta V(t)$, is called retardation, or sometimes recovery, and the response of intensive variables, e.g., $\Delta p(t)$, relaxation. If not, we will say 'relaxation', and compliance or modulus will be called 'susceptibility'. By convention, retardation and relaxation strengths, the step heights ΔJ and ΔG (Figs. 2.2a–b), are understood as amounts with no + or − sign: $\Delta J > 0$, $\Delta G > 0$.

The FDT implies that the fluctuation time for extensive variables is equal to the retardation time τ_R', and the fluctuation time for intensive variables is equal to the relaxation time τ_R,

$$\tau(\text{extensive}) = \tau_R' \quad , \quad \tau(\text{intensive}) = \tau_R \; . \tag{2.9}$$

From the different monotonicity, it follows that $\tau_R' > \tau_R$.

A distinction between *zone* and *state* is important for understanding the glass transition. In the linear region well above T_g, the $J^*(\log\omega)$ and $G^*(\log\omega)$ functions are completely determined by fluctuations in thermal equilibrium, i.e., the whole functions are equilibrium properties of the liquid. It is useful to partition any isothermal equilibrium section parallel to the

▶

Fig. 2.2. Linear response. (**a**) Retardation or creep experiment. Δf program, the perturbation is an intensive variable; Δx answer, the response is an extensive variable. $J(t)$ is the compliance, $J^* = J' - \mathrm{i}J''$ is the dynamic compliance as a function of $\log\omega$, ΔJ is the retardation strength, t the time, τ_R' the retardation time, ω the angular frequency, and inst. is the instantaneous response for relaxation at shorter times than the time window of the apparatus. (**b**) Relaxation experiment exchanges extensive and intensive. $G(t)$ modulus or relaxation function, $G^* = G' + \mathrm{i}G''$ dynamic modulus as a function of $\log\omega$, ΔG relaxation strength, τ_R relaxation time. (**c**) Partitioning an equilibrium isothermal $\log\omega$ section into zones. (**d**) Shear curves in the dispersion zone for the α process (glycerol, T = 192.5 K) [518]. $J_s^0 = 1.8 \times 10^{-9}$ Pa^{-1} (192.5 K) is the steady state compliance. $J_r'' = J'' - 1/\omega\eta$. (**e**) Loss susceptibilities (compliances or moduli) for four models: Debye exponential decay, Havriliak–Negami (HN) equation (2.14), Kohlrausch–Williams–Watts (KWW) equation (2.17) (more details in Figs. 3.7 and 3.8), logarithmic Gauss equation (2.22). (**f**) *Bold lines*: comparison of FWHM (at 0.5000) in base 10 logarithms with equivalent Gaussian dispersion in natural logarithms [exp(−1/2) ≈ 0.6065]. The two compared curves are pinned together at the half maximum value

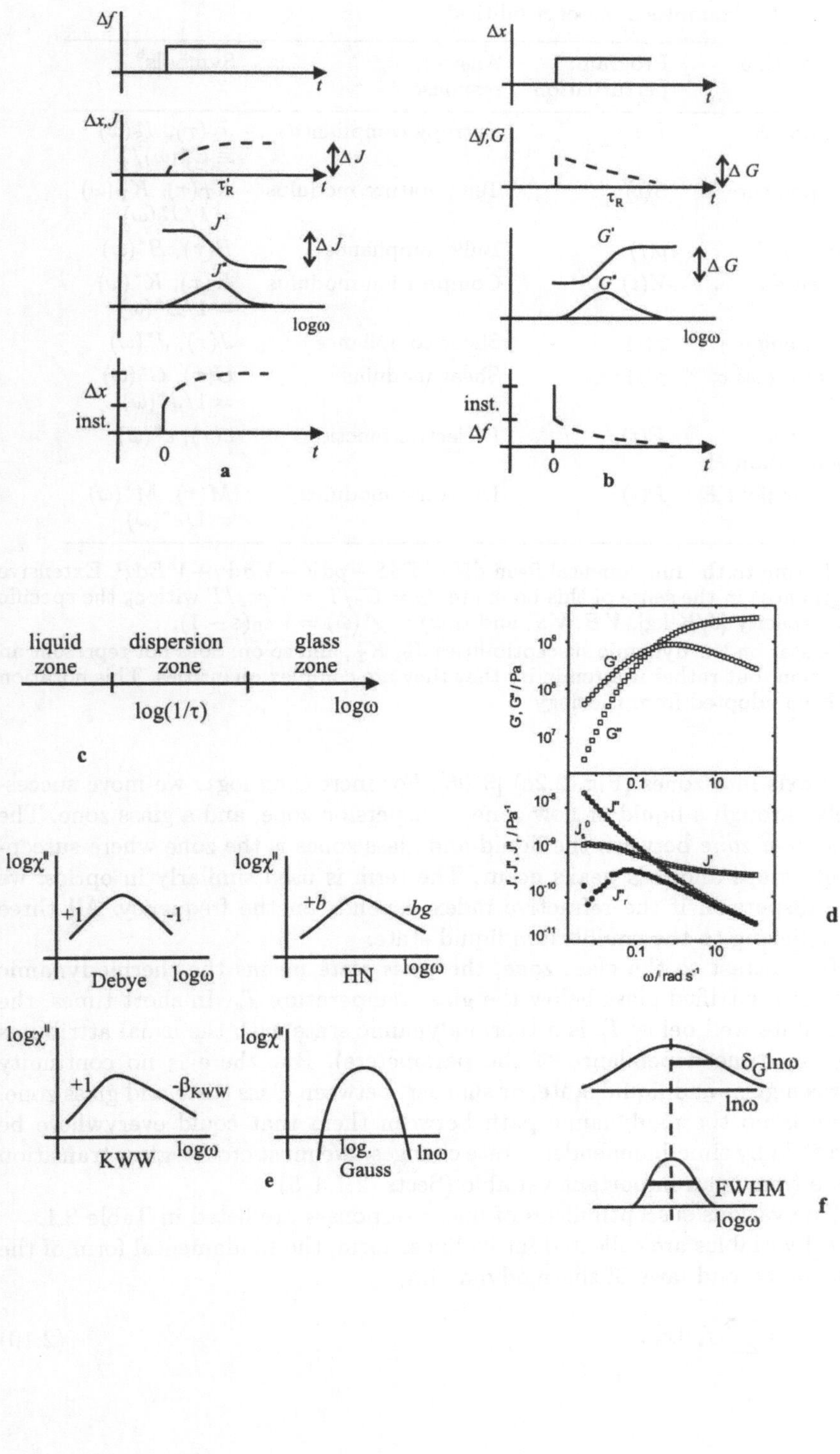
Δf
t
Δx,J
ΔJ
τ'R
J'
J"
logω
Δx
inst.
0
a
Δx
Δf,G
ΔG
τR
G'
G"
inst.
Δf
b
liquid zone
dispersion zone
glass zone
log(1/τ)
logω
c
G, G" / Pa
J', J", J'r / Pa⁻¹
ω/ rad s⁻¹
d
logχ"
+1
-1
Debye
+b
-bg
HN
-βKWW
KWW
log. Gauss
lnω
e
δGlnω
FWHM
f

Table 2.1. Examples of susceptibilities[a]

Fluctuation	Program, perturbation	Answer, response	Symbols[b]
Entropy S	$T(t)$	Entropy compliance	$J_S(\tau)$, $J_S^*(\omega)$ $\equiv C_p^*(\omega)/T$
Temperature T	$S(t)$	Temperature modulus	$K_T(\tau)$, $K_T^*(\omega)$ $= 1/J_S^*(\omega)$
Volume V	$p(t)$	Bulk compliance	$B(\tau)$, $B^*(\omega)$
Pressure p	$V(t)$	Compression modulus	$K(\tau)$, $K^*(\omega)$ $= 1/B^*(\omega)$
Shear angle γ	$\sigma(t)$	Shear compliance	$J(\tau)$, $J^*(\omega)$
Shear stress σ	$\gamma(t)$	Shear modulus	$G(\tau)$, $G^*(\omega)$ $= 1/J^*(\omega)$
Dielectric polarization P	$E(t)$	Dielectric function	$\varepsilon(\tau)$, $\varepsilon^*(\omega)$
Electric field E	$P(t)$	Dielectric modulus	$M(\tau)$, $M^*(\omega)$ $= 1/\varepsilon^*(\omega)$

[a] Referring to the fundamental form $\mathrm{d}U = T\mathrm{d}S - p\mathrm{d}V - V\sigma\mathrm{d}\gamma + VE\mathrm{d}P$. Extensive compliances in the sense of this book are $J_S = C_p/T = V\varrho c_p/T$ with c_p the specific heat capacity [J/K·kg], VB, VJ, and $\alpha(\omega) = \chi^*(\omega) = V\varepsilon_0(\varepsilon - 1)$.
[b] The star on the dynamic susceptibilities J_S^*, K_T^*, and so on, does not represent an operation, but rather is a reminder that they are complex quantities. This notation has been adopted from rheology.

$\log\omega$ axis into zones (Fig. 2.2c) [3, 96]. For increasing $\log\omega$ we move successively through a liquid or flow zone, a dispersion zone, and a glass zone. The *dispersion zone* between the liquid and glass zones is the zone where susceptibility steps and loss peaks occur. The term is used similarly in optics: we have dispersion if the refractive index depends on the frequency. All three zones belong to the equilibrium liquid state.

In contrast to the glass zone, the *glass state* means the thermodynamic state of a vitrified glass below the glass temperature T_g. In short times, the glass state well below T_g is a thermodynamic state with the usual attributes (e.g., no time dependence of the parameters). But there is no continuity between glass and liquid state, or sharper, between glass state and glass zone. There is no thermodynamic path between them that could everywhere be described by time-independent state changes. We must cross a glass transition where time is an important variable (Sects. 2.1.4–5).

The various susceptibilities of linear responses are listed in Table 2.1.

All variables are collected into a linear form, the fundamental form of the first and second laws of thermodynamics,

$$\mathrm{d}U = \sum_i f_i\,\mathrm{d}x_i\,. \tag{2.10}$$

The different terms in (2.10) will be called activities. We thus have a calorimetric ($T\mathrm{d}S$), a volumetric ($-p\mathrm{d}V$), a shear ($-V\sigma\mathrm{d}\gamma$), a dielectric ($VE\mathrm{d}P$), and sometimes other activities. Cross response is also included in the FDT [97]. For instance, the isobaric thermal expansion which is the volume response to a temperature program [98] is determined by the cross fluctuation of volume and entropy $\overline{\Delta V \Delta S}$.

Let us now consider two examples, shear activity and dielectric activity.

Example One. Shear Activity. Shear response in the dispersion zone between the glass and liquid zones of a true liquid differs from what is shown schematically in Figs. 2.2a, b. In the time domain, the shear modulus $G(t)$ tends to zero, and the shear compliance $J(t)$ tends to infinity for $t \to \infty$, with asymptote $J(t) = J_s^0 + t/\eta$. A step is only defined in the real part of J^*, $J'(\log \omega)$ (Fig. 2.2d). The glass zone is characterized by the glass compliance J_g, the transition zone is characterized by the step in J' (or, in logarithms, $\Delta \log_{10} J'$, usually in the 0.3 to 0.8 range), and the liquid zone is characterized by the *steady state compliance* J_s^0 (formerly J_e^0, usually in the 10^{-10} to 10^{-9}/Pa range) and the viscosity η:

$$J_s^0 = \lim_{\omega \ll 1/\tau} \frac{G'(\omega)}{G''^2(\omega)}, \quad \eta = \lim_{\omega \ll 1/\tau} \frac{G''(\omega)}{\omega}. \tag{2.11}$$

The limits in (2.11) are related to the liquid zone (called flow zone in rheology) where the liquid behaves in a Newtonian manner in the linear region,

$$G' \sim 1/\omega^2, \quad G'' \sim 1/\omega. \tag{2.12}$$

Both J_s^0 and η are generated by molecular correlation functions in the time scale of the dispersion zone (Green–Kubo integrals, Sect. 2.2.3).

We have two peaks in the dynamic shear susceptibilities, namely, G'' and $J_r = J'' - 1/\omega\eta$, as a function of frequency (Fig. 2.1d). In the dispersion zone, G' and G'' are of the same order of magnitude. This behavior is called viscoelastic. The fluctuations $\Delta\sigma^2$ in shear stress are intensive and relevant for the modulus G. The fluctuations $\Delta\gamma^2$ in shear angle are extensive and relevant for the compliance J. The symbols $\Delta\sigma^2$ and $\Delta\gamma^2$ mean $(\Delta\sigma)^2$ and $(\Delta\gamma)^2$, and so on.

The average increase in shear angle fluctuation across the dispersion zone of the dynamic glass transition can be calculated from the FDT [18] using

$$\left.\overline{\Delta\gamma^2}\right|_{\alpha \text{ or } a} = k_B T \frac{J_s^0 - J_g}{V_{\mathrm{ref}}}, \tag{2.13}$$

where V_{ref} is the reference volume and the subscripts α and a refer to the α and a processes, respectively. Taking a volume $V_{\mathrm{ref}} = V_\alpha$ of about 10 nm^3, a typical size for a cooperatively rearranging region CRR, the representative subsystem for the α process (Sect. 2.1.3), we get $\overline{\Delta\gamma^2}$ of order 10^{-3} rad^2, i.e., $\overline{\Delta\gamma}$ of order 1°. Extrapolating V_{ref} to the size of one molecule (0.1 nm^3),

Table 2.2. Steady state compliance J_s^0

Substance	T [K]	J_g [$(10^{10}$ Pa$)^{-1}$]	J_s^0 [$(10^{10}$ Pa$)^{-1}$]	$\Delta \log_{10} J'$	Reference
Glycerol	192	2.8	15	0.73	[518]
Salol	220	18	33	0.26	[100]
$Pd_{40}Ni_{40}P_{20}$	590	0.35	0.9	0.41	[101]
$Zr_{65}Al_{7.5}Cu_{17.5}Ni_{10}$	645	0.38	–	–	[101]
Soda-lime-glass	820	–	–	0.42	[102]
PS[a] ($M_w = 1100$)	313	8.3	39	0.67	[103]
PS ($M_w = 700000$)	402	9	35	0.59	[104, 105]
PMPS[b] ($M_w = 5000$)	223	5.4	28.2	0.72	[107]
OTP[c]	241	4.5	11.9	0.42	[106]
6-phenyl ether	248	7.2	17	0.37	[106]
Tri-α-napthylbenzene	342	9.5	28	0.47	[106]
Tri-cresyl phosphate	265	7.76	26	0.52	[106]
Aroclor#1248	223	4.25	10.3	0.38	[106]

[a] PS polystyrene.
[b] PMPS poly methylphenyl siloxane.
[c] OTP orthoterphenyl.

we get a rotational fluctuation of order 10°. Mean rotational jumps of order 10° [99] thus seem necessary to make the liquid flow. Such jumps are perhaps needed for a 'percolation' of rotational jumps of the molecules to generate a shear flow of the liquid. The dispersion zone itself, however, does not [512] seem to have the property of a percolation transition: there is no singularity. Instead, if any, the increase of $\overline{\Delta\gamma^2}$ from the glass zone to the flow zone seems to be a percolation precursor. The transition seems to be smeared or hidden by the flow onset. The steady shear compliance J_s^0 is more an indicator for a threshold of shear flow. We thus have some kind of Lindemann criterion for the flow zone threshold: the mean rotational jumps of molecules must be of order ten degrees.

Experimental values of J_s^0 and J_g for several liquids near T_g are collected in Table 2.2. The temperature dependence of J_s^0 is, compared to $\eta(T)$, not dramatic. J_s^0 increases with increasing temperature. Near T_g, the increase is of order 10% per 10 K.

Example Two. Dielectric Activity. The second example is dielectric spectroscopy. The curves obtained are similar to those in Fig. 2.2a and 2.2b. The compliance is usually expressed by the dielectric function $\varepsilon(t)$ or $\varepsilon^*(\omega)$. The compliance itself is called the dielectric susceptibility $\chi^*(\omega) = V\varepsilon_0[\varepsilon^*(\omega) - 1]$. The dielectric modulus is usually denoted by $M^*(\omega) = 1/\varepsilon^*(\omega)$. The fluctuations that determine linear dielectric spectroscopy are

ΔP^2 for ε and ΔE^2 for M, where P is the dielectric polarization and E an internal electric field for representative subsystems.

The standard interpretation comes from Debye [108] and Tammann [1]. For $\log\omega$ in the liquid zone, i.e., below the dispersion zone, the dipoles can follow the slow periodic perturbations of the external field. The liquid zone has a large spectral density of polarization P fluctuations, and a large compliance. For $\log\omega$ in the glass zone, i.e., above the dispersion zone, the dipoles cannot follow the fast perturbation because their fluctuation relevant for the dynamic glass transition is too slow. The glass zone therefore has a small spectral density of P fluctuations and a small compliance. In the dispersion zone, the dipoles respond to external fields with the same rate as to a fictive, fluctuating internal field that may be imagined to cause the actual fluctuations in the polarization P. The relevant fluctuations are said to be 'functional' to the glass transition. The dielectric intensity $\Delta\varepsilon_\alpha$ = step height at the α process is usually smaller than the switching on of fluctuations of the given number of dipoles when they are assumed to be statistically independent. The cooperative nature of the α process takes advantage of a certain vector compensation of neighboring dipoles.

For substances with electrical conductivity, the relevant dynamics is expressed by a dynamic conductivity $\sigma'(\omega) = \omega\varepsilon_0\varepsilon''(\omega)$. For Ohmic behavior, $\sigma' = \sigma'(0) =$ const. and, therefore, $\varepsilon'' \sim \omega^{-1}$. A dispersion zone for conductivity is indicated by the Jonscher exponent p, $\sigma'(\omega) - \sigma'(0) \sim \omega^p$, $0 < p < 1$ [109]. Two particularities are worth mentioning (details in Sect. 2.2.3). Firstly, the validity of the Jonscher power law is often restricted to a small $\log\omega$ region. In general, a continuous increase is observed in the $\log\sigma'(\log\omega)$ slope. Secondly, the corresponding dispersion zone for conductivity is usually decoupled from the dispersion zone of the dynamic glass transition for the other activities. The isothermal $\sigma'(\omega)$ function for glasses with ionic conductivity far below T_g behaves similarly to $J''(\omega)$ for shear near the dynamic glass transition far above T_g.

The following empirical models are often applied to isothermal linear response curves of the type shown in Fig. 2.2a and 2.2b.

The Havriliak–Negami (HN) Function. For example, for a dynamic compliance in the frequency domain [110],

$$\varepsilon^*(\omega) = \varepsilon_\infty + \Delta\varepsilon\left[1 + (\mathrm{i}\omega\tau_{\mathrm{HN}})^b\right]^{-g} . \tag{2.14}$$

The five parameters mean for dielectrics: ε_∞ the dielectric permittivity at high frequencies above the dispersion zone, $\Delta\varepsilon$ the step height across the dispersion zone, sometimes called the relaxation strength or relaxation intensity (more precisely, for the dielectric compliance would be the retardation strength), τ_{HN} a typical time (HN time, $\tau_{\mathrm{HN}} = 1/\omega_{\mathrm{HN}}$) in the dispersion zone, and two shape parameters which are exponents describing the slopes of the $\log\varepsilon''(\log\omega)$ peak wings [111] (Fig. 2.2e),

$$b = \frac{\mathrm{d}\log\varepsilon''}{\mathrm{d}\log\omega} \quad \text{for} \quad \omega \ll \omega_{\mathrm{HN}}\,, \tag{2.15}$$

and

$$-bg = \frac{\mathrm{d}\log\varepsilon'}{\mathrm{d}\log\omega} \quad \text{for} \quad \omega \gg \omega_{\mathrm{HN}}\,. \tag{2.16}$$

A physical interpretation of the wings will be discussed in connection with Fig. 2.15c. The special case $b = 1$ is white noise at low frequency, and $g = b = 1$ is the dynamic Debye relaxation in the frequency domain, corresponding to a Cauchy distribution for spectral density (Sect. 2.2.4). After Fourier transformation, the Debye relaxation is an exponential decay in the time domain. Any HN function with $0 < b, g < 1$ decays 'more slowly' than the exponential.

For $g = 1$ the HN loss function is symmetric in $\log\omega$. HN is called the Cole function for $g = 1$ and $0 < b < 1$. Its loss peak over $\log\omega$ is also wider than the Debye loss peak. Asymmetry is produced by $0 < g < 1$, and HN is called the Cole–Davidson function for $b = 1$ and $0 < g < 1$.

The Kohlrausch or Kohlrausch–Williams–Watts (KWW) Function. For example, for a compliance in the time domain [112, 113],

$$-[J(t) - J_{\mathrm{e}}] = \Delta J \exp\left[-\left(\frac{t}{\tau_{\mathrm{KWW}}}\right)^{\beta_{\mathrm{KWW}}}\right]\,, \quad 0 < \beta_{\mathrm{KWW}} \le 1\,. \tag{2.17}$$

The four parameters mean: J_{e} the long-time or equilibrium compliance, ΔJ the retardation strength, τ_{KWW} a typical (KWW) time in the dispersion zone, and one shape parameter β_{KWW}, the KWW exponent, describing the stretched exponential of (2.17). The case $\beta_{\mathrm{KWW}} = 1$ corresponds to Debye relaxation and exponential decay. In the frequency domain, $\beta_{\mathrm{KWW}} = -\mathrm{d}\log J''/\mathrm{d}\log\omega$ is the logarithmic slope at high frequency, whilst the slope is 1 at low frequency, corresponding to white noise of the spectral density. More details are shown in Figs. 3.7 and 3.8. $\beta_{\mathrm{KWW}} > 1$ would mean some resonance contribution not appropriate for dynamic glass transition below the terahertz frequency range.

The formula $\exp(-t^{\beta_{\mathrm{KWW}}})$ is often used for moduli, too. This does not exactly correspond to $J^*G^* = 1$ [see (2.8)] in the frequency region if (2.17) is used simultaneously for the compliance. The errors resulting from a simultaneous application of KWW for modulus and compliance are discussed in [114] (Fig. 3.8d).

A useful integral is

$$\int_0^\infty t^{\nu-1} \exp\left(-t^{\beta}\right)\,\mathrm{d}t = \frac{1}{\beta}\,\Gamma(\nu/\beta)\,, \quad \nu > 0\,. \tag{2.18}$$

The average dimensionless relaxation time $\bar{\tau}$ is then

$$\bar{\tau} = \frac{\int t \exp(-t^{\beta})\, dt}{\int \exp(-t^{\beta})\, dt} = \frac{\frac{1}{\beta}\,\Gamma(2/\beta)}{\frac{1}{\beta}\,\Gamma(1/\beta)} . \tag{2.19}$$

From these integrals, we obtain

$$\frac{\bar{\tau}}{\tau_{\mathrm{KWW}}} = 6 \quad \text{and} \quad \frac{\overline{(\tau-\bar{\tau})^2}}{\tau_{\mathrm{KWW}}^2} = \frac{\beta\Gamma(3/\beta) - \Gamma^2(2/\beta)}{\Gamma^2(1/\beta)} = 24 \, ,$$

for $\beta = \beta_{\mathrm{KWW}} = 0.5$, for instance. For the Debye relaxation ($\beta = 1$, putting $\tau_{\mathrm{KWW}} = 1$), we obtain $\bar{\tau} = 1$ and $\overline{(\tau - \bar{\tau})^2} = 1$.

The stretched exponential in (2.17) corresponds to the characteristic function, which is the Fourier transform of the symmetric Levy distribution density $p(x)$ of probability theory. The Levy exponent is usually denoted by α. The Levy distribution is playing an increasing role in understanding the spectral density of fluctuations for the dynamic glass transition, $p = p(\omega)$, $\mathrm{d}P = p(\omega)\mathrm{d}\omega$. We use the KWW exponent β_{KWW} if the retardation compliance is fitted as a function of time over many time decades. We use the Levy exponent α, instead of β_{KWW}, if an interpretation of the high-frequency wing of the loss compliance by means of a Levy distribution is intended. For the HN function (2.16), $\alpha = bg$, if $bg < 1$. Since high frequency corresponds to short times, it is the short-time behavior that is interesting for Levy, not the long-time tail of the stretched exponential.

There is no closed analytical formula for the HN function in the time domain, nor for the KWW function in the frequency domain. The frequencies of the loss peak maxima $\omega_{\max}$ are different from $1/\tau_{\mathrm{HN}}$ and $1/\tau_{\mathrm{KWW}}$, respectively:

$$\omega_{\max} \cdot \tau_{\mathrm{HN}} = \left[\sin\left(\frac{\pi b}{2g+2}\right) \Big/ \sin\left(\frac{\pi b g}{2g+2}\right)\right]^{1/b} , \tag{2.20}$$

and [115]

$$\ln(\omega_{\max} \cdot \tau_{\mathrm{KWW}}) \approx 0.60607\,(\beta_{\mathrm{KWW}} - 1) \, . \tag{2.21}$$

In this book, we take $\tau = 1/\omega_{\max}$ as the typical fluctuation time.

The Logarithmic Gauss Function. This is given by

$$G''(\ln\omega) \quad \text{or} \quad J''(\ln\omega) \propto \exp\left[-\frac{[\ln(\omega/\omega_{\max})]^2}{2(\delta\ln\omega)^2}\right] . \tag{2.22}$$

The parameter $\delta\ln\omega$ means the dispersion of the Gaussian bell curve J'' over the $\ln\omega$ axis. There are no definite wings in a $\log J''$–$\ln\omega$ graph and no other shape parameters. All experience with the dynamic glass transition shows, however, that loss peaks do have logarithmic wings with definite exponents. Moreover, the real part $J'(\ln\omega)$ of Gauss has some overshots for $\delta\ln\omega \lesssim 2$, as for resonances. Only the central part of the Gaussian curve can therefore be used for the dynamic glass transition. This central part, however, is of phenomenological interest when $\delta\ln\omega$ is compared to dispersions of thermodynamic fluctuations via local time–temperature equivalence Sects. 2.1.3, 2.4.5, and 3.5).

Approximate half widths (FWHM = full width at half maximum) and equivalent Gaussian dispersions for the model functions are compared in Table 2.3. Note a useful coincidence: the equivalent Gaussian dispersion in natural logarithms, $\delta_G \ln\omega$, is about 97% of FWHM in base ten logarithms, $\Delta\log_{10}\omega$ (Fig. 2.2f).

2.1.3 Dynamic Glass Transition

The term 'dynamic glass transition' seems a useful shorthand for a longer phrase: the $\log\omega$–T–p (mobility–temperature–pressure) or $\log\tau$–T–p (relaxation time–temperature–pressure) behavior of the main process is the sum of a and α processes above T_g (see [14] and Fig. 2.3a). We used this concept in the Introduction and in the preceding section on the basis of a preliminary definition. We take the boundaries of the equilibrium dispersion zone from Fig. 2.2c for different temperatures T at constant pressure p and transfer them into an *Arrhenius diagram* (logarithm of the frequency, $\log\omega$, vs. reciprocal temperatures, T^{-1}). This defines a stripe that guides the shift of the loss peaks with temperature (Fig. 2.3b), and describes the slowing down of thermal fluctuations in the equilibrium material, e.g., the liquid. The relaxation or retardation times and, equivalently by (2.9), the fluctuation times are of order $1/\omega$ there. This stripe is called the *dynamic glass transition*. Its position in the diagram depends on the pressure p. In the $\log\omega$–T–p space, the dynamic glass transition is a shell of approximately hyperbolic shape (Fig. 1.4, [7]). The region above, i.e., with larger frequencies, was called the glass zone. The region with lower frequencies was called the liquid or flow zone (Fig. 2.2c).

The frequency ω in this stripe takes on an independent molecular meaning. Thermodynamic fluctuations are measured by susceptibilities, via the fluctuation–dissipation theorem FDT. As there is no change of frequency in such experiments (ω identity, Sects. 2.2.4 and 3.4.5), the frequencies in the stripe directly indicate the frequencies of the molecular fluctuation which are relevant for the susceptibilities. We will therefore call $\log\omega$ in the stripe the *molecular mobility* as relevant for the dynamic glass transition.

Along the stripe, falling temperatures correspond to decreasing frequencies or increasing times. When the dispersion zone arrives at a preparation time of order t_g = 1000 s, corresponding roughly to $\nu_g \approx 1$ millihertz or $\omega_g \approx 10^{-2}$ rad/s, the material freezes in, i.e., it vitrifies (Sect. 2.1.4). Such a

Table 2.3. Approximate half widths (FWHM = full width at half maximum in $\log_{10}\omega$) and equivalent Gaussian dispersions $\delta_G \ln\omega$ for the four responses of Fig. 2.2e

	Debye	HN[a]	KWW[c]	log Gauss
FWHM: $\Delta\log_{10}\omega$	1.144	$\approx 0.85\,(1/b + 1/bg - \ldots \ldots - 1/2g) - 0.12$	$\approx 1.14/\beta_{KWW}$	$1.02\ \delta\ln\omega$
$\delta_G \ln\omega$[b]	1.12	0.97 FWHM	$\approx 1.07/\beta_{KWW}$	$\delta\ln\omega$

[a] From [116].
[b] In natural logarithms, the two curves are pinned together at the half maximum.
[c] The figures depend slightly on β_{KWW}. For $\delta_G \ln\omega$, we obtain from the graphs of [115]: 1.04 for $\beta_{KWW} \to 0$, 1.07 for $\beta_{KWW} = 0.5$, and 1.12 for $\beta_{KWW} = 1$.

preparation time may be defined by a cooling rate of order 10 K/min. The transformation from an equilibrium liquid state to a nonequilibrium glass state is called the *thermal glass transition.* The transformation is characterized by a transformation temperature (glass temperature T_g, Sect. 2.1.4) and a surprisingly small *transformation interval* (Fig. 1.2). The latter corresponds to the temperature width of our stripe, i.e., of the dynamic glass transition. The transformation interval increases with the frequency dispersion of the susceptibility loss peak, $\delta\ln\omega$ (Table 2.3 in Sect. 2.1.2), and decreases with the fragility.

We assume that the shape of the Fig. 2.2a and b loss peaks does not dramatically change in a temperature interval corresponding to the width of the dynamic glass transition. Then the variables temperature T and $\log\omega$ (or if we prefer times, $\log\tau$) are locally equivalent (Fig. 2.3b). Geometrically, this property is a trivial consequence of our assumption. Across the dynamic glass transition, the isothermal and isochronous sections of a uniform wall as formed by the loss susceptibilities are similar. This property is called *local time–temperature equivalence* or, more simply, time–temperature superposition.

This superposition is a great help for the extrapolation of susceptibilities to neighboring temperatures, frequencies, or times (see for example [3]). The reduced susceptibility curves obtained from a shift along the dynamic glass transition are called 'master curves'. The shift factor is usually denoted by a_T so that the reduced frequency or time variables are $\log(a_T\omega)$ or $\log(\tau/a_T)$. Using the VFT or WLF equations [see (2.2) or (2.23)] for the temperature dependence of a_T, the master-curve construction is also called VFT or WLF scaling. Globally, for larger $\log\omega$ or T intervals, the superposition is restricted to those intervals where changes in intensity and shape are small (thermorheological simplicity).

It is the logarithm of time or frequency in Fig. 2.3b that is locally equivalent to temperature (Sects. 3.4, 4.2, and 4.3). We prefer the word 'equivalence' when we wish to stress the situation that the shape or position of the chosen loss peak, i.e., the choice of an activity, either does not or only slightly influences the success of a master curve construction, and in particular, of getting the same shift factor a_T [118]. Then $\log\omega$ gets a certain independence and characterizes the dispersion zone by molecular mobility.

Note an important difference. The shift factor is defined along the dispersion zone of the dynamic glass transition and $\log\omega$ increases with T, whereas the Fig. 2.3b equivalence or master curve is defined across the dispersion zone and $\log\omega$ decreases with T.

We now move from stripes to curves in the Arrhenius diagram. Consider a specific susceptibility, and a specific point on its isothermal response function. We will often use the maximum of a loss function, e.g., the maximum of dielectric loss, $\varepsilon''(\log\omega)$, at frequency $\omega_{\max}(T)$ for temperature T. Putting the $\log\omega_{\max}$ values at different temperatures T into the Arrhenius diagram, we get a *trace* of symbols in the dispersion zone (Fig. 2.3c). Such a trace can be considered as a certain representative of the dynamic glass transition. A formula often used for this trace in a certain temperature interval [84], e.g., between T_g and T_c for the α process, or between T_c and T_A for the a process, is the WLF (Williams–Landel–Ferry) equation [118]. The pristine formula is, e.g., for $\omega \equiv \omega_{\max}(T)$,

$$\log\frac{\omega_B}{\omega} = \frac{c_1^B(T-T_B)}{c_2^B-(T-T_B)}\ . \tag{2.23}$$

We have three positive parameters: the two WLF constants, c_1^B and c_2^B, and the trace frequency ω_B at a reference temperature T_B. All three parameters depend on T_B. Common choices for the α process are $T_B = T_g$ or $T_B = T_g + 50$ K [3].

The WLF equation is symmetrical for $\omega_B \leftrightarrow \omega$, $T_B \leftrightarrow T$. This symmetry is better expressed in a $\log\omega$–T diagram than in an Arrhenius $\log\omega$–T^{-1} diagram (Fig. 2.3d). It is well known that the WLF equation is then a hyperbola $xy = x_By_B =$ const. for $x = T - T_0$ and $y = \log\omega - \log\Omega = -\log(\Omega/\omega)$:

$$(T-T_0)\log\frac{\Omega}{\omega} = (T_B-T_0)\log\frac{\Omega}{\omega_B} = B_{\mathrm{WLF}} = \text{const.} \tag{2.24}$$

We have, similarly to the VFT equation (2.2), two asymptotes $\log\Omega$ and T_0, and a third parameter B_{WLF} decreasing with the fragility $m(T_g)$ for the α process:

$$m(T_g) = \log\frac{\Omega}{\omega_g} + \frac{T_0}{B_{\mathrm{WLF}}}\log^2\frac{\Omega}{\omega_g}\ . \tag{2.25}$$

The arithmetic of the WLF equation is described in detail in Sect. 4.3, (4.70)–(4.89). The relationship with the pristine WLF constants is

$$c_1^B = \log(\Omega/\omega_B)\ ,\quad c_2^B = T_B - T_0\ . \tag{2.26}$$

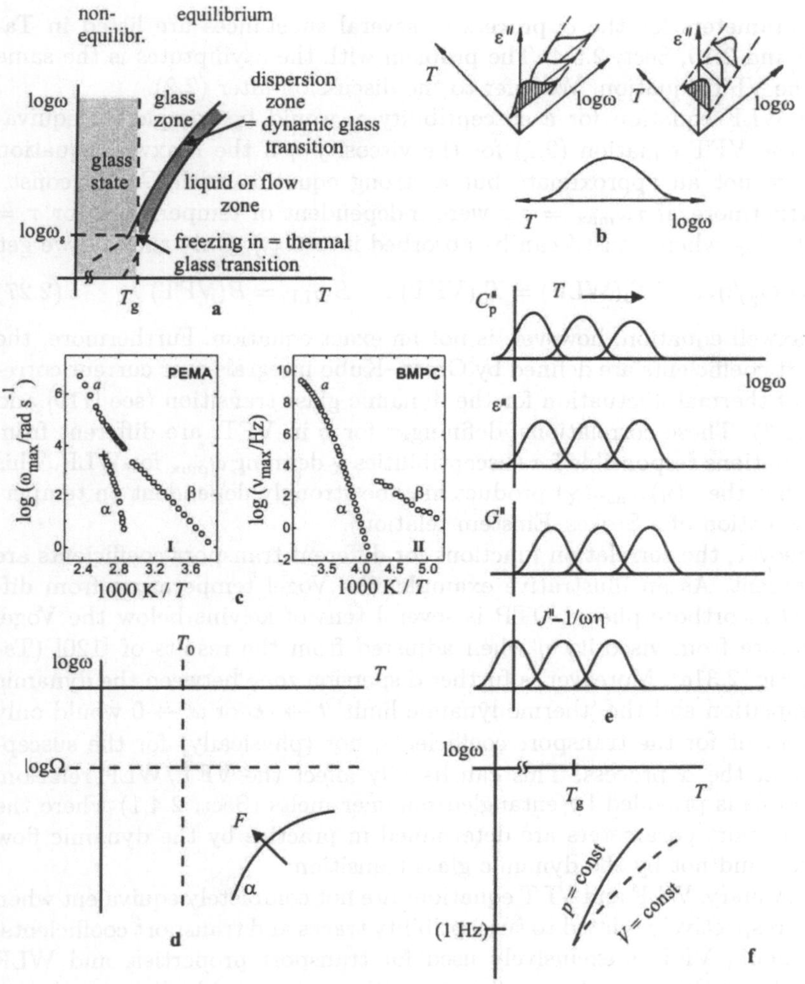

Fig. 2.3. Dynamic glass transition. (**a**) Partition of the isobaric mobility–temperature ($\log \omega$–T) plane for a glass-forming liquid into zones and glass state. (**b**) Local time–temperature equivalence for the example of dielectric-loss $\varepsilon''(T, \log \omega)$. (**c**) Dielectric traces for dynamic transitions in an Arrhenius diagram for two crossover scenarios I, II (Sect. 2.2.1). α is the cooperative α process, a the high-temperature a process, and β the local Johari–Goldstein process. Scenario I refers to polyethyl methacrylate PEMA [237], whilst scenario II refers to bis-methoxy-phenyl-cyclohexane BMPC [243]. (**d**) WLF trace for the α process in a $\log \omega$–T diagram. *Broken lines* are the asymptotes. The *arrow* marked F indicates increasing fragility. (**e**) Examples of loss parts for different activities (schematic). $C_p'' = TJ_S''$ the dynamic heat capacity (Nagel et al.) with J_S'' the entropy compliance, ε'' the dielectric function (a compliance), G'' the shear modulus, and J'' the shear compliance. (**f**) Comparison of dynamic glass transitions in polyvinylchloride PVC [124] 'starting' at T_g for normal pressure, for isobaric ($p =$ const.) and isochoric ($V =$ const.) increase in temperature (schematic)

WLF parameters for the α process of several substances are listed in Tables 2.9 and 2.10, Sect. 2.5.4. The problem with the asymptotes is the same as for the VFT equation. We refer to the discussion after (2.2).

The WLF equation for a susceptibility χ would be completely equivalent to the VFT equation (2.2) for the viscosity η if the Maxwell equation (2.1) were not an approximate but a strong equation with G_∞ = const., and furthermore, if $\tau\omega_{\max} = \tau\omega$ were independent of temperature (or τ = const.'$/\omega_{\max}$, where const.' can be absorbed into G_∞). Only then do we get

$$\Omega = G_g/\eta_\infty\,, \quad T_0(\mathrm{WLF}) = T_0(\mathrm{VFT})\,, \quad B_{\mathrm{WLF}} = B(\mathrm{VFT})\,. \qquad (2.27)$$

The Maxwell equation, however, is not an exact equation. Furthermore, the transport coefficients are defined by Green–Kubo integrals over current correlations of thermal fluctuation for the dynamic glass transition (see [119] and Sect. 2.2.3). These correlations, defining τ for η in VFT, are different from the correlations responsible for susceptibilities χ defining $\omega_{\max}$ for WLF. This means that the $\tau(\eta)\omega_{\max}(\chi)$ product may be strongly dependent on temperature (violation of a Stokes–Einstein relation).

Moreover, the correlation functions for different transport coefficients are also different. As an illustrative example, the Vogel temperature from diffusion D in orthoterphenyl OTP is several tens of kelvins below the Vogel temperature from viscosity η when adjusted from the results of [120] (Table 2.8, Fig. 2.31e). Moreover, a further dispersion zone between the dynamic glass transition and the 'thermodynamic limit' $t \to \infty$ or $\omega \to 0$ would only be important for the transport coefficients, not (physically) for the susceptibilities of the α process. This can heavily affect the VFT/WLF relation. An example is provided by entangled polymer melts (Sect. 2.4.1) where the VFT transport parameters are determined in practice by the dynamic flow transition and not by the dynamic glass transition.

In summary, WLF and VFT equations are not completely equivalent when they are respectively related to susceptibility traces and transport coefficients. In this book, VFT is exclusively used for transport properties, and WLF exclusively for dynamic glass or flow transitions as traced by linear response susceptibilities.

Another point is the concept of *different activities* as defined after (2.10). Consider different responses, e.g., different loss susceptibilities in glass-forming liquids for three given temperatures (Fig. 2.3e). Not only shape and positions of compliances are different from those of the corresponding moduli (e.g., J'' and G'' for shear), but also shape and position of the different compliances (C_p'', ε'', J'') are different. In principle, even confining ourselves to loss maxima, we have six traces for the three activities in the same dispersion zone, e.g., for the α process: shear (J'', G''), dielectric (ε'', M''), and calorimetric (C_p'', K_T''). For the α process of a given polymer, for instance, WLF equations seem appropriate for each trace. Their Ω parameters are different, but their Vogel temperatures T_0 are the same within the limits of experimental uncertainty [121]. This behavior differs from the behavior of diffusion

D and viscosity η mentioned above. The different 'transport activities' have different Vogel temperatures. Different activity behavior can restrict the local and global aspects of time–temperature equivalence.

Different responses for different activities do not seem possible for a simple-activation rate process, even with a barrier distribution. Different activities point to a certain complexity in the molecular process whose parts can be evaluated differently by different perturbations.

The dramatic slowing down of the main transition with falling temperatures is a difficult problem. The curvature of the dynamic glass transition in the Arrhenius diagram, or, equivalently, the finite Vogel temperatures in the VFT or WLF equations, indicates that this increase cannot be understood by a rate theory with a constant transition state, in particular, with a given activation energy. The present physical view of this problem will be described in Sect. 2.5 and Chap. 4. The physicochemical view in the 1960s can be characterized by three fruitful concepts for the α process: free volume [3, 26, 27], cooperativity [122], and Glarum defect diffusion [123].

The dynamic situation on molecular scales is described by at least two frequency regions. Firstly, there is a high-frequency region for local movements, of the same order as the intermolecular distance divided by the thermal molecular velocity, that is, d/v_T of order 0.1 nm / 100 ms^{-1} $\approx 10^{12}$ hertz. Secondly, there is a low-frequency region for molecular mobility, i.e., for Fourier components of fluctuations in the dispersion zone for the much slower dynamic glass transition. There is no a priori reason for any molecule being excluded from either of the two regions.

The *free volume* describes the sensitivity of this situation to all changes that are equivalent to small density changes. A small increase in average intermolecular distances shifts the molecular mobility towards the insensitive high frequencies. The dynamic situation is 'not stable' against a small decrease in density, e.g., of order 5%. We may consider temporary, more or less local *breakthroughs of molecular mobility* in the direction of the high frequencies, organized by density fluctuations. The mobility is sensitive to small changes in intermolecular distances because it is disordered contacts between usually nonspherical molecules that are optimized by free energy in cold liquids. Consequently, free volume is always a small fraction of the total volume, and the density contrast connected with large local mobility variations is correspondingly small.

This picture is not restricted to isothermal density changes. Not only an isobaric (p = const.), but also an isochoric (V = const.) increase in temperature is, in a way, equivalent to a decrease in density. Both changes diminish effects due to core diameters of steep intermolecular repulsion potentials and enlarge the room available for movement. This means that both enlarge the free volume. Accordingly we find WLF type equations for both conditions (Fig. 2.3f, [124]). In modern terms, the isobaric condition corresponds to a more fragile glass transition than the isochoric one [125]. Isobaric conditions, permitting thermal expansion, offer more possibilities for molecular rearrangements to increase the mobility. For the fluctuation aspect of free volume, it is important that even the isochoric–isothermal condition (and any other restriction) of a macroscopic sample does not suppress the relatively large density fluctuations of its small representative subsystems [126] (Sects. 3.1 and 3.2).

The organization of molecular contacts in cold liquids is described by *molecular cooperativity*. The basic reason for cooperativity is the insight that local molecular motion is not sufficient to maintain molecular mobility for very low free volume. The molecules do not relax independently of one another. The motion of a particular molecule depends to some degree on that of its neighbors [123]. The rearranging movement of one particle is only possible if a certain number (N_α) of neighbor particles are also moved. A *cooperatively rearranging region (CRR)* is, according to Adam and Gibbs [122], defined as a subsystem which, upon a sufficient thermal fluctuation, can rearrange into another configuration independently of its environment. The average size of a CRR is, therefore, determined by a spatial aspect of statistical independence of thermal fluctuations.

Note again that all relevant molecular movements are counted in linear response only with respect to their contributions to Fourier components of thermal fluctuation in the dispersion zone of the dynamic glass transition. The definition of a CRR is therefore related to the dynamic glass transition. We will say that a CRR is a 'functional' subsystem [18], i.e., functional to the α process. For instance, a slow temperature fluctuation is attributed to the CRR because it stems from the contributions of the relevant spectral density across the α dispersion zone. This temperature fluctuation cannot be quenched by a large thermal conductivity stemming from the phonons at high frequency (functional independence, see Sect. 3.5).

The average size of CRRs is generally estimated to be a few nanometers [45] and is assumed, from the very beginning, to increase for decreasing temperatures. We expect a wide distribution of CRR sizes if the average size is small, e.g., about one nanometer at higher temperatures. The formulas for this size depend on the treatment that follows the CRR definition via thermal fluctuations. A direct fluctuation treatment by means of the FDT, alternative to [122], will be carried out in Sect. 2.2.5.

In their pioneering paper [122], before Goldstein's famous paper on the energy landscape, Adam and Gibbs used statistical independence to do statistics with the CRRs. They then used the fluctuations to overcome local (landscape) barriers inside the CRRs. This treatment introduced the dynamic relevance of configurational entropy $S_c = S(\text{liquid}) - S(\text{crystal})$. The relationship with dynamics, the Adam–Gibbs correlation, is obtained as

$$\tau = \tau_B \exp \frac{C}{TS_c}\,, \quad TS_c \propto T - T_0\,, \tag{2.28}$$

with $C \approx$ const. calibrated at a reference time τ_B. This yields a WLF type equation for the relaxation time τ.

Guided by (2.28), after an earlier discussion about the general value of the configurational-entropy concept [127], Angell [128] later translated his fragility diagram Fig. 2.1d from viscosity η to reciprocal configurational entropy: $\log[\eta(T)/\eta(T_g)] \longmapsto S_c(T_g)/S_c(T)$. For $\tau_B = \tau_g$, the corresponding fragility measure based on (2.28) is

$$m(T_g) = -\frac{1}{2.3}\frac{d}{d\ln(T/T_g)}\frac{T_g S_c(T_g)}{T S_c(T)} \quad \text{at} \quad T = T_g\,. \tag{2.29}$$

The fragility ranking from configurational entropy S_c is approximately the same as that from viscosity η.

Although the Adam–Gibbs approach is based on the concept of CRRs, their size and concrete temperature dependence cannot be calculated directly from the formulas in the paper [122] (Adam–Gibbs dilemma), because the connection with S_c is obtained from too intimate a connection between internal and external treatment of CRRs. The determination of a CRR size in the nanometer range from the configurational entropy S_c demands some molecular model for the extensive S_c variable [129, 130], since the original formulas gave sizes too far below 1 nm [129, 131].

Glarum [123] introduced the *defect diffusion model* to describe spatial details of the intermolecular organization in a glass-forming liquid. A defect is a spot with low local density, i.e., a concentration of free volume corresponding to a spot with large local mobility. When the defect, via fluctuations, reaches a dipole, for example, the latter relaxes more easily than otherwise. In the extreme model case, and only then, it relaxes completely and instantly. This corresponds to a diffusion of the defect, if the free volume is balanced, e.g., inside the CRR, and if the spot picture is preserved, i.e., the free volume is not homogeneously distributed. If we consider the CRRs as the representative α subsystems, the diffusion length in the fluctuation time of the dynamic glass transition then corresponds to the size of a CRR.

We thus get a spatiotemporal dynamic heterogeneity at least for the larger CRRs. Diffusing islands of higher mobility [132] can be equated with diffusing defects in cells of CRR size, with fluctuating walls of lower mobility. This dynamic heterogeneity is characterized by an average CRR size expressed, for example, by a 'cooperativity' N_α, a mobility dispersion $\delta \ln \omega$ (Table 2.3), and a very small density contrast that will be estimated in Sects. 2.2.6, 2.5.1, equation (2.150), and Sect. 4.3.

I believe that the Glarum defect diffusion model results from stabilization of the mobility breakthroughs described above. The large mobility in the islands is regulated by a stable Levy distribution for the mobilities of partial systems in a CRR (Sect. 4.3).

Partition of the dynamic glass transition into a and α processes (Fig. 2.3c) indicates that different molecular mechanisms may be responsible for the high-temperature a process and the cooperative α process. Physicochemical experience in the 1960s was largely based on the α process.

2.1.4 Thermal Glass Transition. Glass Temperature. Partial Freezing

The thermal glass transition, freezing(-in), and vitrification will be used synonymously to refer to the transition from the equilibrium liquid state to the nonequilibrium glass state by cooling, densification, or extraction of a solvent. A thermodynamic cycle such as liquid $\rightarrow$ glass state $\rightarrow$ liquid is thermoreversible in the sense that the liquid before and after is the same, but it is not reversible, i.e., it is irreversible, in the sense that we have entropy production in the cycle if the thermal or the dynamic glass transition is crossed.

To prepare for the next few sections, the following subjects will also be discussed in the context of the glass transition: structure, nonergodicity, and the third law of thermodynamics.

Thermal Glass Transition. The thermal glass transition was briefly compared with the dynamic glass transition in the third paragraph of Sect. 2.1.3 (Fig. 2.3a). If we cool down a glass-forming liquid with a given cooling rate $\dot{T} = \mathrm{d}T/\mathrm{d}t$, at the glass temperature T_g, the thermal fluctuations become too slow to establish their contribution to thermodynamic variables (Fig. 2.4a). This starts at the slow, flow-zone side of the dispersion zone. The compliances, e.g., the heat capacity $= \dot{Q}/\dot{T}$ as a function of T, take a downward step. As a function of temperature, this step seems similar to the entropy compliance step in $C_p'(\log\omega) \sim J_S'$ as a function of $\log\omega$ across the dynamic glass transition (Fig. 2.2a). According to the local time–temperature equivalence, high frequencies correspond to low temperatures across the dispersion zone. Let us stress again that it is only the Fourier components of this dispersion zone that freeze at T_g.

There is a fundamental difference between the glass zone for the dynamic glass transition and the glass state for the thermal glass transition (Sect. 2.1.2). In the glass zone, the probe frequency is too high for a response from typical molecular rearrangements: they cannot follow the perturbation. The description of this zone will be continued below. In the glass state, molecular rearrangements cannot be observed by any probe frequency. They give no contribution to C_p, because the experimental time available in and below the transformation interval is too short for their realization. The sample in the glass state is no longer in equilibrium. If we wait several hours at given T at the lower end of the transformation interval, as mentioned in the Introduction, fluctuations would come into play and realize its contribution to C_p. This means that the entropy, for instance, depends on time and such a state of affairs is not compatible with equilibrium.

In principle, there are two possible primary concepts for the thermal glass transition. Firstly, the primary concept may be a continuous slowing down of

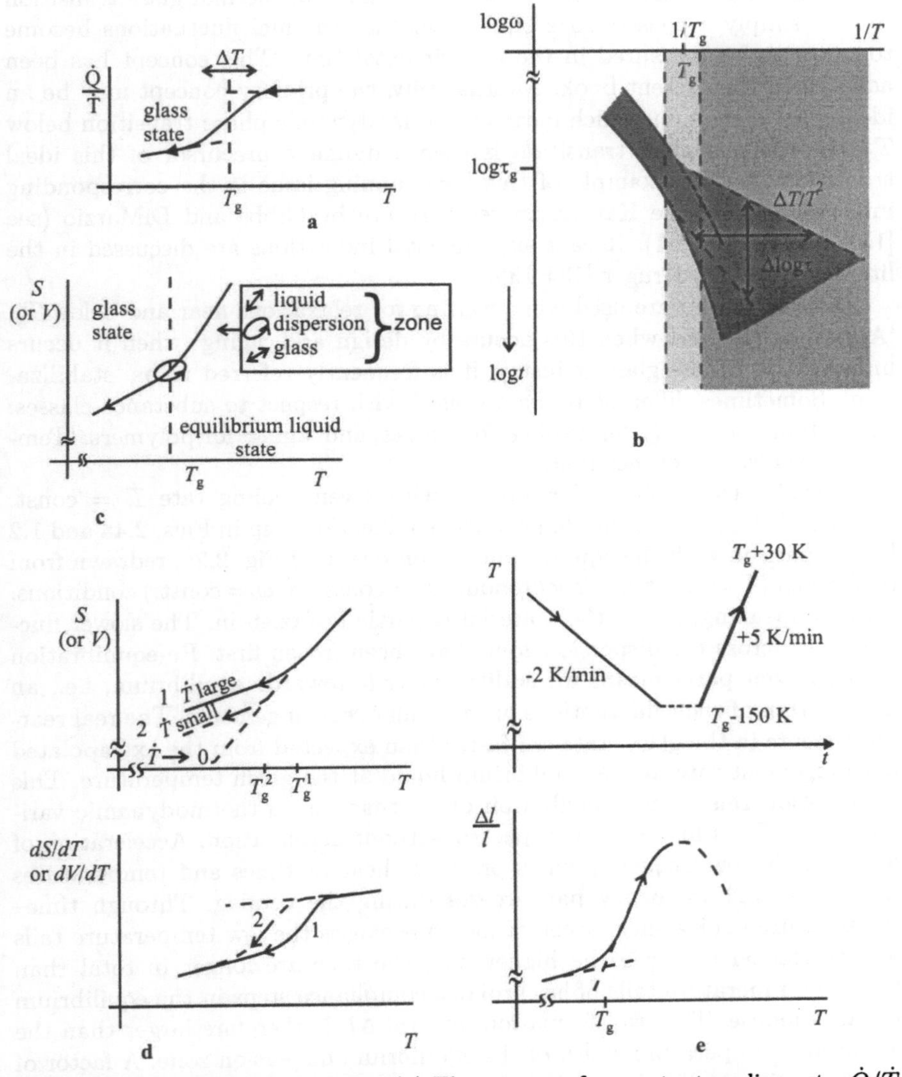

Fig. 2.4. Thermal glass transition. (a) Thermogram for constant cooling rate. $\dot{Q}/\dot{T}$ is the 'dynamic' heat capacity, T_g the 'calorimetric' glass temperature, ΔT the transformation interval (Fig. 1.2). (b) Acceleration effects in the glass state. (c) Thermal glass transition in an entropy–temperature S–T diagram. The *arrowed closed curves* are thermodynamic cycles of a stationary periodic response experiment. Details are explained in the text. (d) Thermal glass transition for two different cooling rates $\dot{T} = |dT/dt|$. (e) T–t program and onset construction of T_g from dilatation as used by glassmakers

the dynamic glass transition in the cold liquid; the thermal glass transition is then simply a consequence of the fact that thermal fluctuations become too slow to be measured in the experimental time. This concept has been adopted in the present book. Alternatively, the primary concept may be an ideal glass transition, which means a thermodynamic phase transition below T_g; the thermal glass transition is then a dynamic precursor of this ideal transition. An old example of this long-running issue is the corresponding interpretation of the Kauzmann temperature by Gibbs and DiMarzio (see [133] and Sect. 2.5.1). Recent experimental indications are discussed in the literature e.g., by Krüger [134,135].

Different terms are used when waiting for relaxations near and below T_g. 'Annealing' is used when this occurs by design and 'aging' when it occurs undesirably. In the glass industry, it is frequently referred to as 'stabilization'. Sometimes different terms are used with respect to substance classes: 'relaxation' for liquids, 'annealing' for glasses, and 'aging' for polymers. 'Tempering' is also sometimes used.

Consider the cooling of a sample with given cooling rate $\dot{T}$ = const. Nonequilibrium affects the shape of the vitrification step in Figs. 2.4a and 1.2 by comparison with the equilibrium response step of Fig. 2.2a, redrawn from isothermal (T = const.) to isochronous ($\dot{T}$ = const., or ω = const.) conditions. Strictly speaking, at T_g, the material is partially frozen in. The slower fluctuations, across the dispersion zone, have been frozen first. Re-equilibration of the frozen parts means an additional drift towards equilibrium, i.e., an *acceleration* of their fluctuations or molecular rearrangements. The real rearrangements in the glass state are faster than expected from the extrapolated rearrangement rate in the equilibrium liquid at the given temperature. This acceleration raises the contribution of fluctuations to thermodynamic variables, i.e., C_p is higher than expected without acceleration. Acceleration of 'thawing' in low-frequency parts produces heat at times and temperatures where the high-frequency part freezes during the cooling. Through time–temperature equivalence, acceleration thus causes the low temperature tails of vitrification curves to be higher, i.e., the tails are longer in total than the low temperature tails of isochronous compliance steps in the equilibrium linear response. The transformation interval ΔT is therefore larger than the isochronal temperature width of the equilibrium dispersion zone. A factor of about two is usually observed.

Let us rephrase this complicated matter in other terms (Fig. 2.4b). Compare the relaxation in the glass state with the extrapolated equilibrium response in the glass zone after a long wait at the same temperature. Because of acceleration, the typical relaxation time is shorter than the response time τ at T, but of course longer than the relaxation time τ_g at T_g. Because of differential effects, the widths of the relaxation region (ΔT, $\Delta \log \tau$) in the glass state are larger than expected from the extrapolated equilibrium dispersion zone. Additionally, effects from fast precursors of the glass transition become important in the glass state, e.g., the Andrade zone A1 in Figs. 2.28a and b.

The difference between glass state and glass zone can also be discussed by small cooling–heating cycles in S–T (entropy–temperature) diagrams (see Fig. 2.4c). We partly follow McKenna [136]. In the dispersion zone of the equilibrium liquid, we have ellipses in the linear region. Their area is proportional to the average entropy production contributed by calorimetric activity,

$$\dot{S} = \frac{1}{2}\omega C_p'' \left(\frac{\Delta T}{T}\right)^2 , \tag{2.30}$$

where $C_p'' = \rho V c_p''$ is the loss part of the extensive heat capacity of the sample at ω and T, and ΔT is the temperature amplitude of the cycle (Sect. 3.3). Outside the dispersion zone ($C_p'' = 0$), the ellipses degenerate to short double arrows ($\nearrow$). Their slope in the glass zone is smaller than in the liquid or flow zone. In the glass state, the cycle is affected by the equilibration acceleration, absent in the equilibrium liquid.

Glass Temperature. The glass temperature T_g depends upon several conditions. Consider higher cooling rates. Crossing the dispersion zone with higher cooling rate means shorter experimental time. This corresponds to higher frequency ω in the $\log\omega$–T plane of Fig. 2.3a. The time needed for thermodynamic establishment of the faster thermal fluctuations is shorter at higher temperature. The time available for establishment is of the order of $\Delta T/|\dot{T}|$, with ΔT the transformation interval. This means (Fig. 2.3a, $\omega \sim 1/\tau$) that the material vitrifies at higher temperatures for higher cooling rates (Fig. 2.4d). Correspondingly, the glass temperature depends on the cooling rate, T_g increasing with $|\dot{T}|$. More precisely, the T_g shift ΔT_g is proportional to $\Delta \log \dot{T}$, because of the logarithm of time in the time–temperature equivalence:

$$\Delta T_g / \Delta \log_{10} \tau = T_g / m , \tag{2.31}$$

with $m = m(T_g)$ the fragility. The glass temperature is not therefore an absolute material property but depends on process parameters such as $\dot{T}$. The old term 'glass transition temperature' was more precise in this regard but is now often substituted by the simpler 'glass temperature' T_g.

Any definition of T_g needs three conventions, besides the composition, pressure p, and history of the sample. We must:

(i) fix the cooling and heating rates since the thermal glass transition is stuck to the dynamic glass transition (Fig. 2.3a),
(ii) choose a susceptibility since the activities are different (Fig. 2.3e),
(iii) fix which point of the smooth vitrification step to use, since there is no jump, cusp, or sharp bend in the curve (Fig. 2.4a).

Most physico-chemists use a 'calorimetric' T_g as follows:

(i) $\dot{T} = 10$ K/min in the heating stage after cooling with the same $|\dot{T}|$,
(ii) heat capacity from dynamic calorimetry,
(iii) an equal-area construction for $\dot{Q}/\dot{T} = C_p$ of Fig. 2.4a, corresponding to a tangent construction in the S–T diagram of Fig. 2.4c. The latter construction induces a T_g uncertainty of about 1 K, because the $C_p(T)$ lines in the glass state and liquid state are not always straight.

Glassmakers have their own standards for getting comparable and reproducible glass temperatures [137]. For instance, they use dilatometry (Fig. 2.4e) after relatively slow cooling – this sharpens the transition at heating – and an onset tangent construction. Further, they require a reproducibility of $\Delta' T_g < 3$ K after some days to guarantee the stability of the material. We expect all three T_gs, from calorimetry, from dilatometry, and from viscosity, $\log_{10} \eta$ (Pa s) $= 12.3$, to be different since the three methods refer to different activities.

A dynamic response experiment (frequency ω) near and below T_g is thus characterized by three times:

(1) a *relaxation time* τ_R in the dispersion zone,
(2) a *probing time* for linear response ($\approx 1/\omega$),
(3) a *preparation time* τ_{prep}.

The relaxation time (1) is measured linearly by a clock, next to the apparatus used to measure a response in the time domain. The probing time (2) corresponds to time intervals as used to measure a correlation function of a thermal fluctuation, or as used, in principle, to measure a frequency ω of a spectral density of this fluctuation. The fluctuation time $\tau(\mathrm{Fl})$ used earlier stems from the probing time concept. The preparation time (3) is related to the transformation interval and depends on the kind of experiment. We may determine τ_{prep} by an annealing or waiting time t_w, by a minimal time interval needed for the experiment, τ_{exp}, by a reciprocal cooling rate reduced by the transformation interval, $\tau_{prep} \sim \Delta T/\dot{T}$, or in some other way. Preparation time and experimental time in the transformation interval will often be used synonymously.

In an equilibrium linear-response experiment, only the first two times are needed. They become equivalent via the FDT. The third time complicates the situation near T_g. If the fluctuation time $\tau(\mathrm{Fl})$ is longer than the preparation time τ_{prep}, $\tau(\mathrm{Fl}) > \tau_{prep}$, then, as mentioned above, the fluctuations cannot establish their equilibrium susceptibilities. They become partially frozen. The borderline between $\tau(\mathrm{Fl})$ and τ_{prep} will be estimated in Sect. 4.5.1, and will be discussed qualitatively in Sect. 2.2.6, (2.79). Waiting or raising the temperature increases the compliances. The recovery of extensive variables by 'thawing' of the partially frozen state is called structural relaxation. As a rule, this is a nonlinear process which will be described in the next section (Sect. 2.1.5).

To avoid any major misunderstandings, we need a convention concerning the way in which the concepts 'structure' and 'nonergodicity' will be used.

Structure. The word 'structure' will only be used in relation to an experimental time τ_{exp}. A *structure* is the spatial arrangement of particles in the material, i.e., a spatial molecular correlation that lives longer than the experimental time chosen for its determination. Let us consider two examples.

- The 'static' structure as determined by the structure factor from X-ray or neutron scattering, for example. The experimental time is given by the reciprocal beam wave frequency, for X rays of order $\tau_{\text{exp}} = 10^{-18}$ seconds, and there is practically no molecular movement in such a short time interval. (Remember that even the frequency of visible light is of order 500 terahertz.) The structure factor $S(Q)$ comes from interference of Huygens waves (2.89), i.e., from snapshots averaged over a large ensemble of equivalent subsystems.
- The structure concept of dynamic scattering (Sects. 2.3.1 and 3.4.2). This example is more complicated. Roughly, we have a 'structure function' $S(Q, \omega)$ which is the Fourier transform of a (better named) 'intermediate scattering function' $S(Q, t)$. Scaled by the structure factor, we get a reduced scattering function

$$f_Q(t) = \frac{S(Q,t)}{S(Q)} . \qquad (2.32)$$

This behaves as a correlation function which 'decays' with time, $|f_Q(t)| < f_Q(0)$. Details will be described in Sects. 2.3.1, 2.3.2, and 3.4.2.

A problem arises in the following situation. On a trial basis, put the experimental time equal to t in $S(Q, t)$. Cross the dispersion zone of the dynamic glass transition from short times (large ω) to long times (small ω). For short times, in the glass zone, $f_Q(t) = 1$, or more precisely (Sect. 2.3.1), $f_Q(t) = \text{const.} \equiv f_Q(T) < 1$, $S(Q, t) = f_Q(T)S(Q)$, where $f_Q(T) < 1$ reflects fast relaxations that have died away in the glass zone near the dynamic glass transition. We then have a true structure that lives longer than the glass-zone times t. For large times, in the flow zone, $S(Q, t) \to 0$, all spatial correlations are extinguished by motional averaging and the liquid is 'homogeneous'. In our trial we would obtain, for flow-zone times t, a liquid without any structure. This means that t in $S(Q, t)$ is more like a probing time than an experimental time for structure as defined in the preceding paragraph.

In the dispersion zone, using our trial, we would observe a 'decay of structure', the transformation of a 'long-lived' liquid structure into 'short-lived' correlations of a homogeneous liquid. The problem is to find, functional to the dynamic glass transition or the α process, a practicable concept that describes the relevant spatial inhomogeneities by saving the structure concept with the short experimental time such that glass, dispersion, and flow zones have the same structure.

We will now discuss dynamic heterogeneity, a possible medium between structure and correlation, as mentioned near the end of the Introduction. There is a major problem. It will be shown in Sect. 2.5.2 that the dynamics in the Arrhenius diagram ($\log \omega$ vs. $1/T$) is general when compared to the individual, multifarious structures of different glass-forming liquids. Moreover, (2.32) for the correlation function $f_Q(t)$ does not depend on Q as much as $S(Q)$ does. We usually only observe some kind of diffusion with a 'structureless' dispersion law, e.g., $\tau_{\mathrm{KWW}} \propto Q^{-2/\beta_{\mathrm{KWW}}}$ (2.96). Constructing spatiotemporally fluctuating distributions, or even a pattern, of molecular mobility for the α process, e.g., a stochastic mobility field $\log \omega(\boldsymbol{r})$ or $\log \tau(\boldsymbol{r})$, we are therefore confronted with low concomitant density contrast [see (2.150)]. The suggestion that we should still construct such $\log \omega(\boldsymbol{r})$ distributions [7, 138] will now be called 'dynamic heterogeneity' (Fig. 2.5a).

A more microscopic formulation of dynamic heterogeneity refers only to independent subensembles with different dynamics [289]. These authors believe that such a formulation can even do without spatial reference. The disadvantage of such a formulation is the absence of any interface with a mechanical picture. Mechanics must reflect the multifarious structures of different glass formers. A chaotic mechanical system with molecular cooperativity must be heterogeneous in such a sense and can be characterized by individual 3-, 4-, or more time correlations. This is, of course, interesting. But I think it would be more appropriate to the aim of finding a general explanation for the general liquid dynamics to develop a definition of dynamic heterogeneity more closely related to the measurement of thermal fluctuations, including scattering with its 2-time correlations. One could start with sections of the $f_Q(t)$ functions, e.g., $f_Q(t) = 1/2$, giving a spatial Fourier transformation of the mobility field $\log t(Q)$ or, related to $S(Q, \omega)$, $\log \omega(Q)$ [139, 140]. The discussion of dynamic heterogeneity will be continued in Sect. 2.2.5.

▶

Fig. 2.5. Thermal glass transition (continued). (**a**) Dynamic heterogeneity of dynamic glass transition. (**b**) Discussion of nonergodicity above and below glass temperature. f_c is the nonergodicity parameter. See text for details. (**c**) Dynamic heat capacity $C_p^*(\omega) = C_p' - \mathrm{i}C_p''$ from periodic calorimetry is only influenced by the underlying cooling rate $\dot{T}$ at temperatures where the heat capacity $C_p(\dot{T})$ from DSC indicates partial freezing-in (schematic) [145]. (**d**) Third Law of thermodynamics for glasses (connection with the equilibrium liquid). $S_c(T)$ is the configurational entropy, $\Delta S'(T)$ the excess entropy when glass and crystal are compared via the equilibrium liquid: $\mathrm{d}\Delta S'/\mathrm{d}\ln T = C_p^{\mathrm{glass}}(T) - C_p^{\mathrm{crystal}}(T)$, $\Delta S' \to \Delta S_0'$ for $T \to 0$. $\Delta S'(T)$ and $\Delta S_0'$ depend on the cooling rate $\dot{T}$. (**e**) Third Law for glasses (Nernst–Simon formulation). The excess entropy S_{exc} [see (2.34)] for an arbitrary example of a single individual glass is put to zero, $S_{\mathrm{exc}} = 0$ for $T \to 0$, and is not therefore related to the liquid by crossing the glass transition (schematic). Both pictures are slight variations of overheads from a lecture given by Johari 1999. (**f**) Kauzmann construction for definition of T_2 = Kauzmann temperature

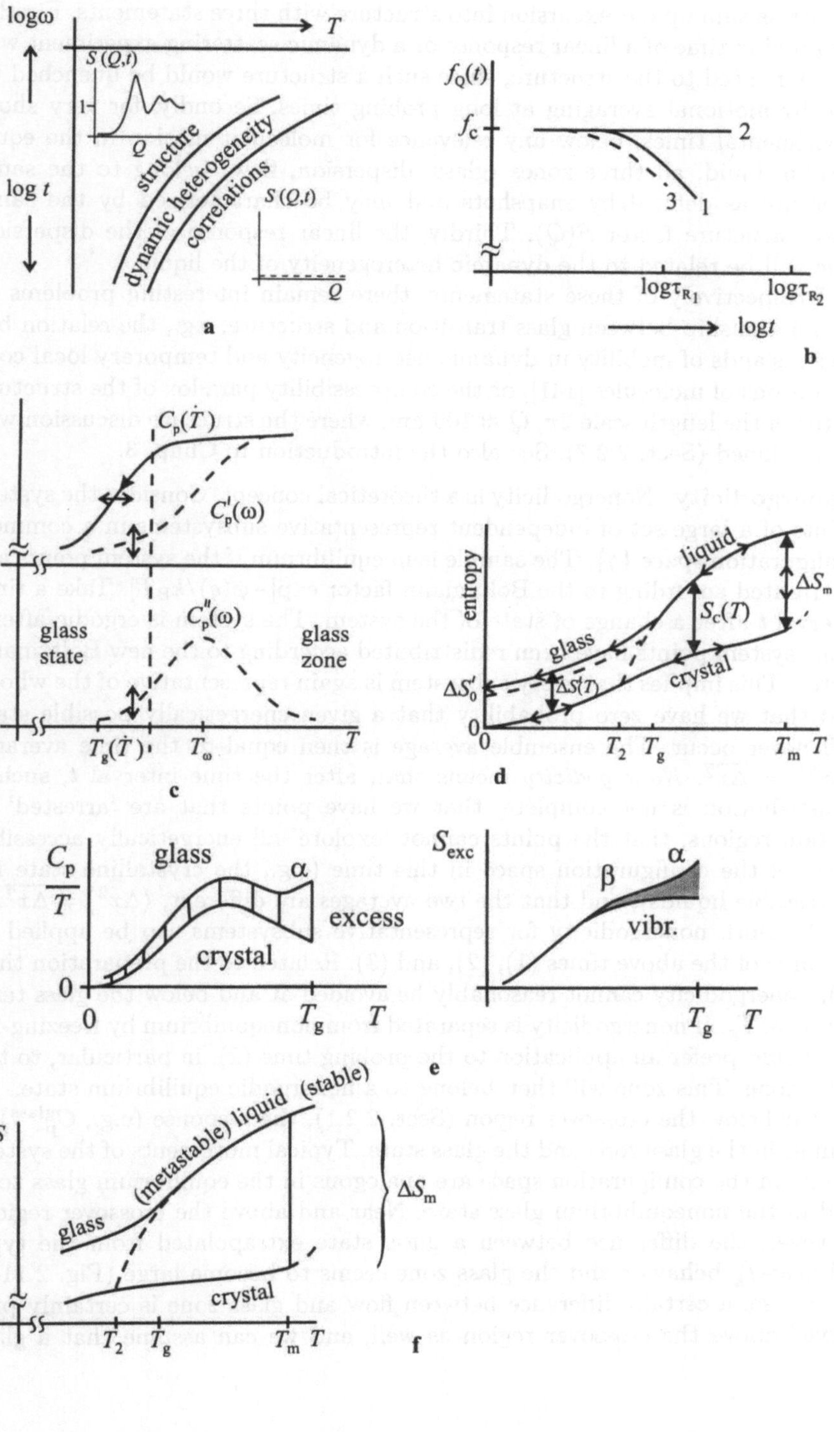
logω
T
S (Q,t)
1
Q
structure
dynamic heterogeneity
correlations
log t
S (Q,t)
Q
a
$f_Q(t)$
f_c
2
3
1
$\log\tau_{R_1}$
$\log\tau_{R_2}$
log t
b
$C_p(\dot{T})$
$C_p'(\omega)$
$C_p''(\omega)$
glass state
glass zone
0
$T_g(\dot{T})$
T_ω
T
c
entropy
liquid
ΔS_m
$S_c(T)$
glass
crystal
$\Delta S_0'$
$\Delta S(T)$
0
T_2
T_g
T_m
T
d
$\frac{C_p}{T}$
glass
α
excess
crystal
0
T_g
T
S_{exc}
β
α
vibr.
T_g
T
e
S
(metastable) liquid (stable)
glass
ΔS_m
crystal
T_2
T_g
T_m
T
f

Let us sum up our excursion into structure with three statements. Firstly, the probing time of a linear response or a dynamic scattering experiment will not be related to the structure, since such a structure would be quenched to zero by motional averaging at long probing times. Secondly, for very short experimental times, below any relevance for molecular motion in the equilibrium liquid, all three zones (glass, dispersion, flow) belong to the same structure as defined by snapshots and may be characterized by the same static structure factor $S(Q)$. Thirdly, the linear response of the dispersion zone will be related to the dynamic heterogeneity of the liquid.

Irrespectively of these statements, there remain interesting problems in the relationship between glass transition and structure, e.g., the relation between islands of mobility in dynamic heterogeneity and temporary local configurations of molecules [141], or the compressibility paradox of the structure factor in the length scale $2\pi/Q \approx 100$ nm, where the structure discussion will be continued (Sect. 2.2.7). See also the introduction to Chap. 3.

Nonergodicity. Nonergodicity is a theoretical concept. Consider the system points of a large set of independent representative subsystems in a common configuration space $\{q\}$. The sample is in equilibrium if the system points are distributed according to the Boltzmann factor $\exp[-\varphi(q)/k_B T]$. Take a time interval t after a change of state of the system. The system is ergodic after t if the system points have been redistributed according to the new Boltzmann factor. This implies that every subsystem is again representative of the whole, and that we have zero probability that a given energetically possible state will never occur. The ensemble average is then equal to the time average, $\langle \Delta x^2 \rangle = \overline{\Delta x^2}$. *Nonergodicity* means that, after the time interval t, such a redistribution is not complete, that we have points that are 'arrested' in certain regions, that the points cannot 'explore' all energetically accessible parts of the configuration space in this time (e.g., the crystalline state for metastable liquids), and that the two averages are different, $\langle \Delta x^2 \rangle \neq \overline{\Delta x^2}$.

A priori, nonergodicity for representative subsystems can be applied to all three of the above times (1), (2), and (3). Related to the preparation time (3), nonergodicity cannot reasonably be avoided at and below the glass temperature T_g. If nonergodicity is separated from nonequilibrium by freezing-in, we should prefer an application to the probing time (2), in particular, to the glass zone. This zone will then belong to a nonergodic equilibrium state.

Far below the crossover region (Sect. 2.2.1), the response (e.g., C_p^{glass}) is similar in the glass zone and the glass state. Typical movements of the system points in the configuration space are analogous in the equilibrium glass zone and in the nonequilibrium glass state. Near and above the crossover region, however, the difference between a glass state extrapolated from the typical near-T_g behavior and the glass zone seems to become large (Fig. 2.11c). Of course, a certain difference between flow and glass zone is certainly preserved above the crossover region as well, and we can assume that a glass state quenched hypothetically with large cooling rates above 10^6 K/s has a configuration-space behavior analogous to that of the glass zone there.

Let us end by discussing the nonergodicity problem with respect to the $f_Q(t,T)$ function of (2.32) for dynamic scattering at different temperatures T (Fig. 2.5b). Curves 1 and 2 are for equilibrium, curve 3 for nonequilibrium. Consider two cases:

- T_1 and $T_2 < T_1$ both far above T_g,
- T_1 and $T_2 < T_1$ both below T_g.

In the first case, curve 1 is the equilibrium curve at a temperature $T_1 > T_g$ where τ_{R1} is the relaxation time at T_1. Curve 2 is for T_2 below T_1 but higher than T_g again, $T_1 > T_2 > T_g$, with a much larger relaxation time τ_{R2}. Only when referring to a probing time t_2, $\tau_{R1} \ll t_2 \ll \tau_{R2}$, may curve 2 be called nonergodic, although in equilibrium. The plateau value of curve 2, $f_Q(t) = f_c$, is then called the 'nonergodicity parameter'. In the glass zone, $f_Q(t)$ is arrested at f_c. In the second case, curve 3 is an actual curve for a temperature T_1 below T_g, $T_1 < T_g$. Curve 1 is the equilibrium curve after a long waiting time, and the difference between curve 1 and curve 3 indicates the shift of structural relaxation in the glass state because of the acceleration in nonequilibrium. Curve 2 is the curve after a long waiting time at the lower temperature T_2, with the relaxation time τ_{R2}. Referring to the preparation time, curve 3 would be the only one related to nonergodicity.

Glass Zone. In my experience, it is in the glass zone that most problems of understanding arise. (An example is the Prigogine–Defay ratio of Sect. 2.5.1.) In the following the glass zone will first be distinguished from the liquid or flow zone in equilibrium (Fig. 2.2c), and then from the nonequilibrium glass state. Let us stress once more that all relevant things here are related to measurements of Fourier components of slow α process fluctuations.

Firstly, if we relate nonergodicity to the probing time, i.e., to the reciprocal frequency of the corresponding zone, then the glass zone is the nonergodic equilibrium frequency region, the liquid or flow zone is the ergodic equilibrium frequency region, and the dispersion zone defines the transition between the two, i.e., the dynamic glass transition. The properties of the glass zone are then the nonergodicity properties listed at the end of the first nonergodicity paragraph above. In anticipation of the FDT (Sects. 2.2.4 and 3.4), especially in its integral form (3.151)–(3.152), the glass zone susceptibilities do not contain the integral of spectral densities or correlation functions over the dispersion zone of the dynamic glass transition – the FDT integration is always over the higher frequencies or shorter times – whilst the liquid or flow zone susceptibilities do contain these integrals. The dispersion zone forms the steps between the two neighboring zones.

Secondly, the glass zone is distinguished from the glass state by the nonequilibrium of the latter, with the acceleration mentioned above and with a violation of the FDT (Sect. 4.5.2). There is no additional acceleration in the glass zone. In anticipation of terms for structural relaxation (Sect. 2.1.5), the fictive temperature T_f in the glass zone is equal to the average thermodynamic temperature, $T_f = T$, there is no partial freezing and no nonlinearity for temperature jumps ΔT smaller than the average temperature fluctuation δT of cooperatively rearranging regions (CRR), $\Delta T < \delta T$. In the glass state, $T_f \neq T$, we have partial freezing, and we can also have nonlinearity for small $\Delta T < \delta T$, down to ΔT of order 0.1 K [142] which is only a few per cent of the temperature fluctuation δT in CRRs.

The borderline between liquid zone and glass state is the thermal glass transition, in the time domain determined by the preparation time, e.g., $\tau_{\mathrm{prep}} \sim \delta T/\dot{T}$ (Sect. 4.5.1). It can even be shifted to temperatures above T_g by means of large heating or cooling rates $\dot{T}$, e.g., by a temperature or density jump after long (several days') tempering and equilibration near and below T_g (Ritland and Kovacs recovery experiments, Sect. 2.1.5). However, the occasional application of glass state concepts such as fictive temperature T_f or mixing parameter x to the glass zone [143, 144] is not appropriate to equilibrium response experiments for the dynamic glass transition above T_g. This application negates the fundamental difference between the time concepts of probing time (2) and preparation time (3). It does not contribute to the distinction between glass zone and glass state.

An illustrative example is Schick's comparison of $C_p(T)$ curves from dynamic and temperature modulated calorimetry near the glass temperature (Fig. 2.5c) [145]. Consider a periodic temperature–time program with a small temperature amplitude A_T (e.g., $\omega = 0.1$ rad/s, $A_T = 0.5$ K) and with an underlying cooling rate (e.g., $\dot{T} = -10$ K/min). The experiment gives a dynamic response $C_p^*(\omega) = C_p'(\omega) - iC_p''(\omega)$ and a freezing-in curve $C_p(\dot{T})$, with all three curves functions of temperature T. If we now vary the cooling rate $\dot{T}$ and look to see where the $C_p^*(\omega)$ curves are affected by this variation (double arrows in Fig. 2.5c), the result is that the $C_p^*(\omega)(T)$ functions are only affected at low temperatures where freezing-in is indicated by the decrease of the $C_p(\dot{T})(T)$ curves, i.e., on the left hand side of the vertical broken line. This line is a map of the preparation time τ_{prep} for the onset of freezing in the thermogram. The line shifts to the right (higher temperatures) for increasing cooling rate $\dot{T}$ of the freezing experiment and, relatively, for decreasing frequency ω of the dynamic experiment. The $C_p^*(\omega)(T)$ functions are not affected to the right of this line. Extrapolating this result to relatively small $\dot{T}$ or relatively large ω, we find that the equilibrium linear response, especially in the glass zone, is not affected by the nonlinearity characterized by $T_f \neq T$ and $x \neq 1$ as typical for the glass state.

Third Law of Thermodynamics. The complicated time-state problems below the glass temperature also generate problems with the Third Law of thermodynamics. They were analyzed by Simon [23] and resulted in the famous Nernst–Simon formulation of this law [146]. Consider a set of glasses vitrified at different cooling rates $\dot{T}$. Obviously, the products have different structures [147], and it follows from the upper part of Fig. 2.4d that they have different entropies, $S^{\text{cryst}}(T) + \Delta S'(T)$, if these are related to a given entropy of the equilibrium liquid above T_g (Fig. 2.5d). They also have different entropies at $T \to 0$, $\Delta S'_0(\dot{T})$, depending on the parameter $\dot{T}$. This contrasts with the pristine Nernst form of the third law which states that the entropy for $T \to 0$ must not depend on a further parameter. We may ask if the products may be regarded as different substances [148] even though they have identical chemical composition.

We should distinguish between configurational, excess, and relaxation properties.

- The term 'configurational properties' refers, as mentioned above, to the difference between liquid and crystal:

$$\Delta C_p(\text{conf}) = C_p^{\text{liquid}} - C_p^{\text{cryst}}\,, \quad S_c = \int^{\text{equil}} \frac{\Delta C_p(\text{conf})}{T} \mathrm{d}T\,. \tag{2.33}$$

  For use below T_g, liquid properties must be extrapolated to the lower temperatures.
- The term 'excess properties' refers to the difference between glass and crystal:

$$\Delta C_p(\text{excess}) = C_p^{\text{glass}} - C_p^{\text{cryst}}\,, \quad S_{\text{excess}} = \int_{T=0} \frac{\Delta C_p(\text{excess})}{T} \mathrm{d}T\,, \tag{2.34}$$

  where $S_{\text{excess}} = 0$ for $T = 0$ is usually assumed (Fig. 2.5e). If, on the other hand, entropies are compared via the equilibrium liquid, we obtain $S_{\text{excess}}(T) = \Delta S'(T)$ and $S_{\text{excess}}(0) = \Delta S'_0$ of Fig. 2.5d. For use of S_{excess} above T_g, glass properties can either be extrapolated from the glass state to the higher temperatures or, alternatively, can be taken from the real part of the dynamic heat capacity in the glass zone. The difference between the $T > T_g$ alternatives comes from entropy recovery of the glass state and is small near T_g, but can become large near the crossover region, usually far above T_g (Figs. 2.11c and 2.31h).
- The term 'relaxation intensity', referring to the step height of the linear response ΔC_p as used so far, was defined between the liquid and glass zones,

$$\Delta C_p = C_p^{\text{liquid}} - C_p^{\text{glass}} = \Delta C_p(\text{conf}) - \Delta C_p(\text{excess})\,. \tag{2.35}$$

  The second equality in (2.35) holds if C_p^{glass} is extrapolated from the real part C'_p of the glass zone.

The definition of configurational properties by (2.33) is more or less formal [149]. The configurational entropy at T_g is influenced by the excess difference in vibrational contributions between glass and crystal, the contribution of the boson peak and the Johari–Goldstein β process, perhaps by a combinatorial effect embracing an increasing number of β processes [150, 511], and the contribution of the high-frequency tail of the α process near but below T_g. The latter will be increased by the vault effect in Sect. 2.2.6. As an example, the frozen free volume in the glass provides more room for vibrations so that C_p(vibr. glass) > C_p(vibr. crystal) can be expected. The situation is sketched in Fig. 2.5e. In our book, however, for simplicity we always refer to (2.33) when configurational properties are used.

Table 2.4 lists the entropy data for 33 substances from Johari's collection [89, 149].

The entropy problem of the first paragraph was solved by Simon. Thermodynamic paths are not only linear sets of 'static' states but they must also be dynamically passable, e.g., by overlapping of the individual fluctuations between neighboring state points. This is not the case for our set of glasses with different structures far below T_g.Their structures are frozen-in and cannot be changed to overlap at low temperatures. Simon therefore added a corresponding restriction to the Nernst formulation of the Third Law: the entropy change associated with any isothermal reversible process of a condensed system approaches zero as the temperature approaches zero (*Nernst–Simon form of the Third Law of thermodynamics*).

The advantage of this formulation is that the usual thermodynamics including the Third Law can be applied to any individual glass far below T_g, i.e., for any glass without structural relaxation in the experimental time. Individuality can be named by the cooling rate $\dot{T}$ at a given pressure, or a compression rate at a given temperature, and so on. In particular, we can put $S \to 0$ for $T \to 0$ for each individual glass. We can then compare the different glasses by their individual excess entropies as a function of temperature, $S_{\mathrm{excess}}(T; \dot{T})$, as long as the fluctuation overlap between neighboring glasses can be excluded (Fig. 2.5e). On the other hand, the comparison can be started above T_g, and we have individual excess entropies $\Delta S'(T; \dot{T})$ and $\Delta S'_0(0; \dot{T})$ values (Fig. 2.5d). However, the determination of configurational properties using (2.33) only requires measurements in the equilibrium liquid and crystal states, not in the glass state.

A long-running issue is the 'equilibrium glass'. Other names are the 'dense amorphous state' [26] or 'ideal glass'. The first question is whether an infinitesimally slowly cooled liquid can produce a glass that is accessible by equilibrium thermodynamics or statistical physics [23]. The second question is whether an ideal glass has an 'ideal glass transition', e.g., near the Kauzmann temperature (see below), which has more than merely dynamic reasons. This question will be discussed in Sect. 2.5.1. It is assumed that S (ideal glass) $\to S$ (crystal) $\to 0$ for $T \to 0$. One way of producing an ideal glass in finite times may be to cross the Π or the Θ asymptote in the pressure–temperature plot (Fig. 5.2d).

Table 2.4. Entropy parameter [89, 149]. $S_c(T_g)$ the configurational entropy at the glass temperature T_g, $\Delta S_0'$ that at $T = 0$ K, $S_{exc}(T_g)$ the excess entropy at T_g

Material	T_g [K]	$S_c(T_g)$ [J/mol K]	$\Delta S_0'$ [J/mol K]	$S_{exc}(T_g)$ [J/mol K]
1-butene	60	19.8	12.8	7
1-pentene	70	27	17.8	9.2
Iso-pentane	65	19	14.1	4.9
3-methylpentane	77	25.1	23.6	1.5
Toluene	117	12.2	7.9	4.3
Ethylbenzene	115	23.5	10.0	13.5
Iso-propylbenzene	126	21	12	9
2-methyltetrahydrofuran	91	23	13.6	9.4
o-terphenyl	240	21.5	15	6.5
Tri-α-naphthylbenzene	342	50.2	33	17.2
d, l-propylene carbonate	160	17.5	9	8.5
Triphenylphosphite	200	37.8	21.5	16.3
3-bromopentane	107	21.6	13.8	7.8
Triphenylethene	248	45.4	18.6	26.8
Ethanol	97	13.6	8.9	4.7
Ethylene glycol	153	17	9	8
1-propanol	100	20.3	11.3	9
1, 3 propanediol	146.8	22	13	9
Glycerol	185	26.9	23.4	3.3
2-methyl-1-propanol	112	20.5	9	11.5
2-butanethiol	92	18.8	16.2	2.6
2-methyl-1-propanethiol	96	17.4	14	3.4
3-methyl-1-butanethiol	100	21.5	13.9	7.6
Salol	220	27.1	14.3	12.8
Butyronitrile	97	12.6	5.4	7.2
Diethylphthalate	180	23	20	3
d, l-lactic acid	204	10.2	8.5	1.7
P-M siloxane	167	30.5	22.9	7.6
$H_2SO_4.3H_2O$	157	34.8	24.7	10.1
B_2O_3	521	24.5	9.9	14.6
Selenium	300	6.3	3.7	2.6
PBBA (smectic, $T_{g,2}$)	185	29.2	9.4	19.8
HBBA (smectic, $T_{g,2}$)	207	36.9	7.5	29.4

The time dependence of thermal properties (e.g., structural relaxation below T_g or heat release at $T < 1$ K, Sect. 2.4.3) touches a general question of thermodynamics: What is measured, in comparison to fluctuations, that can possibly be calculated by statistical physics? This concerns the violation of the FDT and will be discussed in Sect. 4.5.2. Briefly, I think that a certain, small 'part of fluctuation' is needed to drive the acceleration towards equilibrium, whilst the large 'remainder' can be measured as a susceptibility, for example. Finally, thermodynamics has to do with the susceptibilities accessible to laboratory experiments.

An interesting example in relation to Fig. 2.5d is the Kauzmann paradox (see [151] and Fig. 2.5f). The question here is not nonequilibrium in the glass state but a guess about how the equilibrium liquid configurational entropy should be extrapolated to low temperatures $T < T_g$. As the melting entropy is positive, $\Delta S_m > 0$, the liquid entropy is larger than the crystal entropy at the melting temperature T_m. If the heat capacity of the liquid is larger than that of the crystal, $\Delta C_p(\text{ conf}) > 0$, the liquid $S(T)$ slope is larger than the crystal $S(T)$ slope. If the liquid slope is extrapolated in a straightforward manner, with progressively increasing liquid entropy slope, but without a smooth Third Law requirement $S_{\text{liq}} \to S_{\text{cryst}}$ for $T \to 0$, i.e., without an additional input at small configurational entropy, we get an intersection of the extrapolated equilibrium liquid curve and the crystal curve. The intersection temperature is called the *Kauzmann temperature* T_2. The extrapolated situation below T_2 is considered to be paradoxical, because intuitively the classical liquid is connected with a larger entropy than the crystal. This paradox, also called the entropy crisis, has become famous because [72,122,152] the Kauzmann temperature is often near the Vogel temperature for the α process obtained from an extrapolation of the dynamic glass transition by the WLF equation, or of the viscosity by the VFT equation,

$$T_2 \approx T_0 \ . \tag{2.36}$$

This finding supports the Adam–Gibbs approach (2.28) which connects the slowing down of the liquid mobility, i.e., the dynamic glass transition, with the configurational entropy.

The Kauzmann construction Fig. 2.5f, however, is not the key to the glass transition, although the paradox is interesting as such. (We may ask whether there is an additional phase transition in the liquid at T_2 that could serve as an ideal glass transition (Sect. 2.5.1) and that could thereby dissolve the paradox.) There are two points:

- for several strong glasses, the liquid and glass $S(T)$ curves are nearly parallel so that the construction fails [34];
- there are glasses without any metastability of the state above T_g [132,154] (Table 2.5). Metastability is not a precondition for the glass transition. The use of 'supercooled liquid' as a synonym for a 'glass-forming substance' is therefore misleading.

Table 2.5. Glass temperatures T_g of glassy crystals from Suga's collection [153]

Metastable phase		T_g [K]	Stable phase		T_g [K]
Thiophene		37	Thiophene		42
2,3-dimethylbutane		76	Buckminsterfullerene		87
Isocyanocyclohexane		55	β-cyclodextrin$\cdot 11H_2O$		150
		130	Ethylene oxide$\cdot 6.86H_2O$		85
		160	Tetrahydrofuran$\cdot 17H_2O$		85
$CFCl_2$–$CFCl_2$		60	Acetone$\cdot 17H_2O$		90
		90	CO		18
		130	RbCN		30
Ethyl alcohol		90	$CsNO_2$		42
Cyclohexene	I	92	$TlNO_2$		60
	II	93	$SnCl_2 \cdot 2H_2O$		150
Cycloheptane	I	100	$SnCl_2 \cdot 2D_2O$		155
	II	100	H_2O	hexagonal	105
	III	93		cubic	140
Cycloheptatriene		106	D_2O	hexagonal	115
Cycloheptanol		135	Pinacol$\cdot 6H_2O$		155
Cyclohexanol		150	H_3BO_3		290
$Cs_{0.7}Tl_{0.3}NO_2$		48	D_3BO_3		298
			Lysozyme		ca. 150
			Myoglobin		ca. 170

Rather than the melting temperature T_m, a better distinction with regard to the glass transition is the crossover temperature T_c separating cold ($T < T_c$) and warm ($T > T_c$) liquids (Sect. 2.1.1). An Adam–Gibbs correlation (2.28) adjusted to the cold liquid does not fit the warm liquid [72]. More details about the Kauzmann construction have recently been reviewed by Johari [155].

2.1.5 Structural Relaxation. Landscape. Nonlinearity

Structural relaxation. The *structural relaxation* of a glass refers to the change of structure whilst waiting at a temperature T near or below the glass temperature T_g (Fig. 2.6a). A glass vitrified at T_g has a structure which is not identical to that of the extrapolated equilibrium liquid at the waiting temperature $T \lesssim T_g$. We observe a relaxation of structure, volume, entropy, and so on in direction of the extrapolated equilibrium values. In engineering terms, the structural relaxation is accompanied by a *recovery* of extensive variables such as volume, entropy, and so on. In a wider sense, structural relaxation means the behavior of such quantities during and after a temperature–(pressure–composition–) time program near and below T_g.

Compliances decrease, moduli increase, and molecular mobility decreases by structural relaxation during waiting. The first two facts ('hardening') follow from Fig. 2.4d. Waiting means that the compliance steps shift to lower temperatures. The mobility decrease is determined by the objective of structural relaxation – the equilibrium mobility at $T < T_g$ is lower than that at T_g – and by acceleration due to the nonequilibrium (Sect. 2.1.4). The acceleration decreases when equilibrium is approached. Inverse tendencies in the compliances, observed after certain perturbation programs [156], are called 'softening'.

As mentioned in Sect. 2.1.4, structural relaxation is an irreversible process since we observe entropy production, i.e., entropy recovery. On the other hand, it is a thermoreversible process because further heating above T_g would give the same equilibrium liquid properties as before, if there were no chemical reaction in the waiting stage.

An application of the term structural relaxation to the dynamic glass transition is misleading, especially if it refers to the linear response in the main dispersion zone of the equilibrium liquid. In the glass, the structure is improved. The structure factor $S(Q)$ gets more contrast and more details during waiting. In the liquid, the structure before and after response to a pulse perturbation in the dispersion zone is the same. Motional averaging as observed in dynamic scattering from a liquid, e.g., represented by the decay of the intermediate scattering function, $S(Q,t) \to 0$ for $t \to \infty$, is not related to nonequilibrium of the glass state but only to nonergodicity of the equilibrium glass zone (Sect. 2.1.4). It makes no sense to call this decay a structural relaxation. Structural relaxation aims in the other direction, not towards a decay but rather towards the small improvements in structure mentioned above. Nevertheless, some people still use this term for the main transition in the liquid. The term 'structural glasses' is sometimes used to distinguish conventional glasses from orientational or spin glasses (Sect. 2.2.8).

Strictly speaking, we also find a certain improvement in structure during a relaxation experiment for linear response after a jump perturbation in the time domain. But the aim is different. For linear response, we are interested in equilibrium fluctuations, whilst for structural relaxation, we are interested in the change of properties.

The terms 'physical aging' and 'glass stabilization' (Sect. 2.1.4) are sometimes restricted to structural relaxation far below T_g where the typical time scale is much longer than the 'vitrification time' of about 1000 seconds at T_g.

Landscape. Simon [23] concluded from Fig. 2.6a that the structural relaxation $B \to A'$ cannot be described by overcoming a common barrier for correlated particles. He identified the structures at $T(B)$ and $T_g(A)$, and identified the structure change for the structural relaxation $B \to A'$ and for the thermodynamic equilibrium path from $T_g \to T < T_g$, $A \to A'$. The latter is a succession of infinitesimally close equilibrium states which certainly cannot be realized by an activated transition state with a common barrier.

To save the picture of local, molecular saddles on an energy surface we must go to the high-dimensional phase space or configuration space $\{q\}$ of representative subsystems. Any typical one-dimensional picture (Fig. 2.6b, top) of molecular glass-transition movements in the configuration space was later called an *energy landscape*. This is often used for illustrative discussions (see for example [41, 72, 122, 157–160]).

The *landscape* concept is not only applied to one dimension ($d = 1$) and to structural relaxation at $T \lesssim T_g$. For protein folding, two-dimensional ($d = 2$) landscape funnels [161] are used to discuss routes to the native state. The latter is thought to be near the tube of the funnel ('landscape paradigm'). In more recent theoretical work [158, 162–164], the full surface of potential energy or free energy in the configuration space $\{q\}$ for representative subsystems is referred to as a landscape, as it is also for molecular motion in warm liquids far above T_g. In this section, the term landscape will be restricted mainly to one-dimensional pictures and evaluations.

Let us assume that the intermolecular energy φ depends only on the coordinates q of the configuration space, $\varphi(q) = E_{pot}(q)$, and that all glass-transition properties can be derived from such a $\varphi(q)$ (Sects. 2.5.1 and 3.1). The $\varphi(q)$ surface on $\{q\}$ does not depend on the temperature. On the other hand, because of the Boltzmann factor $\exp[-\varphi(q)/k_BT]$, different thermodynamic states sample and weight different parts of the surface differently. The one-dimensional route across the landscape along a 'collective coordinate' should be typical for the slow dynamics of the glass transition in a given state, e.g., at a given temperature. The collective coordinate can be a spatial coordinate q or (e.g., [166]) a time coordinate t along a typical trajectory for the system point in the configuration space. To draw simple one-dimensional pictures means to draw different landscapes that can be 'observed' at different temperatures by the moving point that is representative of the whole subsystem. Briefly, the one-dimensional landscape depends on the state $(T, p, \dots)$, whereas the potential energy surface $\varphi(q)$ in the high-dimensional configuration space $\{q\}$ does not. The evaluation of the representative subsystem should be functional, i.e., only related to the Fourier components of fluctuations in the dispersion zone of the dynamic glass transition.

In the middle part of Fig. 2.6b, landscapes are suggested (partly following Angell [159]) that could perhaps be typical for different temperature regions along the main transition (Fig. 2.1c). Much experience seems necessary to work with these pictures. Anticipating some concepts of new computer simulations (Sect. 2.3.4), I will make three personal remarks.

First Remark. The one-dimensional landscape for the α process is usually characterized by the term 'ruggedness'. I think, alternatively (Fig. 2.6b), that molecular cooperativity of the α process means there are many competing possibilities which can only be distinguished by small energy differences. This in turn means that the effective roughness of the landscape (as a measure of its ruggedness) must be small below the crossover region, i.e., in the cold liquid. If such roughness is connected with the Vogel temperature (Sect. 2.4.1, Fig. 2.24f), we conclude from the bottom part of Fig. 2.1c that the roughness of the α process landscape is actually smaller than that of the high-temperature a process landscape above the crossover region, i.e., in the warm liquid.

In warm liquids, there is no need for cooperativity because the rough landscape corresponding to the many equivalent escaping-the-cage processes for the a process (Sect. 2.3.2) is easily accessible at high temperatures. In cold liquids, the rugged escaping landscape is not accessible, and subsystem points must use the comparatively few (but nevertheless numerous!) cooperative possibilities for the α process, which had only relatively small weight in warm liquids.

In the geometric $d = 3$ space, the cooperatively rearranging regions CRRs are 'separated' (from one another, Sect. 3.1) by statistical independence. The question seems to be how this independence can be mapped in the $\{q\}$ space and, from there, into the landscape. Of course, energetic decoupling implies statistical independence. So one could try to mark the CRRs by higher energy barriers (as indicated by a question mark in Fig. 2.6b). Energetic decoupling is, however, not a necessary condition for statistical independence. The query is chosen because such a barrier is not therefore necessary for definition of the CRR, and is only related to the functionality of the main transition. High barriers are used to separate craters [41] in the landscape. Their identification with CRRs is thus questionable.

A further problem arises with Goldstein's suggestion [158] that the α process "is dominated by potential barriers high compared to thermal energies". In this case, the query means: are the barriers so local that they

▶

Fig. 2.6. Structural relaxation. (**a**) Simon's picture. The 'molecular order' increases during a waiting period below T_g. (**b**) Energy landscape. *Top*: the collective coordinate q is the one-dimensional path in the high-dimensional configurational space that is thought to be representative for the slow mobility and the state of the system point. $\varphi(q)$ is the corresponding section of the energy hypersurface. The system prefers local minima, and attacks local barriers with a certain success-to-attempt rate (schematic). *Middle*: landscape types for the main transition. **1**. Temperature regions. **2**. Relevant processes. **3**. Relation to the excitation profile. **4**. Ruggedness. *Question mark*: at high temperature, a larger ruggedness could easily be overcome. In the cold liquid, the 'more rugged' variant is Angell's suggestion [159]. I think of the less rugged alternative as resulting from α process cooperativity (see text). **5**. Landscape suggestions. The *question mark* in the alternative is related to the question of whether statistical independence for CRRs must really be symbolized by a high barrier (see text). The *question mark* at the ideal glass transition means that the realization refers to Angell's suggestion of larger ruggedness. Higher cooperativity in the alternative would correspond to a flat landscape at the ideal glass transition. *Bottom*: excitation profile after local quenching into internal structures [160]. N is the nonergodicity for an example with too large a cooling rate

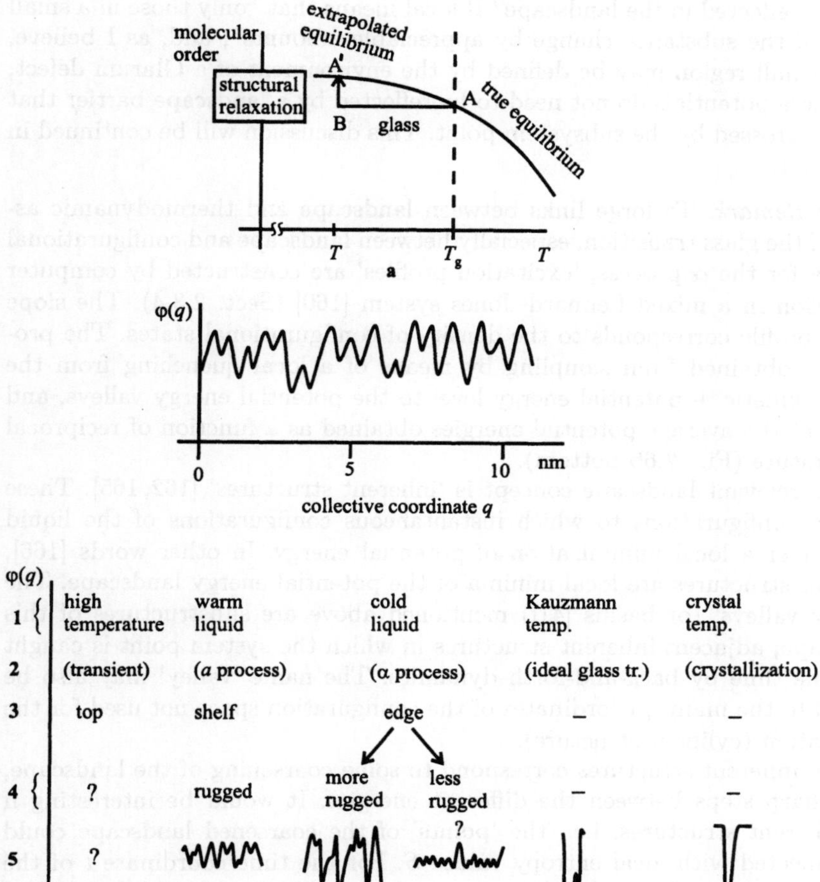
molecular order
extrapolated equilibrium
A'
structural relaxation
A
B
glass
true equilibrium
T
Tg
T
a
φ(q)
0
5
10
nm
collective coordinate q
φ(q)
1
high temperature
warm liquid
cold liquid
Kauzmann temp.
crystal temp.
2
(transient)
(α process)
(α process)
(ideal glass tr.)
(crystallization)
3
top
shelf
edge
–
–
4
?
rugged
more rugged
less rugged
–
–
5
?
?
?
collective coordinate q
Epot
N
?
Tc-1
Tg-1
T-1
b

must be reflected in the landscape? If local means that "only those in a small region of the substance change by appreciable amounts", and, as I believe, that a small region may be defined by the environment of a Glarum defect, then these potentials do not need to be reflected by a landscape barrier that must be crossed by the subsystem point. This discussion will be continued in Sect. 2.2.1.

Second Remark. To forge links between landscape and thermodynamic aspects of the glass transition, especially between landscape and configurational entropy for the α process, 'excitation profiles' are constructed by computer simulation in a mixed Lennard–Jones system [160] (Sect. 2.3.4). The slope of this profile corresponds to the density of configurational states. The profiles are obtained from sampling by means of a local quenching from the total = kinetic + potential energy level to the potential energy valleys, and represent the average potential energies obtained as a function of reciprocal temperature (Fig. 2.6b bottom).

The relevant landscape concept is 'inherent structures' [162, 165]. These are the configurations to which instantaneous configurations of the liquid map under a local minimization of potential energy. In other words [166], inherent structures are local minima of the potential energy landscape. The 'energy valleys' (or basins [41]) mentioned above are substructures of this landscape, adjacent inherent structures in which the system point is caught for some time by back-and-forth dynamics. The name 'valley' may also be related to the many q coordinates of the configuration space not used for the localization (cylinder structure).

The inherent structures correspond to some coarsening of the landscape, with sharp steps between the different energies. It would be interesting if the inherent structures, i.e., the 'points' of the coarsened landscape could be connected with local entropy values S. For the time coordinate t of the landscape, we would get a stationary stochastic $S(t)$ function that could be used to calculate $\overline{\Delta S^2} = k_B C_p$ fluctuations and $\Delta S^2(t)$ correlation functions, at least for longer time intervals. From a comparison with $C_p = (\partial H/\partial T)_p$ and $\overline{\Delta H^2}$ [(3.29) and (3.41)], we can perhaps decide between Gibbs and von Laue treatments of thermodynamics (Sects. 3.2.1 and 3.2.2).

The Fig. 2.6 profile is approximately linear in the α process range and bends to a smaller slope above the crossover temperature T_c. The simplest connection with glass transition dynamics is made by identifying the profile energy with the relevant part of the internal energy U, and driving the system by configurational entropy S_{conf} via minimization of free energy $F = U - TS \approx E_{pot} - TS_{conf}$. Here, S_{conf} is linked to dynamics by the Adam–Gibbs correlation (2.28).

Third Remark. The map from the high-dimensional configuration space $\{q\}$ onto the common three-dimensional real space reflects a concept of quantum physics. The configuration space for the energy surface $\varphi(q)$ of representative subsystems is the same as the configuration space for quantum mechanics.

In quantum mechanics, the reduction to the real space is mediated by the quantum mechanical (QM) experiment, that is, the QM measurement process characterized by a collapse or reduction of the wave function. The QM experiment may be more than the substitution of $E(q)$ into the Gibbs distribution of statistical physics. I think that this *reduction map* from configuration statistics to thermodynamics of subsystems can be described [167] by the fluctuation–dissipation theorem FDT for small perturbations in the equilibrium. The question here is whether the FDT contains temperature fluctuations not allowed by the Gibbs distribution. This discussion will be continued in Sects. 3.2–3.4.

The reduction map reduces correlations and nonlinearities of the mechanical chaos behind thermodynamics to linear response, and the many-time mechanical correlation functions in the configuration space to two-time correlation functions of thermodynamic fluctuations for subsystems. From this point of view, the different landscapes are illustrative representations of the reduction map. Since this map is a mathematical homomorphism, the pull-back to the configuration space, and the following map to the landscape, is always arbitrary to a certain degree, especially as long as the mechanical problems are not solved. Experimental methods for exploring the underlying mechanics are 3- or 4-dimensional NMR methods [168, 169], for example. The discussion of the relationship between mechanics in individual substances (with their multifarious structures) and general dynamics will be continued in Sects. 2.2.5 and 2.2.6. (The keyword here is 'disengagement'.)

Nonlinearity in Structural Relaxation. This subsection describes the nonlinearity in the experimental approach to structural relaxation. Nonlinearity for the response in equilibrium will be discussed in the next section. Briefly, the nonlinearity in structural relaxation is large, of first order in $\tilde{\Delta T}/\delta T$, because the time-reversal symmetry of the equilibrium is broken. This symmetry causes nonlinearity in the equilibrium to be small, in fact of second order $(\Delta T_0/\delta T)^2$. In these estimates, $\tilde{\Delta T}$ is the amplitude of a certain 'temperature deviation from equilibrium' (2.43), ΔT_0 is the temperature amplitude of the experiment, e.g., of temperature-modulated DSC, and δT the average temperature fluctuation of a cooperatively rearranging region CRR near T_g.

Consider isothermal volume or entropy recovery after a jump *from an equilibrium state* (see Fig. 2.7 and [27, 170]). After such a jump from equilibrium at higher temperature to the experimental temperature in the transformation interval, we observe structural relaxation by volume contraction. After a jump from equilibrium at lower temperature (we have to wait a long time for equilibration there!) to the same temperature in the transformation interval, we observe structural relaxation by volume expansion. Even for the same temperature-jump amplitudes, recovery curves are not symmetrical as long as the temperature deviations $\tilde{\Delta T}$ are larger than several tenths of one kelvin, e.g., for polyvinylacetate PVAC, which means several per cent of the transformation interval. The expansion is then slower than the contraction. The recovery amplitudes for contraction and expansion are different. These effects are called *nonlinearity*.

The six parts of Fig. 2.7 give detailed information about the famous Ritland–Kovacs [27, 170] experiments. The temperature–time programs in Fig. 2.7a are symmetric, whilst volume recovery shown in Fig. 2.7b is asymmetric, i.e., nonlinear. The example in Fig. 2.7c is for PVAC with $T_g = 35°C$ and jump amplitudes $|\Delta T| = 5$ K, over a logarithmic time scale (order hours), and uses reduced volumes δ and local effective retardation times τ_{eff} :

$$\delta(t) = \frac{V(t) - V(t=\infty)}{V(t=\infty)} , \quad \tau_{\text{eff}} = -[\mathrm{d}\delta(t)/\delta(t)\mathrm{d}t]^{-1} . \tag{2.37}$$

The effective time is defined as a retardation time which is locally equivalent to an exponential decay. The use of $\mathrm{d}\ln|\delta(t)|$ liberates us from the initial expansion/contraction asymmetries. The difference in the recovery amplitudes for short times results from faster parts of the structural relaxation (which will be discussed in Sect. 2.2.6).

The situation immediately after the jump (Fig. 2.7d) is characterized schematically by a sudden shift of the borderline between glass state and glass zone in the direction of higher temperatures. This holds for both heating and cooling jumps, $\pm\Delta T$, and is caused by freezing-in due to the short preparation time, $\tau_{\text{prep}} \sim \delta T/\dot{T}$, short for steep jumps with large $|\dot{T}|$. Both expansion and contraction start then with partially frozen material, irrespective of the previous equilibration. During structural relaxation, this borderline 'relaxes' to lower temperatures.

The later stages of structural relaxation, where $|\delta|$ is small, of order a few times 10^{-3}, are traditionally represented in a Kovacs plot, $\log\tau_{\text{eff}}$ vs. δ (Fig. 2.7e). The asymmetry in this plot is dramatic. A *Kovacs expansion gap* is sometimes observed. This means that the formal extrapolation of τ_{eff} from the $\delta = -10^{-3}$ range to $\delta \to -0$ yields different τ_{eff} values for different perturbation jump amplitudes ΔT. No indication for such a gap in contraction ($\delta > 0$) has yet been observed. The more or less horizontal line parts, $\tau_{\text{eff}} \approx$ const., mean a local exponential decay independent of δ in a certain δ interval within the $\delta \approx -10^{-3}$ range.

In the final stage, for $|\delta|$ of order 10^{-4} and smaller, we observe a steep and symmetric increase in the effective time. This comes from equilibrium nonexponentiality. Exponential decay in equilibrium would mean $\delta =$ const., independently of δ. However, for a Kohlrausch function (2.17), we have [171]

$$\tau_{\text{eff}} = \frac{\tau_{\text{KWW}}}{\beta_{\text{KWW}}}\left[\ln\frac{\delta_0}{\delta(t)}\right]^{(1-\beta_{\text{KWW}})/\beta_{\text{KWW}}} = \frac{\tau_{\text{KWW}}}{\beta_{\text{KWW}}}\left[\frac{t}{\tau_{\text{KWW}}}\right]^{(1-\beta_{\text{KWW}})} , \tag{2.38}$$

with δ_0 the start value after the jump (Fig. 2.7f). Equation (2.38) describes an equilibrium response that is symmetrical for $\delta \leftrightarrow -\delta$. The final stage of jump experiments is equilibrium retardation of the long-time tails of nonexponential decays. All frozen-in parts are 'thawed' for $|\delta| \lesssim 10^{-4}$ and recovery is described by a linear compliance.

Nonlinearity in Equilibrium. To understand the term nonlinearity for structural relaxation, e.g., for the asymmetries of the Kovacs plot, we should also briefly discuss nonlinearity in equilibrium response at higher temperatures $T > T_g$. The situation is characterized by large perturbation amplitudes in a material with no frozen parts. The first example is 'shear thinning' for large shear rates $\dot{\gamma}$, a typical effect of a 'nonlinear viscoelastic phenomenon in shear' at the flow transition (FT) in polymers. Good reviews can be found in [172, 173], for example. Unfortunately I have not found in the literature analogous experiments at the α relaxation of small-molecule glass formers. The flow transition in polymers, however, can also be considered as some kind of glass transition (Sect. 2.4.1). (This transition is rather sensitive to polymer chain individualities, so that in exceptional cases 'shear thickening' will be observed instead.)

We start with a picture that stems wholly from linear response (upper part of Fig. 2.8a). Define a complex viscosity $\eta^* = G^*/\mathrm{i}\omega$ via the shear modulus G^* and consider the viscosity modulus $|\eta^*|$ [or the reduced real part, $\eta' = G''/a_T\omega$ (2.59)] as a function of reduced frequency $a_T\omega$. This function is often similar to the nonlinear picture for shear thinning, where the reduced experimental viscosity η/η_0, with η_0 being $\eta = \sigma/\dot{\gamma}$ for $\dot{\gamma} \to 0$, is drawn as a function of a reduced shear rate $\dot{\gamma}\eta_0$ (lower part of Fig. 2.8a). This similarity, with $\omega = \dot{\gamma}$, is called the Cox–Merz rule [174]. I think, in case of fulfillment, that both graphs reflect the same physical situation. Consider shear oscillation $\gamma = \gamma_0 \sin(\omega t)$. We then have a combined scaling. The shear rate $\dot{\gamma} = \gamma_0\omega\cos(\omega t)$ can be raised by increasing the frequency ω for given amplitude γ_0 – corresponding to an isothermal shift in the Arrhenius diagram towards the FT dispersion zone – and by increasing the shear amplitude γ_0 for given frequency – corresponding to no shift. This example shows that caution and experience are required for a discussion of nonlinearities at the dynamic glass transition.

The second example of nonlinearity without freezing is periodic calorimetry (Temperature Modulated DSC = TMDSC) with large temperature amplitudes ΔT_0, due to Schick et al. [175]. The standard of comparison is δT, the average temperature fluctuation of the relevant representative functional subsystem, the cooperatively rearranging regions CRR (Sect. 2.1.3). The value of δT will be estimated in (2.139) as $\delta T \approx 0.65\, T_g/m$, decreasing with fragility m. Nonlinearity is thus expected for

$$\Delta T_0 \gtrsim \delta T \approx 0.65\, T_g/m\,, \tag{2.39}$$

i.e., especially for large fragility m where δT is small. The problem is to know the relevant order of $\Delta T_0/\delta T$.

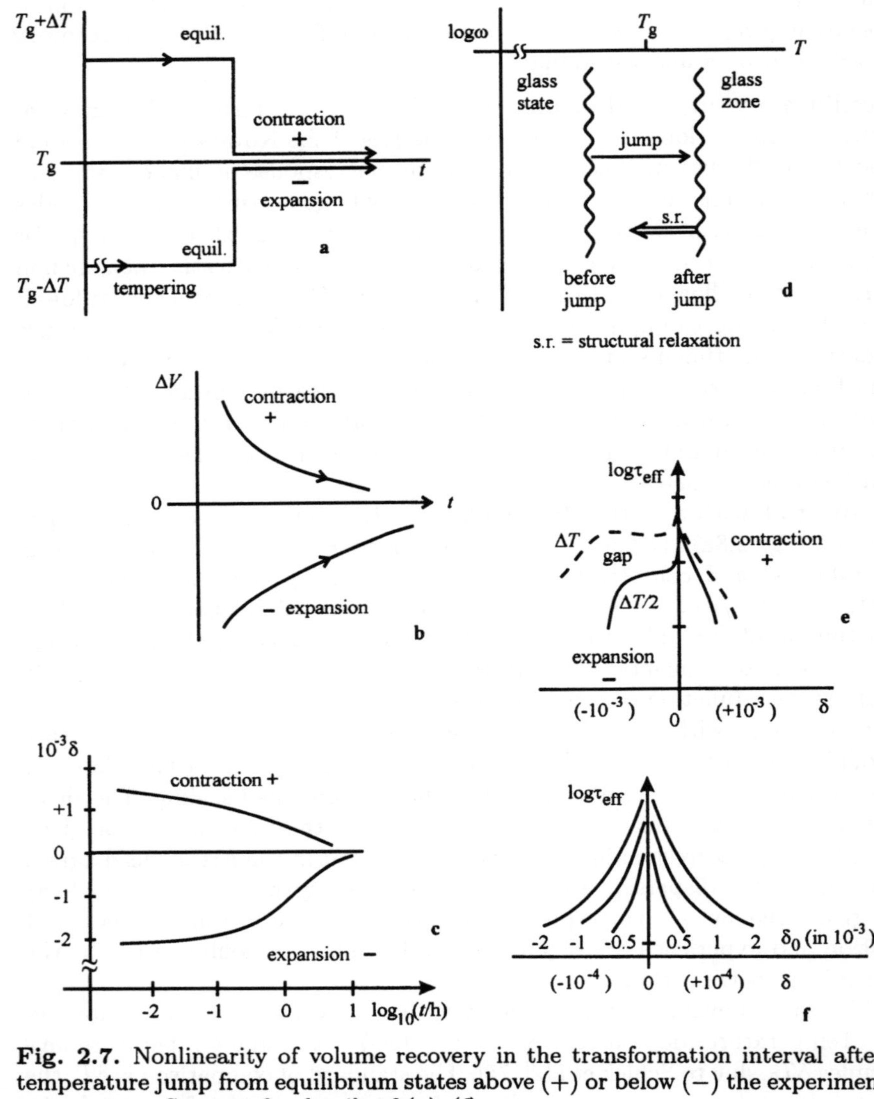

Fig. 2.7. Nonlinearity of volume recovery in the transformation interval after a temperature jump from equilibrium states above (+) or below (−) the experimental temperature. See text for details of (**a**)–(**f**)

The calorimetric responses [175] must first be corrected for the trivial nonlinearity due to the sigmoid shape of the ΔC_p step at the dynamic glass transition, and should then be reduced by ΔC_p to compare them with the glass-transition part of the heat capacity. The first order is the response itself. The second order is quenched by time reversal symmetry [175]. The result of the TMDSC experiments is that nonlinearity is small for $\Delta T_0 < \delta T$ and that the third-to-first order ratio of response amplitudes, reduced by $(\Delta C_p/C_p)^{3-1} = (\Delta C_p/C_p)^2$, is of order several times ten percent for $\Delta T_0 \approx 2\delta T \approx 5$ K in the case of PVAC. Compared to the first order nonlinearity of structural relaxation, the nonlinearity of equilibrium is second order.

The equilibrium nonlinearity is thus much smaller than the structural-relaxation nonlinearity. For the PVAC example near T_g, at $\Delta T_0 \approx \tilde{\Delta T} \approx 0.5$ K, the former is only a few tenths of a percent, whereas the latter is a significant effect. Anticipating the physics behind (2.45), we can argue as follows. A large part of the nonlinearity comes from the exponentialization of the logarithm of time in the local time–temperature equivalence. This is explicitly needed in the first order to connect the temperature T and linear time t in the recovery programs, but it is absorbed in the mobility measure $\mathrm{d}\log\omega$ (logarithmic frequency measure) for susceptibility in the linear response (2.75). The nonlinearity for structural relaxation is then given by the inequality

$$e^{\pm y} \neq 1 \pm y \quad \text{with} \quad y \approx \tilde{\Delta T}/\beta_{\mathrm{KWW}}\delta T \,. \qquad (2.40)$$

Phenomenological Concepts for Structural Relaxation. In order to estimate details for the nonlinearity of structural relaxation, we need some effective phenomenological concepts such as fictive temperature T_{f}, material time ζ, and mixing parameter x.

Tool's *fictive temperature* [176] maps the molecular order of Fig. 2.6a, or the extensive variables of recovery experiments, to a formal temperature T_{f} (Fig. 2.8b). The advantage of this map is that it provides two comparable quantities for two different things. The actual temperature T describes the heat bath of the experiment, and the fictive temperature T_{f} describes the progress of structural relaxation. In an isothermal recovery experiment at a given temperature T near T_g, T remains constant and T_{f} changes, with $T_{\mathrm{f}} \to T$ for large times. In more complicated experiments we have a program history $T(t')$ and a response history $T_{\mathrm{f}}(t') \neq T(t')$. The actual state $T_{\mathrm{f}}(t)$ at $t > t'$ is considered to be affected by both histories.

How can their effects be combined [170, 176]? Consider the state during the contraction experiment of structural relaxation at temperature T (volume recovery). The volume is larger than the equilibrium volume $V(t = \infty) = V_{\mathrm{equil}}(T)$. This means that the free volume is also larger than the equilibrium free volume. We have an additional free volume that is described by T_{f}. Since the equilibrium state at T can also be imagined to affect the relaxation by free volume, both the actual volume state T_{f} and the actual temperature T act in the same manner. It is therefore only necessary to combine them by their part

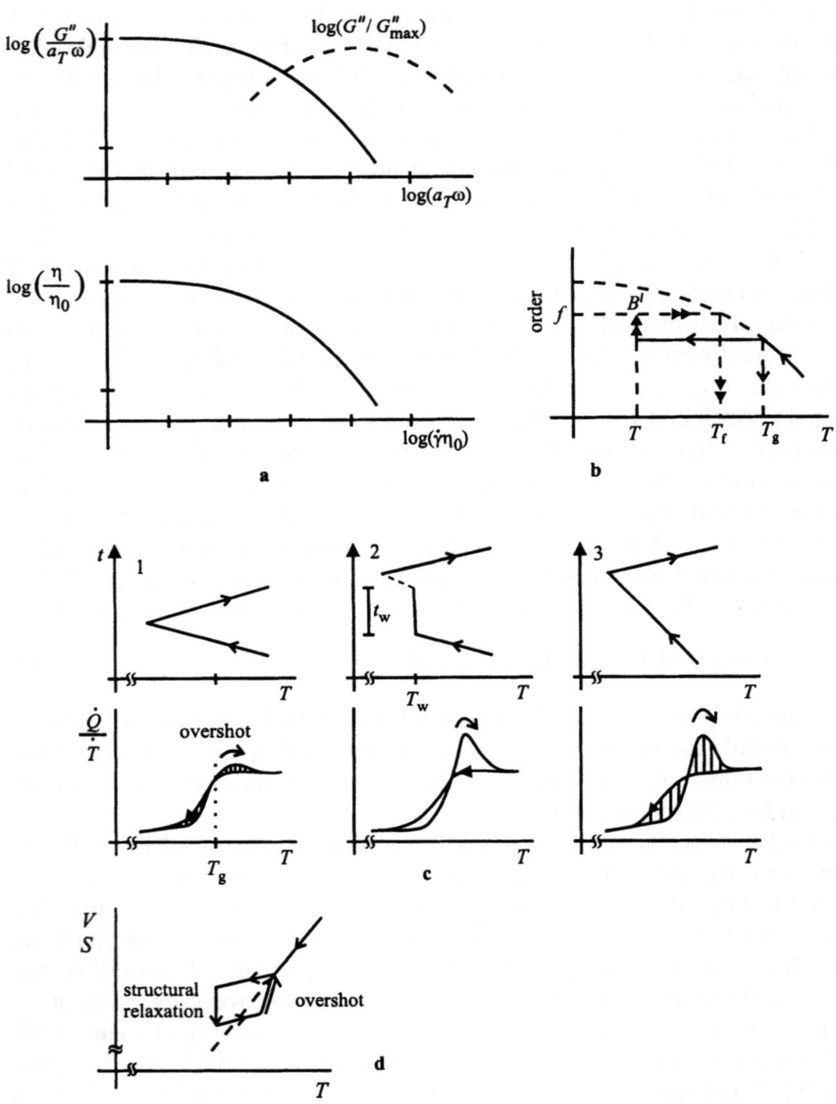

Fig. 2.8. Nonlinearity and structural relaxation (continued). (**a**) Comparison of the real part of linear complex shear viscosity as a function of reduced frequency (*top*) with the experimental shear viscosity as a function of shear rate (*bottom*). (**b**) Construction of Tool's fictive temperature for the B' state. (**c**) Thermograms for different temperature–time programs, from Vol'kenshtein et al. (**d**) Explanation of large overshots after structural relaxation

in the free volume, to mix their weights by some mixing parameter x without additional coupling. For contraction, the larger free volume leads to acceleration that is then expressed by $T_f - T > 0$. The expansion experiment can be described by the same combination. The volume is less than $V_{\text{equil}}(T)$, which leads to deceleration rather than acceleration. This deceleration is described by $T_f - T < 0$.

The phenomenological Narayanaswamy formula for structural relaxation is described by a functional $\mathfrak{F}$,

$$T_f(t) = \mathfrak{F}\{T(t'), T_f(t')\}\ , \qquad t' < t\ , \tag{2.41}$$

where the temperature history $T(t')$ and the volume history $T_f(t')$ must also be mixed without additional coupling (Sect. 4.5.2).

The medium where T and T_f is mixed is the *material time* ζ [177,178]. It is measured by a 'material clock' counting the progress of structural relaxation. Its 'second' is the time interval between properly defined events whose succession is representative of structural relaxation. The material clock is regulated by free volume, i.e., it is faster at higher true temperature and at higher fictive temperature. More precisely, the material time ζ' in the history is considered as a function of true and fictive temperature: $\zeta' = \zeta(T(t'), T_f'(t'))$ [170]. The weights are determined by a *mixing parameter* x, $0 < x < 1$ [179], for example, by

$$\text{rate} \propto \exp\left\{\tilde{\tilde{B}}[T - T_g]\right\} \to \exp\left\{\tilde{\tilde{B}}[x(T - T_g) + (1 - x)(T_f - T_g)]\right\}\ . \tag{2.42}$$

Here, we have approximated by using a tangent for a WLF curve at T_g, with slope $\tilde{\tilde{B}} = |\mathrm{d}\ln\tau/\mathrm{d}T|_{T_g} > 0$. The weight of the actual temperature T is x, and the weight of the fictive temperature T_f is $(1 - x)$.

The core of the Narayanaswamy formulas [179] is that the actual state is convoluted from the history in a similar manner to the linear response. This convolution, however, does not quench the first order of nonlinearity caused by the exponentialization equations (2.40) and (2.42). Nonexponentiality is inserted by using a Kohlrausch function [180, 181] for the retardation function (Narayanaswamy–Moynihan model). The nonexponentiality is proved to be the same as for the underlying equilibrium at T [182]. In this way, $T(t')$ characterizes the slowness of equilibrium fluctuations at T, and $T_f(t')$ describes acceleration ($T_f - T > 0$) or deceleration ($T_f - T < 0$) caused by partial freezing in the nonequilibrium. The full Narayanaswamy formulas will be described in Sect. 4.5.2.

Note that neither the mixing of T and T_f nor their effect on the rate of structural relaxation (2.42) needs any additional coupling constant. This general fact will be called 'functional independence' of the different dispersion zones (Sect. 3.5). For structural relaxation, it means that the main effect can be described by properties of the α process alone [183].

Consider as an example a Ritland–Kovacs jump experiment. Assume a phonon temperature diffusivity of $a = 10^{-7}\mathrm{m}^2/\mathrm{s}$ and a CRR size of 3 nm. The quenching time of a temperature difference would be $\tau \approx (3\,\mathrm{nm})^2/a \approx 10^{-10}\mathrm{s}$, whereas the structural relaxation lasts hours or longer. Similarly, the slow temperature fluctuations δT of CRRs cannot be quenched. Even a sample of mm size is 'equilibrated' in $\tau \approx 10$ s by the phonons, with no further influence on structural relaxation. For given temperature, the fictive temperature is not 'accessible' to the phonons.

Note further that the fictive temperature is not an intensive variable like the temperature but represents the recovery of an extensive variable such as volume V or entropy S. Since their retardations are different activities, we expect the fictive temperature to depend on the activity used for its definition, i.e., $T_\mathrm{f}^{(V)} \neq T_\mathrm{f}^{(S)} \neq \ldots$, where $T_\mathrm{f}^{(V)}$ is defined by volume recovery, $T_\mathrm{f}^{(S)}$ by entropy recovery, and so on. A more recent attempt to bring fictive temperature into contact with the true temperature is the interpretation of the violation of FDT during structural relaxation. The effective temperature that can be so defined there (Sect. 4.5.2) may also depend on the activity for which the fluctuation–dissipation ratio is evaluated. The gain from constructions leading to a set of different temperatures in a sample seems to be limited.

The weakness of the material time concept is obvious. A partly frozen mobility spectrum over several frequency decades, as for a larger dynamic heterogeneity (Sect. 2.2.6), cannot be described by a single material time. This seems inappropriate when structural relaxation after different, complicated temperature–time programs is to be described using a small set of parameters obtained, for example, from a simple cooling–heating cycle. Further, a recent improvement [184] aims at the final state accessible to structural relaxation. Practically [185], the usually constant ΔC_p value in the formulas for entropy recovery is made smaller and temperature dependent, $\Delta C_p(T) < \Delta C_p$. This can also be explained by a short-time, 'instantaneous' relaxation [182] before the recovery time window (Fig. 2.2a). This is indicated, as mentioned above, by the asymmetry of the initial recovery amplitudes in Fig. 2.7c and it may be an indicator for the vault effect (Sect. 2.2.6, Fig. 2.16f). In addition, this asymmetry cannot be described by the pristine Narayanaswamy formulas. Further, another improvement [186] aims at an application of the mixing parameter (x_s) to a configurational entropy variant $S_\mathrm{c}(T, T_\mathrm{f}) = x_\mathrm{s} S_\mathrm{c}(T) + (1 - x_\mathrm{s}) S_\mathrm{c}(T_\mathrm{f})$.

Nonlinearity after Jump Experiments. A nonlinearity condition for structural relaxation after simple temperature–time programs can be estimated from (2.40) and (2.42). The deviation from equilibrium is described qualitatively by

$$\tilde{\Delta} T(t) = (1 - x)[T_\mathrm{f}(t) - T(t)] \,. \tag{2.43}$$

Using (2.39), the nonlinearity condition $|\tilde{\Delta} T(t)| \gtrsim \delta T$ gives

$$|(1-x)(T_\mathrm{f}-T)| \gtrsim 0.65\ T_\mathrm{g}/m\ . \tag{2.44}$$

The $(1-x)$ factor sometimes correlates [183] with the configurational entropy, and therefore with the fragility m. The increase in nonlinearity with fragility m is therefore a general trend.

Let us now give some details. The main nonlinearity for structural relaxation comes from the logarithm of time in the local time–temperature equivalence [(2.40), Fig. 2.3b in Sect. 2.1.3, and also (2.138) and Fig. 2.28h in Sect. 2.4.5, and (2.31)]. Linear temperature–time programs $T(t)$ in the Narayanaswamy response, linear with respect to material time [(4.142) and (4.143)], become exponentialized [170] by this equivalence. The main nonlinearity is therefore generated when the exponential can no longer be linearized:

$$\text{rate} \propto \exp\left[\tilde{\tilde{B}}\tilde{\Delta}T(t)\right] \neq 1 + (1-x)\tilde{\tilde{B}}[T_\mathrm{f}(t) - T(t)]\ , \tag{2.45}$$

consistently with (2.44). A second source of nonlinearity is the effect described by the equilibrium TMDSC experiments above. The latter is of the same order of magnitude in the exponent and its first order is not quenched here (contrary to the equilibrium). We shall only use the condition (2.45) for the following estimations.

The nonlinearity condition (2.45) will now be applied to different stages of increasing waiting time after the jump, analogously to the discussion of jump experiments in Fig. 2.7.

(i) Starting Stage. Consider a jump generated by large $|\dot{T}|$ rates (Fig. 2.7d). The borderline between glass zone and glass state is defined by the preparation time $\tau_\mathrm{prep} \sim \delta T/\dot{T}$, or for rates, reciprocally, by a glass frequency

$$\omega_\mathrm{g} \approx \dot{T}/5.5\,\delta T\ . \tag{2.46}$$

This equation will be derived in Sect. 4.5.1.

Consider a positive temperature jump from equilibrium at T with a typical fluctuation frequency $\omega(T)$, possibly rather different from ω_g. Immediately after the positive jump, the $T_\mathrm{f} - T$ difference is of order

$$T_\mathrm{f} - T \approx \frac{1}{\tilde{\tilde{B}}} \ln\frac{\omega_\mathrm{g}}{\omega} \qquad \text{as long as} \quad \omega_\mathrm{g} > \omega\ . \tag{2.47}$$

Hence, large $|\dot{T}|$ also induces nonlinearity above the glass temperature T_g, as long as $\omega_\mathrm{g} > \omega$. In the PVAC example, $\omega_\mathrm{g} \approx 10$ rad/s for $\dot{T} = 100$ K/min and $\delta T = 2$ K, which can be compared with $\omega \approx 10^{-2}$ rad/s at the glass temperature, i.e., for $T = T_\mathrm{g}$. As $1/\tilde{\tilde{B}} \approx 3$ K/decade, we find nonlinearities e.g., after a positive jump, even starting from equilibrium at $T = T_\mathrm{g} + 5$ K. Curiously, even long 'exchange times' can be found in the Narayanaswamy iterations for such programs above T_g [187]. Generally, nonlinearity decreases for higher starting temperatures T above T_g

since ω increases in (2.47). For maximal jump rates usually accessible by dynamic calorimetry, the nonlinearity is restricted to be below $T_g + 10$ K for samples with parameters similar to those of PVAC.
Negative jumps near T_g are similar.

(ii) Evolution Stage. Roughly speaking, the subsequent recovery rate is determined by the evolution of the recovery $T_f(t) - T$ difference itself in the exponent in (2.45). The temperature T is now the recovery temperature. [The details are complicated, because the full iteration scheme for $T_f(t)$ is nonlinear (Sect. 4.5.2).] The contraction after a positive jump is faster than the expansion after a negative jump (Fig. 2.7b and c). The higher fictive temperatures T_f for the contraction yield the higher rates. The exponential function (2.45) is larger than one for $T_f > T$, as for contraction, and smaller than one for $T_f < T$, as for expansion. Related to the actual temperature T = const. after the jump, we therefore have acceleration for contraction and deceleration for expansion after jumps [171]. The explanation of the extreme nonlinearity of the Kovacs diagram Fig. 2.7e probably requires the full iteration scheme of the functional in (2.41) (see Sect. 4.5.2).

(iii) Final Stage. The final stage of recovery is, as mentioned above, monitored by nonexponentiality of the equilibrium response after complete thawing (Figs. 2.7e and f). Nonlinearity is not yet finished when $\tilde{\Delta T}(t)$ becomes smaller than δT during the evolution. The release of even a small amount of volume or heat has a large effect when the retardation itself arrives at small $\delta(t)$ values. The time when the recovery is finally dominated by symmetric equilibrium nonexponentiality usually corresponds to $|\delta|$ of order 10^{-4} (Fig. 2.7e). Since $|\delta|$ is of order $|T - T_f|/T_g$ it follows that $|T_f - T|$ is of order 0.1 K or smaller [142]. The time Δt where $\delta(t)$ reaches a given value for equilibrium linear response can be estimated backwards from the KWW decay equation (2.17) to give

$$\Delta t = \tau_{\mathrm{KWW}}\{\ln[\delta_0/\delta(t)]\}^{1/\beta_{\mathrm{KWW}}} . \tag{2.48}$$

This increases with decreasing $\delta(t)$ and increasing nonexponentiality $1/\beta_{\mathrm{KWW}}$. Example: $\Delta t/\tau_{\mathrm{KWW}} = 21.2$ for $\delta(\Delta t) = 10^{-4}$, $\delta_0 = 10^{-2}$ (corresponding to an initial jump of about 3 K), and $\beta_{\mathrm{KWW}} = 0.5$. Let us recall that these long times, across the dispersion zone, correspond to thawing of the cooperativity shell around Glarum defects.

Overshots. Structural relaxation during cooling–heating cycles is displayed by the classical entropy recovery experiments of Vol'kenshtein et al. [188–191] (see Fig. 2.8c). They represent the concerted action of several aspects discussed above. Consider three cycles of a DSC (Differential Scanning Calorimetry) experiment.

- Cooling–heating cycle with equal rates $|\dot{T}|$. The heating curve in the thermogram is somewhat steeper than the cooling curve and displays a small peak ('overshot') for some substances. After correction of all heat transfer effects in the apparatus, the net area between heating and cooling is zero (First Law). This was used above to define the calorimetric glass temperature T_g.

- Cooling–heating cycle with a waiting stage in the transformation interval (temperature T_w, time t_w). The overshot increases, considerably under certain conditions, and the net area is positive, compensating the heat production during waiting.
- Cooling–heating cycle with slower cooling than heating. The overshot is observed again but the net area remains zero as long as no heat invisible to DSC must be compensated. The intersection temperature is different from the value of T_g obtained in the first cycle (above) for the same heating rate.

These thermograms can be described qualitatively using the Narayanaswamy formulas (Sect. 4.5.2). Large overshots result from large structural relaxation during the waiting stage or at slow cooling (Fig. 2.8d). More entropy than without structural relaxation must be recovered in the time interval of devitrification. The effective temperature interval for recovery is small, because the difference $T_f - T$ must become small for it [see (2.45)]. Large heat release is then concentrated there in a cumulative fashion (double arrow in Fig. 2.8d). An 'undershot' can also be explained by (2.40) and (2.45). For $T_f < T$, $|T_f - T| < \delta T$, the exponential is small. The effect starts when $T_f \approx T$ is reached. Many examples concerning applications of structural relaxation are collected in Scherer's book [9].

2.1.6 Final Remarks

The classical understanding of the glass transition will be concluded with some remarks. The matter is conceptually difficult and the reader is not recommended to skip the main definitions and conventions above. As a reminder, we may contemplate what would classically be referred to as the 'three canonical features'.

(1) Non-Arrhenius. In the Arrhenius diagram, the traces of the dynamic glass transition curve downwards for $\log \omega$ (Fig. 2.3c) or, correspondingly, upwards for $\log t$ (Fig. 2.1c).
(2) Nonexponential. Relaxations are usually more stretched than Debye's exponential decay [e.g., (2.14) and (2.17), or Table 2.3]. This corresponds to broader dynamic loss peaks (Fig. 2.2e).
(3) Nonlinear. The classical recovery experiments near the transformation interval, starting from equilibrium above or below the experimental temperature, are differently influenced by partial freezing and are not therefore symmetrical (Fig. 2.7).

As a rule, for different substances, the first feature when restricted to the α process depends on one parameter, such as the fragility, the second on two, such as the wing exponents b and bg of the Havriliak–Negami function (2.14), and the third on three or more parameters, such as fragility, mixing parameter, and possible influence of the $A1$ process and a short-time relaxation. This means that only the first expresses glass-transition similarity in the sense of Fig. 2.1f. The second and third are more or less individual.

Other effects are sometimes superposed on the glass transition. We may consider the example of a metastable glass. Heating it from below, near T_g, the mobility gained may immediately be used to realize a phase transition, e.g., a crystallization. This can significantly modify the thermograms. For separation of glass transition and phase transition it may be useful to have other properties in mind – T_g increases with the rate $|\dot{T}|$ and no kind of nucleation is necessary for glass transition – and to use suitable methods, e.g., Temperature Modulated Differential Scanning Calorimetry (TMDSC) [192].

2.2 Serious Problems

In the 1970s, it was generally felt that only a few details were lacking from a complete classical picture that would lead to a closed theory of the glass transition. This section aims to describe several serious problems which destroyed those hopes. The issue is approached through a series of questions that are listed in Sect. 2.2.2, once the important phenomenon of crossover region has been introduced. The descriptive style should not induce the reader to believe that all the points raised here have been brought to a stage of full understanding.

2.2.1 Crossover Region and β Process

Typically, the viscosity as a function of temperature (Fig. 2.1c) and the trace of the dynamic glass transition in the Arrhenius diagram (Sect. 2.1.3) consist of three parts (A, a, α): an Arrhenius process or molecular transient (A) at very high temperatures, corresponding to very high frequencies and very short relaxation times, a high-temperature process (a process) in the warm liquid, at high temperatures, and a cooperative process (α process) in the cold liquid, at low temperatures. The latter usually includes the glass temperature T_g, corresponding to millihertz frequencies and relaxation times of order 1000 seconds. The transition from a to α is called the *crossover region*. This region is characterized by a crossover temperature T_c (Fig. 2.1c) usually defined by extrapolations from both sides (Table 2.10 in Sect. 2.5.4).

Crossover data was first collected for polymers. The Arrhenius diagram of dielectrical (Fig. 2.3c), mechanical, certain NMR, and other traces for polymers shows that a third process (β) usually shares in the crossover. Figure 2.9a contains the traces for polymers known in 1965. They are taken from

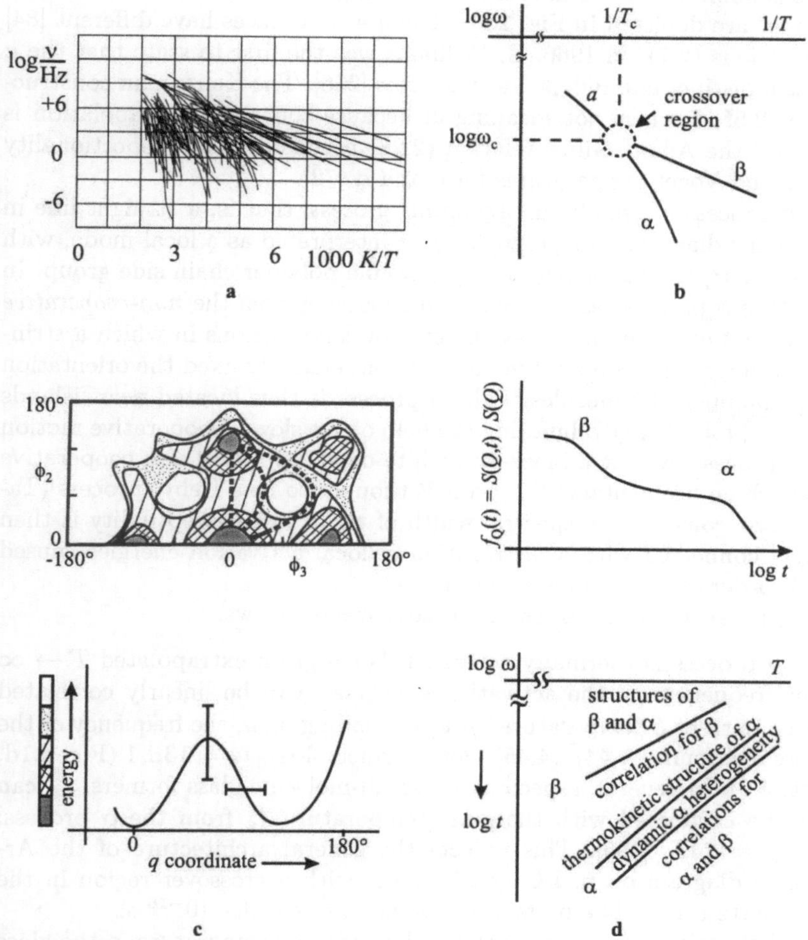

Fig. 2.9. Crossover region. (**a**) Compilation of all traces in the Arrhenius diagram for polymers known in about 1965. (**b**) Arrhenius diagram for a given polymer near the crossover region. a is the high-temperature process, β the local Arrhenius process (later called the Johari–Goldstein process), and α the cooperative process. (**c**) *Bottom*: construction of a 'landscape' for n-pentane from sampling of typical paths (*top*) in a two-dimensional conformational energy map (energy mountain) on the plane for the two inner angles ϕ_2, ϕ_3 corresponding to given outer angles. *Bars* indicate schematically the possible landscape variations from the different paths. (**d**) *Top*: decay of the reduced scattering function $f_Q(t) = S(Q,t)/S(Q)$ at the retardation times for the β and α processes. *Bottom*: attempt to characterize the corresponding zones in a $\log\omega$–T diagram by using the terms structure, correlation, and dynamic heterogeneity

the famous book by McCrum, Read, and Williams [2]. The traces for one single polymer are depicted in Fig. 2.9b. The a and α traces have different [84] WLF equations (2.7). In 1966, G. Williams was the first to state that the a process is a distinct and independent process [206]. The Kauzmann construction (Fig. 2.5f) for a is not meaningful because too much extrapolation is needed, and the Adam–Gibbs relation (2.28) needs different proportionality constants and Vogel temperatures for a and α [72].

The β process is usually an *Arrhenius process*, that is, a straight line in the Arrhenius diagram. For polymers, β is interpreted as a local mode, with or without [193] the participation of a possible polymer chain side group. In general, "the β process can be envisaged as arising from the *non-cooperative* rearrangement of some molecules engaged by large regions in which a stringent requirement of a cooperative motion has relatively fixed the orientation of a large number of molecules." The β process is thus located near 'islands of mobility' (Johari [194]). Ignoring the help of the slower cooperative motion for the β process, we get a black-and-white depiction (local yes, cooperative no) in which an 'elementary' β process is thought to be a Debye process (Table 2.3). The considerable spectral width of the β loss susceptibility is then intuitively connected with a distribution of local activation energies caused by the disorder of amorphous structure.

The general properties of the β process are as follows.

(i) The β process is thermally activated. For a given extrapolated $T \to \infty$ limit frequency ν_0, the activation energy ε_A can be linearly correlated with a current β temperature $T_\beta(\nu)$ depending on ν, the frequency of the β loss maximum (4.44)–(4.45). For polymers, $\log_{10} \nu_0 = 13 \pm 1$ (Fig. 4.1d, Sect. 4.2). In general, especially for small-molecule glass formers, ε_A can even be correlated with the glass temperature T_g from the α process: $\varepsilon_A/k_B \approx 24T_g$ [195]. This reflects the general architecture of the Arrhenius diagram on p. IX, or Fig. 2.9b, with a crossover region in the megahertz range. The prefactor $1/\nu_0$ is then of order 10^{-16} s.

(ii) The dielectric intensity $\Delta\varepsilon_\beta$ weakly depends on temperature in the glass state $T < T_g$, but increases above the glass temperature T_g. The increase is often discussed in terms of larger average amplitudes due to more free volume, decompensation of 'antiparallel' dipole arrangements, and increasing spatial density [196] of elementary β processes. In the glass state, tempering decreases $\Delta\varepsilon_\beta$ correspondingly, whereas quenching into the glass states increases $\Delta\varepsilon_\beta$.

(iii) There are many substances where the dielectric intensity remains small up to the crossover region, or even seems to be completely missing. (These are A glasses in the terminology used by Rössler [197], whilst those with a dielectric intensity $\Delta\varepsilon_\beta \neq 0$ are called B glasses. This will be further discussed below).

Leaving the black-and-white depiction, we may ask: is the β process organized or at least helped by the dynamic heterogeneity of the slow α process?

Four aspects of this serious problem will be commented. It will serve also to illustrate some concepts introduced in Sect. 2.1.

(i) Mechanics. The concrete molecular mechanism, as confirmed for instance by NMR [198], such as "a $180 \pm 10°$ flip of the ester unit accompanied by a restricted main chain rearrangement with a $\pm 20°$ rms amplitude", as for poly(ethyl methacrylate) PEMA, should be complemented by the role of all neighboring molecules or monomeric units. It is general experience, for example in the context of polymer solutions [199], that the neighborhood contributes about 50% to the activation energy, as a rule of thumb.

(ii) Landscape (Fig. 2.6b). The two-dimensional energy mountain Fig. 2.9c is the conformational energy surface for the two inner angles (ϕ_2, ϕ_3) of n-pentane with fixed outer angles ($\phi_1 = \phi_4 = 0$, [200]). Many paths are possible for the 'subensemble' point, especially when distributions of kinetic energies, outer angles ϕ_1 and ϕ_4, and environments are considered. This results in a distribution of barrier heights and shapes for a typical one-dimensional landscape picture (bars in the lower part of Fig. 2.9c). Although overcoming a single saddle in the landscape corresponds to about 0.5 nanometers in real geometry, the definition of the β process in a landscape based on more degrees of freedom is not easy [201]. Short modes of about 0.5 nm may belong not only to the β but also to the α process with its wide mode length spectrum [202]. Moreover, the representation of the α process as a boundary for β depends on the contrast that can be connected with the definition of cooperatively rearranging regions (CRRs) by statistical independence from their environment (Sect. 2.2.5). It seems difficult, especially near the crossover, to find two landscapes at the same temperature, one functional to α and the other functional to β.

(iii) Thermokinetic Structure. Let us recall the structure discussion concerning Fig. 2.5a in Sect. 2.1.4. The picture must be doubled now (Fig. 2.9d). The intermediate scattering function, reduced by the structure factor $f_Q(t)$ of (2.32), decays in two steps: first β and then α. The structure contribution of the β process decays at the β trace to a β correlation in space and time. In the time zone between β and α we have, however, a quasi-fixed α structure contribution. Here lives the β process. This α contribution may be called a 'thermokinetic structure' [7,18]. It is a temporary distribution of locally α-optimized real-space arrangements after motional averaging of β. A landscape concept for it would be certain inherent α structures. It serves as a quasi-mechanical, larger-scale [203] 'boundary condition' for the β process correlations.

(iv) Dynamic Heterogeneity. The α contribution to the correlation function $f_Q(t)$ finally decays when the α trace is crossed in the direction of larger times. The spatial aspects of the decay in the α dispersion zone were characterized by the concept of dynamic heterogeneity, e.g., a spatial distribution of α mobilities (pattern of mobility field). We expect dynamic heterogeneity to reflect the thermokinetic α structure in some way, irrespective of the low contrast of the concomitant density differences: we expect the local β process to be located near the Glarum defects, i.e., near the islands of α mobility in the mobility field. Sampling all NMR mobilities indicates [204] that more or less all molecules participate in β, but to different degrees. Large jumps may be linked with the defect vicinity, i.e., with the activated β process of the black-and-white depiction, and small-angle fluctuations with the assisting cooperativity shell.

Summarizing the four aspects, the β process is organized by the α process, the α process is a precondition for the β process (no β without α), and the two processes are therefore not independent, but a sufficient mobility difference between them below the crossover region permits a black-and-white depiction of the β process as a low-dimensional activation process.

The β process usually survives vitrification. It is first and foremost the cooperativity shell that vitrifies near the glass temperature T_g. This aspect will be discussed in Sect. 2.2.6. An interesting proof for the local nature of the β process results from the pressure and temperature dependence of its spectral width, investigated in polyvinylchloride PVC by Koppelmann [205]. Extrapolations to high temperature and to negative pressure lead to a Debye process for equivalent decreases in density, of order a few per cent. The missing barrier spectrum there indicates a local chain-mode by purely conformational variations, a so-called 'local mode' of polymer chains with no mobile side groups.

From such investigations we expect variations of free volume to have only a minor effect on the position of the β trace in the Arrhenius diagram. An extreme example is the series of homologous poly(n-alkylmethacrylates) (Fig. 2.10a). More free volume shifts both the a and the α process to higher mobility, whereas the local β process is not (or is less) affected. Free volume can be varied by pressure [206], by polymer chain length for molecular mass below 10 kg/mol [207], or by the length of the n-alkyl group in the side chain [208]. Increasing pressure and increasing chain length diminish the free volume, the latter because of the decreasing number density of the disturbing end groups. Since the β process for these homologous series is nearly independent of such free volume variations and, therefore, fixed in the Arrhenius diagram, the crossover region is shifted to low frequencies (Fig. 2.10a). The hertz range is reached for the n-hexyl member [209].

Let us now leave the polymers. Johari and Goldstein showed in a series of papers [210] that the Arrhenius diagram of Fig. 2.9b type is not restricted to polymers. Instead, as a rule, this diagram topology is typical for almost all glass formers. The dielectric β intensity depends on the substance. It varies from the same order as that for α down to dielectric loss ε'' of order 10^{-4} for orthotherphenyl OTP, for instance [211]. The calorimetric β intensity ΔC_p is usually weak [154], but there are exceptions such as for some biopolymers [212]. Small ΔC_p means that the entropy fluctuation due to the β process is small. [The reader should not confuse the retardation intensity ΔC_p(= liquid

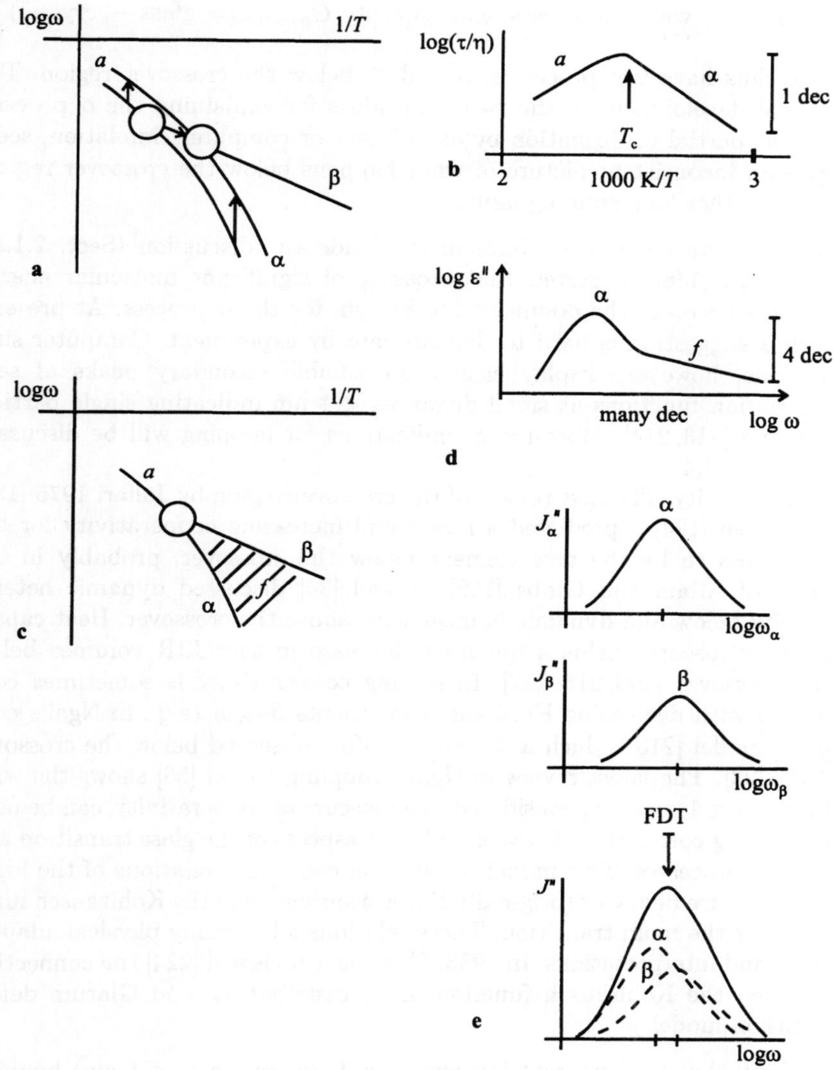

Fig. 2.10. Crossover region (continued). **(a)** Shift of the crossover region in an Arrhenius diagram caused by increasing free volume for fixed β trace (example poly(n-alkyl methacrylates). **(b)** Example for the behavior of the Stokes–Einstein relation near T_c, for 1,3,5-tri-α-naphthyl benzene. See text for details. **(c)** An independent f process between α and β. **(d)** Wing from the f process in dielectric loss for α below the crossover temperature. **(e)** Suggestion for a decomposition of overlapping dynamic α and β compliances in the crossover region, using elements of probability theory

– glass zone) with the excess heat capacity $C_{p\,\text{excess}}$(= glass – crystal) of Fig. 2.5e.]

We thus have two processes, α and β, below the crossover region. The historical development of the two main ideas for explaining the α process, and their partial confirmation by experiment or computer simulation, seem to give an inconsistent picture of what happens below the crossover region. This is another long-running issue.

- Hopping. As mentioned above in the landscape discussion (Sect. 2.1.5), Goldstein [158] suggested that crossing of significant molecular energy barriers becomes the dominant mechanism for the α process. At present, such a suggestion is hard to demonstrate by experiment. Computer simulations, however, display clearly identifiable secondary peaks of self-correlation functions at small distances < 1 nm indicating single particle hopping [213, 214]. More recent indications for hopping will be discussed in Sect. 2.3.4.
- Cooperativity. The first review of the crossover region by Johari 1975 [132] – see also [194] – predicted a larger and increasing cooperativity for the α process to be the new element below the crossover, probably in the sense of Adam and Gibbs [122]. Angell [34] discussed dynamic heterogeneity below and dynamic homogeneity above the crossover. Heat capacity spectroscopy yields a quadratic increase in the CRR volumes below the crossover [209, 215–217]. Increasing cooperativity is sometimes connected with decreasing Kohlrausch exponents β_{KWW} (e.g., in Ngai's coupling model [218]). Such a decrease is often observed below the crossover [219–223]. The latest review of Ngai's coupling model [55] shows the wide field where $1-\beta_{\text{KWW}}$, considered as a measure of cooperativity, can be used for finding correlations between different aspects of the glass transition and related processes. This model is based on continuity relations of the high-temperature behavior (originally the A transient) and the Kohlrausch functions for the main transition. These relations allow many physical adaptations and interpretations. In 1988, Shlesinger reviewed [224] the connection between the Kohlrausch function, Levy distribution, and Glarum defect diffusion model.

I think that the apparent inconsistency between local (< 1 nm) hopping and cooperativity over scales of a few nanometers may be resolved by the formation of Glarum defect diffusion below the crossover region, for the α process. The local elements of α and β are located near the defect: α hopping is connected with a defect diffusion step, β-process motion with no diffusion step, and the cooperativity shell is necessary for assisting the two local elements and for the defect diffusion. This discussion will be continued in Sects. 2.2.5–2.2.6, 2.5.3 (physical picture) and Sect. 4.3 (Glarum–Levy defects).

Although it is historically not quite correct, the β process is usually called the *Johari–Goldstein process*. It would be an act of fairness and compensation towards G. Williams to call the high-temperature a process the *Williams–Götze process*: Williams for his 1965 statement about the a process quoted above [206], and Götze for his mode-coupling theory which in fact describes the main features of the a process (Sect. 2.3.2).

The structure of the Arrhenius diagram with the a, α, and β traces (Fig. 2.9b) implies certain relationships between them, e.g., the β process time at T_g, $\log \tau_\beta(T_g)$, and properties of the α process there, e.g., the fragility $m = m(T_g)$, defined in (2.3). For substances with crossover frequency in the MHz range, $\log \tau_\beta(T_g)$ increases with decreasing m [225], as expected if the β process slope increases with the α process slope in the crossover. If m increases with the α spectral width at T_g [77, 226], then there is also a correlation between $\log \tau_\beta(T_g)$ and $\delta \log \omega(T_g)$ or, equivalently (Table 2.3), $1/\beta_{KWW}$. A correlation between the β process activation energy ε_A and the α process glass temperature T_g was mentioned above. Furthermore, if the crossover frequency goes down, as for the poly(n-alkyl methacrylates), then $\log \tau_\beta(T_g)$ also goes down for $T_c \to T_g$, of course.

The temperature dependence of *Stokes–Einstein relations* (or ratios) also changes at the crossover (see Fig. 2.10b and [227]). In general, this relation will be used for the ratio of characteristic times from different activities in the same dispersion zone, e.g., a time from light scattering vs. a time from viscosity [or, according to (2.1), vs. the viscosity itself]. Historically, the Stokes–Einstein ratio compares translational diffusion and shear viscosity, and the Stokes–Debye ratio rotational diffusion and viscosity; use of a combination of the two is sometimes called the SED relation. As a rule, at least one transport coefficient (Sect. 2.2.3) is included in the general Stokes–Einstein relation. The details of the temperature dependence largely depend on the kind of activities and on the individual substance investigated [228, 229].

The temperature dependence of many other liquid properties also changes significantly at the crossover temperature T_c. An example is the amplitude of terahertz cage rattling on the angstrom scale [230].

A specific relaxation process f (Fig. 2.10c) seems to be confined between the α and β processes. It is often indicated by a 'high-frequency tail' of dielectric loss on the high-frequency side of the α process, the *Nagel wing* (see Fig. 2.10d and [231]). It seems to be an isothermal precursor of the α process and most likely corresponds to an Andrade law in the shear compliance of liquids, $J(t) \sim t^{1/3}$ [232]. This precursor is important for physical aging below T_g, where the structural relaxation can no longer be described as caused only by thawing of the partially frozen α process alone [233]. For a determination of β process parameters, we must decide whether the Nagel wing is restricted to the zone between α and β or whether it forms the basement for β so that $\varepsilon^* = \varepsilon^*$ (wing)$+\varepsilon^*(\beta)$ [197]. This reference also contains an example in which the bend between α and f can be observed between α and β (3-fluoroaniline). The discussion of relaxation between dispersion zones will be continued in Sect. 2.4.5.

The details of the crossover region (within the circles of Figs. 2.9b, 2.10a, and c) have recently been investigated by linear response and dynamic neutron scattering [209, 234]. The data evaluation for comparable α and β relaxation intensities depends on an assumption about the way α and β can be distinguished in the crossover where their frequencies are of the same order and do overlap [235].

Let us analyze the $\alpha\beta$ separation problem at issue using probability theory. To begin with, the Second Law for linear response is equally valid for extensive or intensive variables. Both $\omega J''(\omega)$ and $\omega G''(\omega)$ are proportional to heat release = power spectrum = entropy production, as described by (3.78)–(3.81). Since the spectral density of any thermal fluctuation is positive, it can be interpreted as a probability density with measure $d\omega$. Because of the fluctuation–dissipation theorem FDT, the imaginary parts of both the dynamic compliances $J''(\omega)$ or the dynamic moduli $G''(\omega)$ can be interpreted as probability densities with logarithmic frequency measure $d \log \omega$ [see (2.75)]. We thus have a probability density $P = J''$ or $P = G''$ and a probability space $\log \omega$. Thinking of a histogram, P would be the ordinate and $\log \omega$ the abscissa. Details will be described in Sect. 3.1 in the discussion of Figs. 3.1c and d.

The most simple α and β distinction would be to draw two histograms, one for P_α vs. $\log \omega_\alpha$ and the other for P_β vs. $\log \omega_\beta$ (Fig. 2.10e). This corresponds to the assumption that the α process and the β process can be distinguished a priori. An analysis of linear response experiments on this simplified basis has to face two obstacles.

- The experiment based on the FDT cannot distinguish between $\log \omega_\alpha$ and $\log \omega_\beta$ (ω identity of the FDT, Sects. 2.2.4 and 3.4.5). The two probability spaces must be identified and the common measure is $\log \omega$ without any index.
- The additivity of probabilities, $P = P_\alpha + P_\beta$, should be valid for the loss aspects of both compliances and moduli,

$$J'' \stackrel{?}{=} J''_\alpha + J''_\beta \quad \text{and} \quad G'' \stackrel{?}{=} G''_\alpha + G''_\beta \,.$$

  The additivity of response, however, cannot simultaneously be valid for both susceptibilities: $J^* = J^*_\alpha + J^*_\beta$ is not compatible with $G^* = G^*_\alpha + G^*_\beta$ because of (2.8), $G^* J^* = 1$. This means we have to decide whether J^* or G^* is considered to be additive.

The simplest decision is to connect additivity with compliances [236], since they correspond to thermodynamically extensive variables. To remain inside our a priori assumption, it therefore suffices to make two decisions: the distinguishing determination of what α and β is (e.g., each can be represented by its own Havriliak–Negami formula, so that α and β are then distinguished by different parameter sets), and the choice of which variable is additive (e.g., the compliance). This approach works down to a difference of maximum loss frequencies of only one decade, and down to maximum intensity ratios of 0.1 [237]. The approach is only connected with the additivity of probability measures and the distinguishability of α and β (the mutual exclusion, i.e., the 'either/or' of elementary probability theory). Any reference to statistical independence ($P = P_\alpha \cdot P_\beta$, the 'as well as') would be an additional restriction here and is usually biased [235,238]. More theoretical details will be described in Sect. 3.1.

We should remark that, in contrast to spectral densities, their Fourier transformations, which are correlation functions, cannot generally be interpreted as probability densities. Correlation functions can become negative, which is impossible for probability densities. Instead, the correlation functions are 'characteristic functions' of probability densities, namely of $x^2(\omega)$ and $f^2(\omega)$. Moreover, the problem of distinguishing α and β will not automatically be solved by dynamic neutron scattering. Since the spatial scales of β and the high-frequency wing of α relaxation do also overlap, an additional distinction criterion is also needed for scattering in the crossover [202] although a new parameter, the scattering vector Q, is available. As an aside, scattering quantities like $S(Q, \omega)$ or $S(Q, t)$ are in principle extensive variables.

Using the additivity of compliances, we obtain, in a certain generalization of experimental findings by Murthy, Schönhals and others [208,217,239–241], two scenarios for the dielectric traces in the crossover region (Figs. 2.11a and 2.3c). The *splitting scenario*, or *scenario I*, displays a separate α onset in the Arrhenius diagram, an $a\beta$ continuity with a bend, and a linear increase of the dielectric and calorimetric α intensity. The *merging scenario*, or *scenario II*, displays a quasi-$a\alpha$ continuity, a usually weak β intensity which does not allow speculation about an onset, and the dielectric (and probably the calorimetric) α intensity does not tend to zero.

The details of scenario II [239,242] are difficult to measure if the β intensity is very small [211,243].

It seems that scenario II is more frequent than scenario I. The latter is, besides the poly(n-alkyl methacrylates) [117,208,209,244,245], also observed for example in DGEBA epon [246, 247] and PPGE epon [248], and there are indications for it in substances such as toluene [249] and polybutadiene PB [234]. For these four substances the crossover frequencies are above the megahertz range. Differences in the molecular pictures which may discriminate between I and II will be discussed in Sect. 2.5.3. Briefly, it is assumed that the a process cage above the crossover for II is formed by a larger number of particles than that, as conjectured for I, needed for the first configuration shell.

The general crossover topology of Fig. 2.9b is stable against substitution of temperature by pressure [206, 251] or substitution of mobility $\log \omega$ by homologous substance variation [252] (Fig. 2.11b).

Nondetectable dielectric β intensities [197, 225] do not mean that there is no crossover for such substances. The four examples of [197], propylene carbonate, propylene glycol, glycerol, and tricresyl phosphate, do have a crossover region that is well defined by different WLF curves for the dielectric a and α process traces [84]. The small dielectric β intensity for orthoterphenyl OTP can be enlarged below T_g by quenching with a large cooling rate [211]. A possible explanation may be smaller vaults (Sect. 2.2.6) in the glass structure which means more islands of mobility, supported by more free volume at the higher T_g for quenching. Since the calorimetric β intensity is often small, heat capacity spectroscopy usually avoids the $\alpha\beta$ separation problem. We can directly study the crossover from a to α. In poly(n-hexylmethacrylate), an example for scenario I, we observe a saddle in the $C_p''(\log\omega, T)$ contour map: the a process dies out and the α process sets in separately (Fig. 2.31a).

Consider now the α process for increasing temperatures approaching the crossover from below. The disappearance of the calorimetric retardation intensity ΔC_p at T_c, $\Delta C_p \sim (T_{on} - T)$, $T < T_{on}$, $T_{on} \approx T_c$, in the splitting scenario I would lead to a thermodynamic dilemma, if ΔC_p were related to the configurational difference (2.33). The configurational entropy would then become independent of temperature, which does not seem reasonable [253]. This dilemma is resolved by comparison with (2.35). The retardation intensity ΔC_p is not between liquid and crystal but between liquid and glass zone, which makes a great difference in the crossover (Fig. 2.11c).

Heat capacity spectroscopy data in the crossover region indicate that α cooperativity increases quadratically below the onset for both [209, 217] scenarios:

$$\left.\begin{array}{lll} N_a \text{ of order } 1 & \text{for} \quad a\,, & T > T_{on}\,, \\ N_\alpha \sim (T_{on} - T)^2 & \text{for} \quad \alpha\,, & T < T_{on}\,, \end{array}\right\} \tag{2.49}$$

►

Fig. 2.11. Crossover region (continued). (**a**) Details of dielectric traces and intensities in the crossover region. Two different scenarios have been observed so far (I, II). *Upper parts*: Arrhenius diagrams. *Lower parts*: dielectric intensities $\Delta\varepsilon$. The α onset is near the intersection of α with the extrapolation of the β slope. (**b**) Stability of the general crossover region topology Fig. 2.9b under change of variables: pressure p on the left (T = const.), side-chain length on the right [$\log\omega$ = const., C-number of the alkyl group in the side chain of poly(n-alkyl methacrylates)]. Schematic representation. (**c**) Heat capacity C_p of scenario I for the liquid (l), glass (g, zone and state), and crystal (c). The *double arrow* ↕ is the l–g relaxation step ΔC_p, quite different from configurational (l–c) and excess (g–c) differences in the crossover region. The extrapolation of the liquid state to temperatures below T_g was first experimentally pursued in [250]. The extrapolation of the glass state to temperatures above T_g (*dotted curve*) may result in differences between glass state and experimental values for the glass zone (*continuous curve*) from heat capacity spectroscopy. The differences come from entropy recovery in the glass state (Figs. 2.8b, 2.4d, 2.5e)

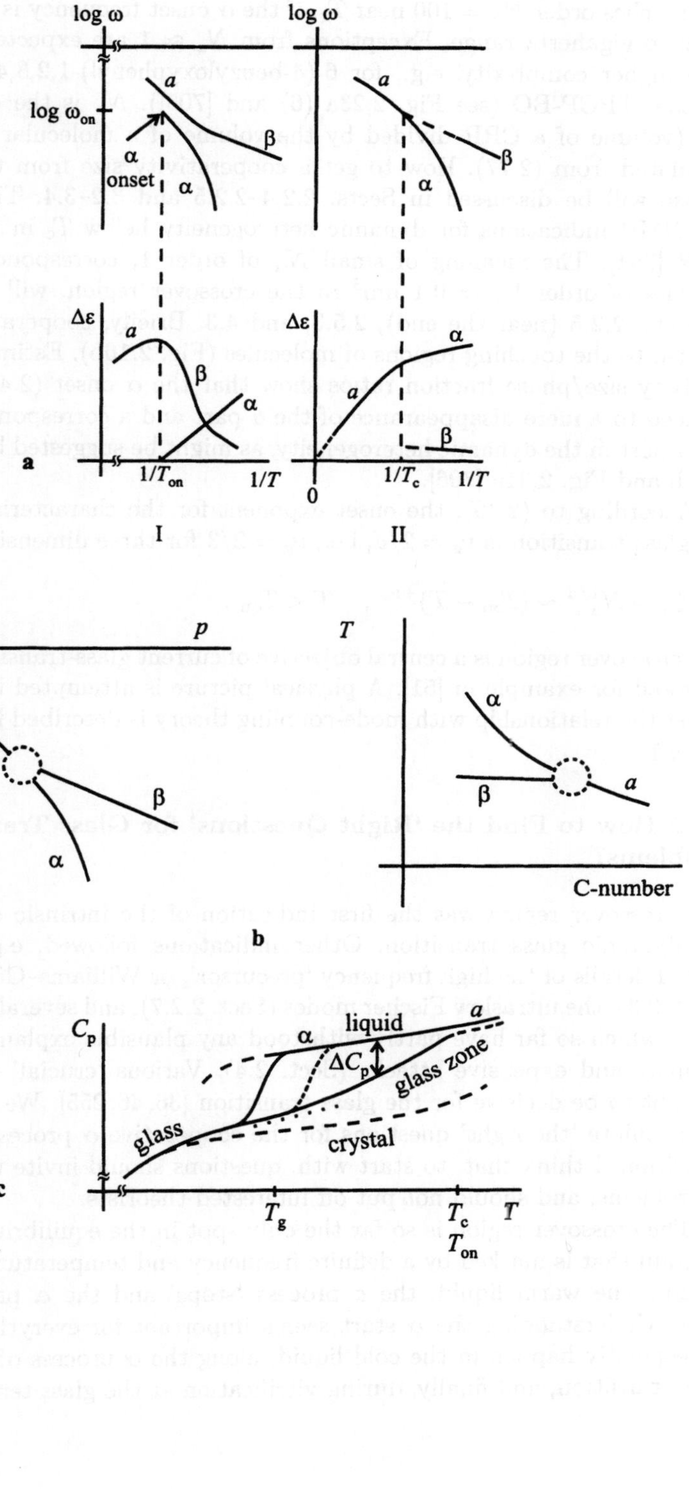

log ω
log ω_on
onset
a
α
β
Δε
1/T_on
1/T
I
log ω
Δε
0
1/T_c
1/T
II
logω
p
T
C-number
b
C_p
liquid
ΔC_p
glass zone
glass
crystal
T_g
T_c
T_on
T
c

and reaches order $N_\alpha = 100$ near T_g, if the α onset frequency is in the megahertz to gigahertz range. Exceptions from $N_a \approx 1$ are expected for liquids with higher complexity, e.g., for 6-(4-benzyloxyphenyl)-1,2,3,4 tetraphenyl fulvene, TPCP-BO (see Fig. 2.23a (6) and [705]). N_α is the cooperativity size (volume of a CRR divided by the volume of a molecular unit) and is calculated from (2.77). How to get a cooperativity size from the linear response will be discussed in Sects. 2.2.4–2.2.5 and 3.2–3.4. There are also ^{15}N NMR indications for dynamic heterogeneity below T_c in the salt melt CKN [254]. The meaning of small N_α of order 1, corresponding to CRR volumes of order $V_\alpha \approx 0.1$ nm^3 in the crossover region, will be discussed in Sects. 2.2.5 (near the end), 2.5.3, and 4.3. Briefly, cooperativity will be applied to the touching regions of molecules (Fig. 2.16b). Estimates of cooperativity size/phase fraction ratios show that the α onset (2.49) cannot be reduced to a mere disappearance of the a part and a corresponding onset of the α part in the dynamic heterogeneity, as might be suggested by Figs. 2.34a and b and Fig. 2.31a [706].

According to (2.49), the onset exponent for the characteristic length of the glass transition is $\nu_g = 2/d$, i.e., $\nu_g = 2/3$ for three dimensions ($d = 3$):

$$\xi_\alpha \sim N_\alpha^{1/d} \sim (T_{on} - T)^{+\nu_g}\,, \quad T < T_{on}\,. \tag{2.50}$$

The crossover region is a central objective of current glass-transition research, reviewed for example in [51]. A physical picture is attempted in Sect. 2.5.3, whilst the relationship with mode-coupling theory is described in Sects. 2.3.2 and 4.1.

2.2.2 How to Find the 'Right Questions' for Glass Transition Problems?

The crossover region was the first indication of the intrinsic complexity of the dynamic glass transition. Other indications followed, e.g., the unexpected details of the high frequency 'precursor', or Williams–Götze a process (Sect. 2.3), the ultraslow Fischer modes (Sect. 2.2.7), and several long-running issues which so far have partly withstood any plausible explanation in spite of many and expensive attacks (Sect. 2.4). Various 'crucial' questions are thought to be decisive for the glass transition [36, 46, 255]. We attempt here to formulate 'the right' questions for the cooperative α process of the glass transition. I think that, to start with, questions should invite the resolution of problems, and should not put off interested theorists.

The crossover region is so far the only spot in the equilibrium Arrhenius diagram that is marked by a definite frequency and temperature ($\log\omega_c, T_c$). Cooling the warm liquid, the a process 'stops' and the α process 'starts' there. Understanding the α start seems important for everything that will subsequently happen in the cold liquid, along the α process of the dynamic glass transition, and finally, during vitrification at the glass temperature T_g.

(i) Why does the high-temperature a process stop or at least change at the crossover region? Are the possibilities of the cooperative α process also 'exhausted' somewhere at low temperatures, and also for the extrapolated equilibrium?
(ii) What are the important new elements below the crossover? How do they evolve below the crossover? Which properties of the α process are the 'boundary conditions' for the β process?

Several attempts to formalize these problems by means of a theoretical framework (Chap. 3) will be described in Chap. 4.

(iii) What are the consequences of the fact that a definite pattern of dynamic heterogeneity in the cold liquid vitrifies at the glass temperature T_g? How can a glass be distinguished from an amorphous solid produced without any glass transition?

In this book, I will look for an answer to these questions in terms of molecular cooperativity, spatial aspects of dynamic heterogeneity, and their characteristic length.

Another problem is the enormous breadth and multifariousness of the substance class that has glass transitions with general properties characterized, for example, by the three canonical features mentioned in Sect. 2.1.6. This means that very abstract models must be constructed. On the other hand, if we accept the general feeling that typical molecular cooperativities can be organized by a few dozen particles in 'strong interaction', corresponding to characteristic lengths in the range of a few nanometers, then we are confronted with the suspicion that individual properties of the molecules should dominate the transition. Consider the parameter space after the map described by Fig. 2.1f.

(iv) For which glass-transition properties and for which temperature, frequency, pressure, concentration, or parameter regions is similarity observed for glass transitions in different substances, and where do molecular individualities shine through?

This question is related to the most imposing mystery of the glass transition: why do we get *general* Arrhenius diagrams with a clear architecture (see the figure on p. IX and Fig. 2.32) for the *multifarious* molecules, molecular interactions, and structures of the different substances that form glasses? Let us recall once more that the dynamic glass transition and the related processes in these figures include the complete relaxation dynamics of all liquids.

2.2.3 Different Transport Properties

The spreading of traces for different transport properties in the Arrhenius diagram is expected to be much larger than for the various activities of the linear response as listed in Table 2.1. This effect will be called *different transport activities*. In the linear response, fluctuations in such variables as entropy, volume, and temperature ($\Delta S, \Delta V, \Delta T$) are either strongly coupled, so that $\overline{\Delta S \Delta T} = k_{\mathrm{B}}T$, or coupled by susceptibilities which behave similarly, e.g., $\overline{\Delta S \Delta V} = k_{\mathrm{B}}T(\partial V/\partial T)_p$, the volume expansivity, even if $\overline{\Delta V \Delta T} = 0$ [see (3.37)–(3.40)]. The transport coefficients, on the other hand, differ greatly from the very beginning. Let us start with two examples.

First Example. Translational diffusion is different from spin diffusion (see Fig. 2.12a), because no particle need be translated for the redistribution of spins. The orientational diffusion of molecules is defined similarly to spin diffusion, although the molecule can be translated during a change of orientation.

Second Example. Consider a set of charged particles distinguished by the index $i = 1, 2, 3, \ldots$ (e.g., ions). Their full mean square displacement correlation function

$$\overline{\boldsymbol{r}^2}(t) = \sum_i \overline{\boldsymbol{r}_i^2}(t) + \sum_{i \neq j} \overline{\boldsymbol{r}_i \boldsymbol{r}_j}(t) \tag{2.51}$$

is related to the electrical conductivity, whereas the self-diffusivity is only related to the first term, the self term, on the right-hand side of (2.51). The difference can be understood by referring to Fig. 2.12b. Moreover, the mobility of the ions may differ greatly from the molecular mobility of the host molecules.

The connection between transport coefficients and molecular mobilities is mediated by correlation functions of velocity fluctuations for the particles. Near thermal equilibrium, the reference to thermal velocity is given by $k_{\mathrm{B}}T$ alone (Einstein relation), characterizing the 'microbrownian' character of particle mobility. The final relations contain the molecular mobilities or time scales of the dispersion zone under consideration, i.e., of slow dynamics such as α or ϕ processes, and are a special case of the fluctuation–dissipation theorem, viz., Green–Kubo integrals. These will be described after two preliminary remarks.

First Point. The equation for translational diffusivity,

$$D = \frac{1}{6t}\left\langle \overline{\boldsymbol{r}_i^2}(t) \right\rangle , \tag{2.52}$$

is valid if the average squared displacement is proportional to time. The reduced (auto) velocity correlation function $\psi(t)$ is defined by

$$\psi(t) = \frac{\overline{\boldsymbol{v}_1(0)\boldsymbol{v}_1(t)}}{\overline{v_T^2}} , \tag{2.53}$$

where v_T is the thermal velocity of order $(k_{\mathrm{B}}T/m)^{1/2}$ and m the particle mass. Partial integration yields

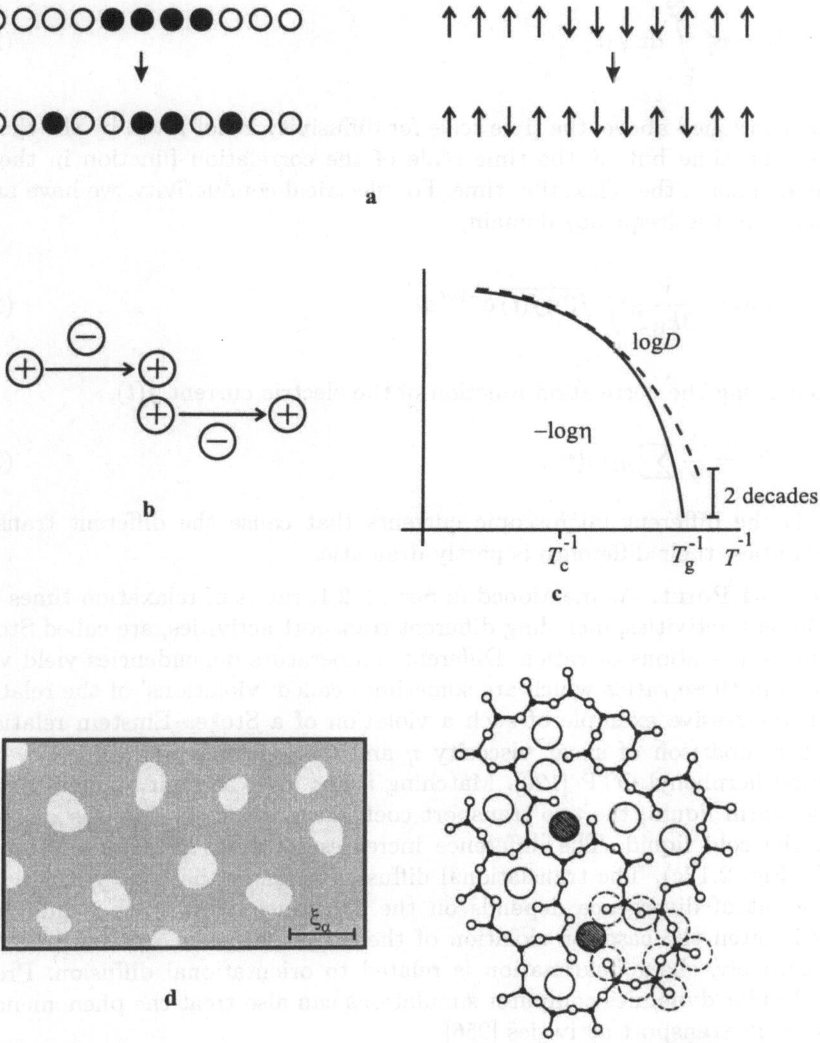

Fig. 2.12. Transport properties. (**a**) Translation (*left*) and spin diffusion (*right*). No particle translation is necessary for the latter. (**b**) Difference between ionic conductivity and self-diffusivity of ions. (**c**) Divergence of reciprocal viscosity η^{-1} and self-diffusivity D at low temperatures (below the crossover temperature T_c). (**d**) Dynamic heterogeneity for the α relaxation represented by a Glarum defect diffusion model. *White*: islands of mobility, *gray*: cell walls (cooperativity shells) of lower mobility. ξ_α = characteristic length = average size of a CRR = average distance between defects. The pattern fluctuates in space and time. (**e**) Structure model of sodium–calcium silicate glass showing the Si–O network [17]

$$D = v_T^2 \int_0^\infty \mathrm{d}t \, \psi(t) \, . \tag{2.54}$$

As mentioned above, the time scale for diffusivity is not given by the thermal collision time but by the time scale of the correlation function in the dispersion zone, the relaxation time. For electrical conductivity, we have analogously in the frequency domain,

$$\sigma(\omega) = \frac{V}{3k_\mathrm{B}T} \int_0^\infty \overline{\boldsymbol{j}(0)\boldsymbol{j}(t)} \, \mathrm{e}^{-\mathrm{i}\omega t} \mathrm{d}t \, , \tag{2.55}$$

containing the correlation function of the electric current $\boldsymbol{j}(t)$,

$$\boldsymbol{j}(t) = \frac{1}{V} \sum q_i \boldsymbol{v}_i(t) \, . \tag{2.56}$$

It is the different microscopic currents that cause the different transport activities; their difference is partly dramatic.

Second Point. As mentioned in Sect. 2.2.1, ratios of relaxation times from different activities, including different transport activities, are called Stokes–Einstein relations or ratios. Different temperature dependencies yield variations in these ratios which are sometimes called 'violations' of the relations. An impressive example of such a violation of a Stokes–Einstein relation is the comparison of shear viscosity η and translational self-diffusivity D of orthotherphenyl OTP [120]. Matching η and D^{-1} at high temperatures, in the warm liquid, the two transport coefficients diverge below the crossover, in the cold liquid. The difference increases to about a factor of 100 near T_g (Fig. 2.12c). The translational diffusion is faster than the viscosity. The amount of divergence depends on the substance, that is, it is individual, as is often the case for violation of the Stokes–Einstein relation [228, 229]. Intuitively, shear deformation is related to orientational diffusion. Present molecular dynamics computer simulations can also treat the phenomenon of different transport activities [256].

Different transport activities indicate dynamic heterogeneity. The general Green–Kubo formulas – an integral form of the FDT – connect the transport coefficients χ with microscopic spatial and temporal integrals of specific current density correlations:

$$k_\mathrm{B}T\chi \propto \int \mathrm{d}\boldsymbol{r} \int \mathrm{d}t \, \overline{\boldsymbol{j}_\chi(0,0)\,\boldsymbol{j}_\chi(\boldsymbol{r},t)} \, . \tag{2.57}$$

The space integration is actually confined to the representative subsystems, the CRRs for the α process, statistically independent from the environment and functional to the dispersion zone. To explain this statement, let us discuss (2.57) for three situations: (i) for critical states with large correlation lengths

ξ, (ii) for local processes with small typical lengths < 1 nm, and (iii) for dynamic heterogeneity, as in the cooperative α process, with characteristic lengths of a few nanometers.

(i) Consider χ = thermal conductivity [257]. Consider the x component of heat current density $j_\chi(\boldsymbol{r}, t)$. Its slowly-varying component for critical slowing down is given by $\delta s(\boldsymbol{r}, t) v_x(\boldsymbol{r}, t)$, where $\delta s(\boldsymbol{r}, t)$ is the local specific entropy fluctuation and $v_x(\boldsymbol{r}, t)$ the x component of the local velocity correlation. For large correlation lengths ξ, the microscopic quantity $j_\chi(\boldsymbol{r}, t)$ in fact contains products of macroscopic variables like δs and v_x. Assuming mutual statistical independence of δs and v_x, the right-hand side of (2.57) is then calculated using hydrodynamics appropriate for large length scales of order several tens of nanometers. The result, by mode-coupling theory MCT, is expressed in terms of a weakly diverging critical viscosity and strongly diverging heat capacity at constant pressure: thermal conductivity behaves near criticality roughly as the diverging correlation length ξ.

(ii) If the current correlation is confined to a small particle cage then χ can only reflect different averages of a time distribution inside the dispersion zone. The integral (2.57) is then sensitive to different moments of the distribution. This approach is discussed in detail in the review [50]. As an extreme example, we have averages of t or of $1/t$, where t stands for rotational and $1/t$ for translational correlations. The first would stress large times, and the second short times. The ratio would be of order $\exp(2\delta \ln \omega)$, where $\delta \ln \omega$ is the width of the α dispersion zone (Table 2.3, e.g., $\exp(2 \times 1.1/\beta_{\mathrm{KWW}}) \approx 23$ for $\beta_{\mathrm{KWW}} = 0.7$). We expect only a small dependence on individual substance parameters beyond $\delta \ln \omega$ for small cages below 1 nm.

(iii) Consider a cooperatively rearranging region CRR as a representative for the dynamic heterogeneity of the α process. Outside a CRR, the correlation in (2.57) tends to zero because the CRRs are statistically independent of their environment. Compared to the local modes of point (ii), larger variations between different transport activities can now be produced by the integration over spatial differences in the current correlations.

For instance [258], rotational correlation time may again weight the least mobile environments, $\tau_c = \langle \tau_i \rangle$, and translational diffusion may weight the most mobile environments, $D_T \propto \langle 1/\tau_i \rangle$. Anticipating a dispersion law (general scaling, Sect. 2.2.5), namely that larger times correspond to larger modes, rotational diffusion is connected with larger modes in a CRR, and translational diffusion with shorter modes. Referring to the Glarum defect diffusion as the model for α process cooperativity, rotational diffusion is connected with the cooperativity shell, and translational diffusion to the defects, i.e., to the islands of mobility. Substance individuality of transport properties is then caused by the spatial inte-

gration in (2.57) if larger modes in a CRR (of order ξ_α) reflect molecular individuality.
Conversely, the actual observation of such individualities indicates that the larger cooperativity modes are in fact sensitive to molecular parameters. It seems that substance individualities shine through here. A universal hydrodynamic treatment as for critical states (i) does not seem reasonable for length scales ξ_α of a few nanometers.

Self-diffusion being faster than viscosity (Fig. 2.12c) is a special case of a more general phenomenon called *translational enhancement*. Comparing different diffusivities, e.g., translational and orientational [259], we find that translational diffusivity is faster. Enhancement is also observed for tracer diffusion some 5 K above T_g (Ediger et al. [260]). This problem is serious. Let us attempt a qualitative explanation by dynamic heterogeneity in real space.

Consider again the Glarum defect diffusion model introduced at the end of Sect. 2.1.3. On the scale of Fig. 2.12d, the Glarum model yields a dynamic heterogeneity pattern represented by a fluctuating cell structure with cell walls of low mobility and islands of high mobility. As mentioned above, larger spatial scales inside a CRR, as for the walls, correspond to slower modes, and shorter spatial scales, as for the islands, correspond to faster modes. The general length scale is given by the characteristic length ξ_α of a few nanometers (average diameter of a CRR), and the general temporal scale by the α-process relaxation time and its dispersion $\delta \ln \omega$. Orientational diffusion and shear viscosity can only be defined in a larger ensemble of neighboring molecules. They depend on the cooperativity of the neighbors, i.e., they 'get stuck' in the cooperativity shells of the islands, the slow cell walls. On the other hand, translational diffusion is defined by moves of a single molecule that changes its neighborhood. It may thus be dominated by fast flips across the islands of mobility. The flips may make only a small contribution to the orientational diffusion.

We may anticipate the Levy treatment of the Glarum defect diffusion model (Sects. 2.2.4, 3.6 and 4.3). The Glarum defect corresponds to the preponderant Levy component. The diffusion corresponds to a random occurrence of defects in space and time in the general CRR time and length scales mentioned above. Since a defect must be constructed from about 30 particles (Sect. 2.4.3), the random movements of defects inside a CRR containing about 100 particles are connected by the random but quasi-continuous path of a spot with extraordinary free volume. Flips across the islands of mobility can thereby 'add' to fast movements across the CRRs without much affecting orientations. Fast translation can thus 'decouple' from the slow orientational diffusion necessarily involving the cell walls. Conversely, a characteristic length of about 3 nm was concluded from the translation enhancement of tracers with different size and shape [260].

It seems that translation enhancement is again embedded in the more comprehensive phenomenon of *translational decoupling*. Decoupling of ionic electrical conductivity from typical time scales of structural relaxation is often observed in alkali silicate glasses (and others) above and even below T_g (Moynihan [261], Angell [262]). The positive alkali ions are small, in the diameter range from 0.16 nm for lithium to 0.33 nm for caesium, smaller than the elementary units of the Si–O network structures (Fig. 2.12e). Charge neutrality is ensured by negative non-bridging oxygen ions. The positive ions are spheres not directly tied to orientational rearrangements that dominate the structural relaxation. The ions can use 'paths' in the network, as indicated in the upper part of Fig. 2.12e. All ion sites are equivalent for glasses with only one ion type: exchanging equal ions at a given site does not require larger-scale and, therefore, time-consuming rearrangements.

The translation decoupling process can mark its own dispersion zone [86, 263], which itself is similar to a dynamic glass transition. It may be different from the usual dynamic glass transition of the material and its continuation below T_g. The latter defines boundary conditions for the decoupling. The conditions can vary with temperature (varying thermokinetic structures or degree of vitrification) and time (structural relaxation below T_g). The new transition will be called a *decoupled* or *embedded glass transition*. Decoupling means that the embedded trace in the Arrhenius diagram has another mobility $\log \omega$ than that for the original transition. In the Glarum model, the decoupled transition is probably related to the islands of mobility. This idea is developed in the description of the mixed alkali effect (Sect. 2.4.4).

In a certain simplification, the increase in dynamic ionic conductivity across the dispersion zone can be compared with the increase in fluidity across the dynamic glass transition of small-molecule glass formers. Compare the a.c. conductivity of ions $\sigma(\omega, T)$ with the shear response behavior $\eta'(\omega, T)$ (Figs. 2.13a–d).

First Point. In the overview given in Fig. 2.13a, the temperature dependence of σ at low frequency (the 'static' or d.c. conductivity) is compared for three different substances: a good conductor, a semiconductor, and an insulator (from top to bottom). The dynamic conductivity of a given substance for different frequencies as a function of reciprocal temperature has bends in the decoupling zone (Fig. 2.13b). The bend 'points' $\{\log \omega, 1/T\}$ form the trace of the dynamic decoupled glass transition in this Arrhenius diagram which can, as a rule, be continued below T_g. The low temperature activation energy E_a is smaller than the apparent activation energy of the trace E_σ (Fig. 2.13b). The ratio increases with the glass transition dispersion, $\delta \ln \omega \approx 1/\beta_{\mathrm{KWW}}$ [264–266],

$$\beta_{\mathrm{KWW}} E_\sigma / E_a \approx 1 . \tag{2.58}$$

The decoupled glass transition is often an Arrhenius process ($E_\sigma = \text{const.}$) with a reasonable frequency prefactor of order 100 terahertz. This corresponds to a strong dynamic glass transition.

Let us repeat that, as a rule, such decoupled glass transitions are observed in the glass state, i.e., below the conventional glass temperature of vitrification.

The dynamic behavior of $\sigma(\omega, T)$ as a function of $\log \omega$ is obtained by conductivity spectroscopy covering today the frequency range from 10^{-3} to above 10^{14} Hz, including the vibration modes (see Fig. 2.13c and [267]).

Second Point. The behavior of the electrical conductivity $\sigma = \sigma'$ near the bend is analogous to that of the shear 'fluidity' $1/\eta'$ at a normal glass transition (Fig. 2.13d). Note the definitions from the linear response

$$\eta' = G''/\omega \qquad \text{of the shear modulus ,} \tag{2.59}$$

$$\sigma = \sigma' = \omega\varepsilon_0\varepsilon'' \qquad \text{of the 'conductivity term' in } \varepsilon^*(\omega) \text{ .} \tag{2.60}$$

The three zones indicated in Figs. 2.13d, for conductivity and shear, are analogous ($\Longleftrightarrow$) in the following way.

(i) The flow zone FZ with $\eta(T) \Longleftrightarrow$ to the zone of d.c. conductivity $\sigma_{\mathrm{d.c.}}(T)$, with the higher activation energy E_σ.
(ii) The dispersion zone DZ $\Longleftrightarrow$ to the bend of $\sigma(\log \omega)$ (straight line with slope 1 in Fig. 2.13c). The bend range is sometimes called the Jonscher regime [109], where the frequency dependence can, in a relatively small $\log \omega$ range, be approximated by

$$\sigma \text{ (ii) } \propto \omega^{\beta_{\mathrm{KWW}}} \text{ .} \tag{2.61}$$

(iii) The glass zone GZ with decreasing (!) η' viscosity $\Longleftrightarrow$ to a low loss regime in dynamic electrical conductivity with the lower activation energy E_{a}.

In summary, we have approximately the following analogy between the normal dynamic glass transition for shear and the decoupled dynamic glass transition for ionic conductivity:

$$\text{FZ} - \text{DZ} - \text{GZ} \Longleftrightarrow \text{(i)} - \text{(ii)} - \text{(iii)} \text{ .} \tag{2.62}$$

The ionic conductivity increases with frequency in the glass zone beyond the decoupled glass transition.

For ionic conductivity, the whole $\sigma(\log \omega)$ curve can be described by a unique formula with a unique model, irrespective of the partition into three zones [268–271]. A high-frequency conductivity $\sigma_{\mathrm{hf}}(T)$ is defined after subtracting the vibration wing with slope 2 (Fig. 2.13c). The master curve is given by

$$\omega_{\mathrm{onset}}/\omega = E_1(\ln[\sigma(\omega)/\sigma_{\mathrm{d.c.}}]) - E_1(\ln[\sigma_{\mathrm{hf}}/\sigma_{\mathrm{d.c.}}]) \text{ ,} \tag{2.63}$$

where ω_{onset} is a suitably defined reference frequency in the dispersion zone and $E_1(x)$ is the exponential integral

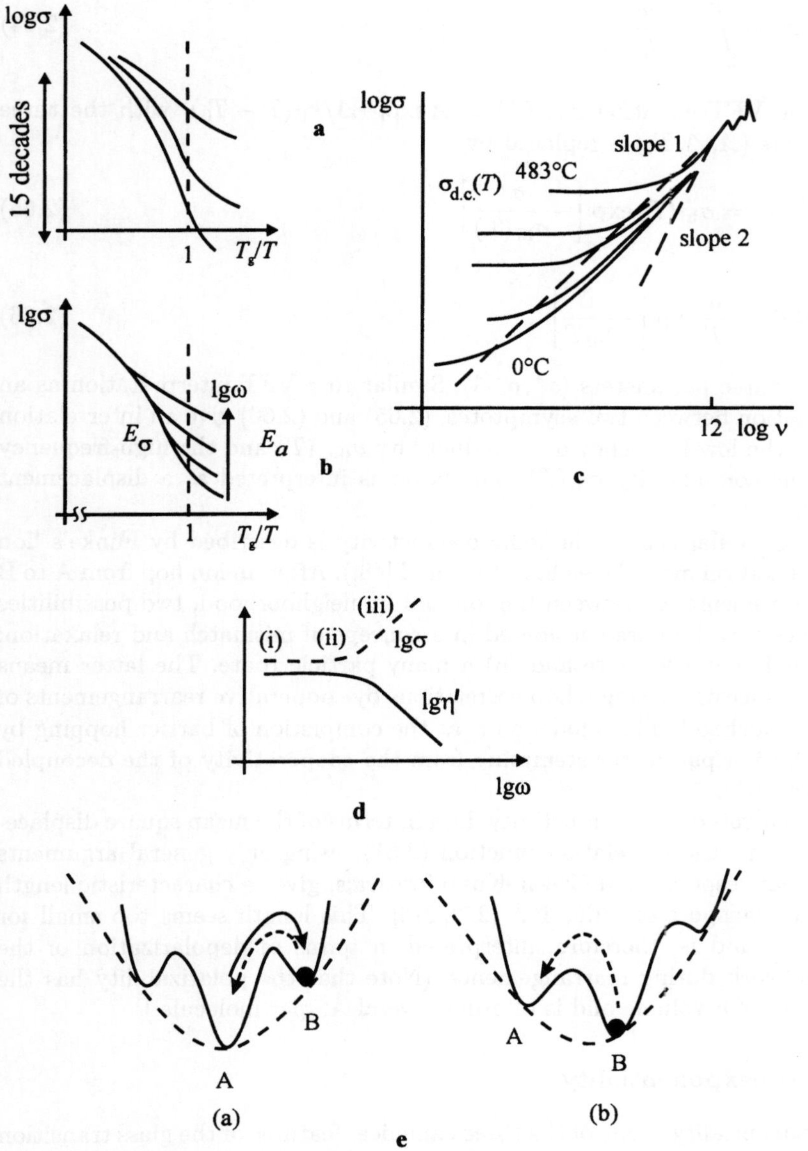

Fig. 2.13. Ionic conductivity. **(a)** Comparison of good, medium, and poor amorphous electrical conductors. **(b)** A.c. conductivity σ as a function of reciprocal temperature for several given frequencies ω. **(c)** Conductivity spectroscopy (schematic). Temperatures are related to the sodium silicate glass $Na_2O{\cdot}3SiO_2$, $\nu = \omega/2\pi$ [709]. **(d)** Electric-to-shear analogy. See text for details. **(e)** Funke's ion jump relaxation model. See text for details

$$E_1(x) = \int_x^\infty e^{-\xi} \frac{d\xi}{\xi} . \tag{2.64}$$

The ionic VFT equation $\sigma_{\mathrm{d.c.}}(T) = A\exp[-\Delta/k_B(T - T_0)]$ with the three parameters (A, Δ, T_0) is replaced by

$$\sigma_{\mathrm{d.c.}}(T) = \sigma_{\mathrm{hf}}(T) \exp\left[-\frac{\sigma^*}{\sigma_{\mathrm{hf}}(T)}\right] , \tag{2.65}$$

$$\sigma_{\mathrm{hf}}(T) = \frac{a}{T} \exp\left[-\frac{\Delta}{k_B T}\right] , \tag{2.66}$$

with the three parameters (σ^*, a, Δ). Similar to a VFT interpretation as an interpolation between two asymptotes, (2.65) and (2.66) give an interrelation between the low-frequency d.c. conductivity $\sigma_{\mathrm{d.c.}}(T)$ and the high-frequency Arrhenius conductivity $\sigma_{\mathrm{hf}}(T)$. The latter is interpreted as a displacement current.

On molecular scales, the ionic conductivity is described by Funke's 'ion jump relaxation model' (see Fig. 2.13e and [48]). After an ion hop from A to B creating a mismatch between the ion and its neighborhood, two possibilities for a backward hop are considered in a concept of mismatch and relaxation: (a) a single particle route and (b) a many particle route. The latter means the assistance of the single hop correlations by cooperative rearrangements of the neighborhood. This model aims at the completion of barrier hopping by a distribution parameter stemming from the cooperativity of the decoupled glass transition.

An analysis of the conductivity data in terms of the mean square displacement of ions, the correlation function (2.51), using only general arguments from linear response and Green–Kubo integrals, gives a characteristic length for ionic transport of order 1 Å [272, 273]. This length seems too small for ion moves and is, therefore, interpreted in terms of depolarization of the glass network during rearrangements. (Note that the polarizability has the dimension of a volume and is of order several Å^3 per molecule.)

2.2.4 Nonexponentiality

Nonexponentiality is one of the three canonical features of the glass transition (Sect. 2.1.5). It means that the relaxation curves over logarithmic time or frequency are wider than a *Debye relaxation* (Table 2.3). The latter is defined by an exponential decay in the time domain corresponding to a so-called 'Lorentz line' or 'Lorentz curve' in the logarithmic frequency domain. We have, for a modulus,

$$G(t) = \Delta G \exp(-t/\tau) , \quad G^*(\omega) = \frac{\Delta G}{1 + i\omega\tau} . \tag{2.67}$$

The Lorentz line for $G''(\omega) = \operatorname{Im} G^*(\omega)$ is defined in this book by

$$G''(y) = \frac{\Delta G\, y}{1+y^2}\,. \tag{2.68}$$

Its characteristic shape is revealed by plotting of G'' as a function of $\log(\omega\tau) = \log y$ (Fig. 2.14a). Changing the relaxation time τ shifts the $G''(\log\omega)$ peak to another mobility $\log\omega$ without changing its shape. [Such behavior, also for other shapes of $G^*(\log(\tau\omega))$ or $J^*(\log(\tau\omega))$, is called thermorheological simplicity, where τ is considered as a function of temperature, $\tau = \tau(T)$ (or $p, x, \ldots$). The shift is usually characterized by a *shift factor* a_T, the ratio of a characteristic time or frequency at temperature T to time or frequency at a reference temperature.] The aim of this section is to discuss serious problems with the Debye relaxation itself and with its extension to nonexponentiality.

The conventional motivation for a Debye relaxation comes from thermodynamics. The relaxation rate is assumed to be proportional to the deviation $\Delta f \neq 0$ from equilibrium ($\Delta f = 0$):

$$\frac{\mathrm{d}\Delta f}{\mathrm{d}t} = -\frac{\Delta f}{\tau}\,. \tag{2.69}$$

This immediately yields the modulus $G(t)$ of (2.67). Such an argument is well suited to the linear thermodynamics of irreversible processes or to hydrodynamic treatments, operating with fluxes and forces [274–276]. However, the characteristic length of 3 nm for the molecular glass transition seems too small for a reasonable application of fluxes and driving forces. The collective character of thermodynamic fluxes and forces does not seem appropriate to the cooperative molecular motions intended for the glass transition.

A better transformation from microscopic to macroscopic phenomena of the glass transition is the fluctuation–dissipation theorem FDT [94,277] mentioned several times above. It reads the same for intensive variables (f, modulus G) and extensive variables (x, compliance J):

$$G''(\omega) = \frac{\pi\,\omega\, f^2(\omega)}{k_B T}\,, \quad J''(\omega) = \frac{\pi\,\omega\, x^2(\omega)}{k_B T}\,, \tag{2.70}$$

where $x^2(\omega)$ and $f^2(\omega)$ are the spectral densities of fluctuations for extensive and intensive variables, respectively. In the following, we consider the slow contributions, from the α functional subsystems, i.e., from CRRs for the α process of the glass transition (Fig. 2.10e in Sect. 2.2.1, and Sects. 3.2–3.4).

For a Debye relaxation in a 'solid' ($G_{\mathrm{equil}} > 0$), both $G''(\log y)$ and $J''(\log y)$ are Lorentz lines. This follows from (2.8), viz., $J^* \cdot G^* = 1$. They are mutually displaced (Fig. 2.14b) by $\Delta \log y = \log(G_{\mathrm{glass}}/G_{\mathrm{equil}})$, i.e., by the logarithm of the glass-zone to liquid-zone ratio of G or J across the dispersion zone. The *modulus–compliance displacement* $\Delta \log y$ increases with the relaxation intensity, e.g., $\Delta G = G_{\mathrm{glass}} - G_{\mathrm{equil}}$.

The FDT equation (2.70) can be applied to each frequency separately: there is no frequency change between microscopic spectral density and macroscopic susceptibilities ('ω identity' of the FDT). This results from the indistinguishability of quanta $\hbar\omega$ used in any linear response experiment (Fig. 3.6b, Sect. 3.4). Testing, for example, the situation Fig. 2.14b by a dynamic linear response experiment with the frequency ω means simultaneously testing the right $J''(\omega)$ wing and the left $G''(\omega)$ wing. This means that the Debye relaxation with finite intensity is not elementary for the dynamic glass transition. In fact it consists of many Fourier components of fluctuations which actually contribute differently to modulus and compliance. The spectral densities in the FDT equation (2.70) are composed of these fluctuation components and are related to an average CRR as representative subsystem.

The representation of a nonexponential relaxation in the time domain by a spectrum of exponential decays, i.e., Debye relaxations (Sect. 3.3), must therefore be considered with caution. Such a spectrum is very useful in practice because it corresponds to a highly qualified sharpening of peaks and steps by means of a differentiation or a Tichonov regularization [278]. Microscopically, however, a spectrum is not appropriate to the cooperative glass transition. A certain compromise seems to be implied by the partition of a non-Debye relaxation into Lorentz lines with infinitesimally small amplitudes. The modulus/compliance displacement vanishes for them, $\Delta \log y \to 0$, but the thermodynamic aspects are retained. These lines will be called *cooperativity modes*. They correspond to independent Fourier components with a thermodynamic halo (Sect. 4.5.1).

A second argument against the use of Debye relaxation with finite intensity as spectral elements for understanding nonexponentiality comes from the Levy distributions for cooperativity (Sects. 2.1.2, 2.2.7, 3.6, and 4.3–4.5). The Debye relaxation corresponds to a Levy exponent $\alpha = \beta_{\mathrm{KWW}} = 1$ of the KWW stretched exponential (2.17). Consider first the compliance $J(t)$. According to the FDT, in the time domain,

$$-[J(t) - J_{\mathrm{equil}}] = \frac{x^2(t)}{k_{\mathrm{B}}T}\,, \quad G(t) - G_{\mathrm{equil}} = \frac{f^2(t)}{k_{\mathrm{B}}T}\,, \tag{2.71}$$

the compliance is proportional to the self-correlation function $x^2(t)$ of x fluctuations of the representative subsystems, the CRRs.

The correlation function and spectral density are mutual Fourier transforms, viz.,

$$x^2(t) = 2\int_0^\infty \mathrm{d}\omega\,\cos(\omega t)\,x^2(\omega)\,, \quad x^2(\omega) = \frac{1}{\pi}\int_0^\infty \mathrm{d}t\,\cos(\omega t)\,x^2(t)\,. \tag{2.72}$$

The same transformations connect $f^2(t)$ and $f^2(\omega)$. Consider the spectral density $x^2(\omega)$, the 'power spectrum', as probability density for x^2 to occur at a given ω. Then $x^2(t)$ and $J(t) - J_{\mathrm{equil}}$ are proportional to the characteristic

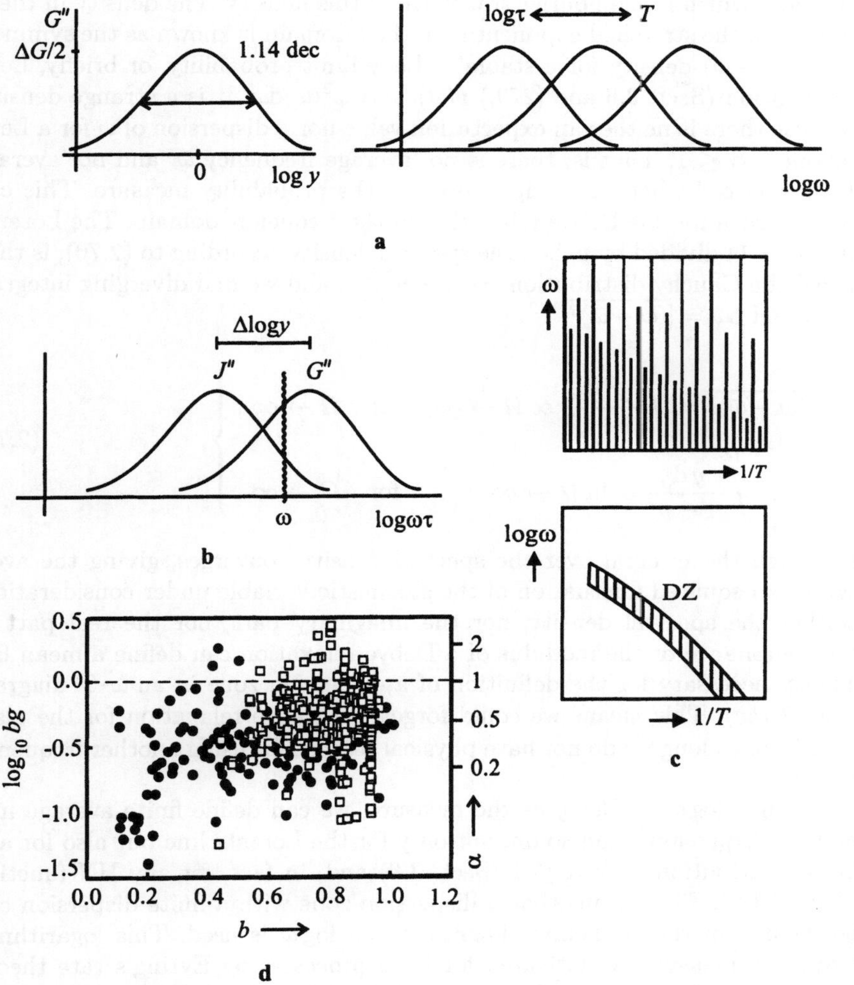

Fig. 2.14. Nonexponentiality. **(a)** *Left*: Lorentz line for a Debye relaxation, $y = \omega\tau$. *Right*: shift of the Lorentz line for increasing relaxation times $\tau(\leftarrow)$ or increasing temperature $T(\rightarrow)$. **(b)** Testing extensive fluctuation by a compliance J, and intensive fluctuation by a modulus G of a dynamic linear response experiment with given frequency ω. **(c)** A dispersion zone DZ cannot be defined by fluctuation with measure $\mathrm{d}\omega$, but can be defined by susceptibility with logarithmic measure $\mathrm{d}\log\omega$, as in an Arrhenius diagram. **(d)** No correlation between the wing parameters of dielectric Havriliak–Negami functions (2.14) for different substances. □ polar liquids, • polymers (many are covered by □ [283]). The b exponent is for low frequencies, and bg for high frequencies. The similarity at high frequency is expressed by the Levy exponent $\alpha = bg$ for $bg \leq 1$. From theory: $0 < \alpha \leq 1$ is for Levy and $\alpha = 2$ is for Gauss; further $0 < b \leq 1$, and $b = 1$ corresponds to white noise

function, which is the Fourier transform of this density. The density in the ω domain of the stretched exponential in the t domain is known as the symmetric ($\omega \leftrightarrow -\omega$) density for a stable or Levy limit probability, or briefly, Levy distribution (Sect. 3.6 and [279]) $p(\omega)\mathrm{d}\omega \propto x^2(\omega)\mathrm{d}\omega$. It is a strange density, because there is neither an expectation value nor a dispersion of ω for a Levy exponent $\alpha \leq 1$. That is, there is no 'average frequency $\bar{\omega}$' and no 'average dispersion $\delta\omega$' when $\mathrm{d}\omega \propto \mathrm{d}y$ is used as the probability measure. This can even be seen for the Debye relaxation in the frequency domain. The Lorentz line ($\alpha = 1$) divided by y, i.e., the spectral density according to (2.70), is then called the Cauchy distribution, $\sim (1+y^2)^{-1}$, and we find diverging integrals for $\bar{\omega}$ and $\delta\omega = [\overline{(\omega-\bar{\omega})^2}]^{1/2}$:

$$\left.\begin{aligned} \overline{(\omega-\bar{\omega})^2} &\propto \int^{\Omega} \frac{y^2\,\mathrm{d}y}{1+y^2} \propto \Omega \to \infty \quad &&\text{for} \quad \Omega \to \infty\,, \\ \bar{\omega} &\propto \int^{\Omega} \frac{y\,\mathrm{d}y}{1+y^2} \propto \ln \Omega \to \infty \quad &&\text{for} \quad \Omega \to \infty\,. \end{aligned}\right\} \tag{2.73}$$

Although the integral over the spectral density converges, giving the average mean squared fluctuation of the stochastic variable under consideration, neither the spectral density, nor the imaginary part, nor the real part of the compliance or the modulus of a Debye relaxation can define a mean frequency necessary for the definition of a dispersion zone in an ω–T diagram (Fig. 2.14c). This means we could forget the Debye relaxation for the glass transition as long we do not have physical reasons for using another frequency measure.

Using $\mathrm{d}\log\omega = \mathrm{d}\log y$ as the measure, we can define finite average mobilities, dispersions, and so on, not only for the Lorentz line but also for any Levy distribution with $\alpha \leq 1$ (Sect. 3.6) and, in fact, for any HN function (Sect. 2.1.2). This means that a dispersion zone with a finite dispersion can be defined in the Arrhenius diagram when $\log\omega$ is used. This logarithmic frequency measure is motivated for local processes by Eyring's rate theory using a transition state with an activation energy E_{A},

$$\omega \propto \exp(-E_{\mathrm{A}}/k_{\mathrm{B}}T)\,, \tag{2.74}$$

and, generally, by the FDT as the transform of microscopic fluctuation to macroscopic susceptibility,

$$x^2(\omega)\,\mathrm{d}\omega \propto k_{\mathrm{B}}TJ''(\log\omega)\,\mathrm{d}\log\omega\,, \quad f^2(\omega)\,\mathrm{d}\omega \propto k_{\mathrm{B}}TG''(\log\omega)\,\mathrm{d}\log\omega\,. \tag{2.75}$$

The frequency measure for the spectral density is $\mathrm{d}\omega$ and the frequency measure for the susceptibility is $\mathrm{d}\log\omega$. In this sense it is the linear response experiment, the FDT, which defines a dispersion zone and which ensures that any trace in the Arrhenius diagram can be interpreted by an average mobility $\overline{\log \omega}$ with an average dispersion $\delta \log \omega \propto \delta \ln \omega$. In the time domain, such a sharp analysis is hidden because the time average $\bar{t}$ and time dispersion δt of KWW functions do exist for all exponents $0 < \alpha \leq 1$ (2.18).

The macroscopic clock of the measuring device determines a time that is conceptually different from the many time differences necessary for the definition of a correlation function. There is no one-to-one map of a small-scale microscopic process $x(t)$ to the response $J(t)$, and the 'quadratic' Wiener–Khintchin theorem (Sect. 3.4.6) [280] must be used because of the stochastic nature of $x(t)$. A surviving sharp mark of the microscopic picture is only the frequency ω finally defined by the Planck condition $\Delta E = \hbar\omega$, for the many quantum jumps. The clock time is not a sharp mark for microscopic statistics because, as mentioned above, time differences must be used to define correlation functions. This means that a microscopic description of cooperativity or the corresponding dynamic heterogeneity has primarily to do with mobility $\log \omega$ and with Fourier components as 'modes' ω, not with a spectrum of local ($< \xi_\alpha$) Debye relaxations with different τ values.

This view exemplifies the idea that the quantum mechanical description including the measurement process may be used for a better understanding of a classical situation. For $\hbar\omega / k_B T \ll 1$, the time domain and the frequency domain are entirely equivalent. But there is no simple and general classical way to understand why $\mathrm{d} \log \omega$ should be used to define a dispersion zone with high-frequency wing slopes $|\mathrm{d} \log \chi'' / \mathrm{d} \log \omega| \leq 1$.

The nonlocal nature of cooperativity requires the mobility picture to be completed by spatial aspects, or relations to subsystems or subensembles. Conversely, $\beta_{\mathrm{KWW}} < 1$ is not, in general, a sure characteristic of molecular cooperativity. Such exponents are also observed for the high-temperature a process that is considered rather to be local, both by experiment [281] and by theory [282].

As an example in the $\log \omega$ domain, we shall discuss the Havriliak–Negami analysis of dielectric compliances for a large set of glass formers (Fig. 2.14d) [242, 283, 284]. The wing slopes of the HN equation, b and $-bg$ [see (2.15) and (2.16), Fig. 2.2e], are not coupled. Similarity was defined in Sect. 2.1.1 as dependence on only one substance parameter. The dielectric compliances, however, depend on these two substance parameters. The dielectric nonexponentiality cannot be reduced either to fragility m or to a global Levy or KWW exponent $\alpha = \beta_{\mathrm{KWW}}$ alone. Instead the diagram Fig. 2.14d reflects an individuality of glass transitions.

The question is: are both wings individual, or is there still a similarity about one of the wings? We seek arguments supporting the idea that it is the high mobility wing which is similar and the low mobility wing which remains individual. This would correspond to the individuality analysis made for transport coefficients (Sect. 2.2.3).

At present, this question cannot be decided without some theoretical input. We have two fundamental ideas: hierarchy or democracy of modes. In a way, hierarchy is connected with a serial, and democracy with a parallel diagram. *Hierarchy* means [285] that the slower mobility of larger parts is governed by the faster mobility of smaller parts. The degrees of freedom are divided into a sequence of levels such that those in level $n+1$ are locked except when some of those in level n find the right combination to relax them, thus representing a hierarchy of constraints in real systems. *Democracy* means that all parts of a partition are entirely equivalent, and that a coarser partition and a finer partition are related by a non-hierarchic scaling. An example is a partition by a multiplicative law in the domain of attraction of the Levy distribution (Sect. 4.3). It is one of the ironies of glass-transition history that followers of both ideas tried to find an explanation of the experimental KWW function.

Intuitively, hierarchy connects similarity with the larger regions. An example, although surely not relevant for the glass transition, is critical scaling by Kadanoff decimation. The resulting critical slowing down is observed at low frequencies, i.e., large times, corresponding to large spatial scales. On the other hand, democracy connects similarity between different substances with the smaller regions. The laws of limit distributions are best detected for finer partitions and best observed in the corresponding tail (wing) at high frequencies.

I shall discuss further only the democratic variant with Levy exponent α. It is then the high-frequency wing that is the carrier of glass-transition similarity with the single parameter $\alpha = bg \leq 1$, and it is the low-frequency wing that is the carrier of molecular individuality with parameter b. In the KWW function (2.17) of the time domain, after Fourier transformation, it is the short-time part that is responsible for substance similarity, not the long-time tail.

According to a dispersion law (or general scaling principle Sect. 2.2.5) the high frequency wing is related to large scattering vectors Q, i.e., to short modes. This means that the islands of mobility in the Glarum model for the α process (Fig. 2.15b) are the carriers of the Levy similarity, and the cell walls, the cooperativity shells, are the carriers of individuality. This corresponds to our finding for the transport properties (Sect. 2.2.3). We will speak about a *Glarum–Levy defect* of dynamic heterogeneity when the free-volume Glarum defect is connected with the singularity of the democratic Levy distribution. As mentioned in Sect. 2.2.1, the first review of the relationship between Glarum defects and the Levy distribution is due to Shlesinger [224].

As an aside, in mode-coupling theory MCT for the a process, the Levy distribution is also obtained for large Q [282].

2.2.5 Dynamic Heterogeneity. Characteristic Length. Cooperatively Rearranging Regions CRR

I shall start by anticipating a theoretical sketch of Glarum–Levy defects (Sects. 3.6 and 4.3) to give the concept of dynamic heterogeneity an aim.

Consider a fluctuating mobility field $\log\omega(\boldsymbol{r},t)$ similar to Fig. 2.12d for one CRR (Fig. 2.15a, right-hand side). Partition the CRR into partial systems i with local frequency ω_i and local free volume V_i, and put $\omega_i = \omega_i(V_i)$. Assume that the $\omega_i(t)$ of the partial systems are statistically independent stochastic functions and let the cooperativity be organized by the balance of free-volume redistribution between them. Then the spectral density $\Delta V'^2(\omega)$ is a Levy distribution (Sect. 4.3).

Any Levy distribution with exponent $\alpha < 1$ has a preponderant component j with high local mobility $\log\omega_j$ and high local concentration of free volume V_j'. The defects are islands of mobility. The stochastics makes the islands appear and disappear randomly on the spatial scale of a CRR and on the time scale of the α process, in such a way that we have one island per CRR on average. I think that dynamic heterogeneity should be defined in such a way as to reflect this behavior.

Apart from its use in some theoretical developments (Sects. 3.6 and 4.3), this definition of dynamic heterogeneity will be applied to the discussion of the β process (Sect. 2.5.3), glass structure (Sect. 2.2.6), Fischer modes (Sect. 2.2.7), tunnel states at low temperatures (Sect. 2.4.3), and the mixed alkali effect (Sect. 2.4.4).

The reader is now recommended to refer back to Glarum's definition of molecular cooperativity and Adam and Gibbs' definition of cooperatively rearranging regions (CRRs), both at the end of Sect. 2.1.3. Typical length scales of order a few nanometers for the glass transition were first discussed in the early 1970s [286–288]. The characteristic length of the glass transition was defined as the average size of the CRRs [226]. A radical concept of dynamic heterogeneity, beyond linear response for thermal fluctuation but strongly related to manageable experimental tests, was developed in the 1990s in Mainz [289, 290]. The connection between cooperativity and dynamic heterogeneity, although simple in our above sketch, is a serious problem in the Mainz concept. We will also continue the discussion about the disappearance of the intermediate scattering function across the α dispersion zone (see Sect. 2.1.4 and the text accompanying Fig. 2.5a).

The Mainz groups [289] "call a system dynamically heterogeneous if it is possible to select a dynamically distinguishable subensemble by experiment or computer simulation." This radical definition was helpful in inventing new and powerful experiments and really can do *without* spatial aspects. Doing without any spatial aspects, however, would mean neglecting the relationship with cooperativity. Moreover, an additional map on thermal fluctuations is sometimes needed since they are accessible by linear response or dynamic scattering.

The methodological point of the Mainz definition is 'selectability'. The idea is to find subensembles that have different time behavior, and then find correlations between them. This concept can be applied to micromechanics even in the configuration space $\{q\}$. NMR experiments, for example, can be applied to measure higher-dimensional time correlations [291]. To link such correlations with cooperativity, we must also find spatial correlations in the 3-d space [292]. If we use linear response for defining cooperativity [122,226], we need a map from $\{q\}$ mechanics to linear response. The general methodological problem in defining dynamic heterogeneity therefore concerns the question of whether it is defined at a kinematic or a mechanical level in the $\{q\}$ space or at a thermodynamic, linear response level.

In other words, the length determination can depend on the activity used for its definition. Let us consider some examples. Kinematic definitions based on molecular dynamics computer simulations [293–295] are related to correlations between molecular displacements (Sect. 2.3.4). These 'kinematic lengths' are expected to be larger than a 'characteristic length' defined by the pattern of entropy fluctuation in space and time [296], because for instance, parallel displacements such as in molecular strings are observed as kinematic correlations but are not active in the entropy pattern. This discussion will be continued in Sect. 2.3.4 [Fig. 2.23a (6)]. NMR correlations tested by spin diffusion [292] determine a 'heterogeneity length' as relevant for the size of less mobile regions. Thinking of a space-filling pattern, the less mobile regions are only a fraction of a CRR that also contains the island of mobility. The heterogeneity length is therefore expected to be smaller than the characteristic length.

According to the concept discussed in the first paragraphs of this section, I think we should refer to linear response and dynamic scattering concepts when dynamic heterogeneity is defined. Linear response and the FDT can be concentrated in dispersion zones, i.e., in Fourier components of slow thermodynamic fluctuations. Since a characteristic length such as the size of representative subsystems can be determined from these fluctuations, we can include spatial aspects from the beginning.

Such a characteristic length ξ_α has nothing [or little, see (2.150)] to do with correlation lengths from structure. This is evident in the crossover region where ξ_α reaches 0.5 nm, the size of the high-temperature a process cage for scenario I. The length for liquid structure is larger there. The definition of a CRR by statistical independence is different from that by direct and indirect energetic coupling of particles in the structure. This can be understood if there is some abstract and functional medium where statistical independence can be related to the dispersion zone under consideration. We choose the mobility field $\log \omega(\boldsymbol{r}, t)$ and, for our present aims, we intend to connect dynamic heterogeneity with the fluctuating pattern of this field. Since the general dynamics is largely decoupled from the multifarious molecular structure (Sect. 2.5.2), it seems unnecessary to refer directly to the structural heterogeneity of the material. As an example consider the large order parameter for polymer chain segments, $S(t) = (1/2)\left\langle 3\cos^2\theta - 1\right\rangle_t$, observed in polyethylmethacrylate [297]. This may influence the glass or crossover temperatures, T_g or T_c, but seems unimportant for the decrease in $\xi_\alpha(T)$ when T_c is approached.

On the other hand, relationships between the dynamic heterogeneity and dynamic structure factor (Sect. 2.1.4) are not a priori excluded. Whether or not dynamic heterogeneity is really reflected in the one-dimensional $S(Q)$ function is related to the concomitant density contrast of a dynamic heterogeneity defined in real space (2.150).

We can summarize the discussion in three pedagogical steps.

First Step. We use a constructive definition of *dynamic heterogeneity*. We partition a representative subsystem and connect each partial system i with its own mobility $\log \omega_i$ with frequency ω_i or, alternatively with its own relaxation time τ_i (Fig. 2.15a), all related to the slow dynamics of entropy fluctuation for the α dispersion zone. The first connection is in the frequency domain [7,138,298] and the second in the time domain [290]. Giving parts of the system their own mobility reflects, in a way, the construction in Fig. 2.10e (Sect. 2.2.1) for distinguishing α and β mobilities in the crossover region. Since the CRRs are statistically independent of the environment and functional to the α process, the representative subsystem is the CRR. The partition is inside each CRR. Considering a larger part of the system with several CRRs, the inner distributions give a pattern of dynamic heterogeneity, fluctuating in space and time (Fig. 2.15b is for the time domain). New aspects enter the scene in large patterns, for many CRRs. We then get an external Levy situation (Sect. 3.6) that leads to the Fischer modes (Sects. 2.2.7 and 4.4).

Second Step. The discussion is related to a definite model. For ideological purposes, we should distinguish the terms 'collective' and 'cooperative'. On the relevant time scale, *collective* means that all particles do the same, as in a wave or in a cage of nearest neighbors, where all particles are equivalent. On the other hand, *cooperative* means that the particles do not do the same. In the Glarum model, for example, the molecular movement in the islands of mobility is distinct from those in the cell walls. In the time domain, one could imagine that relaxation might be fast near the islands and slow in the cell walls (see Fig. 2.15b and [290]).

Third Step. We should ask if there is some general relation between the time and length scales across the dispersion zone. In the frequency domain, we may discuss a relation between frequency ω and wave vector Q from scattering. The corresponding picture for dielectric loss is Fig. 2.15c. We use here a general *dispersion law*, well known from solid state physics. In general, high frequencies correspond to large Q vectors, i.e., after Fourier transformation, to short modes. The converse is also true. For crystals and quantum liquids with their collective quasi-particles, the dispersion law can be expressed by a graph with a sharp line. In amorphous materials, however, we should think about a tendency expressed, formally, by $\mathrm{d}\log\omega/\mathrm{d}Q > 0$ for cooperativity. This tendency is best expressed by diffusivities. Beyond the dispersion zone, self-diffusion has the dispersion law $\omega = DQ^2$, with D the self-diffusivity. In the dispersion zone, dynamic neutron scattering indicates a steeper dispersion law, $\omega \propto Q^{2/\alpha}$ with $\alpha < 1$ the corresponding Levy exponent [281] [see (4.124)].

The term *general scaling* is also used to express the fact that, across the dispersion zone, the larger modes are the slower ones. In the frequency domain, we have the general dispersion law: larger frequencies ω for the spectral density, and larger mobilities ($\log \omega$) for the susceptibility, correspond to shorter modes with larger Q values. In the time domain, the larger-scale spatial correlations of the pattern for dynamic heterogeneity are related to the larger times for the correlation function of cooperativity.

From the standpoint of general scaling, it is rather trivial that we have islands of mobility, and not of immobility. The islands correspond to short modes with large mobility, and the continuous cell walls correspond to long modes with low mobility. On the other hand, without general scaling, and if the relaxation times τ_i are only pinned at points, both possibilities can a priori be discussed [168].

A wide mobility distribution plus a dispersion law (general scaling) does not, of course, contradict the stronger Mainz definition of dynamic heterogeneity. On the other hand, it is of great value if the Mainz subensembles can actually be observed or computed, e.g., by exchange NMR as in the Spiess

▶

Fig. 2.15. Dynamic heterogeneity. (**a**) Definition of dynamic heterogeneity by partition of a CRR, represented in the time and the frequency domain. (**b**) Fluctuating pattern formed by several CRRs in the time domain. This corresponds to a similar mobility pattern in the frequency domain. *White*: islands of mobility with larger local values of $\log \omega$. The average distance between the islands ξ_α, i.e., the average size of a CRR, is called the characteristic length if the pattern is related to entropy (or density) fluctuation. (**c**) General scaling in the frequency domain. Q = scattering vectors. (**d**) Peaks, areas, and steps related to the average temperature fluctuation contribution of the dynamic glass transition (α process). $G_T^* = G_T' + \mathrm{i}G_T''$ is the dynamic temperature modulus of linear response, and $\Delta T^2(\omega)$ is the spectral density of temperature fluctuation. The relevant δT^2 for (2.77) is equal to the twofold area *hatched* in the middle part. The separation of $\Delta T^2(\omega)$ for α from that of other dispersion zones is analogous to the construction Fig. 2.10e for the separation of α and β. (**e**) Time–temperature equivalence gives the equation $(\mathrm{d}\log\omega/\mathrm{d}T)_{\mathrm{along}} = \delta \log\omega/\delta T$. (**f**) A burned hole at frequency $\Omega = \omega$ is a result of shifting spectral parts of $\Delta T^2(\omega)$ along the α dispersion zone. See text for discussion of regions 1 and 2 in the dispersion zone. (**g**) Logarithm of cooperativity $\log_{10} N_\alpha$ and characteristic length ξ_α at T_g for different glass formers, including low-molecular weight substances, polymers, one orientational glass transition in a plastic crystal, and one metallic glass [299]. □ homologous series of poly(n-alkylmethacrylates) including one random copolymer with styrene. m is the fragility

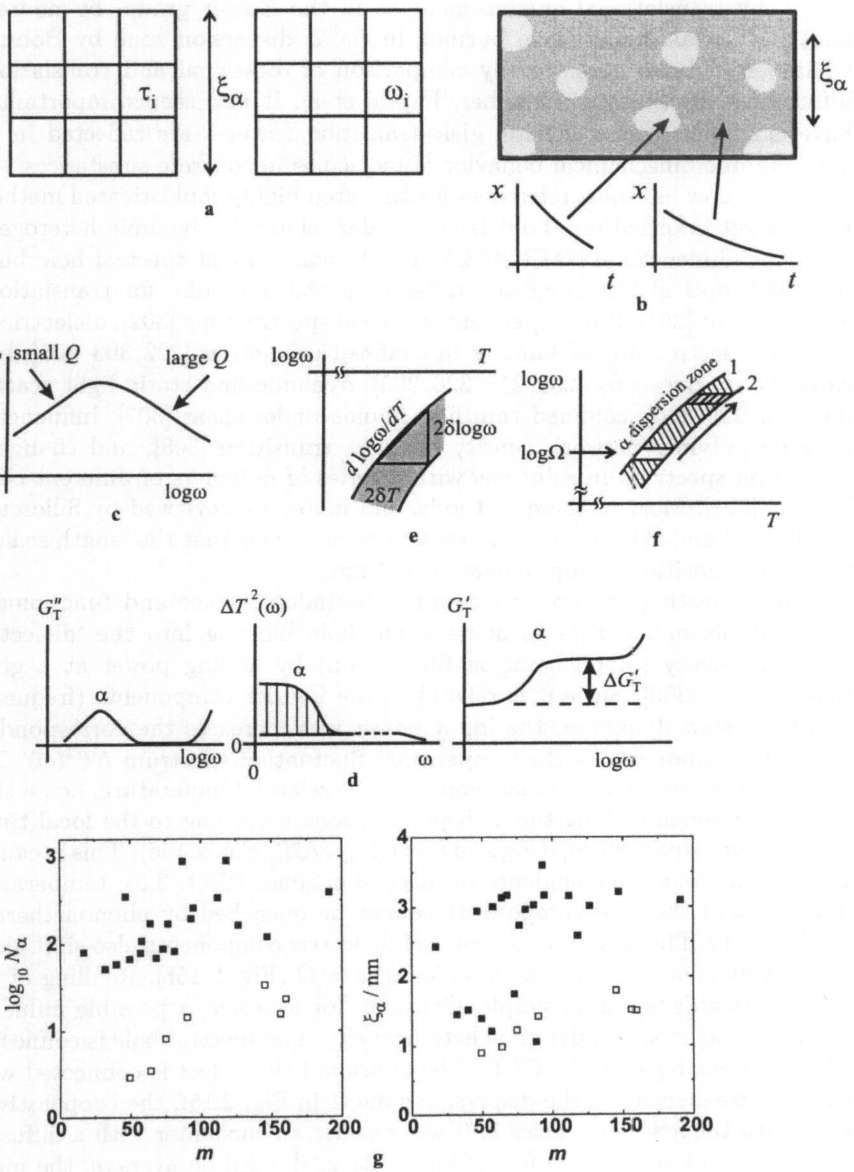
τi
ξα
ωi
a
ξα
x
t
x
t
b
logε″
small Q
large Q
logω
c
logω
T
d logω/dT
2δlogω
2δT
e
logω
logΩ
α dispersion zone
1
2
T
f
G″T
α
logω
ΔT²(ω)
α
0
0
ω
d
G′T
α
ΔG′T
logω
log10 Nα
3
2
1
0
0
50
100
150
200
m
g
ξα / nm
4
3
2
1
0
0
50
100
150
200
m

group, by translational enhancement as in the Ediger group, by nonresonant (i.e., relaxational) hole burning in the α dispersion zone by Böhmer, Chamberlin, Loidl et al., or by comparison of rotational and translational diffusivities by Sillescu, Fleischer, Fujara et al. It also seems important to have examples of how general glass-transition features are reflected in the concrete micromechanical behavior of molecules in concrete substances.

Let us now list some references for the often highly sophisticated methods used to get information about length scales related to dynamic heterogeneity: multi-dimensional NMR [168, 254, 292], nonresonant spectral hole burning [300], optical bleaching and different probe molecules for translational enhancement [301], time dependent solvation spectroscopy [302], dielectric relaxation spectroscopy of samples in confined geometries [222, 303–305], heat capacity spectroscopy [209, 215, 226, 299], dynamic and static light scattering [143, 229, 306], confined thin film liquids under shear [307], influence of growing polymer network density on glass transition [308], and change of relaxation spectrum in solutions with a series of polymers of different chain lengths [309]. Most of these methods, and more, are reviewed by Sillescu et al. [45, 289] and Ediger [50]. The present consensus is that the length scale of the glass transition is approximately 1–4 nm.

An interesting example concerning the independence and functionality of the α dispersion zone is nonresonant hole burning into the 'dielectric' spectral density (of polarization fluctuation) by adding power at a given frequency Ω [300]. Since it is related to the Fourier components (frequency ω) of the slow dynamics, the input power also increases the corresponding ($\omega = \Omega$) components of the temperature fluctuation spectrum $\Delta T^2(\omega)$. The only possible response is enhancement of the related temperature, i.e., a shift of the components along the α dispersion zone according to the local time–temperature equivalence, $\mathrm{d}\log\omega/\mathrm{d}T = \delta\log\omega/\delta T$ (Fig. 2.15e). This is caused by the functional independence of dispersion zones (Sect. 3.5): temperature deviations of the slow components cannot be quenched by phonon thermal conductivity. This means that powered dielectric components also shift along the α dispersion zone leaving a 'hole' at $\omega = \Omega$ (Fig. 2.15f). Refilling of the hole afterwards is not so simple. Consider, for example, a possible influence of the spatial aspect of dynamic heterogeneity. The spectral hole is connected with different regions of a CRR. The Glarum–Levy defect is connected with the high frequencies of the dispersion zone, 1 in Fig. 2.15f, the cooperativity shell with the low frequencies 2. If we consider an encounter with a diffusing Glarum defect as required for refilling [123, 224], then on average, the mean diffusion time, equal to the mean α relaxation time, will become important for refilling.

The characteristic length taken as the average CRR size defines the spatial scale of dynamic heterogeneity in the sense of our approach (Fig. 2.15b). [If there is a wide distribution of CRR sizes, as expected in the crossover region (Sect. 4.3), then the parameters of the length distribution are also important for the description of dynamic heterogeneity. The mobility pattern generated by dynamic heterogeneity is expected [202] to be heavily affected by such a distribution.]

The question of whether the dynamic heterogeneity pattern is connected with concomitant density differences is not trivial [310]. The latter are expected to be very small, because the free volumes regulating the relations between mobility and state variables are – irrespective of their precise definition – very small quantities. The mobility contrast of the Fig. 2.15b pattern is large, of order $\delta \ln \omega$, whereas the concomitant density contrast is very small. A further problem is the density contrast for small characteristic lengths of order $\xi_\alpha \approx 0.5$ nm, i.e., of order the molecule diameter, as expected in the crossover region. The discussion of these important issues will be continued in Sects. 2.5.1–3.

The characteristic length ξ_α can be calculated from heat capacity spectroscopy data by means of a fluctuation formula [7, 226], because the length is related to the entropy fluctuation pattern. Formulas of this type belong to the very small group of formulas from which definite lengths for dynamic heterogeneity can actually be calculated at the present time. Having data at different temperatures, temperature dependencies can also be deduced, viz., $\xi_\alpha(T)$. Let us consider how the formula can be found.

The characteristic length can be calculated from linear response if, for a pair of intensive and extensive variables (f, x), the mean fluctuation δf and the glass transition step height Δx can be determined (Sects. 3.2–3.4). The point is that, by means of the FDT, the fluctuation as recalculated from the response is representative for the smallest independent subsystem, i.e., the CRR. Since the fluctuation amplitude depends on the CRR size, the latter can be recalculated from the response.

Consider the following representativeness *gedankenexperiment*. Divide the sample in two equal parts and measure the peak of specific dynamic heat capacity $C_p''(\log \omega)$ at the dynamic glass transition in one part. The result is, of course, the same as for the original sample. Repeat division and measurement on the chosen half as long as the result remains the same. [For small systems, experimental scattering can in principle be diminished by sampling over long times, because the fluctuations are ergodic with respect to the sampling time in the equilibrium above T_g, and also for the glass zone.] The result will change, however, if the part reaches the size of a CRR. Mutual dependence of subsystems would destroy representativeness. This means that the macroscopic linear response gives the same $C_p''(\log \omega)$ peak as for one CRR of average size. The susceptibility at the dynamic glass transition is a representative for fluctuation of one average CRR. The average cooperatively rearranging region CRR is the smallest, or minimal *representative subsystem* for the α process. As mentioned above, since the amplitude of fluctuation depends on the subsystem size, we may conversely determine this size from fluctuation, via the FDT from linear response susceptibilities. The consequences of the representativeness concept are far-reaching and will be described in Sect. 3.1.

The representativeness concept cannot be applied to crystals. The translational symmetry has an a priori meaning of some large scale entity. The elementary unit of an atomic or molecular lattice is not a smallest representative subsystem. Fluctuations related to translational symmetry are not important, and thermodynamics can be completely based on Gibbs distributions (Sect. 3.2). Disordered aspects of crystals, e.g., molecular orientations in plastic crystals, are, however, subject to the representativeness concept. We can define CRRs in orientational glasses.

Robertson was one of the first to call attention to the representativeness concept [311]. Conversely, in a way, after an analysis of local polymer chain conformations, he concluded that molecular rearrangements seem to be completely contained in regions as small as 2–3 nm.

Let us now recall that the CRR was defined by statistical independence of thermal fluctuations from the environment, and that these thermal fluctuations are functional to the α process, i.e., defined exclusively by the slow Fourier components in the dispersion zone for the α process.

From calorimetry (x = entropy, f = temperature), we first obtain [226] a characteristic volume,

$$V_\alpha = \xi_\alpha^3 = \frac{k_B T^2 \Delta(1/C_V)}{\rho \delta T^2} , \tag{2.76}$$

from which a 'cooperativity' N_α can be calculated,

$$N_\alpha = \frac{RT^2 \Delta(1/C_V)}{M_0 \delta T^2} , \tag{2.77}$$

where R is the molar gas constant,

$$\Delta(1/C_V) = (1/C_V)_{\text{glass zone}} - (1/C_V)_{\text{liquid zone}} ,$$

C_V is the specific heat at constant volume, M_0 is the molar mass of what is considered as a particle, and δT is the average temperature fluctuation of an average CRR. The above gedankenexperiment shows that it is the fluctuation of a CRR which is represented by an α spectral density of temperature or entropy fluctuation in the α dispersion zone (Fig. 2.15d). The α labels in the pictures of Fig. 2.15d illustrate what precisely is meant by saying that the CRR is an α *functional subsystem*. The cooperativity N_α of (2.77) is the average number of particles in a CRR of volume $V_\alpha = N_\alpha/n$, with n the particle number density, and $\xi_\alpha = V_\alpha^{1/3}$ the characteristic length.

As an aside, the same consideration applied to the volume–pressure pair, for example, instead of the entropy–temperature pair, results in

$$V_\alpha = \frac{k_B T \Delta(1/B_S)}{\delta p^2} , \tag{2.78}$$

where B_S is the entropic (adiabatic) compressibility, $B_S = -(1/V)(\partial V/\partial p)_S$, and δp the average pressure fluctuation of a CRR [compare with the fluctuation formulas (3.37) in Sect. 3.2]. The experimental expense in getting V_α from (2.78) is larger than in getting it from (2.76). The average CRR sizes from calorimetry and compressibility are the same since the entropy and volume fluctuation pattern are coupled by thermal expansivity [see (3.38), Sect. 4.3].

Equations (2.76)–(2.78) are based on Gaussian distributions for the relevant fluctuations. Consistency with the intended Levy distribution treatment will be discussed in Sect. 2.4.5.

Since, at present, neither the temperature fluctuation δT nor the temperature modulus G_T can be measured directly, δT is estimated from the time–temperature equivalence $\delta T/\delta \log \omega = (\mathrm{d}T/\mathrm{d}\log \omega)_{\text{along trace}}$ (see Fig. 2.15e). Details of this situation will be discussed in Sect. 2.4.5, (2.138), and Fig. 2.28h, but see also Fig. 2.3b. It will be proved in Example 1 of Sect. 3.3 that the dispersion δT of a Gaussian fit for the isochronal ($\omega =$ const.) $C_p''(T)$ peak from heat capacity spectroscopy is equal to the average temperature fluctuation of a CRR [299]. Furthermore, C_V may be approximated by C_p with a mean 25% uncertainty [299] as estimated from O'Reilly's thermodynamic data in [312,581] (Table 2.11 in Sect. 2.5.4).

The existence of a temperature fluctuation δT rests on the assumption that the FDT can be phenomenologically formulated for any term in the energy form defined by the First Law of thermodynamics, (2.10) and Table 2.1, i.e., including T and S in $T\mathrm{d}S =$ heat (see Sect. 3.2–3.4 and [126,313,314]).

The calculated cooperativities N_α and characteristic lengths ξ_α at the glass temperature T_g do not correlate in a one-to-one manner with the fragility m (see Fig. 2.15g and [299]). Apart from a large scatter outside the confidence intervals of order 50–200% for N_α, we only see a certain trend of N_α with m and, for the poly(n-alkylmethacrylates) mentioned in Sect. 2.2.1, a trend of N_α with increasing distance of T_g from the crossover temperature T_c. As expected from the two preceding sections, the scatter of characteristic lengths reflects the molecular individuality of the various glass formers.

The N_α data from calorimetry indicate a strong temperature dependence of cooperativity [216]. For decreasing temperature, the cooperativity increases from $N_\alpha \approx 1$ in moderate liquids near and above the crossover region $T \gtrsim T_\mathrm{c}$ up to $N_\alpha \approx 30$–300 near the glass temperature $T = T_\mathrm{g}$ [see (2.49), (2.50), and (2.142)]. As radical changes of structure are not observed in this temperature range, this means that the mobility pattern disengages itself from the nearest neighbor shells of molecular structure. Summarizing the relevant remarks from earlier in this section, we have to explain what a cooperativity of order $N_\alpha = 1$ near T_c could mean, perhaps in terms of the mobility field; what role is played by a wide distribution of CRR sizes near T_c; and, near T_g, what can be said about extremely small density contrasts (2.150) when the mobility contrast remains large.

2.2.6 Glass Structure

In liquids, the large temperature dependence of cooperativity must be compared with merely subtle structural changes in large temperature ranges. The latter have been reviewed recently in [315]. I think that the conventional questions about 'structure–property relations' are not appropriate for the cooperativity problem in liquids.

In glasses, we should ask whether or not the freezing-in of dynamic heterogeneity near the glass temperature has consequences for the glass structure below T_g that make a difference between glasses and amorphous solids prepared without a glass transition.

The structure of liquids and glasses is disordered, multifarious, and irregular. Nevertheless, experiments and computer simulations result in a typical picture for the structure factor $S(Q)$. This map from three-dimensional real space onto the one-dimensional reciprocal Q space is mainly dictated by the steep repulsion potentials. We usually find a sharp main peak at a wave vector Q of order the 2π reciprocal intermolecular distances, the 'principal maximum of $S(Q)$', and a weaker and broader second peak from second order diffraction and, possibly, from intramolecular details (Fig. 2.16a). In mixtures and substances with different atom types, different main peaks corresponding to different coordination partners can be observed, e.g., Si–O, Si–Si, and O–O in silicate glasses. Sharpness and details of $S(Q)$ increase with lower temperatures. The changes in $S(Q)$ with structural relaxation of the glass state are small and are, in principle, understood. They are directed towards the equilibrium structure at the annealing temperature (arrow in Fig. 2.16a). This figure is also typical for amorphous solids not fabricated via a glass transition.

There is sometimes a prepeak almost at the half Q value of the main-peak Q (not shown in Fig. 2.16a). This prepeak implies a nanophase separation [316] which has nothing to do with the glass transition.

▶

Fig. 2.16. Vitrification of dynamic heterogeneity. (**a**) Structure factor $S(Q)$ of amorphous solids. The *arrow* indicates the tendency for falling temperature. (**b**) Schematic diagram for touching of molecules living with very little free volume. (**c**) Partial freezing of a susceptibility peak. *Hatched*: frozen. Is there a local relationship between a glass frequency ω_g and the cooling rate $\dot{T}$? (**d**) Schematic diagram of a molecular vault. (**e**) Glarum–Levy shade of dynamic heterogeneity with frozen cell walls (*gray*) and partly surviving defect mobility (*white*). (**f**) Additional enthalpy retardation for given waiting time (e.g., t_w = 10 min) as a function of waiting temperature T_w, after cooling (vault effect, *left-hand side*). Location of the vault effect in the Arrhenius plot (*right-hand side*). (**g**) Boson peak in low-temperature heat capacity of amorphous solids. (**h**) The same in the vibration energy state density as a function of frequency ν. (**i**) The plateau in thermal conductivity $\kappa(T)$

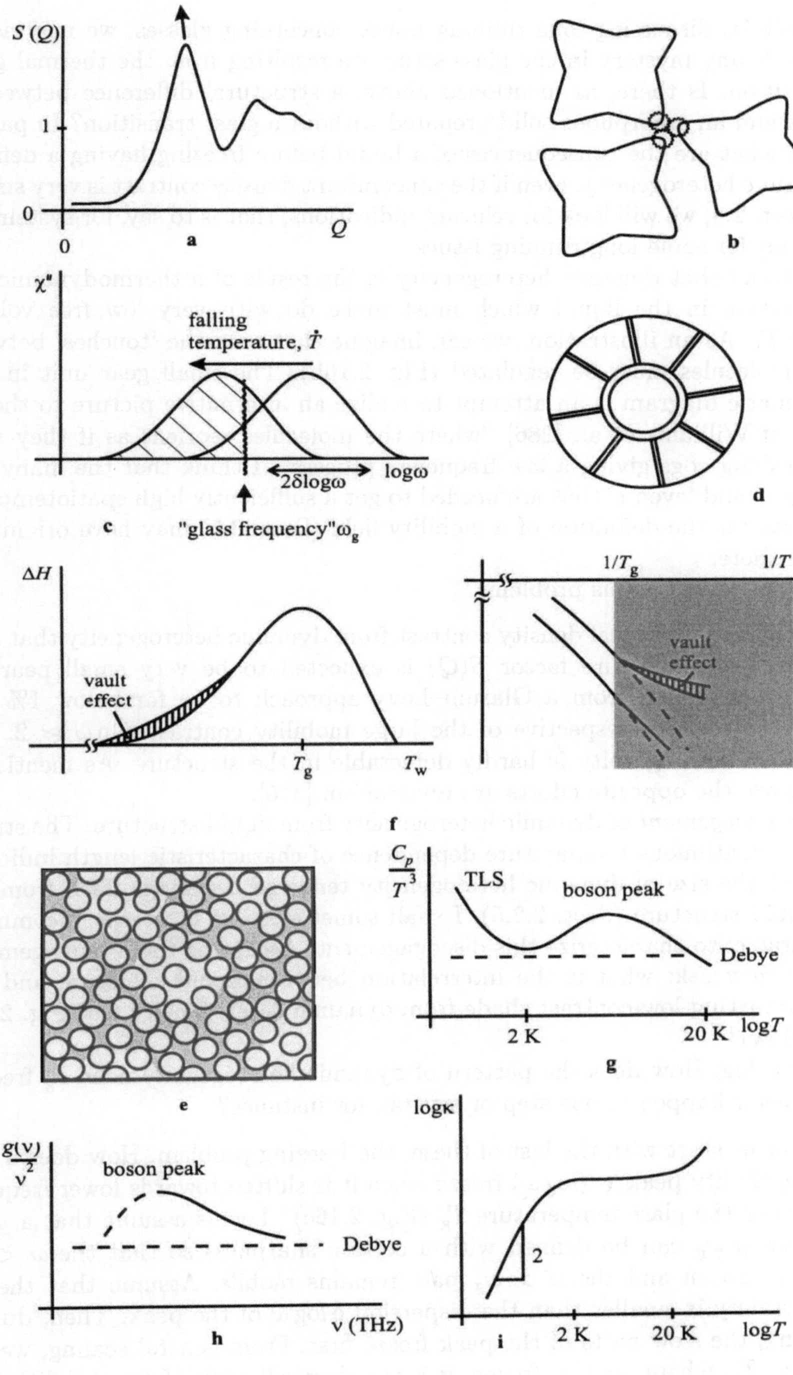

S (Q)
1
0
0
Q
a
b
χ"
falling
temperature, $\dot{T}$
2δlogω
logω
c
"glass frequency"ω_g
d
ΔH
vault
effect
T_g
T_w
$1/T_g$
$1/T$
vault
effect
f
e
$\frac{C_p}{T^3}$
TLS
boson peak
Debye
2 K
20 K
logT
g
$\frac{g(\nu)}{\nu^2}$
boson peak
Debye
ν (THz)
h
logκ
2
2 K
20 K
logT
i

Whilst discussing long-running issues concerning glasses, we may ask if there is any mystery in the glass structure resulting from the thermal glass transition. Is there, as mentioned above, a structural difference between a glass and an amorphous solid prepared without a glass transition? In particular, what are the consequences of a liquid before freezing having a definite dynamic heterogeneity, even if the concomitant density contrast is very small? In Sect. 2.4, we will look for relevant indications, that is to say, for systematic reasons for some long-running issues.

Recall that dynamic heterogeneity is the result of a thermodynamic optimization in the liquid which must make do with very low free volume near T_g. As an illustration, we can imagine that only the 'touches' between the molecules must be regulated (Fig. 2.16b). The small gear unit in this schematic diagram is an attempt to realize an alternative picture to the vision of Williams et al. [286] "where the molecules reorient as if they were enmeshing cogs giving a low frequency process". I think that the many 'attempts' and 'events' that are needed to get a sufficiently high spatiotemporal density for the definition of a mobility field (Sect. 3.5) may have originated from there.

I see three serious problems.

- Contrast. The local density contrast from dynamic heterogeneity that may affect the structure factor $S(Q)$ is expected to be very small near T_g. It is estimated from a Glarum–Levy approach to be far below 1% [see (2.150)] [296] irrespective of the large mobility contrast $\delta \ln \omega \approx 2$. Dynamic heterogeneity is hardly detectable in the structure. As mentioned above, the opposite efforts are reviewed in [315].
- *Disengagement* of dynamic heterogeneity from liquid structure. The strong and continuous temperature dependence of characteristic length indicates that the size of dynamic heterogeneity tends to disengage itself from the liquid structure (Sect. 2.2.5). I shall sometimes use the term 'incommensurable' to characterize this disengagement. Accepting the disengagement, we may ask: what is the interrelation between liquid structure and the fluctuating low-contrast shade from dynamic heterogeneity (see Fig. 2.16e below)?
- Freezing. How does the pattern of dynamic heterogeneity near T_g freeze? Does it happen in one step or several, for instance?

Let us start with the last of these, the freezing problem. How does a loss susceptibility peak $\chi''(\log \omega)$ freeze when it is shifted towards lower frequencies near the glass temperature T_g (Fig. 2.16c)? Let us assume that a *glass frequency* ω_g can be defined with a certain sharpness so that the $\omega < \omega_g$ part is frozen and the $\omega > \omega_g$ part remains mobile. Assume that the ω_g uncertainty is smaller than the dispersion $\delta \log \omega$ of the peak. Then, during cooling, the slow parts of the peak freeze first. From general scaling, we see that at T_g, where 'half' is frozen, it is the slow cell walls of the mobility pattern (Fig. 2.15b) that are frozen, and it is the islands of mobility that remain mobile.

The calculation of a glass frequency ω_g is conceptually difficult since cooling is primarily given by a cooling rate $\dot{T} = |dT/dt|$, not by a frequency. A general expression for a calorimetric ($\chi'' = C_p''$) *vitrification rate*,

$$a = \frac{\dot{T}}{\omega_g \delta T} \approx 5.5 \,, \tag{2.79}$$

will be derived by means of cooperativity modes in Sect. 4.5.1. The cooling rate $\dot{T}$ is to be reduced by the average temperature fluctuation δT of a CRR. The large number 5.5 (compared with the first estimate $a \approx 1$ [7]) follows from comparison of the logarithmic frequency measure for susceptibilities, here their dispersion $\delta \log \omega$, with the linear cooling rate $\dot{T}$. As a rule, experimental values for a lie between 5 and 7 [317].

It seems to be an open problem whether the glass transition is related to a general physical problem: can (2.79) be fully understood via classical statistical mechanics? We have a new time concept, the preparation time $\tau_{prep} \sim \delta T/\dot{T}$ (Sect. 2.1.4), a certain functionality and independence of the α process (Figs. 2.15d and 2.16c, and Sect. 3.5), and we think that fluctuations that are too slow cannot be thermodynamically established (like $C_p = \overline{\Delta S^2}/k_B$). Is there room for insertion of a Levy distribution treatment in addition to the omnipresent Gaussian distribution in statistical mechanics, and does temperature really fluctuate? We may ask whether these questions can be answered solely by computer simulation with the usual averages. Do we need additional information about a natural, i.e., quantum mechanical measurement on and with subsystems [318,615]? This discussion will be continued in Sect. 3.4 and in particular Sect. 3.4.7, where the fluctuation–dissipation theorem FDT is put forward as an equation for such a measurement.

The above contrast and disengagement problems of dynamic heterogeneity will now be combined into the question as to whether a certain island mobility can survive the further cooling below T_g. Assume that a CRR at T_g is large, i.e., that it contains many more molecules than needed for the first coordination shell of liquid structure. A vitrified molecular cooperativity shell can then act like a *vault* that withstands the negative pressure evolving through the higher thermal contraction of the mobile islands inside (Fig. 2.16d). The mortar for the vaults may be the three-groups of blocked gear wheels in the schematic diagram Fig. 2.16b. The higher contraction of the islands generates additional free volume when compared to a situation without vaults. This may help the island mobility to survive. The small density contrast of dynamic heterogeneity in the liquid is amplified by freezing of cooperativity shells and survival of island mobility in the glass. It may be that, near T_g, we can detect a new length defined by the amplified contrast of the island-to-shell ratio.

The integral free-volume effect will be estimated from the formula $\Delta f = \Delta\alpha\,(T_g - T_0)$, where $\Delta\alpha = \alpha^l - \alpha^g$ is the step height in the thermal cubic expansion coefficient at T_g, and T_0 the Vogel temperature. The α^l coefficient then stands for the island, and α^g for the vault. For polymers and silicate glasses, we obtain the order $\Delta f = 1\%$ which corresponds to the amount of conventional free volume at T_g. This means that at least parts of the islands retain their mobility far below T_g. This seems important for the ionic conductivity of glasses. Preconditions are, as mentioned above, that the cooperativity at T_g is large enough for vault formation, e.g., $N_\alpha > 15$ particles [319], and that we really have islands of mobility and not the opposite dynamic heterogeneity, with islands of immobility (Fig. 2.16e).

A more or less direct proof for increased mobility near the islands below T_g would be an enhanced enthalpy retardation so far below T_g that Narayanaswamy modelling (Sects. 2.1.5 and 4.5.2) would give small or zero retardation there (Fig. 2.16f). This calorimetric 'vault effect' is located near the high-frequency boundary of the dispersion zone. The pristine Narayanaswamy model is not expected to be able to model such a consequence of dynamic heterogeneity, so that this model may be used to separate the vault effect from the low temperature tail of enthalpy retardation or, correspondingly, from the short-time part of the underlying Kohlrausch retardation function. Systematic experimental indications for a vault effect are described in [182]. The first steps in structural relaxation (Fig. 2.2a), reflected by an initial asymmetry in the recovery curves (Fig. 2.7c), also observed by adiabatic calorimetry [320], may be interpreted as a temperature dependence of the vault effect. Earlier indications (Fig. 2.5e, reviewed by Johari [149]), may also be interpreted in terms of the vault effect in so far as they are labelled by α in Fig. 2.5e.

An order-of-magnitude estimate for the remaining size of mobility islands at very low temperatures can be obtained from the FDT for shear, provided that the islands are effectively independent subsystems. Take the operating two-level tunneling states TLS at $T_0 = 1$ K (Sect. 2.4.3) as indicators for the finally remaining free volume part stemming from the systematic vault effect. [A smaller nonsystematic effect is expected from randomness of packing, as for amorphous solids without glass transition (Sect. 2.4.3).] The FDT gives

$$V_{\text{island}} \approx \frac{\tan\delta \cdot k_B \cdot T_0}{\pi \cdot n_{\text{TLS}} \cdot G' \cdot \overline{\Delta\gamma^2}}\,, \tag{2.80}$$

where $\tan\delta$ is the shear loss factor G''/G', n_{TLS} the island fraction per particle (e.g., $n_{\text{TLS}} \approx 10^{-5}$), G' the shear modulus, and $\overline{\Delta\gamma^2}$ the mean shear angle fluctuation taken as relevant for rotational tunneling. Using values for typical two level states TLS ($\tan\delta \approx 5 \cdot 10^{-4}$ as for polymethylmethacrylate, PMMA) we obtain, taking $\overline{\Delta\gamma^2} \approx 1$ for large rotational tunnel 'jumps', volumes of order V_0, the volume of one particle. This corresponds to the core of a Glarum defect in our picture (Fig. 2.16e).

The $N_\alpha > 15$ condition for vault formation reflects a trivial aspect of the incommensurability mentioned above when describing the disengagement problem. The disengagement of the continuous temperature dependence of dynamic heterogeneity from the molecular shell structure in amorphous solids can be tested for series of substances in which the $N_\alpha(T_g)$ change crosses the $N_\alpha(T_g) = 15$ mark. The structure discussion will be pursued with an application of the vault picture to several examples (tunneling systems in Sect. 2.4.3, and mixed alkali effect in Sect. 2.4.4).

Dynamic heterogeneities may be important for heterophase equilibria with glasses. A nonequilibrium glass can be a partner of partial equilibria with respect to those fluctuations (e.g., entropy for heat, volume for work, concentration for diffusion) that can, for given times and temperatures, participate in exchange of heat for thermal equilibrium, work for mechanical equilibrium, and particles for phase equilibrium (including electrochemical equilibria). Some interesting examples are discussed by Rehage [321, 322], and the decoupled glass transitions (Sect. 2.2.3) should also be mentioned here.

We conclude this section with an example of an amorphous structure phenomenon where the need for a thermal glass transition or vaults remains an open question, namely the *boson peak* [323]. This peak is observed in the C_p/T^3 vs. $\log T$ diagram between 2 and 20 K (Fig. 2.16g) and is considered as typical for any amorphous solid, and not only for glasses. The Sokolov ratio, $R =$ (peak height/minimum height) in the C_p/T^3 curves, varies between $R = 0.2$ and 0.9, and R increases with fragility [324]. This peak is related to a peak of vibration state density divided by ν^2, $g(\nu)/\nu^2$, in the terahertz region for frequency ν (Fig. 2.16h) and can be deduced from Raman scattering or dynamic neutron scattering, for example. The high frequency wing indicates the transition to molecular scales, and the low frequency wing the transition to the Debye continuum. The peak can thus be interpreted in terms of the additional vibrations possible in amorphous structures. Computer simulations indicate [325] that the corresponding modes involve strangely branched chains of about 20 particles with small mode amplitudes of angstrom order. Possible chain 'links' are random arrangements with shorter distances between the particles, missing in crystals. Stringlike movements are also observed in computer simulations for the liquid [293]. We may ask, among other things, to what extent such chains are modified by residual island mobility.

Let us mention a few further ideas for the boson peak. The thermal conductivity κ as a function of temperature has a plateau of order $\kappa \approx 0.1$ W/Km, approximately in the temperature range of the boson peak (Fig. 2.16i). Translated to the frequency domain, this is the region where the sound waves cease to contribute to κ. A length scale for the boson peak can be estimated from a typical frequency and the sound velocity where sound waves no longer contribute [326]. The result is about two nanometers, i.e., of order the CRR size at T_g. As a speculation, motions of molecules within the frozen CRR may be correlated, resulting in a localized motion within the CRR [327]. A boson peak in amorphous solids without CRRs could not be explained by this speculation. Further ideas concern transverse [328], anharmonic [39], and soft mode [329] contributions.

I think that boson peak and CRR are not interrelated. Frozen-in higher poly(n-alkyl methacrylates), where it is the high-temperature a process with $N_a \approx 1$ that is frozen, also manifest the boson peak [319]. Moreover, too many simple models of the amorphous solid, with no relevance to any dynamic or thermal glass transition, also manifest the boson peak. For example, the boson peak is also a characteristic feature of disordered systems of coupled oscillators [330].

2.2.7 Fischer Modes

Debye and Bueche [331] observed an excess light scattering from PMMA and two silicate glasses caused by spatial correlations in the $\xi_D = 100$ nm range. Fischer [306] clarified the dynamic nature of this phenomenon by dynamic light and partly dynamic X-ray [332] scattering in many other glass formers, classified it as a general feature of glass formers, and explained this scattering by equilibrium density fluctuations of extraordinary intensity ('clusters') in a new ultraslow dispersion zone 5–8 decades below the dynamic glass transition: 'ultraslow modes' or Fischer modes (ϕ in the figure of p. IX). The wave vector range is $Q_\phi \approx 2\pi/\xi_\phi \approx 0.1$/nm. The structure factor $S(Q)$ between Q_ϕ and the main peak from intermolecular distances (at $Q \approx 2\pi/(0.5$ nm$)$ ≈ 10/nm) has a low plateau value of order $S(Q) \approx 0.01$–0.1. This section aims to describe the plateau scattering and Fischer modes. I believe that these modes are typical for all glass-forming liquids and seek a general dynamic explanation.

The term 'ultraslow' is used in various ways by theorists. They call the relaxation behavior of a and α processes ultraslow because the exponentials connected with VFT or WLF type equations become slower than any power-law slowing down by approaching a critical point. The use made by theorists may be confusing because they sometimes associate the Vogel temperature with a presumed hidden phase transition.

▶

Fig. 2.17. Fischer or ultraslow modes ϕ and plateau scattering. (**a**) Levy statistics for explaining plateau scattering and Fischer modes. The preponderant component for the CRR scale is the Glarum–Levy defect (*left*); the preponderant component for the ϕ scale is called speckle (*right*). (**b**) Structure factor $S(Q)$ as a function of scattering vector Q (schematic). (**c**) Excess scattering in the Q range of Fischer modes. *Arrow*: increasing temperature. (**d**) Particle fluctuation in the plateau Q range from X-ray scattering (ensemble average) and from thermodynamic compressibility (time average). (**e**) Temperature behavior of the Stokes–Einstein–Debye ratio obtained from depolarized light scattering (τ_{ls}) and viscosity (η) for DGEBA epon [339]. (**f**) Typical example for the temperature dependence of the length scale $\xi_\phi(T)$ of Fischer modes

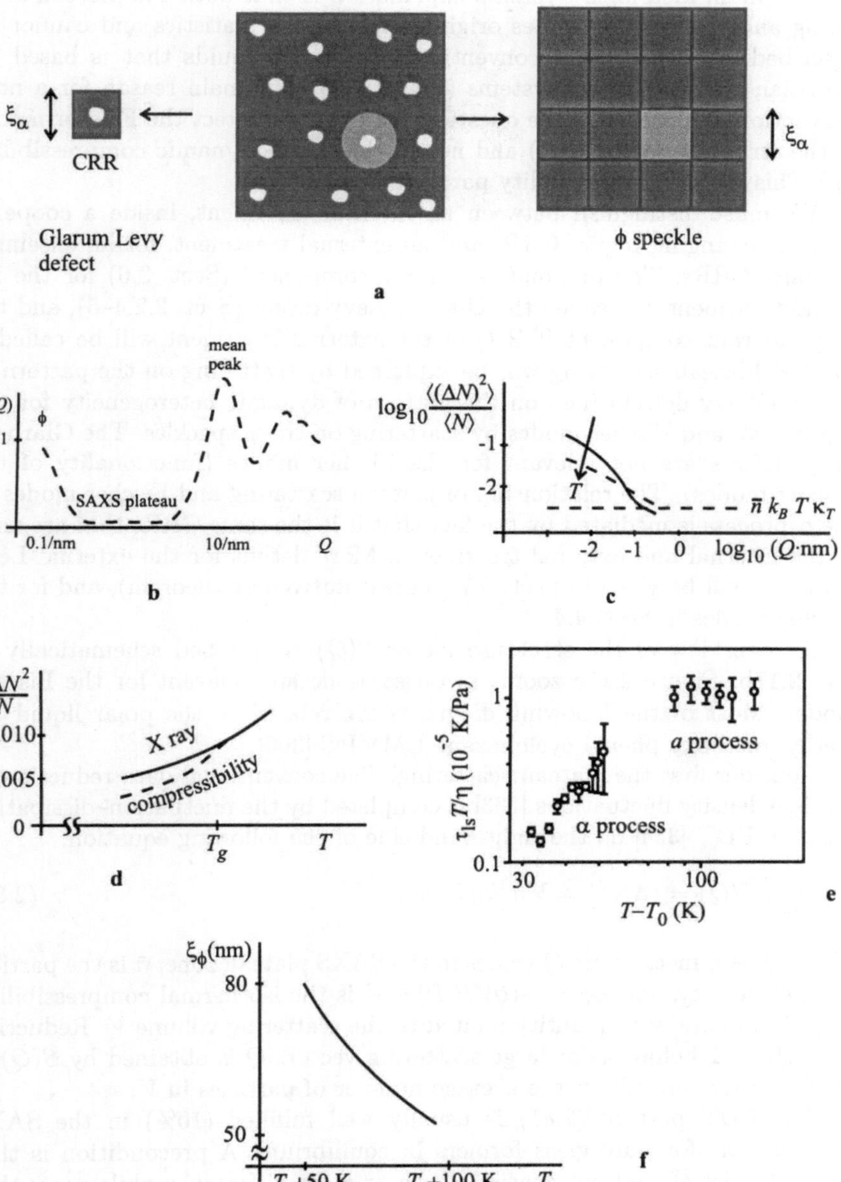
ξα
CRR
Glarum Levy
defect
ϕ speckle
a
mean
peak
S(Q)
ϕ
1
SAXS plateau
0.1/nm
10/nm
Q
b
log10 ⟨(ΔN)²⟩/⟨N⟩
-1
-2
T
n̄ kB T κT
-2
-1
0
log10 (Q·nm)
c
ΔN²/N
0.010
0.005
0
X ray
compressibility
Tg
T
d
τls T/η (10⁻⁵ K/Pa)
1
0.1
a process
α process
30
100
T−T0 (K)
e
ξϕ(nm)
80
50
Tg+50 K
Tg+100 K
T
f

The main idea for a dynamic explanation is that both the plateau scattering and the Fischer modes originate from Levy statistics and cannot be described, therefore, by a conventional theory of liquids that is based on Gaussian statistics of subsystems (Fig. 2.17a). The main reason for a non-conventional approach is the question: why can we detect the Fischer modes in the structure factor $S(Q)$ and not in the thermodynamic compressibility κ_T? This is the compressibility paradox (see below).

We must distinguish between an internal treatment, inside a cooperatively rearranging region CRR, and an external treatment, for an ensemble of many CRRs. The preponderant Levy component (Sect. 3.6) for the internal treatment was called the Glarum–Levy defect (Sect. 2.2.4–5), and the preponderant component (CRR) of the external treatment will be called ϕ *speckle*. Plateau scattering will be explained by scattering on the pattern of Glarum–Levy defects (i.e., on the pattern of dynamic heterogeneity for the α process), and Fischer modes by scattering on the ϕ speckles. The Glarum–Levy defects are not relevant for the Fischer modes (functionality of the Fischer modes). The relationship of plateau scattering and Fischer modes to the α process is mediated by the fact that it is the same CRRs that are used in the internal and external treatments. More details for the external Levy situation will be given in Sect. 3.1 (representativeness theorem), and for the Fischer modes in Sect. 4.4.

An overview of the structure factor $S(Q)$ is sketched schematically in Fig. 2.17b. Figure 2.17c zooms in on some details relevant for the Fischer modes. Most of the following diagrams are related to the polar liquid bis methyl methoxy phenyl cyclohexane BMMPC [306].

Consider first the plateau scattering. The conventional data reduction to particle density fluctuations [333] is completed by the fluctuation–dissipation theorem FDT [334] on the right-hand side of the following equation,

$$\lim_{Q\to 0} \tilde{S}(Q) = \langle \Delta N^2 \rangle = V\bar{n}^2 k_{\mathrm{B}} T\, \kappa_T\ , \qquad (2.81)$$

where $Q \to 0$ means here Q values in the SAXS plateau zone, $\bar{n}$ is the particle number density, and $\kappa_T = -(\partial V/\partial P)_T/V$ is the isothermal compressibility. $\tilde{S}(Q)$ is an extensive quantity related to the scattering volume V. Reduction to $S(Q) \to 1$ behavior for large scattering vectors Q is obtained by $S(Q) = \tilde{S}(Q)\,/\,\langle N \rangle$ where $\langle N \rangle$ is the average number of particles in V.

The FDT part of (2.81) is usually well fulfilled (10%) in the SAXS plateau zone for pure glass formers in equilibrium. A precondition is that the molecules should not consist of two or more different, mobile parts that could simulate concentration fluctuations. An example are the voluminous n-alkyl side groups of higher poly(n-alkyl methacrylates) [335–337]. $\langle \Delta N^2 \rangle$ calculated from scattering is then higher than calculated from compressibility, not fully reflecting such concentration fluctuations. Since molecule parts are not usually thermodynamically compatible, the parts tend to separate as far as possible on the molecular length scale of a monomeric unit (nanophase separation [252, 316, 707]).

Below the glass temperature T_g, the $\overline{\Delta N^2}$ particle fluctuation calculated from compressibility, i.e., the time average, is smaller than $\langle \Delta N^2 \rangle$ calculated from scattering, i.e., the ensemble average (see Fig. 2.17d and [338]). This follows from the nonergodicity of the glass state. Not all ensemble fluctuations can in fact be realized in the experimental time for compressibility measurements below T_g.

The size of $S(Q)$, or $\Delta N^2/N$, in the SAXS plateau zone may be estimated from the assumption that we have an ideal gas of scattering Glarum–Levy defects for $Q < 2\pi/\xi_\alpha$, where ξ_α is the characteristic length of the glass transition of order a few nanometers. Scaling the gas formula $S(Q) = 1$ with the particles as original scatterers gives

$$S(Q) \approx 1/N_\alpha \, , \tag{2.82}$$

where N_α is the number of particles per defect, i.e., per CRR, since there is one defect per CRR. N_α is the cooperativity of the α process. A precondition for a gas treatment is a sufficiently large distance between defects, i.e., large cooperativity $N_\alpha \gg 1$. This restricts the estimate to a temperature range sufficiently far below the crossover temperature. The details depend on the form factor of the defects, as yet unknown.

Further information about cooperativity N_α (or CRR volume V_α) can be obtained from a comparison of relaxation times τ_{ls} for dynamic light scattering with viscosity η [229]. Both light scattering (Tyndall effect from the Glarum-Levy defects) and dielectric spectroscopy are sensitive to rotational dynamics of molecules. The time τ_{ls} corresponds to the maximum frequency of the optical susceptibility $\chi''(\log \omega)$. Above the crossover temperature, $T > T_c$ for the a process, the transport activities are similar because N_a is of order 1. The Stokes–Einstein–Debye relation for $T > T_c$ is therefore independent of temperature:

$$\tau_{ls} = \frac{V_h \eta}{k_B T} \, , \tag{2.83}$$

where V_h is a volume of molecular size (of order 0.1 nm^3). Below the crossover temperature, $T < T_c$ for the α process, the comparison is governed by the difference between the two activities that are intuitively both related to rotational dynamics. The effect expected is therefore smaller than the translation enhancement between self-diffusion and rotational diffusion. Data for depolarized light scattering in the case of DGEBA epoxy resin (DGEBA epon, diglycidyl ether of bisphenyl-A, [339]) indicated a decrease of about one decade (Fig. 2.17e) and were correlated by the ansatz

$$\tau_{ls} \propto \frac{\eta}{V_\alpha T} \, . \tag{2.84}$$

The damping below T_c reflects the increase in V_α (or N_α) with decreasing temperature. A simple explanation would be to connect the viscosity with an extensive variable, a compliance, and the dynamic light scattering with an intensive variable, a modulus, although both are related to rotational dynamics. I expect an influence of molecular individuality on (2.84), as for all formulas related to the characteristic length $\xi_\alpha = V_\alpha^{1/3}$.

Let us now consider the Fischer modes ϕ. The large increase in $S(Q)$ or $\langle \Delta N^2 \rangle / \langle N \rangle$ for $Q \lesssim 2\pi/\xi_D \approx 1/10$ nm, now well documented by dynamic X-ray scattering [332], seems to contradict the FDT equation (2.81) because there is no comparable, but only a small ($< 1\%$ [340]) increase in the liquid compressibility in typical ϕ times (*compressibility paradox*). This means that an additional order parameter ϕ can be introduced [306] for the fluctuations on the spatial scale of the Debye–Bueche correlation length ξ_ϕ,

$$\lim_{Q \to 0} S(Q) = \frac{1}{\bar{n}^2} \left(\frac{\partial \bar{n}}{\partial \phi} \right)^2 \langle \Delta \phi^2 \rangle \ . \tag{2.85}$$

The connection between this order parameter and the extraordinary density fluctuations related to the Levy speckles of Fig. 2.17a is a serious problem. It indicates that, for the nonconventional approach, beyond the Gaussian distribution, scattering and compressibility are different activities. This will be discussed at the end of the section.

The length scale of the Fischer modes, ξ_ϕ, is experimentally determined from an analysis of the $S(Q)$ increase at small Q values (Fig. 2.17c). This length decreases with temperature. An example is shown in Fig. 2.17f (for BMMPC after adjustment by an Ornstein–Zernike equation [306]).

The time scale τ_ϕ of the Fischer modes can be determined by Photon Correlation Spectroscopy PCS (Fig. 2.18a). A nearly exponential decay has been observed [306]. This is compatible with a hydrodynamic treatment, with forces and fluxes from irreversible thermodynamics. Correspondingly, a diffusion–dispersion law can be derived from the dependence on the scattering angle in the Q range from 0.01 to 0.03 nm^{-1},

$$Q^2 \tau_\phi = \text{const.} \ , \tag{2.86}$$

where τ_ϕ is the ϕ relaxation time and Q the length of the wave vector or *scattering vector*,

$$Q = \frac{4\pi}{\lambda} \sin \frac{\theta}{2} \ . \tag{2.87}$$

Here, θ is the scattering angle and $\lambda = \lambda_0/n$ the wavelength in the medium, with n the refractive index. Equation (2.86) indicates hydrodynamic ϕ *diffusion modes* (Fig. 2.18b). Extrapolating (2.86) to the α times of the dynamic glass transition we arrive at a Q_α scale whose length corresponds to the characteristic length of the dynamic glass transition ξ_α = defect distance = CRR diameter, $2\pi/Q_\alpha \approx \xi_\alpha$ [306].

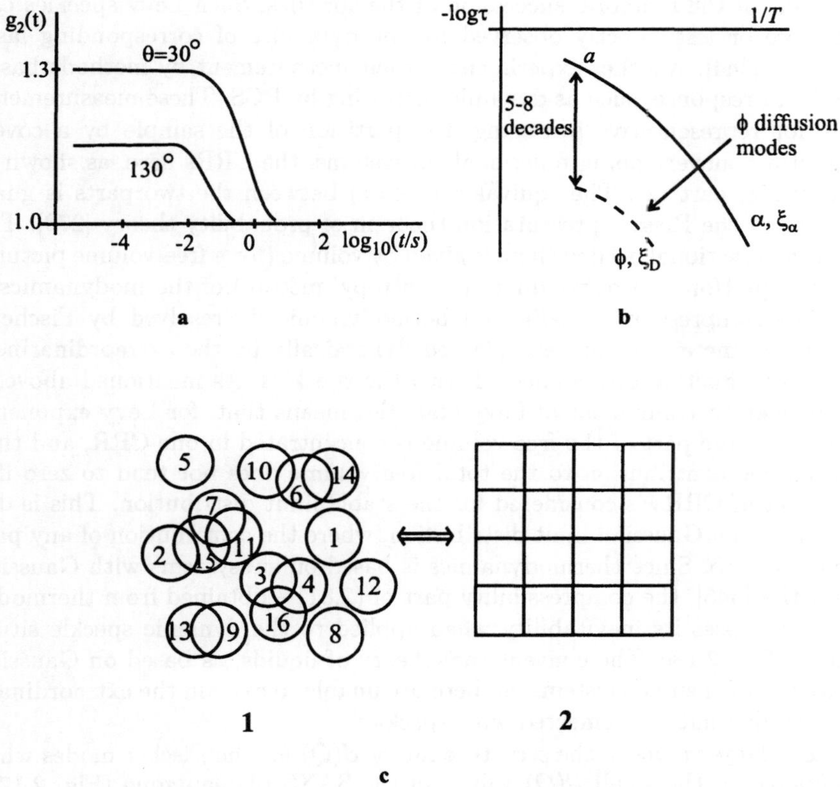

Fig. 2.18. Fischer modes (continued). (**a**) PCS normalized density autocorrelation function for two scattering angles θ, about 100 K above T_g, in the case of BMMPC (10 K above bulk melting temperature). (**b**) Schematic Arrhenius diagram showing the relationship between the dynamic glass transition (a and α processes) and Fischer modes (ϕ). (**c**) Levy speckles (1) and additive partition in CRRs (2) for a dynamic explanation of the compressibility paradox and the hydrodynamic character of ϕ diffusion. See text for details

The conventional explanations suggested so far for the Fischer modes have been based on thermodynamics. They use a free energy for optimization of a two-liquid model [306], for optimization of frustration-limited clusters [341], or for the relation between glass clusters and fluid regions in the liquid [342], similarly to the two-phase model in [343]. When we seek a nonconventional dynamic explanation, we should explain how the Levy speckles of Fig. 2.17a can explain Fischer's order parameter (2.85) and the hydrodynamic character of the ϕ diffusion modes.

The problem is illustrated in Fig. 2.18c. Part (1) is a visualization of the random occurrence of Levy speckles (CRRs with preponderant concentrations of free volume) in space and time. The (arbitrary) numbers are an example of the temporal succession of the speckles. Such Levy speckles can be more or less directly observed by the dynamics of corresponding laser speckles [344]. Another experiment is their measurement by methods based on linear response, such as dynamic scattering by PCS. These measurements use, for representative averaging, the partition of the sample by a covering with nonoverlapping functional subsystems, the CRRs here, as shown in Fig. 2.18c, part (2). The equivalence ($\Longleftrightarrow$) between the two parts is guaranteed by the Riesz representation theorem of probability theory [279]. The positive functional needed there is given by volume (for a free-volume picture) or entropy (for a corresponding 'free-entropy' picture) of thermodynamics.

The compressibility paradox (thermodynamically resolved by Fischer's order parameter ϕ) can be explained dynamically by the extraordinariness of density fluctuations connected with the speckles. As mentioned above, a preponderant component of Levy statistics means that, for Levy exponents $\alpha < 1$, a large part of the free volume is concentrated in one CRR, and that its relative contribution to the total free volume does not tend to zero if a large set of CRRs is considered for the stable limit distribution. This is different from a Gaussian limit distribution, where the contribution of any part tends to zero. Since thermodynamics is based on subsystems with Gaussian statistics [345], the compressibility part of (2.81) – obtained from thermodynamics – loses its inevitability when applied to the dynamic speckle situation of Fig. 2.18c. The conventional theory of liquids, as based on Gaussian statistics of their subsystems, is therefore unable to explain the extraordinary density fluctuation connected with speckles.

The large values of the structure factor $S(Q)$ for the Fischer modes when compared to the small $S(Q)$ values in the SAXS plateau zone (Fig. 2.17b) can be estimated by scattering from a gas of speckles. The scaling by particle scatterers leading to the low fluctuation plateau of (2.82) is not relevant here since the preponderant CRRs (the speckles themselves) are now the scatterers. This gives $S(Q)$ of order 1 as actually observed.

The ϕ diffusion as detected by PCS must also be explained in the nonoverlapping covering of Fig. 2.18c, part (2). Mapping the speckles from part (1) into the covering does not mean that CRRs must actually be moved. Instead, the diffusion can be visualized as some kind of spin diffusion without any translation. The CRRs of the covering are randomly in the state "having extraordinary concentration of free volume". Calling this state 'spin 1' and the others 'spin 0', we can try to describe the spin 1 fluctuation by a spin diffusion. As the spatial scales are large, we can apply hydrodynamic methods which means, phenomenologically, Debye relaxation ($\beta_{\mathrm{KWW}} = 1$) and classical diffusion (2.86), although the speckles are regulated by a Levy exponent $\alpha < 1$. Experimental diffusion coefficients and computer simulation for the a process indicate (Fig. 2.23a (5)) [295] that distances ξ_ϕ can be covered by the particles in times τ_ϕ. The corresponding particle diffusion (Fig. 2.23a (2)) [398] can also be understood by hydrodynamic methods.

Speckle diffusion therefore differs from Glarum–Levy defect diffusion inside the CRR. Since a typical CRR has $N_\alpha \approx 100$ particles and since a defect (island of mobility) requires about 15 particles for its definition, the random occurrence of Glarum–Levy defects can practically always be described by some diffusion of the defect. A hydrodynamic treatment does not seem possible inside the small CRR scale ($N_\alpha = 100$, $\xi_\alpha = 3$ nm), and another diffusion law is expected, e.g., $\omega \sim Q^{2/\alpha}$, as observed by dynamic neutron scattering (see [281] and Sect. 4.5).

The proof that an ensemble of representative CRRs obeys Levy statistics with $\alpha \leq 1$ is described in Sect. 3.1 (representativeness theorem). Length (ξ_ϕ) and time (τ_ϕ) scales for the Fischer modes are estimated theoretically in Sect. 4.4.

2.2.8 Similarity in Different Substances

The aim in this section is to return to the surprising similarity of dynamic glass transitions against the background of structural multifariousness in glass-forming substances.

The *a* process in warm liquids can certainly be observed in all liquids above the melting temperature T_m. The observability of the cooperative α process in cold liquids and of the corresponding thermal glass transition at T_g depends on an 'accident', namely that crystallization does not intervene. In this sense all liquids have a glass transition with the three canonical features mentioned in Sect. 2.1.6. The dynamic glass transition (a, α) and its associated components (β, ϕ, b, c) represent the typical dynamic behavior of liquids (see figure on p. IX).

There is growing evidence that all materials with molecular disorder have a glass transition in this sense, if crystallization or another phase transition can be prevented. This expectation was first stated by Tammann in his famous 1933 book [1].

The multifariousness of molecular and liquid structures is demonstrated by the observation of glass transitions in different molecular and atomic liquids, melts, and mixtures (e.g., selenium, small-molecule polar and non-polar liquids, Lennard–Jones mixtures, silicate glasses, magmas, chalcogenides, salts, metals [42]), polymers (with a certain multiplicity of two, three or four glass transitions, see Sect. 2.4.1 below), biomaterials such as proteins [346–348], and others. There are also glass transitions in plastic crystals (positions ordered, orientations disordered) and in liquid crystals (positions disordered, orientations ordered) [349, 350], in the latter usually with a twofold multiplicity, $\parallel$ and $\perp$. The dynamics of some special materials, such as spin glasses [351], and of charge density waves [352], e.g., in orthorhombic TaS_3, is partly related to glass transition behavior, but possibly superposed by other features.

Exceptional glass formers [353] are outside the scope of our book. Examples of such glass formers are amorphous water, silicon SiO_2 and BeF_2, and proteins. They are characterized by the formation of new glass types by first order transitions from liquids, or by the property of not having a typical glass transition at all, or by nano-multiphase transitions, respectively, or by something with other peculiarities. We concentrate our attention on 'typical' glass formers, or better 'moderate liquids', that will be defined near the end of this section.

It seems that only two features are sufficient for the occurrence of a glass transition: a certain disorder and the absence (or remoteness) of a phase transition, both related to the same molecular degrees of freedom.

We have again some terminological delicacies. The molecules in plastic crystals exhibit rotational motion on a lattice formed by their centers of mass. Orientational glasses are always mixed crystals of this sort. In other words, a disordered crystal is a *plastic crystal* if the disorder is self-induced and, below its T_g, is a frozen-in orientational disorder with no a priori given random field (e.g., cyclohexanol), whilst a disordered crystal is an *orientational glass* if the disorder is given as an alloying-induced disorder, a chemical disorder, or a quenched compositional fluctuation, i.e., if we have a static heterogeneity (e.g., a spin glass, or CKN as in Sect. 2.4.3). The antonym of the two is a *structural glass*, having self-induced disorder in positions and other variables.

Orientational glasses (Sect. 2.4.3) have not so far been able to help in elucidating the basic dynamic problems of 'structural' glass transitions. Self-induced disorder seems decisive for the latter, although spin glass concepts are revived by discussion of the crossover region as a hidden phase transition when a new dimension is included by a coupling parameter. The problem of whether thermodynamics, within the framework of Gibbs distributions, is ultimately the reason for the glass transition will be discussed in the introduction to Sects. 2.5 and 2.5.1. The alternative is a purely dynamic reason. Colloidal glass transitions will be discussed in Sect. 2.3.3. Macroscopic dynamics such as traffic before a jam, or changing success in society (Sect. 4.3), is sometimes also discussed in terms of the glass transition.

We defined similarity as dependence on a single parameter (Sect. 2.1.1). To understand similarity of structural glass transition in materials with such multifarious particles and potentials requires an abstract model or theory. Considering the small characteristic length of order a few nanometers for the α process, we must also consider a certain degree of concomitant individuality (Figs. 2.14d, 2.15g). I concluded from the Glarum model (Sects. 2.2.3–5) that it is the short and fast modes that should be similar: similarity is related to the dynamic behavior at the Glarum defects. Let us now discuss again the general reasons for this hypothesis. Concrete modelling by Levy distribution will be described later (Sects. 3.6, 4.1, and 4.3).

The free volume operates between the slow main transition (α and a processes) and the much faster terahertz process c (for α and a) or the Johari–Goldstein β process (for α). If the molecular configurations open a small spatial region with somewhat smaller local density, e.g., 10%, then a lot of fast

Fourier components for thermal fluctuation evolve – a mobility breakthrough – ranging in principle from the α or a frequencies to the β or terahertz frequencies (Sect. 2.1.3).

As seen from the liquid structure, the number of directly participating particles forming a cage of nearest neighbors is about 10. As seen from the dynamic heterogeneity or cooperativity, we may imagine a localization of the mobility breakthrough on the escaping doors of a cage. For the a process, this is directly supported by the experimental $N_a \approx 1$ values from calorimetry for moderate liquids. The two views were made consistent within the framework of disengagement and incommensurability concepts developed for the relation between structure and dynamic heterogeneity (Sects. 2.2.5–6).

The localization of mobility for both the a and α processes generates a large number of events when compared with the periods of the slow a and α Fourier components. It is the concentration of free volumes on the escaping cage doors that produces there an event field with high density in space and time which can be treated by the continuous methods of probability theory. Such a mobility field is a precondition for disengagement and incommensurability between dynamic heterogeneity and structure. Such a field may also explain the similarity of the dynamic glass transition above the multifariousness of the underlying molecular and liquid structure.

Control of this localized dynamic breakthrough is provided by the mathematical fact that all probability distributions coming in the domain of attraction of a limit distribution – as a consequence of a democratic, increasingly finer partition of the event field – must tend to a Levy distribution (Sect. 3.6). This stable distribution is considered in the mode-coupling theory [282] and in an appropriate model for the Glarum defect (Glarum–Levy defect, Sect. 4.3). In this way, the similarity parameter is the Levy exponent α. For $0 < \alpha < 1$, we find α at the high frequency wing of loss susceptibilities,

$$\alpha = -\frac{\mathrm{d}\log \chi''}{\mathrm{d}\log \omega} \quad \text{for} \quad \text{large } \omega \,. \tag{2.88}$$

This α exponent may be different not only for different substances, but also for different activities, even for the same substance at the same temperature. The physical interpretation of the Levy exponents is a serious problem, and indeed one of the major problems of the glass transition. It will also be discussed in Sect. 4.3.

Let us again compare the molecular basis for critical universality and glass-transition similarity. Kadanoff scaling for larger and larger regions leads to fixed, universal sets of critical exponents for distinct universality classes. On the other hand, partition into smaller and smaller parts leads to a continuous similarity parameterized by the Levy exponent α for the glass transition. Since a map between the two pictures seems difficult, we do not expect the concept of universality class to apply to the glass transition, and in particular to specific α values. The large fluctuation and large correlation length in critical states result from the vanishing of thermodynamic forces near boundaries of thermodynamic stability. The large fluctuation for the glass transition results from the localization of dynamic breakthroughs with small characteristic length.

Even for high density of the event field, the ultimate comparison with molecular features is mediated by the Boltzmann constant k_B. Although we have many events, the relative fluctuation is of the order of $1/\sqrt{f}$, where f is the small number of molecular degrees of freedom for the escaping cage door. The relative fluctuation is large but accessible to the continuous methods of probability theory (Table 3.2 in Sect. 3.5).

On the other hand, individuality for the α process of the glass transition is organized by the cooperativity shell of the Glarum defect. This corresponds to slower and larger modes across the dispersion zone (general scaling of Sect. 2.2.5).

The crossover region separates the high-temperature a process without cooperativity shell from the low-temperature α process with cooperativity shell. The question as to whether this crossover region between warm and cold liquids can be connected with a phase transition that belongs to a definite universality class is not the same as the above similarity problem, and will be considered in Sect. 2.5.1.

Obviously, for the α process, the overall similarity expressed by the fragility m mediates between the dynamic breakthrough similarity of the Glarum–Levy defect and the individuality of characteristic length for the cooperativity shell. We find empirically a certain correlation between m and the α exponent [77] (as it contributes to the $\delta \ln \omega$ width of the dispersion zone) and the cooperativity N_α (Fig. 2.15g). This is the special reason for the WLF homomorphism [see Fig. 2.1f and (2.4)] that generates the intermediate similarity be means of fragility. For given frequency and temperature asymptotes, $\log \Omega$ and T_0 in (2.24), there is only one parameter B_{WLF} that remains free for similarity.

This book is thus concerned with *moderate liquids* as related to the dynamics. The term 'moderate' is related to moderate complexity between simple and normal liquids with unavoidable crystallization, on the one hand, and exceptional glass formers as mentioned above, on the other hand. Moderate liquids are thus glass-forming liquids of moderate complexity with dynamics characterized by predominance of similarity, limited individuality, and no exceptional features of the kind mentioned above.

Apart from such fundamental considerations, each glass former has its own experimental peculiarities such as spurious crystallization, polyamorphism, hygroscopicity, persistent impurities, special secondary γ relaxation, etc., that require careful consideration. This is the main reason why glass transition research is based on a few favorites: B_2O_3, window glass, Glycerol, OTP, PVAC, PS, the salt mixture 'CKN', true CKN (toxic), and some others.

2.3 High Frequency Relaxation. Not as Simple as Expected

This section describes the processes with frequencies above the crossover frequency (figure on p. IX). The results have come from a good collaboration between experimental physics (dynamic neutron and light scattering, colloidal systems) and theoretical physics (mode-coupling theory MCT) in 1982–92. Several findings were unexpected.

(i) The high-temperature *a* process, or Williams–Götze process in the warm liquid does not tend to a Debye exponential decay but remains more stretched out. Its trace in the Arrhenius diagram is curved, at least for the gigahertz region and below.
(ii) The microscopic transient (A) from vibrational modes bifurcates into the *a* process and a nonactivated pair of terahertz processes called the boson peak (*b*) and cage rattling (*c*).
(iii) The extrapolation of the *a* process to lower temperatures seemed to culminate in a dynamical critical state at T_c in a region that was later identified with the crossover region to the α process (C), usually in the mega-to-giga hertz range.
(iv) Götze's MCT can describe the overall picture as well as several details with an elegant method of statistical mechanics.
(v) No characteristic length larger than a particle diameter has so far been observed by dynamic neutron scattering, and nor is it needed by MCT.

It would have been rather difficult to manage the large raw data streams e.g., from dynamic neutron or light scattering, without a theoretical base. This success tempted the inventors of MCT to hope that the crossover splitting of *a* into the cooperative α and the Johari–Goldstein β process could also be understood in a similar way. They therefore named the high-temperature process α (= 'structural relaxation') and the cage rattling process β. This caused some irritation in the glass transition community, which continues to some extent to this day (Table 2.6). Furthermore, tempted by the elegance of MCT, the inventors sometimes tried to avoid an explanation of their results by simple, non-mathematical molecular models or pictures of physical chemistry. Nevertheless, the synergism of experiment and theory in this field was a huge impetus for the whole field of glass transition research.

The term precursor is used in different ways for the glass transition. Along the main transition, it is the *a* process in the warm liquid that is the precursor of the cooperative α process in the cold liquid. Isothermally, across the α dispersion zone, it is the $A1 = f$ process between α and β (Figs. 2.10c and d) that is the precursor of α.

Table 2.6. Labelling different dispersion zones[a]

	MCT	This book
Cage rattling	β, β_{fast}	c
Boson peak	–	b
Escape from cage	α, 'structural'	a, Williams–Götze
Johari–Goldstein	β_{slow}	β
Cooperative	Not distinguished from α	α
Crossover index	$c =$ critical	$c =$ crossover

[a]The labels α, β, γ, ..., were first introduced in polymer science to mark successively the maxima or shoulders of isochronous ($\omega =$ const.) shear loss factors as a function of temperature, $\tan\delta(T)$, starting from high temperatures. It is only for amorphous polymers that these α and β coincide with the conventional α and β used in this book. The glass transition terminology in semicrystalline polymers was later burdened by more complex symbols, e.g., α_c, $\beta(U)$, and $\beta(L) =$ crystalline, upper (interfacial), and lower (amorphous) glass transition, respectively, for the cooperative processes in polyethylene [28], and γ_{I}, γ_{II}, γ_{III} for more local processes. In liquid crystals LC, the cooperative process has two components which are usually labelled by α ($\|$) and δ ($\perp$).

2.3.1 Giga-to-Terahertz Results

The traces of the originally expected and actually observed processes are compared in the Arrhenius diagrams of Fig. 2.19a. With falling temperatures, the intensity of cage rattling c decreases as expected for an overdamped, more relaxational process, and the intensity of the boson peak b increases as expected for an underdamped, more vibrational process.

▶

Fig. 2.19. Giga-to-terahertz results from neutron scattering and dielectric spectroscopy (schematic). $a =$ high temperature, or Williams–Götze process, $b =$ boson peak, $c =$ cage rattling. (**a**) Comparison of the high-frequency parts of Arrhenius diagrams as expected before 1980 and now observed. $\alpha =$ cooperative process, $\beta =$ Johari–Goldstein process, ⓒ crossover region. $\alpha\beta$ is an old symbol for the a process which does not reflect its distinct and independent nature. (**b**) Information about dynamic heterogeneity from a comparison of iso-Q lines as from dynamic neutron scattering and traces of dielectric susceptibility (∿∿∿) in an Arrhenius diagram (schematic) [140]. (**c**) Intermediate scattering function for the a process as a function of time t. The parameter is temperature, $Q \approx 1.44$ Å^{-1} around the first principal diffraction peak, and the example here is glycerol [360]. (**d**) Reduced plot of coherent and incoherent components of the $S(Q,t)$ function for the a process in the nanosecond range. f_Q is the fraction of $S(Q,t)$ for the c process, and times are reduced by the viscosity η divided by temperature T so that $t = 1$ ns for 300 K. $Q = 1.43$ Å^{-1}, example glycerol [360]. (**e**) Iso-Q traces in an Arrhenius diagram as basis for a search for dynamic heterogeneity, e.g., by a comparison with other traces [234] as in (**b**)

For description of dynamic neutron scattering (Mezei et al., Richter et al. [354,355]), we need some qualitative impression of the concepts used there. Details [356] will be presented in Sect. 3.4.2. Scattering with no energy transfer $\hbar\omega = 0$ is called elastic and is characterized by the *structure factor*,

$$S(Q) = \frac{1}{N}\left\langle \sum_{jk} \exp\left[\mathrm{i}\boldsymbol{Q}\left(\boldsymbol{r}_j - \boldsymbol{r}_k\right)\right] \right\rangle , \tag{2.89}$$

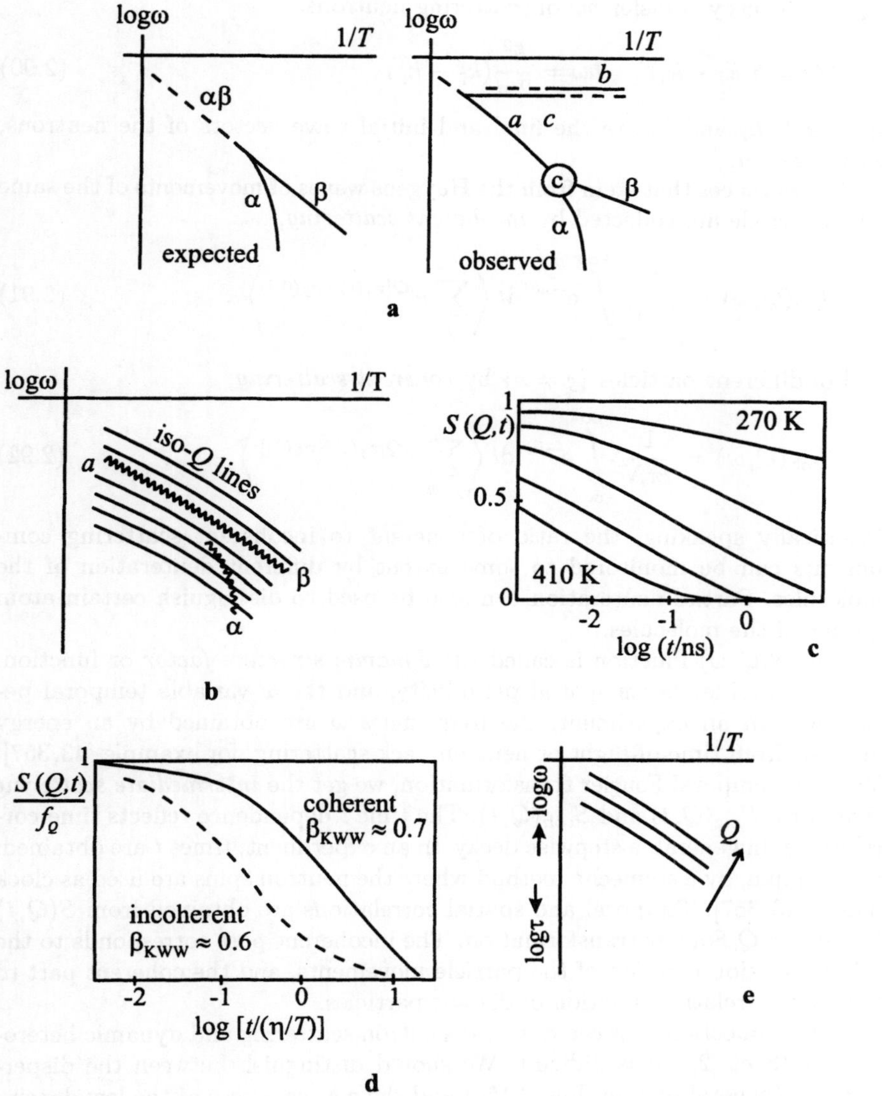

where $\langle\ \rangle$ is the ensemble average over the scattering volume containing N scattering particles. The exponent describes the interference of waves from all pairs of scattering particles superposed via the Huygens principle of optics. $\boldsymbol{Q}$ is the scattering vector of the neutrons, for instance, with length given by (2.87), and $(\boldsymbol{r}_i, \boldsymbol{r}_k)$ are the space positions of the particle pairs at a given time. In isotropic materials, $S(Q)$ depends only on the length of the scattering vector $Q = |\boldsymbol{Q}|$. A careful calibration of all contrast factors is needed to obtain absolute values of $S(Q)$ from scattering intensities.

Dynamic or inelastic scattering means not only momentum transfer $\hbar\boldsymbol{Q}$, but also energy transfer $\hbar\omega$ of scattering neutrons:

$$\hbar\boldsymbol{Q} = \hbar(\boldsymbol{k}_\mathrm{f} - \boldsymbol{k}_\mathrm{i}) \ , \quad \hbar\omega = \frac{\hbar^2}{2m}(k_\mathrm{f}^2 - k_\mathrm{i}^2) \ . \tag{2.90}$$

In (2.90), $\boldsymbol{k}_\mathrm{f}$ and $\boldsymbol{k}_\mathrm{i}$ are the final and initial wave vectors of the neutrons, with mass m.

Interferences that stem from the Huygens waves of movements of the same given particle are collected by *incoherent scattering*,

$$S_\mathrm{inc}(Q,\omega) = \frac{1}{2\pi N}\int_{-\infty}^{+\infty} \mathrm{e}^{-\mathrm{i}\omega t}\mathrm{d}t \left\langle \sum_j \mathrm{e}^{\mathrm{i}\boldsymbol{Q}[\boldsymbol{r}_j(t)-\boldsymbol{r}_j(0)]} \right\rangle \ , \tag{2.91}$$

and of different particles $\{j \neq k\}$ by *coherent scattering*,

$$S_\mathrm{coh}(Q,\omega) = \frac{1}{2\pi N}\int_{-\infty}^{\infty} \mathrm{e}^{-\mathrm{i}\omega t}\mathrm{d}t \left\langle \sum_{j\neq k} \mathrm{e}^{\mathrm{i}\boldsymbol{Q}[\boldsymbol{r}_j(t)-\boldsymbol{r}_k(0)]} \right\rangle \ . \tag{2.92}$$

Practically speaking, the ratio of coherent to incoherent scattering components can be monitored to some extent by different deuteration of the molecules. Partial deuteration can also be used to distinguish certain atom groups of the molecules.

The $S(Q,\omega)$ function is called the *dynamic structure factor* or function. The Q variable checks spatial periodicity, and the ω variable temporal periodicity. In an experiment, the frequencies ω are obtained by an energy analysis from time-of-flight or neutron back-scattering, for example [43, 357]. Without temporal Fourier transformation, we get the *intermediate scattering functions*, $S_\mathrm{inc}(Q,t)$ and $S_\mathrm{coh}(Q,t)$. The time t dependence reflects time correlations; in general, a stepwise decay. In an experiment, times t are obtained, for example, by a spin echo method where the neutron spins are used as clock hands [43, 357]. Temporal and spatial correlations are obtained from $S(Q,t)$ by inverse Q Fourier transformation. The incoherent part corresponds to the self-correlation function of the particle movements, and the coherent part to the pair correlation function of distant particles.

The connection between dynamic neutron scattering and dynamic heterogeneity (Sect. 2.2.5) is difficult. We should distinguish between the dispersion law (general scaling, Fig. 2.15c) and the consequences of the low density contrast of the dynamical pattern (Fig. 2.15b). Dispersion laws can also be applied to molecular motions beyond the dynamic heterogeneity of the glass transition, e.g., for diffusion and wave propagation. A dispersion law should express some relationship between the temporal and spatial variables taken from a well-defined property of scattering, e.g., a relation between ω and Q from $S(Q,\omega)$, or a relation between t and Q from $S(Q,t)$. This may be a typical $\chi''_Q(\omega)$ maximum frequency $\omega_{\max}$ or a typical KWW time from the $S(Q,t)$ decay for the a or α process as a function of the scattering vector, $\omega_{\max}(Q)$ or $\tau_{\mathrm{KWW}}(Q)$ [358], or times from the half step of the $S(Q,t)$ decay as a function of Q, for example. An example will follow later in this section [see (2.96)]. Information about the pattern of dynamic heterogeneity may hardly be expected from possible peculiarities in the t–Q or ω–Q relation beyond some diffusion-like power law relations, since the density contrast of the pattern is so small. More information may be obtained from a comparison of Fig. 2.15c peak traces from other activities, such as dielectric activities, with the t–Q or ω–Q relations from dynamic scattering in an Arrhenius diagram (see Fig. 2.19b and [140]).

Quasi-elastic scattering is restricted to small ω with $|\boldsymbol{k}_{\mathrm{f}}| \approx |\boldsymbol{k}_{\mathrm{i}}| = 2\pi/\lambda_0$, and λ_0 the wavelength of the incoming neutrons. This scattering is often applied for investigating the propagation of thermodynamic perturbations, e.g., diffusion, or thermal and mechanical waves.

For order-of-magnitude estimates, let us recall that for thermal neutrons, λ_0 usually lies in the range of several angstroms. An energy $E = 1.60\times10^{-19}$ J corresponds to an X-ray wave number of 8.07 cm^{-1}, and

$$E = \hbar\omega = k_{\mathrm{B}}T \Longleftrightarrow 1\ \mathrm{meV} = 0.24\ \mathrm{THz} = 11.6\ \mathrm{K}\,. \tag{2.93}$$

The relation between a scattering vector Q and the corresponding distance r may be estimated from $Qr \approx 2\pi$, with large Q corresponding to small r. An exact 2π product is only obtained for the Bragg equation $Qd = 2\pi$, where d is the distance between the reflecting planes in the crystal. For a Lennard-Jones liquid near the triple point, we get $\sigma Q^{\mathrm{peak}} = 6.9 = 1.1 \times 2\pi$ where σ is the molecule diameter and Q^{peak} the Q value of the main peak of the structure factor $S(Q)$. A similar result holds for hard spheres under equivalent conditions [359]. Comparing the $S(Q)$ peak position with the average distance $\bar{d}$ of the spheres gives $\bar{d}Q^{\mathrm{peak}} = 7.3 = 1.16 \times 2\pi$.

Consider the decay of the high-temperature a process in the intermediate scattering function for glycerol at a given Q value of order the position of the main diffraction peak. This decay shifts through the time window from 0.005 ns to 3 ns by lowering the temperature from about 410 K to 270 K (Fig. 2.19c). The time scales with the viscosity η (2.1) or, more precisely, with η/T (2.57).

In general, because of the $\int S(Q,\omega)\mathrm{d}\omega = S(Q)$ normalization (3.143), the dispersion may be scaled by reduction with $S(Q)$. Considering the behavior of the quotient $S(Q,t)/S(Q)$, for example, only a 'sublinear diffusion process' remains as in Fig. 2.19d, with no details from the strong Q dependence of the structure factor $S(Q)$. In other words, $S(Q,t)(T)$ scales with the fraction $f_Q(T)$ that relaxes by a process (c) faster than a, viz.,

$$f_Q(T) = S(Q, t \ll t_a)\,(T)\ , \quad 0 < f_Q \leq 1\ . \tag{2.94}$$

The coherent (distant) and incoherent (self-)scattering functions for the a process do not have the same shape (Fig. 2.19d). They can each be adjusted by Kohlrausch functions (2.17),

$$S(Q,t) \approx f_Q \exp\left[-\left(\frac{t}{\tau_{\mathrm{KWW}}}\right)^{\beta_{\mathrm{KWW}}}\right]\ , \tag{2.95}$$

with different KWW exponents β_{KWW} [360]. Coherent and incoherent scattering are different activities in the sense of Sect. 2.1.3. Consequently, the self part and the distant part of the correlation function for the a process are also different activities. The difference is registered by the different KWW exponents. The exponents are usually in the 0.5–0.8 range and do not significantly depend on temperature for the a process.

Consider now the announced example for testing of dynamic heterogeneity by scattering (Figs. 2.19b and e). Take the characteristic time for given Q as a function of T and put the trace in the Arrhenius diagram (e.g., τ_{KWW} for polyisobutylene PIB, polybutadiene PB, and other polymers [358]). Then compare the iso-Q traces for different Q parameters. Isothermal sections approximately represent the dispersion law for the a process by a power law (*sublinear diffusion* if $\beta_{\mathrm{KWW}} < 1$)

$$\tau_{\mathrm{KWW}} \propto Q^{-2/\beta_{\mathrm{KWW}}}\ . \tag{2.96}$$

The Q range is from (0.2?) 0.3 to 1.0 Å^{-1}, corresponding to spatial regions from about 6 to 20 (33?) Å, from $Qr = 2\pi$, and times t in the range from 0.1 to 100 ns. The question marks indicate a range extension of larger experimental uncertainty. The dispersion law (2.96) describes the cage-escaping paths of the high-temperature a process (Sect. 2.3.2) by a sublinear diffusion for $\beta_{\mathrm{KWW}} < 1$, scaled by the Kohlrausch function across the whole dispersion zone. A linear diffusion $\tau \sim Q^{-2}$ would be obtained for $\beta_{\mathrm{KWW}} = 1$, as for a hydrodynamic regime. The a-process cage hinders free 'microbrownian' diffusion.

No bend or peculiarity could be observed in the experimental dispersion laws, the $t(Q)$ or $\omega(Q)$ functions. The dielectric trace for the a process is parallel to the iso-Q lines (Fig. 2.19b). We conclude that no definite dynamic heterogeneity is observed for the Williams–Götze process by these neutron scattering experiments [139]. All molecules are equivalent for the a process. The kind of non-equivalence predicted by the Glarum model is only expected for the α process, below the crossover region.

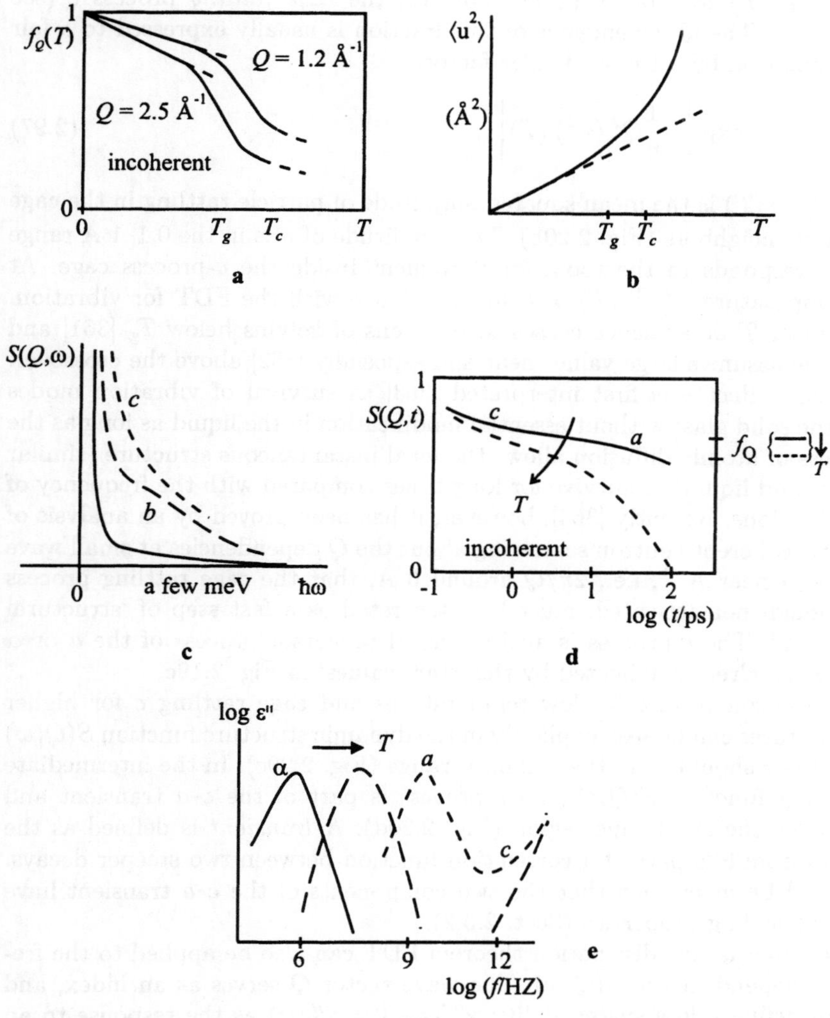

Fig. 2.20. Giga-to-terahertz results (continued). (**a**) c process fraction $f_Q(T)$ as a function of T, parameter Q, example OTP [708]. (**b**) Mean squared c process amplitude as a function of temperature. (**c**) Identification of boson peak b and cage rattling c in the dynamic scattering function $S(Q, \omega)$ for different temperatures. *Continuous curve* $T = 10$ K, *long dashes* near T_g, *short dashes* $T = T_g + 70$ K, $Q = 1.4$ Å^{-1}, example PIB [43]. (**d**) Intermediate scattering function for the c process in the picosecond range. $Q = 1.2$ Å^{-1}, $T = 293$–327 K, example OTP ($T_g = 342$ K). (**e**) Dielectric loss ε'' for frequencies in the mega-to-terahertz range, example glycerol [366]. The problem of constant loss (Sect. 2.4.5) is ignored in this graph

The $f_Q(T)$ fraction of (2.95) indicates the *cage rattling* process c (see Fig. 2.20a). The Q dependence of this fraction is usually expressed to a fair approximation by a Debye–Waller factor,

$$f_Q(T) = \exp\left[-\frac{1}{6}Q^2 \left\langle u^2 \right\rangle (T)\right] , \tag{2.97}$$

where $\left\langle u^2 \right\rangle (T)$ is the mean squared amplitude of particle rattling in the cage of nearest neighbors (Fig. 2.20b). The amplitude of u is in the 0.1–1 Å range and corresponds to the room for movement inside the a-process cage. At low temperatures, $\left\langle u^2 \right\rangle (T) \propto T$ in accordance with the FDT for vibration. This linear T dependence ceases several tens of kelvins below T_g [361] and the slope assumes large values near and especially [362] above the crossover. This cage effect was first interpreted [363] as survival of vibration modes from the solid glass without essential modification in the liquid as long as the slowness of atomic diffusion allows the local instantaneous structure (similar in glass and liquid) to survive for long times compared with the frequency of the vibrations. Recently [364], however, it has been proved by an analysis of inelastic coherent neutron scattering about the Q dependencies at small wave vectors Q near Å^{-1}, i.e., $2\pi/Q$ around 6 Å, that the cage rattling process c, although not activated, must be interpreted as a fast step of 'structural relaxation'. The c process is an isothermal precursor process of the a or α process, as already indicated by the 'start values' in Fig. 2.19c.

The boson peak b for low temperatures and cage rattling c for higher temperatures can be seen explicitly in the dynamic structure function $S(Q,\omega)$ as peaks or shoulders in the 1–2 meV range (Fig. 2.20c). In the intermediate scattering function $S(Q,t)$, the c process is part of the c–a transient and dominates the short-time region (Fig. 2.20d). A *transient* is defined as the moderate middle part of a correlation function between two steeper decays. It should be mentioned that the two components of the c–a transient have different scaling properties (Sect. 2.3.2).

The fluctuation–dissipation theorem FDT can also be applied to the frequency dependence of $S(Q,\omega)$. The wave vector Q serves as an index, and we can define a loss susceptibility $\chi''(Q,\omega) = \chi''_Q(\omega)$ as the response to an external perturbation with frequency ω,

$$\chi''_Q(\omega) \propto \omega S(Q,\omega) . \tag{2.98}$$

In the simplest case, for scattering by density fluctuations in the nanometer range, this χ''_Q corresponds to a Q partition of the dynamic bulk compliance. For refined decorations of the molecules with deuterons, it does not seem such a simple matter to define a corresponding macroscopic perturbation program. Experimentally, dielectric (ε'') and optical [365] compliances are more easily accessible (see Fig. 2.20e and [366]). The a process has a clear ε'' maximum, shifting to higher frequencies $f = \omega/2\pi$ with temperature. The minimum between the a process and the thermal terahertz frequencies, corresponding to the c–a transient in the time domain, would be deeper without the c process. For lower temperatures, the minimum is discussed in terms of a 'constant loss' contribution (Sect. 2.4.5).

The wavelength of light scattering is in the micrometer range, much too large to detect spatial details of any nanometer dynamic heterogeneity. The χ'' peaks for the a process from neutron and light scattering are similar [367]. This indicates that, in the giga-to-terahertz range, spatial scales corresponding to the scattering-vector range between $Å^{-1}$ (neutrons) and 1/10 μm (light) are not important. (The Fischer modes ϕ are obviously not relevant for such χ'' experiments. The ϕ modes are too slow; the few speckles cannot be seen at the high frequencies considered in our frequency range. The ϕ modes are not functional to this range.) This finding, and the factorization of (2.95) for the a process, with f_Q only from the c process, is a further indication that the Williams–Götze process a has no characteristic length larger than a particle diameter. There is probably no density- or entropy-active many-particle molecular cooperativity above the crossover. This was first conjectured by Johari 1975 [132], although with no discussion of different activities. If it exists at all [209], cooperativity is confined to the cage doors for particle escape, having the same size as the particles. Besides the ms rattling amplitude $\langle u^2 \rangle$ for the c process, the only characteristic length scale for the a process is the nearest-neighbor distance for forming the cage. Both c and a processes should therefore be explained by dynamics within and with the cages formed by neighbors that are equivalent to the central particle.

Such an a-process cage is a specific and mathematized (by MCT) form of the old physicochemical concept of a cage for glass-transition cooperativity due to Jenckel [368]. The physicochemical development of the cage concept will be described in Sect. 2.5.3

2.3.2 Götze's Mode-Coupling Theory (MCT)

Particle motion above the crossover region has two components, cage rattling c and escape from the cage a (Fig. 2.21a). This has been confirmed by molecular dynamics simulations [369]. For mode-coupling theory MCT, we partly follow a description by Cummins et al. [370], starting from vibrations in the terahertz range at a temperature far above the crossover (vibration is not shown in Fig. 2.21a). With decreasing temperature and increasing density, each molecule becomes progressively more trapped in the transient *cage* formed by its neighbors, and makes many collisions within the cage (cage rattling c) before finding a way out via cage diffusion (a process). This cage is a self-consistent construction where all participating molecules are completely equivalent. The cage is composed of molecules that are themselves surrounded by cages of their own neighbors. This equivalence is related on the time scale of the a process during which the cage undergoes collective distortions that may open the way for the cage diffusion event, i.e., the escape of a molecule from its cage.

The genesis, selected applications, and future prospects for MCT are reviewed by Kawasaki [371].

Mode-coupling theory comes from the Liouville equation for the dynamics (time t) of the phase point distribution for equivalent subsystems in a representative phase space. A Mori–Zwanzig projection separates a 'systematic' dynamic variable $A(t)$ from a random force $R(t)$ for the other, hidden variables beyond $A(t)$. The hidden variables are transiently lost by the orthogonality of the projection. The systematic variable $A(t)$ is sometimes called 'relevant', and the random force $R(t)$ 'irrelevant'. The systematic $A(t)$ is considered as 'the' variable that is able to describe the a and c processes.

The usual choice is the normalized density–density correlator

$$A(t) = \phi_Q(t) \equiv \frac{\langle \rho_Q^{\mathrm{cc}}(0)\, \rho_Q(t) \rangle}{\left\langle |\rho_Q|^2 \right\rangle} , \tag{2.99}$$

where $\rho_Q(t)$ is the complex Fourier Q component of spatial density correlations that fits the intermediate scattering function $S(Q,t)$, and $\left\langle |\rho_Q|^2 \right\rangle = S(Q)$ is the structure factor as a function of the scattering vector Q. To find $A(t)$, we need to know the equilibrium properties of A and the dynamics of R.

A generalized *Langevin equation*,

$$\frac{\mathrm{d}A(t)}{\mathrm{d}t} = \mathrm{i}\Omega_0 A(t) - \int_0^t \gamma(t')\, A(t-t')\, \mathrm{d}t' + R(t) , \tag{2.100}$$

▶

Fig. 2.21. Mode-coupling theory MCT. (**a**) General model of particle motion in liquids above the crossover region: cage rattling c and the diffusion event, or escape from the cage of equivalent neighbors a (Williams–Götze or high-temperature process of the dynamic glass transition). The point at the end of the *arrow* in the middle part of the bottom row will be called the 'cage door'. (**b**) Overview of the correlation function $\phi(t)$ as obtained from idealized MCT. (1) Ergodic a process for $T > T_c$, (2) nonergodicity for $T < T_c$, T_c the critical temperature, f_c the c fraction at T_c, the nonergodicity parameter of MCT. (**c**) Tails of the c and a processes for the c–a transients. This is the field of the reduction theorem. t_1 and t_2 are differently scaled times. T^- means $T < T_c$, T^+ means $T > T_c$. (**d**) Overlapping of first and second scaling for a T^+ isotherm in the c–a transient. (**e**) Minimum of loss susceptibility from a Fourier transform of the c–a transient into the frequency domain. (**f**) Example of an A2 type fold, the simplest catastrophe: PVT surface representing the equation of state for a van der Waals fluid. The cusp is a projection of spinodals s near the critical point C on the pT plane. In a way, the relevant variable A corresponds to the order parameter V, and the kinetic parameters (c_1, c_2) of the coupling equation (2.103) correspond to the control variables (T, p). The isothermal distance between the spinodals is $\Delta V_s \sim (T_c - T)^{1/2}$, $T < T_c$

is always correct but would be useless without further specification, since we have three unknown functions, viz., $A(t)$, $\gamma(t)$, and $R(t)$, and only one equation (Ω_0 is a frequency constant of terahertz order). The fluctuation–dissipation theorem FDT relates the memory function γ to the random force R and ensures the fluctuational input. In the frequency domain (ω), the theorem reads

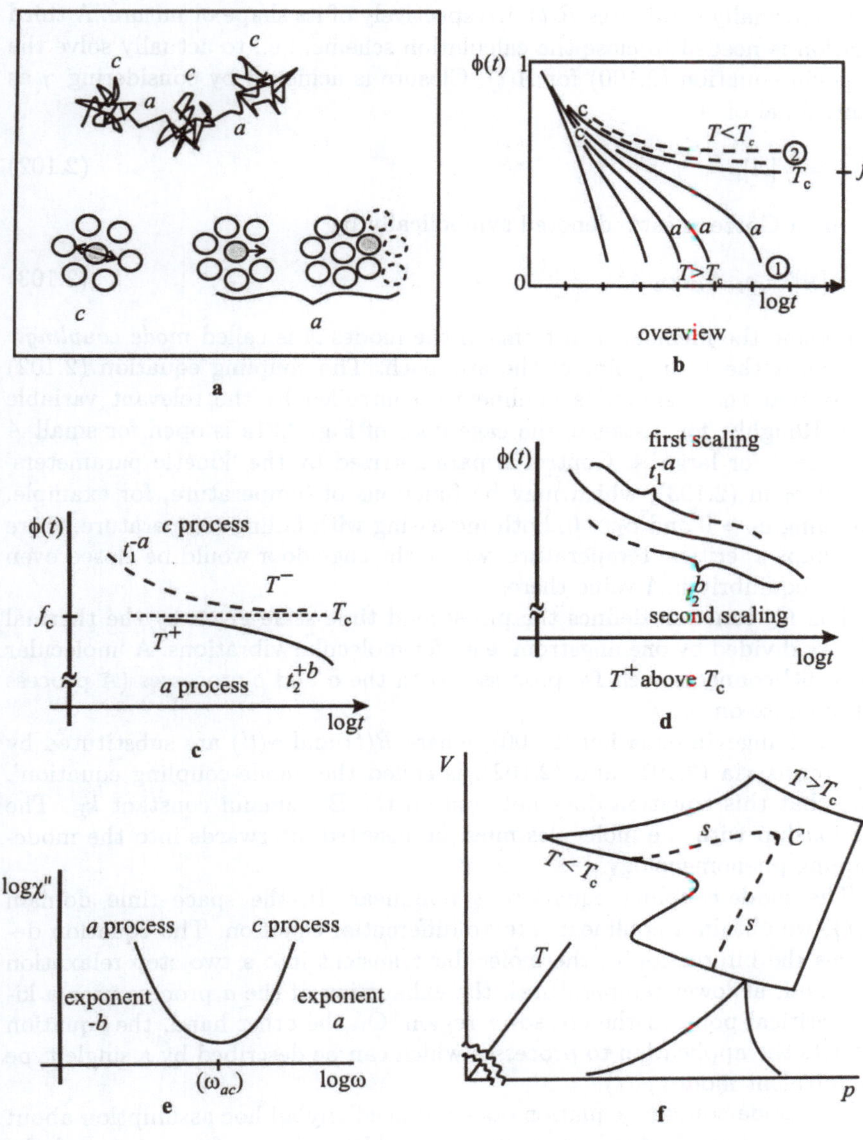

$$\gamma(\omega) = \langle A^2(0) \rangle^{-1} \int_0^\infty \mathrm{d}t\ \mathrm{e}^{-\mathrm{i}\omega t} R^2(t) \propto R^2(\omega) \,, \tag{2.101}$$

where $R^2(t)$ is the correlation function and $R^2(\omega)$ the spectral density of the random force $R(t)$. $\gamma(\omega)$ is the dynamic form of $\gamma(t)$. The FDT equation (2.101) formally eliminates $R(t)$, irrespectively of its shape or nature. A third equation is needed to close the calculation scheme, i.e., to actually solve the Langevin equation (2.100) for $A(t)$. Closure is achieved by considering γ as a functional of A,

$$\gamma = \gamma\,[A], \tag{2.102}$$

e.g., by a Götze ansatz, denoted symbolically by

$$\gamma\,[A] = c_1\,A + c_2\,A^2 \,. \tag{2.103}$$

Expressing the memory γ in terms of the modes A is called *mode coupling.*

This is the main point of the approach. The coupling equation (2.102) means that the memory is nonlinearly controlled by the relevant variable itself. Roughly, for instance, the cage door of Fig. 2.21a is open for small A and closed for large A. Control is parametrized by the 'kinetic parameters' c_1 and c_2 in (2.103), which may be functions of temperature, for example. Assuming $c_1 > 0$ and $c_2 > 0$, both increasing with falling temperature, there is perhaps a 'critical temperature' where the cage door would be closed even for the equilibrium A value there.

The Ω_0 constant defines the picosecond time scale given by the thermal velocity divided by one angstrom, e.g., for molecular vibrations. A 'molecular transient' connects such Ω_0 processes with the a and c processes (A process in the figure on p. IX).

The Langevin equation (2.100), where $R(t)$ and $\gamma(t')$ are substituted by $A(t)$ terms via (2.101) and (2.102), is called the 'mode-coupling equation'. Note that this equation does not contain the Boltzmann constant k_B. The relationship with the molecules must be inserted afterwards into the mode-coupling phenomenology.

The mode-coupling equation is nonlinear. In the space–time domain $\{\boldsymbol{r}, t\}$, we obtain a nonlinear integro-differential equation. The equation describes the bifurcation of the molecular transient into a two step relaxation (a, c) and, at lower temperatures, the exhaustion of the a process near a kinetic critical point in the crossover region. On the other hand, the equation restricts the application to processes which can be described by a single type of equivalent modes $A(t)$.

The mode-coupling equation does not need any ad hoc assumption about the cage motion. Instead, it solves the problem of cage formation and the problem of motion within the cage simultaneously. The equation yields a rich structure of $A(t)$ solutions. Using (2.103) or similar functionals [33,372–375], the equation can model many nonlinear things which can be done with modes $A(t)$ for equivalent, collective movements of a certain number of particles with steep repulsion potentials: cages of neighbors, back-scattering, feedback, and backflow [33, 376]. Equation (2.103) is not a Taylor series expansion. A small parameter for a hypothetical 'linear' series expansion in the nonlinear interplay between angstrom-sized cage rattling and cage-sized diffusion events probably does not exist. The terms in the coupling (2.103) are not selected by principles, but rather by their success. The results of the calculation, however, prove not to be very sensitive to this selection. MCT turns out to be a robust calculation algorithm. Moreover, once selected, the kinetic parameters c_1 and c_2 can be calculated from interparticle potentials. Kinetic parameters are controlled by temperature and pressure: they are the *coupling parameters* that increase along the a process trace in the direction of the crossover.

Mode coupling with one single $A(t)$ variable seems unable to describe a dynamic heterogeneity larger than the cage door of Fig. 2.21a. We may imagine cages constructed from more particles than the nearest neighbors, e.g., from 20 particles, without giving up the equivalence of the particles and their modes and without giving up the particle size of the cage door, at least for moderate liquids. This size is dictated by the diffusion concept: a particle (and not a part of it or a lot of them) changes the particle environment (Sect. 2.5.3). Intuitively, for larger-scaled dynamic heterogeneity, we would need two non-equivalent modes, e.g., a collective mode and a cooperative mode. Two modes seem to be necessary, for instance, to describe the Glarum defect diffusion model for the α process with a collective cage near the defect that can only be opened with the assistance of a slower cooperativity shell in the neighborhood. In the Levy treatment for the α process, the equivalence as for the a process is broken by the preponderant Levy component responsible for the defect. We may ask whether, in general, cooperative movements of particles can be described by equations of the mode-coupling type at all. I think that $A(t)$ is restricted to collective modes. Many trials have so far indicated that this deficiency of (2.100)–(2.102) (*idealized MCT*) cannot be compensated by adding time derivatives $\dot{A} = \mathrm{d}A/\mathrm{d}t$, etc. to $\gamma\,[A]$ (*extended MCT*). Idealized MCT cannot model a characteristic length of the glass transition beyond the cage door dimension (particle diameter ≈ 0.5 nm).

Let us now describe the main success of MCT. We use the correlation function $\phi(t)$ of (2.99), sometimes dropping the Q index. Idealized MCT can reasonably model the overview picture as in fact experimentally observed (Fig. 2.20d) above the crossover, $T > T_\mathrm{c}$ (Fig. 2.21b). For not too high a temperature, high frequency cage rattling c is separated from the a process by a transient, i.e., as mentioned above, a $\phi(t)$ piece with smaller slope. The $\phi(t)$ curves are ergodic (1) above a critical temperature T_c and show the slower Williams–Götze a process as a decay that shifts to longer times for lower temperatures. The a process decay is more stretched out than a Debye exponential decay. Below T_c, the $\phi(t)$ curves are nonergodic (2). The $\phi(t)$ decay is arrested at a finite f_c nonergodicity parameter value, e.g., $\phi(t) \to f_c$ for $T = T_c$. The slowing down of the a process is caused by approach towards arrested dynamics. This arrest tempted early investigators to think that they might later model even the thermal glass transition at $T_g = T_c$.

The elegant core of MCT is an analytical treatment of the long c and a tails in the c–a transient (Fig. 2.21c). For the c process tail, we obtain (Sect. 4.1)

$$c \text{ process}: \qquad \phi_Q(t) - f_Q^c = h_Q\, t_1^{-a} , \tag{2.104}$$

and for the a process tail

$$a \text{ process}: \qquad \phi_Q(t) - f_Q^c = -h_Q\, t_2^{b} , \tag{2.105}$$

where f_Q^c ($= f_c$ in Figs. 2.21b and c) is the critical nonergodicity parameter [i.e., the $S(Q,t)$ fraction of the c process at $T = T_c$, see also Fig. 2.20a], $t_1 = t/t_{c\ \text{process}}$ is a reduced time for c (first scaling law, unexpected before), and $t_2 = t/t_{a\ \text{process}}$ is a differently reduced time for a (second scaling law). Both (2.104) and (2.105) are factorized with respect to Q and t, i.e., in an h_Q amplitude depending only on the wave vector Q, and the differently scaled time factors. Both time factors are power laws with the exponents a and b, respectively. The exponent a belongs to the cage rattling c process, and the exponent b to the Williams–Götze a process (Figs. 2.21c–d). Both scaling times, $t_{c\ \text{process}}$ and $t_{a\ \text{process}}$, diverge at T_c for the tails with given a and b exponents. This follows from $\phi_Q(t) \to f_Q^c = \text{const.}$ for $T \to T_c$ (Fig. 2.21c). The time dependence of (2.105) is called the *von Schweidler law*, and b is the 'von Schweidler exponent'. Fig. 2.21d gives an impression of how the two tails with their different scaling laws overlap in the c–a transient. The temperature dependence of the Q factor is obtained as

$$h_Q(T) \sim |T - T_c|^{1/2} \text{ for } T^+ \text{ and } T^- , \tag{2.106}$$

similarly to a classical critical point.

A further interesting point is [282] that, for large Q (short distances), the a process tends to a Kohlrausch law,

$$\lim_{Q\to\infty} \frac{\phi_Q(t_2)}{f_Q^c} = \exp\left(-t_2^b\right) , \tag{2.107}$$

with a Levy exponent α equal to the von Schweidler exponent, $\alpha = b$. The reason for the Levy distribution there is the participation of many independent Q and Q' values in the vertices (Sect. 4.1) that determine the coupling parameters in the symbolic coupling equation (2.103).

MCT delivers a relationship between the two tail exponents a and b of (2.104)–(2.105) through a coupling constant λ given by

$$\lambda = \frac{\Gamma^2(1-a)}{\Gamma(1-2a)} = \frac{\Gamma^2(1+b)}{\Gamma(1+2b)} \, . \tag{2.108}$$

After Fourier transformation, the transient of Fig. 2.21c corresponds to a minimum of the susceptibility $\chi''(\log \omega)$, Fig. 2.21e. A model or picture of relaxation is usually concerned with the maximum of $\chi''(\log)$. MCT, however, gives a relation for the $\log \chi''(\log \omega)$ wings on either side of the transient minimum at $\omega = \omega_{ac}$,

$$\begin{aligned} b &= -\frac{\mathrm{d}\log \chi''}{\mathrm{d}\log \omega} \quad \text{for} \quad \omega \ll \omega_{ac} \, , \quad a \text{ process} \, , \\ a &= \frac{\mathrm{d}\log \chi''}{\mathrm{d}\log \omega} \quad \text{for} \quad \omega \gg \omega_{ac} \, , \quad c \text{ process} \, . \end{aligned} \tag{2.109}$$

The fit procedures to check (2.108) should use the master curve constructions mediated by the two different scaling laws [377].

The exponents obtained depend on the substance. A typical value for the coupling constant is $\lambda = 0.73 \pm 0.03$, as for toluene [378]. The molecular parameters that control λ are not known. In our terminology (Sect. 2.1.1), the dependence on one parameter λ was called similarity, not universality.

Equations (2.104)–(2.105) and (2.108)–(2.109) result from the so-called reduction theorem of idealized MCT (Sect. 4.1). This is an asymptotic solution of the mode-coupling equation near the c–a transient, i.e., for the situation in Fig. 2.21c, $|\phi_Q(t) - f_Q^c| \to 0$:

$$\phi_Q(t) - f_Q^c \sim h_Q \cdot G(t) + O(G^2) \, , \quad \frac{\chi_Q''(\omega)}{\chi_Q^T} = h_Q \chi''(\omega) \, , \tag{2.110}$$

where χ_Q^T is an FDT reduction factor. The time factor $G(t)$ of asymptotic idealized MCT is determined by

$$\sigma + \lambda G^2(t) \sim \frac{\mathrm{d}}{\mathrm{d}t} \int_0^t G(t-t') \, G(t') \, \mathrm{d}t \, , \tag{2.111}$$

with σ a scaling constant and λ the coupling constant, $1/2 \le \lambda < 1$. Equation (2.108) follows by inserting the tails (Fig. 2.21c) in (2.111). The structure of (2.108) is evident from integrals like

$$\int_0^t (t-t')^{-\mu} \, t'^{-\nu} \mathrm{d}t' = t^{1-\mu-\nu} \frac{\Gamma(1-\mu) \, \Gamma(1-\nu)}{\Gamma(2-\mu-\nu)} \, , \tag{2.112}$$

which result from the asymptotic reduction equation (2.111) for the $t \to \infty$ tails.

In the next order of approximation (2.110), the transient overlap yields a diverging relaxation time τ (critical slowing down) and viscosity η as

$$\tau \sim \eta \sim (T - T_c)^{-\gamma}\,, \quad \gamma = \frac{1}{2a} + \frac{1}{2b}\,, \quad T > T_c\,. \tag{2.113}$$

This indicates the nonergodicity at $T = T_c$ and $T < T_c$.

Idealized MCT is thus 'exhausted' near T_c because a permanent nonergodicity was, of course, never observed in the gigahertz range. The mode-coupling equation from (2.100)–(2.102) is an integro-differential equation of analytical cubic structure in A. From Thom's catastrophe theory [379,380], we see that a critical state of the A_2 fold type will then be reached at T_c. Such a type follows from any critical germ F that can be reduced to $F(A,c) = A^3 + cA$, whilst the integro-differential nature of the equation is not important here.

The A_2 fold is well known for the equation of state of the van der Waals fluid (Fig. 2.21f). In order to actually get such a fold, or the related cusp in the control variables with a typical square root dependence (2.106), the process must be driven by very small differences to critical parameters like $|T - T_c|$ or $|p - p_c|$ for the van der Waals equation. A singularity is observed for molecular critical states near a thermodynamic instability with their small temperature ($C_V \to \infty$) and pressure ($(\partial V/\partial p)_S \to \infty$) fluctuations. Near the glass-transition crossover, however, no critical thermodynamic instability in controllable thermodynamic variables has so far been observed. Nevertheless, the diverging correlation lengths at T_c connected with the fold have been sought using molecular dynamics computer simulations [381–384]. The discussion of a residual thermodynamic particularity in the crossover will be continued in Sect. 2.5.1.

As described in the preceding sections, I think that all relevant fluctuations are large because the representative subsystem becomes small, and that other possibilities such as cooperativity may compete and evolve below T_c. The simplest interpretation is, as mentioned above, that large $A(t)$ modes close the doors for escape paths from the cages, due to the $A(t)$ mode coupling. The diffusion possibilities of idealized MCT are exhausted at T_c. A new type of mode seems necessary for assisting cage opening below T_c.

The addition of time derivatives $\dot{A}(t) = \mathrm{d}A/\mathrm{d}t$ in the extended MCT guides the state path in the kinetic c_1c_2 plane away from the critical point and results in a rounding off of the cusps linked with critical singularities. This picture, with $a\alpha$ continuity in the crossover region, was not confirmed in the (few) cases where the splitting scenario I (Fig. 2.11a) of the crossover region could be investigated in detail [209].

It has sometimes been stated that idealized MCT and its verification on colloidal systems (next section) established a simple standard example for the glass transition that could be analytically extended to the cooperative α process below crossover. Some enthusiasts stated that MCT plays a similar role in glass transition research to the oscillator in quantum mechanics. I would say it does not, because a new, essential element for the distinct α process is missing: cooperativity. Instead, MCT established and fairly describes the Williams–Götze a process as an independent and distinct process in the general dynamic scenario on p. IX. The a process trace is the precursor of the α process trace. The a possibilities seem exhausted in the crossover, and cooperativity causes a more or less separate onset of the α process. For α, the a cage seems to be transferred to the islands of mobility of its dynamic heterogeneity (Sect. 2.5.3).

2.3.3 Colloidal Glass Transition

The glass transition in a suspension of hard spherical colloids can be monitored by increasing the colloid fraction ϕ [385,386]. Larger ϕ in the colloidal systems corresponds to lower temperatures in ordinary glass formers. As the colloid motions are much slower than the molecular motions, we expect to find a colloidal glass transition that corresponds to the a process of the high-frequency relaxation in molecular liquids. The typical features of the a process, as well as the cage rattling process c, are indeed obtained for ϕ in the range between 0.5 and 0.6 (see Fig. 2.22 and [387]). For suspensions with particles of radius $R \approx 200$ nm, the typical c process times are of order 10^{-2} s, and the a times shift up to 10^3 s, for $\phi \approx 0.58$. The colloids are large enough to be able to visualize the cages explicitly [388].

The intermediate scattering function $S(Q,t)$ in the colloid system can be measured by dynamic light scattering. Results are well modelled by idealized mode-coupling theory MCT for hard spheres. The time scale for the colloidal transient of order $t_0 = 10^{-3}$ s is obtained by data adjustment. All other parameters can be taken from the hard sphere model. Since MCT can also model the high-frequency relaxation in molecular liquids for $T > T_c$ (corresponding here to $\phi < \phi_c \approx 0.58$), we may say that colloidal glass transitions are a simple model for the molecular Williams–Götze process (a) when $\phi_c - \phi$ is mapped to $T - T_c$. The question of a colloidal glass transition beyond ϕ_c, that is, a colloidal α process, remains open at present. Investigation of a possible crossover region for the colloidal glass transition seems an interesting task. First attempts to find dynamic heterogeneity [389] and cooperativity [390] have been reported for ϕ ranges where a crossover region is to be expected.

Obviously, in a certain approximation for the a process, the thermal motion of molecules can be substituted by the Brownian motion of the colloids. However, a microscopic time t_0' calculated from the thermal velocity and the radius of the colloidal spheres is smaller than the adjusted colloidal-transient time t_0: $t_0' \approx$ radius R divided by thermal velocity $\sim R \cdot \sqrt{R^3} \sim R^{2.5}$. Comparing $R = 200$ nm with the molecular radius $R_m \approx 0.25$ nm for a given density, we obtain $t_0' \approx 10^{-12}$ s $\times (200/0.25)^{2.5} \approx 1.8 \times 10^{-5}$ s $\ll t_0 = 10^{-3}$ s. The difference is explained by colloidal interaction and hydrodynamics of the viscous background. This difference clearly does not change the essential features of the a process. It seems that the Williams–Götze process a and cage rattling c [391] can be understood on a minimal basis: equivalence of particle movements up to the a time scale (collective modes) and Brownian character of particle motion.

The time ratio between the colloidal transient or the cage rattling, 10^{-3}–10^{-2} s, on the one hand, and the a process near the crossover, 10^3 s, on the other hand, is about 5 orders of magnitude. This corresponds to the time ratio of molecular transient and crossover for the molecular glass transition when the crossover frequency is in the 10 megahertz range, a typical value (Table 2.10).

In other words, and following partly [387], the physical situation for the colloidal glass transition may be characterized as follows. The c process describes the dynamics of localized particle clusters or cages. The first scaling indicates that during the early part of the c process, the dynamics in the liquid and in the glass are indistinguishable. The a process describes the breakdown of the particle cages leading to large-scale particle diffusion and flow. The dynamics of particle cages on the verge of breaking up is shared by both a and c processes and is described by the second scaling with the von Schweidler law.

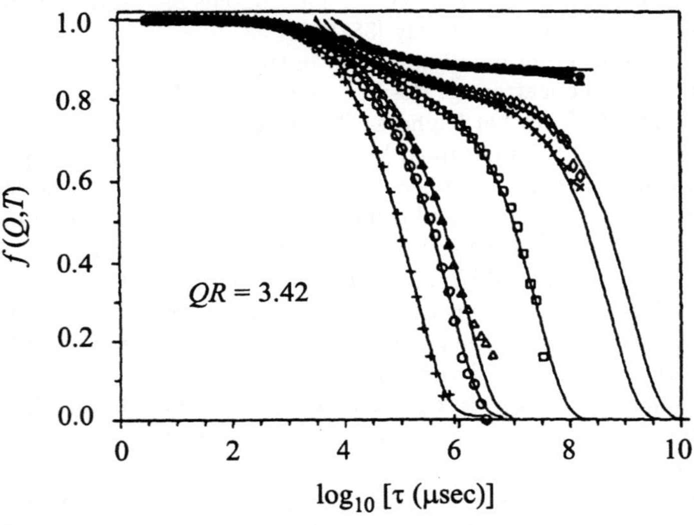

Fig. 2.22. Colloidal glass transition. Intermediate scattering function for a wave vector near the main peak of the structure factor, $QR = 3.42$, $R \approx 200$ nm [387]. The symbols are for suspension volume fractions from $\phi = 0.494$ (+) to $\phi = 0.587$ (•). The *solid curves* are predictions from idealized MCT with an adjusted transient time $t_0 = 10^{-3}$ s

2.3.4 Computer Simulations

Modern computers can help to solve Newton's mechanical equations for a sufficiently large subsystem of N molecules in glass-forming substances, e.g., cold liquids or liquid mixtures. To be interesting for the cooperative α process of the dynamic glass transition, the following conditions should be fulfilled:

(i) Find a molecular structure and intermolecular potentials, or a molecular mixture, so that crystallization can be avoided and that the crossover frequency is as high as possible, e.g., in the 10 megahertz range or higher (Table 2.10).
(ii) Find an integration method that can determine sufficiently slow Fourier components of reasonable observables, preferably several decades below the crossover frequency, e.g., in and below the megahertz range.
(iii) Choose the system size N large enough to find statistically independent subsystems of CRR size with $N_\alpha(T)$ particles, $N \gg N_\alpha(T)$, and choose boundary conditions that do not disturb, or barely disturb, CRR independence on the α time scale (CRR stands for cooperatively rearranging regions).
(iv) Find the map to the observables of thermodynamic linear response and dynamic scattering to compare 'computer experiments' with the relevant real experiments.

Conditions (ii) and (iii) are requirements for finding dynamic heterogeneities evolving below an α onset in the crossover region. If the temperature dependence of cooperativity follows (2.49), $N_\alpha \propto (T_{\text{on}} - T)^2$, $T < T_{\text{on}}$, the simulations must go far below the onset temperature T_{on}, e.g., 20 K or three frequency decades for fragile glass formers. Since the free volume is usually a small quantity (Sect. 4.3), and if (or since) the simulations are based on molecular positions and orientations, a high precision in the megahertz simulation is essential. A dynamic heterogeneity of 1–2 decades is caused by free volume changes of order 0.1% or less.

Condition (iv) seems manageable if the equilibrium fluctuation of thermodynamic variables is exactly described by the Gibbs distribution with its separable momentum $\{p\}$ and coordinate $\{q\}$ variables. Problems arise for large genuine temperature fluctuations in the case of the α process (e.g., $\delta T = 15$ K) that cannot be modelled by this distribution (Sect. 3.2). To compare with the characteristic length from calorimetry or compressibility, the kinematic particle data must be transferred to entropy or density fluctuations in order to construct a pattern from which the characteristic length can be determined.

To compare analytical methods with computer simulations, we partly follow Binder [392]. Disadvantages of analytical methods are: prejudice cannot be excluded by the selection of seemingly important modes or degrees of freedom, the necessarily few parameters are intended to be adjustable, and the necessary mathematical simplifications and approximations cannot easily be surveyed in any case, e.g., the next order cannot be handled. It is possible that a parameter adjustment is also satisfactory with irrelevant models. The advantages of computer simulation are: the number of degrees of freedom is not so limited, pretests can be made on a microscopic basis, one is not so restricted to ideal cases, variables can be determined which are not accessible or are hardly accessible by real experiments, and the reliability of models can be tested by a broad parameter variation. The latter can be a help in finding the truly essential conditions, informative parameters, configurations, etc., for the phenomenon under consideration. Moreover, certain complicated experiments cannot be analyzed without the aid of parallel computer simulations.

Of course, analytical methods also have advantages and computer simulations have disadvantages. One point is what is called understanding. Analytical methods are often closer to experimental concepts than computer simulations.

As indicated in the first paragraph, we shall concentrate on mastering the crossover region and cooperativity. Classical reviews for application of computers to the glass transition are [36, 214, 393–395].

The integration of Newton's equations is called *molecular dynamics* (MD) in computer simulations. By experience, the temporal integration step is of order a few femtoseconds (10^{-15} s). To get the crossover region at 10^{-7} s, we need 10^8 integration steps. This is near the limit of what seems at present achievable, even with parallel computer strategies. At the time of writing (1999), the simulation of 10^{-7} s processes requires one year CPU time, and progress is about one decade per five years. Two decades below this crossover (10^{-5} s) would need 10^{10} steps. A systematic search for a MD simulation system with a crossover region at or above the gigahertz range would, therefore, be interesting (Table 2.10, e.g., similar to epon 828 or to propylene glycol).

Many MD studies in the last few years have been restricted to the α process precursors: the high-temperature a process and cage rattling c. This problem has been solved to some extent [381–384, 396, 397]. An important result has been obtained: dynamic neutron scattering, mode-coupling theory MCT, colloidal glass transition, and MD describe nearly the same picture for the Williams–Götze process a and cage rattling c. In particular [397], MD does provide a good description, in terms of idealized MCT, of the critical temperature T_c, exponent parameters, nonergodicity parameters, and critical amplitudes.

Two examples will illustrate the situation. In the first example, Heuer's group [141] looked for a relation between dynamic heterogeneity and local structural properties, using MD with two concrete molecular models, for polycarbonate PC and phenyl salicate (salol). In the time scale of 100 ps, they found, for dynamic heterogeneity, clusters of slow molecules on the length scale of about 0.8 nm, i.e., no islands of mobility for the a process. The length increases for decreasing temperature, in fact, by about 30% between $T/T_g = 1.8$ and 1.2. Slow molecules are correlated to local structural properties: high local density, high local regularity of the molecular arrangement, and a low local potential (more stable). The declared aim of that paper was understanding of structure–dynamics properties. It corresponds to the subensemble separability of the Mainz approach to dynamic heterogeneity and does not discuss a general origin for dynamic heterogeneity, in particular, a possible need for such a heterogeneity for the a process with equivalent molecules. It does confirm, however, the intuitive connection between low local mobility, small local free volume, and deep local potential energy there.

The second example, from Glotzer's group, is a search for cages in Lennard–Jones mixtures [398]. They find, of course, much more detail than MCT, some of which are represented in Figs. 2.23a (1)–(4) and briefly described in the caption to this figure. They define a 'dynamic entropy' $S(r_a)$ via a reciprocal average first-passage time $\tau(r_a)$ of a virtual sphere with radius r_a,

$$S(r_a) = \frac{1}{\tau(r_a)}\,, \quad \tau(r_a) = \int_0^\infty \mathrm{d}t\; t\, P_{r_a}(t)\,. \tag{2.114}$$

This entropy is induced to provide a bridge between a mechanical definition and a thermodynamic definition. The typical 'cage diameter' r_a^{c} is defined by the properties of the $\tau(r_a)$ function and corresponds rather to a mean square displacement $\langle u^2\rangle$ in the Debye–Waller factor (2.97) than to a geometrical cage diameter as defined by the distance to the nearest neighbors. The r_a^{c} diameter corresponds to typical c–a process transient values of the mean squared displacement, $\langle r^2(t)\rangle_{\text{transient}} \approx (r_a^{\mathrm{c}})^2$. The time distribution $P_{r_a^{\mathrm{c}}}(t)$, for the cage diameter $r_a^{\mathrm{c}} = r_a$, is broad with a half-width of about one decade. This means that the physicochemical cage picture for the a process is condensed from extreme fluctuations. The paper concentrates the rattling itself, not on the structure or dynamic heterogeneity connected with the cage. The dynamic entropy $S(r_a)$ is not compared with an experimental entropy compliance such as from heat capacity spectroscopy which, although measured so far only for a few substances above the crossover region [209, 399], indicates a larger characteristic length than the r_a^{c} diameter.

The first computational investigations of genuine spatially heterogeneous dynamics [160, 256, 293, 294, 381, 384, 400–408] indicate [256, 294, 408] an increase in spatial correlation when the crossover region is approached from above and display, for example, stringlike motions of molecule chains [293]. The latter may prove interesting for an explanation of the boson peak.

Figures 2.23a (5) and (6) show examples for the larger-scale kinematic lengths obtained from MD computer simulations. Figure 2.23a (5) from Heuer's group [295] describes particle motions in a 2-d system for time intervals 50 times longer than the a process time. This picture points to the slower and larger Fischer modes. Figure 2.23a (6) compares two kinematic lengths in 3-d Lennard–Jones systems from Glotzer's group [256, 294, 408] with characteristic lengths from entropy fluctuations as detected by heat capacity spectroscopy [399]: S is from kinematic correlations and L from string lengths. The kinematic lengths for the a process are larger than the two examples of characteristic lengths for the a process measured so far calorimetrically in moderate liquids [poly(n-hexyl) and (n-decyl)methacrylate]. The kinematic analyses include correlations (e.g., parallel displacements) that do not contribute to entropy fluctuations.

To explain the relation between kinematic and characteristic length, we use the term *factual*, i.e., relating to facts. For factual entropy fluctuations, we need large amplitudes of disordered molecular movements that can really change the weights of different configurations responsible for local entropy fluctuation. In this sense, opening the cage door seems entropy active, but collective distortions of the cage, with small molecular amplitudes, also necessary for the a process, do not seem entropy active. For strings, I see two extreme possibilities. Either the string molecules move collectively, as through

►

Fig. 2.23. Computer simulation of the dynamic glass transition. (**a**) (1–4). Example of a search for cages in a Lennard–Jones mixture by molecular dynamics MD [398]. (1) Schematic of a particle trajectory. r_a is the radius of a sphere for definition of first-passage time $\tau(r_a)$. (2) Mean squared displacement of the majority species (reduced LJ units). (3) Typical first-passage time distribution $P_{r_a^c}(t)$ at the cage scale $r_a = r_a^c$. (4) Cage size r_a^c as defined from the maximum value of $\mathrm{d}\ln\tau/\mathrm{d}\ln r_a$ for the particle localization regime. (5–6) Search for kinematic lengths for the a process from MD computer simulations. (5) Particle motions in a 2-d system [295]. (6) Kinematic lengths from correlations (S) and strings (L) of a 3-d Lennard–Jones system [256, 294, 408] compared with characteristic lengths from heat capacity spectroscopy in moderate liquids (*continuous curves*). The *dashed line* is for TPCP-BO, a complicated fulvene derivative which probably does not belong to the class of moderate liquids [399]. For comparison we used the square root $N_a^{1/2}$ of a correspondingly estimated number of particles. The crossover temperature for the computer simulation was put at $T_c = 0.435$, and the Vogel temperature $T_0 = 0.32$, in Lennard–Jones units. (**b**) Distribution of particle displacements from MD. $4\pi r^2 G_{sA}(r,t_1)$ at times t_1 from $\langle r^2(t_1)\rangle_A = 1$, where G_{sA} is the van Hove self-correlation function for the larger particles (A) of a 50:50 Lennard–Jones binary mixture. Reduced units: the distance $r = 1$ corresponds to the average σ diameter, the temperature $T = 1$ to the average ε energy parameter. The MCT critical temperature is estimated here to be $T_c = 0.592$ [561]. (**c**) Bond fluctuation model for Monte Carlo MC computer simulation of the glass transition. Movement $\rightsquigarrow$ of an effective monomer unit from i to f changes the effective bond length from 2 for the initial 3-d bond {2,0,0} (*dashed line*) to $5^{1/2} = (2^2 + 1^2)^{1/2}$ for the final bond {2,1,0} (*continuous line*). (**d**) Fredrickson model in $d = 2$ dimensions with $z = 2$: the particular spin $\otimes$ ($\sigma = 0$ or 1) has a nonzero flip rate only if it has z or more mobile ($\sigma = 0$) spins in its immediate environment. (**e**) Splitting scenario of a $d = 2$ modified Fredrickson model with stochastic barrier and interaction energies. $\{\beta_1, \beta_2, \beta_3\}$ are distinguishable β process components for the modified Fredrickson model. (**f**) Sixteen 50×50 Butler–Harowell pictures for the modified Fredrickson model at reduced temperatures $T/T_{on} = 0.5$, 1.0, 1.5. Mobile regions are *white*

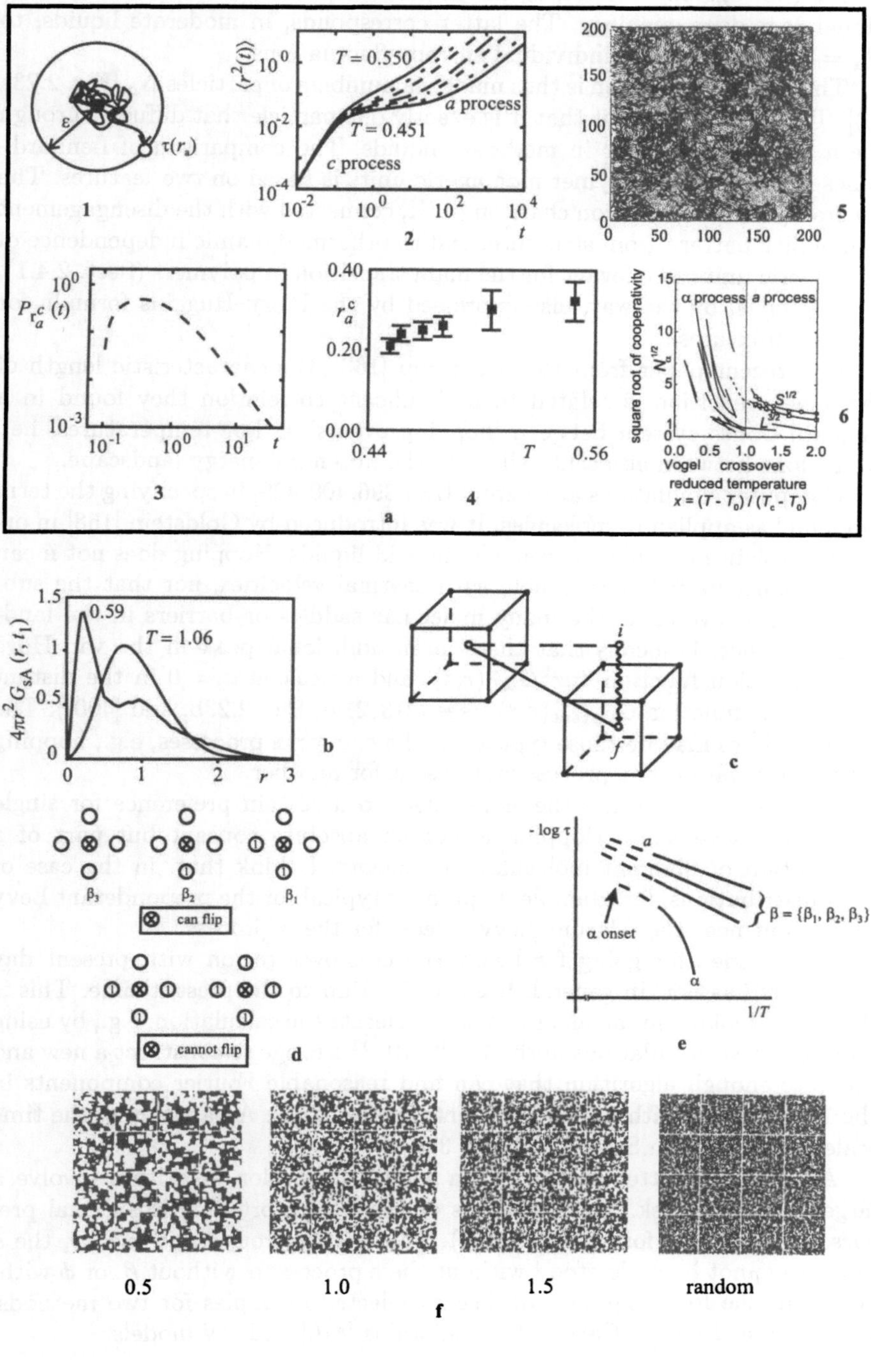

ε
τ(r_a)
1
⟨r²(t)⟩
T = 0.550
a process
T = 0.451
c process
t
2
3
4
a
5
square root of cooperativity
α process
a process
Vogel
crossover
reduced temperature
x = (T - T₀) / (T_c - T₀)
6
0.59
T = 1.06
4πr² G_SA (r, t₁)
b
i
f
c
β₃
β₂
β₁
can flip
cannot flip
d
- log τ
α onset
α
1/T
β = {β₁, β₂, β₃}
e
0.5
1.0
1.5
random
f

a tube (not entropy active), or the molecules move as a succession of individual cage-door openings. The latter corresponds, in moderate liquids, to $N_\alpha = 1$ with the above individual entropy fluctuations.

The above comparison is thus made via numbers of particles N_a [Fig. 2.23a (6)]. This reflects the fact that it is exactly one particle that diffuses through the a process cage door in moderate liquids. The comparison of Lennard–Jones particles with polymer monomeric units is based on two features. The generality of the relaxation chart on p. IX, connected with the disengagement of mobility pattern from structure, and the thermodynamic independence of monomeric units as relevant for the main transition in polymers (Sect. 2.4.1). The latter is, by the way, also expressed by the Flory–Huggins formula for polymer mixtures.

In a recent paper from Heuer's group [166], the characteristic length of the glass transition is related to a significant correlation they found in a Lennard–Jones system between 'hopping events' at low temperatures, i.e., transitions between adjacent valleys in the potential energy landscape.

Computer simulations are helpful [160,396,400,409] in specifying the term 'hopping' as applied to molecules. It was introduced by Goldstein [158] in order to explain molecular movements in cold liquids. Hopping does not mean that a molecule falls into a hole with thermal velocities, nor that the subsystem must overcome the many molecular saddles or barriers in the landscape. Rather, it means that there is an additional peak in the van Hove self-correlation function $4\pi r^2 G_{\text{self}}(r,t)$, and a peak at $r = 0$ in the distant part of this function $G_{\text{distant}}(r,t)$, (see [213, 214], Fig. 2.23b, and [400]). The relevant times may be those typical for the c, a, or α processes, e.g., hopping may be typical for one process and absent for another.

Hopping in this sense therefore refers to a certain preference for single molecular movements. Hopping is not an absolute concept but part of a distribution of different molecular movements. I think that, in the case of Levy distributions, for example, hopping is typical for the preponderant Levy component near the Glarum–Levy defects for the α process.

An old idea for going far below the crossover region with present day computers has not, in general, been realized up to the present time. This is the idea of looking for an algorithm to accelerate the calculation, e.g., by using representative calculations in the 10^{-15}–10^{-11} s range to construct a new and accurate enough algorithm that can find reasonable Fourier components in the 10^{-6} s range with $\Delta t = 10^{-11}$ s integration steps. A partition of the time scale is discussed in Sect. 3.5, Table 3.2.

All empirical attempts to skip a solid acceleration algorithm involve a large element of risk. An example is omission of short-time isothermal precursors (c for a, β for α, or α for ϕ). This is dangerous if, physically, the a process cannot be understood without the c process, α without β, or ϕ without α. In the following we will discuss selected examples for two methods: coarse-grained Monte Carlo MC simulations [410] and toy models.

Monte Carlo methods(MC) are usually applied to simplified coarse-grained models. The idea is to solve the time acceleration problem by omitting very fast degrees of freedom. To consider the chaotic nature of molecular movements, some stochastics must be introduced. Some kind of repeated randomness assumption [411] must be applied after each omission step, e.g., to find clever MC algorithms that efficiently sample the complicated energy landscape of a glass-forming liquid. A local MC [412] movement is a decision between different possibilities for a local change using such transition probabilities that, globally, certain principles are not violated. Examples for such principles are detailed balance to keep the equilibrium, and appropriate probabilities and ensembles to define equilibrium thermodynamic control parameters such as temperature and pressure and, perhaps, their variations by a cooling rate, etc. If temperature fluctuation is important for the phenomenon under consideration, the Gibbs equilibrium distribution is not appropriate [condition (iv) of the first paragraph of this section, and see also Sect. 3.2]. After application to the ensemble, the real time scale must be determined since the thermal velocity scale is eliminated by the omission steps.

Usually, temporal coarsening is accompanied by spatial coarsening using molecular models that omit all details considered to be unimportant for the glass transition. There is a considerable risk of finding degenerate models, e.g., a glass transition with $T_0 = 0$ or even $T_c = 0$, i.e., without Vogel temperature or a cooperativity onset. There are, however, examples where several reasonable properties of the equilibrium fluid close to T_g could be obtained [413]. An advantage of MC is that it can use movements with unphysical dynamics, e.g., allowing the system to equilibrate more effectively at low temperatures.

Since all materials with disorder show a glass transition, it would not seem important to model intermolecular potential functions exactly. Coarsening must, however, guarantee that a possibly special but definite intermolecular and intramolecular potential can, in principle, be mapped onto the coarsened situation in a reasonable manner, and that MC efficiently samples all important slow Fourier components of the movements in this potential. To avoid artefacts, one has to prove that ordering (e.g., crystallization) can definitely be excluded.

An example which is simulated to the end, from Binder's group, is the 'bond fluctuation model' [414]. This is for amorphous polymers which are usually good glass formers. Several chemical bonds along the backbone of a linear macromolecule are coarsened into one effective bond. Depending on the polymer conformation, this bond can be longer or shorter. If we use the lattice spacing of a simple cubic lattice as the length unit (Fig. 2.23c), the bond length b is varied between 2 and $\sqrt{10}$, for example. This model contains 5 bond lengths and 87 bond angles. A mapping (homomorphism) of real polymer potentials into the parameters of effective bonds (lengths and angles) seems possible [415]. An effective repulsion potential is defined via mutually excluded volumes of the chains: each effective monomer blocks the

8 neighbor sites of the elementary cube of the lattice from further occupation. An effective landscape potential is modelled by the transition probability for a random monomer movement:

$$W = \exp \frac{\varepsilon}{k_B T} , \tag{2.115}$$

where ε is the energy change due to the movement. The Hamiltonian puts $\varepsilon = 0$ for the {3,0,0} and $\varepsilon =$ given > 0 for all other bonds (Fig. 2.23c). Only if W exceeds a random number z, uniformly distributed between zero and unity, is the attempted movement $i \to f$ actually performed. This 'Metropolis algorithm' [416] generates a Boltzmann distribution for large times $t \to \infty$, i.e., a Gibbs distribution for the equilibrium.

The model contains an internal frustration. Energetically, all bonds tend to {3,0,0} with $\varepsilon = 0$, but this bond requires a particularly large volume, not available for all bonds at high bond density ϕ.

For suitably chosen volume fractions of occupied sites (e.g., $\phi = 0.533$), there is no detectable tendency towards crystallization. The structure factor of effective monomeric units is qualitatively compatible with experiment. Typical polymer properties such as the Debye function for the single-chain structure factor and reasonable Rouse modes (Sect. 2.4.1) are also obtained. There are indications of a Vogel temperature $T_0 \approx 0.12$–0.13 when the trace for the main transition is fitted by a WLF curve. Extrapolations by means of MCT give a crossover temperature of $T_c \approx 0.15$. The temperatures are reduced by $\varepsilon = 1$ and $k_B = 1$.

The older *Gibbs–DiMarzio model* [133,417] is static and more direct. This model is related to the ideal glass transition by the configurational entropy S_c and the Kauzmann temperature T_2. Polymer chains are distributed on a primitive lattice. Parameters are the hole energy for unoccupied sites (α), volume fraction of holes (ϕ_α), and energy difference(s) between different local chain conformations ($\Delta\varepsilon$). The estimated configurational entropy $S_c(T; \alpha, \phi_\alpha, \Delta\varepsilon)$ has a steep slope with temperature T, and the original extrapolation $S_c \to 0$ defines a Kauzmann temperature T_2 that is identified with a Vogel temperature $T_0 = T_2(\alpha, \phi_\alpha, \Delta\varepsilon)$. The advantage of this model is that many features of the glass temperature T_g for polymers (chain length, stiffness, pressure dependence, copolymerization) can be successfully correlated in analogy with T_2 as a function of the $\{\alpha, \phi_\alpha, \Delta\varepsilon\}$ parameters. As an example, the glass temperature and even the glass transition can be discussed as a function of the hole fraction and the fraction of high-energy bonds [418].

A comparison between the Gibbs–DiMarzio model and the bond fluctuation model is interesting [419]. The original approximation of the Gibbs–DiMarzio model by a quasi-lattice calculation indicated a second order transition at T_2, but a bend at T_2 was soon expected instead. This was proved later [420]. Analogous calculations for the bond fluctuation model yield a similar curve for the configurational entropy $S_c(T)$. Surprisingly, T_2 was extrapolated to be not only above T_0, but also above the MCT critical temperature T_c as calculated dynamically, i.e., $T_2 > T_c$. This means, against any expectation from the Adam–Gibbs approach, that the dynamic behavior of the bond fluctuation model below T_c cannot be correlated with the configurational entropy. This is in contrast with the experience [72] that in molecular liquids such a correlation is reasonable below T_c and fails above T_c. In addition, no ideal glass transition could be detected in a polydisperse hard disk system [421].

Instructive reviews concerning computer glass transitions in polymers are [422, 423], and a general review of the application of computer simulations to the glass transition is [410].

The next step of abstraction are *toy models* for the glass transition. They usually start from a lattice set of entities (such as points and spins) whose movements depend on the multiple state of the environment (empty or occupied sites, up or down sites). It is not possible practically to map reasonable intermolecular potentials on such dependencies. It therefore seems risky to derive consequences for molecular glass transitions from features of toy models. Another problem is to choose initial conditions and algorithms such that ergodic equilibria can be reached. On the other hand, we can perhaps see which kind of properties may be important for this or that property of the cooperative α process. An example for points that actually move with time on lattices is described by Jäckle et al. [424].

Another example for a toy model is due to Fredrickson et al., the 'spin-facilitated kinetic Ising model' (see [425] and Fig. 2.23d). Such an Ising model is a lattice with spins, e.g., $\sigma = 0$ for a more mobile spin, and $\sigma = 1$ for a less mobile spin. The MC flip probabilities are

$$P_i(1 \to 0) \propto \exp \frac{\varepsilon_i}{k_B T} , \quad P_i(0 \to 1) = 1 , \tag{2.116}$$

with the additional 'topological' flip condition, that at least a minimal number z of nearest neighbors must be mobile ($z = 2$ in Fig. 2.23d). The neighboring $\sigma = 0$ mobile spins facilitate the relaxation of the central spin. Assume that all ε_i are equal to a given ε. At high temperatures $T > \varepsilon/k_B$ and for small z, the equilibrium population of mobile spins is sufficient to allow most spins to flip in their immediate surroundings. The strong dynamical constraints imposed by topology are not felt. At low temperatures $T < \varepsilon/k_B$, however, the population of mobile spins is small and the vast majority of spins cannot flip in their immediate surroundings. Relaxation at such temperatures can only occur by cooperative spin flipping events.

The $z = 2$ Fredrickson model in two dimensions (2 SFM, $d = 2$) has a main transition with curved trace in the Arrhenius diagram. The Vogel temperature is zero, $T_0 = 0$, and there is no sharp crossover. The cooperativity, as defined by independent functional subsystems, really increases steeply below $T < \varepsilon/k_B$ [426]. The equilibrium entropy $S(T)$ can be calculated from the equilibrium populations for this model. It corresponds to the configurational entropy $S = S_c$ of glass formers, with an *Adam–Gibbs correlation*, i.e., a linear relation from (2.28),

$$\log \tau_{\mathrm{r}} = \text{const.} \times \frac{\varepsilon}{TS_{\mathrm{c}}}\,, \tag{2.117}$$

where τ_{r} is a reduced model time such that $\tau_{\mathrm{r}} = 1$ for $\varepsilon/TS = 0$. For the model, (2.117) holds for about 6 time decades where ε/TS ranges from zero to about 4 [710].

This 2 SFM, $d = 2$, model was later [202] modified by a special realization of the repeated randomness hypothesis. The energy was completed by interaction energies with nearest neighbors j: $\varepsilon_i \to \varepsilon_i + \sum_j \varepsilon_{ij}$, and both ε_i and ε_{ij} were randomly chosen at each flip from a uniform distribution over 0 to $2\Delta\varepsilon$. This is some kind of dynamic Ising model, not a spin glass with quenched disorder. The idea is that a hypothetical observer sitting on the phase space point in the energy landscape sees an entirely new landscape at each move. This is thought to be an acceleration of the algorithm, and I would expect to get an equilibrium distribution which is different from the Gibbs distribution, with additional fluctuations.

The main result from this modified Fredrickson model is a detailed splitting scenario for the crossover with three continuous β process traces, β_1, β_2, β_3 in Fig. 2.23e, and a curved α trace with a separate α onset about one frequency decade below the slowest β and with a linear increase of an α intensity, $\Delta x_\alpha \sim T_{\mathrm{on}} - T$, $T < T_{\mathrm{on}} = 1.17\Delta\varepsilon/k_{\mathrm{B}}$. The Arrhenius diagram Fig. 2.23e corresponds roughly to the splitting scenario I of Fig. 2.11a. The separation into three β relaxations is a consequence of the three flip possibilities in Fig. 2.23d that cannot be smeared out in $d = 2$ dimensions.

This toy model has a dynamic heterogeneity with islands of mobility below the onset, $T < T_{\mathrm{on}}$, larger for lower temperatures [202]. The corresponding pattern can be visualized by Butler–Harowell pictures [427] giving the faster spins a different gray or color (Fig. 2.23f). A typical length scale at the onset $T = T_{\mathrm{on}}$ is about 7 units and does not decrease in the temperature interval up to $T = 1.5T_{\mathrm{on}}$. The islands of mobility below T_{on}, however, become the more continuous phase; we have more 'islands of immobility' above T_{on} [202]. It is possible that the Butler–Harowell pictures count not only cages, if any, in the model, but also distances between cages. The latter could hardly be linked with entropy fluctuation, so that the characteristic length is expected to be smaller than the typical length from the pictures.

The equilibrium fraction of mobile spins – an indication of free volume – tends to zero at about $T = T_{\mathrm{on}}/2$. This temperature is higher than the estimated Ising critical temperature at about $T_{\mathrm{on}}/4$. The relaxation mechanisms seem to be exhausted at $T_{\mathrm{on}}/2$, because there is too little free volume. The Vogel temperature is estimated to be small, $T_0 \approx 0$.

2.4 Long-Running Issues

Problems actively investigated for decades without breakthrough indicate that one or a few fundamental questions have not been localized. This section aims to discuss three groups of long-running issues that probably have some relevance to such questions about the glass transition.

- The first group are glass transitions under special spatial conditions: in polymers where the chain coil diameter of individual macromolecules is larger than the cooperatively rearranging regions CRR for the α process (Sect. 2.4.1), and in confining geometries of different dimensions (layers $d = 1$, cylinders $d = 2$, and spheres $d = 3$) with spatial scales of order the CRR diameters (Sect. 2.4.2).
- The second group are phenomena in glasses below T_g that may crucially depend on the glass structure as a result of freezing of dynamic heterogeneities at T_g, viz., the tunnel states at low temperatures (Sect. 2.4.3) and the mixed alkali effect in silicate glasses (Sect. 2.4.4).
- The third group seems at first sight to be less important for the glass transition, namely, processes between the traces in the Arrhenius diagram (Sect. 2.4.5). Are details of dynamic heterogeneities important for the rates of crystal nucleation and crystal growth (Sect. 2.4.6)?

In other words, this section is not intended to give a neutral description of the issues mentioned. Instead the text concentrates on special concepts which I believe are crucial for understanding the glass transition: cooperativity and dynamic heterogeneity. I hope that, conversely, these concepts can be illustrated and sharpened by confrontation with the long-running issues.

2.4.1 Glass Transition and Flow Transition in Polymers

Atactic, flexible, long macromolecular chains are usually good glass formers without any tendency to crystallize. In comparison with small-molecule glass formers, we have two additional length scales: the entanglement spacing d_e of order 3–7 nm and the average chain end-to-end distance $\bar{R}$. For example, for vinyl polymers, the latter is $\bar{R} \approx 0.7\sqrt{N}$ nm for $N \gtrsim 20$, where N is the number of monomeric units. The chain diameter corresponds to the molecular diameter of the other glass formers, $\sigma \approx 0.5$ nm. Polymer networks have a further length scale, the average distance between the crosslinks. It seems advantageous to have independent length scales for comparison, if we are interested in the characteristic length for the glass transition.

Consider the dynamic glass transition in a series of amorphous polymers of linear flexible macromolecules with increasing molecular weight M, i.e., with increasing chain length. Examples are atactic polystyrene PS or polyvinylacetate PVAC. For monomers and oligomers, we observe a glass transition similar to that in other small-molecule glass formers, e.g., the α process in cold liquids. Increasing M further, the low frequency wing of loss peaks is modified by Rouse modes R (or normal modes) which will be discussed later. Finally, above an entanglement molecular mass M_e of order 10^4 g/mol, corresponding to about $N_e \approx 100$ monomeric units along the chain, a second dispersion zone ft separates from the α process (Fig. 2.24a). This zone is called the *flow transition* or terminal transition. In the shear storage modulus $G' = \mathrm{Re}\,(G^*)$, we observe a plateau zone between the α process and flow transition ($G' \approx$ const. $= G_N^0$, the plateau zone modulus of order MPa) whose width $\Delta \log \omega$ increases dramatically with M for $M > M_e$,

$$\frac{\mathrm{d}\Delta \log \omega}{\mathrm{d} \log M} = 3.4 \pm 0.2 \,, \tag{2.118}$$

for the $M \gg M_e$ asymptote.

The transition from the glass zone ('solid behavior') to the flow zone ('liquid behavior') for high-mass polymers ($M > M_e$) consists of two components: the main transition and the flow transition ft (Fig. 2.24b). The *main transition* now includes not only the a or the α process, corresponding to the definition on p. IX, but also the Rouse modes R and an entanglement process e (see below). I suggest that the main and flow transitions represent different dynamic glass transitions, 'one upon another', represented by widely separated ($\Delta \log \omega$) Fourier components for spectral densities. The formulation 'one upon another' [18] means that both components are generated by movements of the same molecules. In other words, high-mass polymers show a *multiplicity* of glass transitions [428].

Which component corresponds to the usual glass transition of small-molecule glass formers? The answer 'the α process' is supported by three arguments:

(i) The continuity represented by Fig. 2.24a.
(ii) The position of the dielectric (ε'') and heat capacity (C_p'') loss peaks which are near the G'' shear loss peak of the main transition [121], although the differences between the three peaks may vary by more than one frequency decade [429].
(iii) An additive analysis of shear compliances J^* which is successful upon the condition that the values of steady state compliance J_s^0 for the α process are similar in a series of polystyrenes [104]. Compliances correspond to additive thermodynamic variables; additivity is here related to the independence of different dispersion zones (Fig. 2.24c).

Comparing the steady state compliances for the flow transition, J_N^0 of order $1/G_N^0 \approx 1$/MPa, with that of the proper glass transition, i.e., the α process, $J_s^0 \approx 1$/GPa (Table 2.2), we find that $J_N^0 \approx 10^3 J_s^0$. According to the FDT (2.13), these equilibrium compliances correspond to a threshold of mean shear angle fluctuations $\overline{\Delta\gamma^2}$ necessary for the shear flow that follows: the flow of monomeric units, modified by Rouse modes and confined by entanglements, follows on a threshold that is determined by J_s^0, and the flow of whole polymer chains follows on a threshold that is determined by J_N^0. Extrapolated to monomeric units, the necessary shear angle fluctuation $\Delta\gamma$ for the slow flow after ft is $1000^{1/2} \approx 30$ times larger than $\Delta\gamma$ for the fast flow after α (several per cent for α corresponding to a shear angle of about 10°, Sect. 2.1.2). This means that the full flexibility of chains is required for the slow chain flow. Related to the characteristic length of an independent subsystem (ξ_α for α, d_e for ft), the threshold shear angle fluctuations are much smaller, of order $\Delta\gamma \approx 0.1\%$.

Consider an isothermal section of the Arrhenius diagram Fig. 2.24c below the crossover temperature T_c for increasing relaxation times, i.e., decreasing mobility $\log\omega$. The succession of relaxation processes is as follows. The local Johari–Goldstein modes β come first, and the α process enables the monomeric units to flow; this flow is modified by the polymer chains (Rouse modes R), confined and transiently stopped by entanglements e, which are then loosened at the flow transition ft allowing the diffusion of whole chains. This succession obeys the general scaling (Sect. 2.2.5), i.e., increasing length scales for increasing times or decreasing mobilities: $\beta \approx 0.5$ nm, $\xi_\alpha \approx 2$–3 nm = the CRR diameter, the Rouse modes between ξ_α and d_e, $d_e \approx 3$–7 nm = the entanglement spacing, and finally the 'coil radius' $\bar{R} > d_e$ [430]. Let us follow through this sequence.

The Johari–Goldstein β process in polymers [2, 3] usually has a larger intensity of shear and dielectric response than that in the corresponding monomer. An example are the n-alkyl methacrylates [431]. It seems typical [432] that

- the side group(s), if any, and main chain groups act together,
- that the polymer chain piece is clamped on both ends of the local relaxation, rather like a crankshaft [433].

Note that β relaxations do also occur in polymers without mobile side groups, e.g., in polyvinylchloride PVC. This led to the concept of 'local modes' of chains for the β relaxation.

The α process in polymers is assumed to be a cooperative process of particles like the glass transition in small-molecule glass formers. The existence of chains did not lead to a better understanding. It seems clear that for flexible polymers, the role of particles is played by the monomeric units. Comparable cooperativities ($N_\alpha \approx 100$) and characteristic volumes ($V_\alpha \approx 10$ nm^3) of both substance classes at the glass temperature T_g mean that the α process cannot be understood from the movements of one chain alone. Instead, monomeric units of all chains going through the characteristic volume take part in the cooperativity (e.g., 5 to 10 chain parts in a volume $V_\alpha \approx 10$ nm^3 at T_g corresponding to about 100 monomeric units).

Rouse modes [434] operate between the length scale of minimal Gaussian coiling for flexible macromolecule chains (of order ξ_α for flexible vinyl polymer chains) and entanglement spacing d_e, or a maximal Rouse mode for the whole chain as for molecular weights below the level at which entanglement begins, viz., $M < M_e$. The movements of monomeric units are indicated by arrows in Fig. 2.24d. They are driven by thermal forces ('microbrownian' motions), modified by chain continuity ('entropy springs'), and damped by 'friction' due to the other monomeric units treated as a mean field. The motion is scaled by the Rouse rate

$$W = \frac{3k_{\mathrm{B}}T}{\zeta_0 a^2}\,, \tag{2.119}$$

which combines the three features: the microbrownian motion via $k_{\mathrm{B}}T$, the structure length via a, and the monomeric friction coefficient via ζ_0 [3]. ($\overline{R^2} = a^2N$ describes the Gauss coils of chain parts with N monomeric units, the 'beads' of a bead spring model for Rouse modes; the structure length $a \approx 0.7$ nm for vinyl polymers, as mentioned in the first paragraph of this section.) The mean squared displacement of a monomeric unit in the Rouse regime is

$$\overline{\Delta r^2}(t) \approx a^2(Wt)^{1/2}\,. \tag{2.120}$$

Compared with ordinary diffusion, $\overline{\Delta r^2}(t) \sim t$, Rouse diffusion is an example of *sublinear diffusion.* The corresponding dispersion law is $\omega \sim Q^4$ for Rouse instead of $\omega \sim Q^2$ for normal or linear diffusion. Equation (2.120) also obeys the general scaling, that is, larger length for larger times, and sublinearity indicates a distributed hindering of diffusion. The Rouse modes are the eigensolutions of overdamped mechanical equations [434] for the beads. We have,

▶

Fig. 2.24. Glass transition multiplicity in amorphous polymers with flexible chains. α is the proper glass transition, a the high temperature process, β the local mode or Johari–Goldstein process, R Rouse modes, modified at low temperatures, e the entanglement process, and ft the flow transition. (**a**) Dynamic storage shear modulus G' as a function of frequency for a series of substances from monomer to high-mass polymer. M is the molecular mass. The frequencies are scaled (mastered) by a shift factor a_T so that the α transitions coincide. G_{N}^0 is the rubbery plateau modulus. (**b**) Storage (G') and loss (G'') shear modulus as a function of frequency; ε'' and C_p'' are the peaks of dielectric and calorimetric loss (schematic). (**c**) Traces of the different processes in the Arrhenius diagram. (**d**) Hook picture of chain entanglements. d_{e} is the entanglement spacing. (**e**) Indications of distinct α and e processes in the shear spectra H (from modulus) and L (from compliance). Example: PVAC, units Pa and Pa^{-1}, respectively. (**f**) Pre-averaging of the energy landscape for the slow flow transition ft by the fast main transition α diminishes the landscape roughness: $\varepsilon_{\mathrm{ft}} < \varepsilon_\alpha$. (**g**) Encroachment of the main transition for lower temperatures as caused by the convergence of the e and α traces. (**h**) Extrapolation of main and flow transitions leads to a crossing (X) at low frequency and temperature

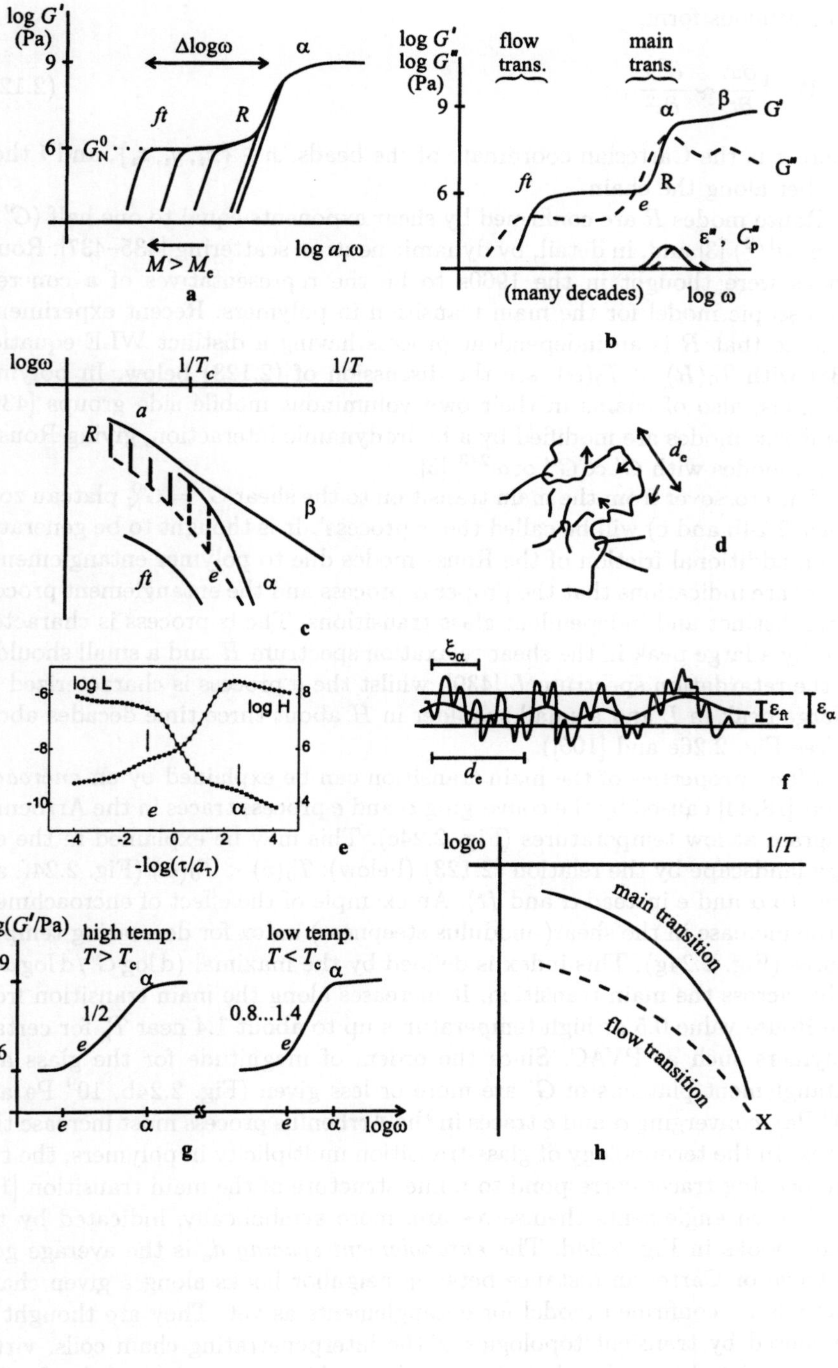
log G'
(Pa)
9
6
Δlogω
α
ft
R
G_N^0
log a_Tω
M > M_e
a
log G'
log G"
(Pa)
flow
trans.
main
trans.
β
G'
G"
e
ε", C_p"
(many decades)
log ω
b
logω
1/T_c
1/T
c
d_e
d
log L
log H
-6
-8
-10
-4
-2
0
2
4
-log(τ/a_T)
ξ_α
ε_{ft}
ε_α
f
log(G'/Pa)
high temp.
T > T_c
low temp.
T < T_c
1/2
0.8...1.4
logω
g
main transition
flow transition
X
h

in continuous form,

$$W^{-1}\frac{\partial x}{\partial t} \sim \frac{\partial^2 x}{\partial i^2}\,, \tag{2.121}$$

where x is the Cartesian coordinate of the beads, $x \in \{x_i, y_i, z_i\}$, and i their number along the chain.

Rouse modes R are confirmed by shear exponents equal to one half ($G' = G'' \propto \omega^{1/2}$) [3] and, in detail, by dynamic neutron scattering [435–437]. Rouse modes were thought in the 1960s to be the representatives of a concrete microscopic model for the main transition in polymers. Recent experiments indicate that R is an independent process having a distinct WLF equation [438] with $T_0(R) < T_0(\alpha)$; see the discussion of (2.123) below. In polymer solutions, also of chains in their own voluminous mobile side groups [430], the Rouse modes are modified by a hydrodynamic interaction, giving Rouse–Zimm modes with $G' \propto G'' \propto \omega^{2/3}$ [3].

The crossover from the main transition to the shear $G' = G_{\mathrm{N}}^0$ plateau zone (Figs. 2.24b and c) will be called the 'e process'. It is thought to be generated by an additional friction of the Rouse modes due to polymer entanglements. There are indications that the proper α process and the entanglement process e are distinct and independent glass transitions. The α process is characterized by a large peak in the shear relaxation spectrum H and a small shoulder in the retardation spectrum L [439], whilst the e process is characterized by a large peak in L and a small shoulder in H about three time decades above α (see Fig. 2.26e and [105]).

Many properties of the main transition can be explained by an *encroachment* [18,44] caused by the converging α and e process traces in the Arrhenius diagram at low temperatures (Fig. 2.24c). This may be explained in the energy landscape by the relation (2.123) (below): $T_0(e) < T_0(\alpha)$ (Fig. 2.24f, applied to α and e instead α and ft). An example of the effect of encroachment is the increase in the shear(-modulus steepness) index for decreasing temperatures (Fig. 2.24g). This index is defined by the maximal $(\mathrm{d}\log G'/\mathrm{d}\log\omega)_T$ value across the main transition. It increases along the main transition from the Rouse value 0.5 at high temperatures up to about 1.4 near T_{g} for certain polymers such as PVAC. Since the orders of magnitude for the glass and entanglement plateaus of G' are more or less given (Fig. 2.24b, 10^9 Pa and 10^6 Pa), converging α and e traces in the Arrhenius process must increase this index. In the terminology of glass-transition multiplicity in polymers, the two encroaching traces correspond to a fine structure of the main transition [18].

The entanglements themselves are, more symbolically, indicated by the chain hooks in Fig. 2.24d. The *entanglement spacing* d_{e} is the average geometrical or Cartesian distance between neighbor hooks along a given chain. There is no confirmed model for entanglements as yet. They are thought to be caused by transient topologies of the interpenetrating chain coils, virtually completed into closed circles, and by packing constraints of the flexible chains with a certain diameter (of order 0.5 nm) [440, 441]. Let us recall that minimal chain lengths N_e are necessary for entanglements (Fig. 2.24a).

Consider now the flow transition process ft. It is thought to be generated by fluctuational disentanglements of whole chains, a presupposition for their diffusion as whole chains. Of course, ft relaxation times increase sharply with chain length,

$$\tau_{\mathrm{ft}} \approx \tau_0 N^{3.4\pm0.2} . \tag{2.122}$$

The theoretical explanation for the exact exponent in (2.122) is still an open problem. Simple chain reptation models give 3.0 for the exponent [442, 443].

Some think that the flow transition relaxation time τ_{ft} is triggered by τ_0, a characteristic time of the main transition, with a $\tau_{\mathrm{ft}}/\tau_0$ ratio independent of temperature. This is wrong. Careful investigations of the ft and α process traces [103, 444, 445] show that $\log \tau_{\mathrm{ft}}$ as a function of temperature T obeys a WLF equation (2.24), but with a Vogel temperature several tens of kelvins below that for the α process,

$$T_0(\mathrm{ft}) < T_0(\alpha) . \tag{2.123}$$

This means that the extrapolated WLF traces for ft and α must converge and eventually, in the extrapolation, intersect at low temperatures and frequencies (marked by a cross in the Arrhenius diagram Fig. 2.24h).

Such a convergence seems to be typical for two glass transitions which occur 'one upon the other'. Encroachment for α and e processes was mentioned above. Two such glass transitions are also observed in liquid crystals [446], where rotation around the long axis ($\|$) is faster than that around the short axis ($\perp$), e.g., for molecules like cigars. The two different molecular rotations are obviously the elements of two distinct glass transitions – indicated by different WLF curves, each with $T_0(\perp) < T_0(\|)$ asymptotes and broad dielectric loss peaks – but with the same molecules. Glass transitions, however, with different molecules or molecular parts (as for micro or nanophase separations [316, 447]) need not follow such a T_0 relation.

The interpretation problem concerning the Vogel temperature T_0 was mentioned above (Sect. 2.1.1). Let us first state that the existence of two different T_0 asymptotes generated by the same molecules is hardly compatible with the idea of interpreting T_0 as an extrapolated phase transition of any kind. The single temperature (or $k_B T_0$ = energy) parameter of the WLF equation should be connected with the actual temperature range of its application. The only energy parameter of the appropriate landscape picture is the roughness ε of the accessible energy surface (Fig. 2.24f). The existence of a faster α process means pre-averaging on the ξ_α scale so that the remaining roughness $\varepsilon_{\mathrm{ft}}$ for the slower flow transition is smaller than ε_α for the faster α transition. If we put $T_0 \propto \varepsilon$ with a given proportionality constant for a given substance, (2.123) then follows from the $\varepsilon_{\mathrm{ft}} < \varepsilon_\alpha$ pre-averaging relation.

I think that the right questions for the flow transition issue are as follows. What kind of dynamic glass transition is the flow transition? What kind of particles are involved? What kind of interaction occurs between them? From Fig. 2.24d, we see that the relevant particles are at least the size d_e of the entanglement spacing, and at most of the size $\bar{R}$ of the coil radius, also called the 'chain coil diameter'. This means that we are somewhere between a molecular and a colloidal glass transition. No crossover region or Johari–Goldstein process related to the flow transition has yet been detected. As there are many entanglement hooks inside the geometric volume of a chain coil which are not directly related to the given chain (Fig. 2.24d), the particles should have the ability to penetrate each other to a certain extent. This means that a particle can have 50 neighbors, say, and that complete interpenetration of too many particles is avoided by outer entanglements and chain diameters, not by 'hard cores' thought of as spheres in the conventional meaning.

It seems that the flow transition is a strange glass transition. We should mention the interesting experiments of Antonietti et al. [448]. They observed that many macroscopic properties previously thought to be inevitably connected with single chains (Rouse and reptation modes) can also be observed in polymer colloid samples without solvent. Polymer chains are quite different from crosslinked polymer cubicles. This indicates once more that the map from molecules to glass transition (Fig. 2.1f) is a homomorphism in the sense that different molecular situations can give similar glass transitions including some similar dependencies on parameters. In this case, the molecular mass of chains $\Leftrightarrow$ the molecular mass of the colloidal cubicles.

Let us also discuss the issue of a so-called *liquid–liquid transition* at a T_{ll} temperature above the glass temperature T_g. After many discussions about small particularities in thermal properties at $T_{ll} > T_g$ [449], the case seemed to have been settled by interpreting T_{ll} as 'the thermal glass temperature' of the flow transition, i.e., the temperature where the typical relaxation time of the flow transition is lowered to about 10^3 seconds, the typical experimental time [450]. This would mean that there is no T_{ll} in small-molecule glass formers. The discussion concerning a liquid–liquid transition, however, has recently been taken up again in connection with the issue of a general thermodynamic marker for the crossover region (Sect. 2.5.1). Are there two transitions, T_{ll} for the flow transition and $T_{l\rho}$ for the crossover region [451]?

Above the crossover temperature T_c, MC computer simulation for the c and a processes [452] reveals a first regime that corresponds to motion of a monomeric unit in its local environment. It is dominated by the cage effect and will be described by the idealized mode-coupling theory. The second, slower regime is governed by the late c–early a process transient. In this regime, the connectivity of the monomeric units begins to interfere with the cage dynamics and finally becomes dominant. It seems that the flow transition ft is not affected by the difference between the a process ($T > T_c$) and the α process ($T < T_c$), i.e., the question of whether monomeric units become mobile without ($T > T_c$) or with ($T < T_c$) main-transition cooperativity. This is an example of functional independence (Sect. 3.5).

2.4.2 Hindered Glass Transition in Confining Geometries

The initial question can be formulated as follows. Suppose that research into the confined glass transition seeks an independent confirmation of the characteristic length of the glass transition in the bulk material without confinement. Assume, as an idealized case, that the molecular motions which cause the dynamic glass transition are not affected by the confining walls as long as all three dimensions are larger than the characteristic length $\xi_\alpha(T)$. Consider a spherical pore with given diameter D and volume V_D of nanometer size (Fig. 2.25a). Consider then the following cooling process. We move along the trace of the dynamic glass transition to lower temperatures. The volume $V_\alpha(T) = \xi_\alpha^3(T)$ of a cooperatively rearranging region CRR increases. In the ideal case, we have the undisturbed WLF curve for bulk at high temperatures $T > T_D$, and a modified trace in the Arrhenius diagram at low temperatures $T < T_D$, where T_D is implicitly given by the coincidence of the two volumes,

$$V_D = V_\alpha(T_D)\,. \tag{2.124}$$

According to Adam and Gibbs, the WLF curvature for the α process is connected with increasing cooperativity, i.e., increasing $V_\alpha(T)$. At $T = T_D$, this increase is stopped and glass transition cooperativity must make do with a constant volume. Assume that this situation is realized by a fixed molecular mechanism with a given activation energy equal to the apparent activation energy of the WLF curve at $T = T_D$ (Fig. 2.25b). The glass transition for given $T < T_D$ in the pore is then faster than in bulk. Having a set of pores with different diameters, we observe a T_D shift: T_D increases with decreasing pore diameter (Fig. 2.25c). Below T_D, we find higher mobility $\log \omega$ in smaller pores, i.e., lower glass temperatures in smaller pores. This idealized case was called the *hindered glass transition* [7].

In preliminary experiments [453], the limitations of the idealization were immediately evident. The glass transition is affected by the pore walls even at much larger diameters than ξ_α, starting at about $D = 10$ nm, and its parameters, including T_g, depend seriously on pore surface properties. T_g variations up to ±15 K were observed for polyvinylacetate PVAC thermally polymerized in initially deoxidized glass pores.

Guided by the relaxation chart, that is, the Arrhenius diagram of semicrystalline polyethylene PE (Fig. 5.1e), where traces like the pore curves of Fig. 2.25b are actually the rule [7], Schick had the idea of using semicrystalline polyethylene terephthalate PET for testing ξ_α against the morphological layer thickness D_a of the amorphous mobile phase between plane crystal lamellas. The foreign boundary of the pores is thus substituted by a self-organized interface in the same material. The layer thickness D_a can be considerably varied by the crystallization regime [454], whereas the variation of T_g against the bulk material was moderate (order $\Delta T_g = 5$ K). This means that the molecular mechanisms in bulk and in the layer, i.e., in a one-dimensional $d = 1$ confined geometry, are quite the same. Systematic experiments [455] (reviewed in [222]) proved that the calorimetrically determined ξ_α values from (2.76) are smaller than the bulk ξ_α values and correlate with the variable layer thickness D_a determined by different methods (Fig. 2.25d). Extrapolating to large D_a values, we find the bulk length near $D_a \approx 2\xi_\alpha$. However, the higher mobilities predicted by the hindered glass transition, were not observed in this 'weak' $d = 1$ confinement.

Systematically decreasing characteristic lengths with decreasing but larger pore sizes were also observed for a small-molecule glass former in a series of porous glasses with pore diameters between 7.5 and 2.0 nm [305]. The subtle independence condition for the CRR definition is sensitive to small perturbations from the distant pore or layer walls. Since the liquid structure in a $2\xi_\alpha$ pore, say, cannot be optimized to the same extent as in the bulk, we have more free volume, and this means smaller CRRs in the pores or layers. We may also ask if the effect from the distant pore or layer walls is mediated by longer kinematic lengths, such as discussed for the a process above the crossover in Sect. 2.3.4.

The formulas (2.76)–(2.78) for determining the characteristic length ξ_α or cooperativity N_α are derived from the von Laue treatment of thermodynamics (Sect. 3.2). An alternative formula from the Gibbs distribution [143, 429, 456] is $N_\alpha = RT^2/\Delta C_V M_0(\delta T)^2$, where δT is not considered as a fluctuation of temperature but as a spatial variation of the glass temperature, δT_g. This formula gives the ✠ symbols in Fig. 2.25d. The characteristic length in the amorphous mobile layers of PET would be much larger than the layer thickness. This seems to be impossible and favors the von Laue approach. Similar discrepancies between the two approaches are obtained for several organic liquids in porous glasses [457].

We can speculate about the small D limit relaxation. In small pores, as mentioned above, the environment for cooperative relaxation is not so extremely optimized, i.e., we have more free volume than in bulk. This would imply that we should find the β process in the small pore if this process can be localized at all. Experiments in small $D < 1$ nm zeolite pores are reported by F. Kremer's group [458]. Krüger's group observed [459] that the α process disappears with decreasing pore diameter. If the high temperature a process has no characteristic length or only a small one, then the T_D shift should stop at the crossover.

Any serious attempt to confirm the bulk characteristic length for the glass transition in confining geometries must try to separate [305, 460] or eliminate effects from the pore or layer interface. Roughly speaking [461], strong interaction between the glass former and the interface material causes the glass temperature T_g to increase, whilst weak interaction causes T_g to decrease. The latter effect, in the ideal case the hindered glass transition, used to be called the 'confinement effect' on the glass transition. The other effect is called the 'surface' or 'interaction effect' (see also [462]). The interaction effect has three parts.

(i) Structure effect. The glass former in the pore may have a structure different from that in the bulk. This may concern the near wall regions (as indicated, for example, by an extra ε'' peak [463, 464]) but also the core regions (as indicated, for example, by different T_g values).
(ii) Density effect. A priori, the density of the glass former in the pores must not be the same as in bulk. Thermodynamically, we have a curved surface so that the pressure inside – and therefore the density – should be different from a material with large plane surfaces (as for $d = 1$ layers [465,466]). The ΔT_g shift would then be proportional to the reciprocal pore diameter $1/D$. The density effect can be avoided by investigating the glass transition in the continuous phase between hard spherical particles of a filler material [467].
(iii) Process effect. At low viscosity, we can imagine observing an isobaric glass transition in the pores as well. But what about at low temperatures, where the viscosity becomes very high? Is the glass transition isochoric there [125]?

The three effects are interwoven in such a way that many people believe there to be an independent phenomenon, the glass transition in confining geometry, that can never be used to confirm or to disprove bulk characteristic lengths. Let us end by discussing some current aspects of this problem.

Lower T_g values for smaller pores were first reported in [465] (see also [468]). The T_g depression reaches the order $\Delta T_g = -10$ K from bulk to pores a few nanometers in size. The depression is proportional to the inverse pore diameter $\Delta T_g \sim 1/D$. This could be interpreted as a density effect.

A T_g depression of about $\Delta T_g \approx -30$ K was also observed for polystyrene PS layers in the thickness range from 100 to 10 nm [469]. T_g was reproduced by a formula $T_g(D) = T_g(\infty)[1 - (A/D)^\delta]$ with an exponent $\delta \approx 1.8$ and a length $A \approx 3.2$ nm of the order of the characteristic length for PS. In a series of free standing polymer films, with thickness between 21 and 200 nanometers, for PS of different molecular masses between 100 and 400 kg/mol, the length A from the above formula for $T_g(D)$ decreases with increasing temperature [470]. Qualitatively, this resembles the analogous decrease in characteristic length $\xi_\alpha(T)$ from calorimetry. A review of the glass transition in thin polymer films is given by Forrest et al. [471].

A lifting of the Arrhenius diagram trace (Fig. 2.25c) was directly observed for small pores by dielectric methods [463,472]. For other substances, however, ΔT_g was positive [473]. The dielectric loss curve $\varepsilon''(\log \omega)$ often consists of two components, the proper α process and, usually at lower frequencies, the α process modified by the structure of the surface (see Fig. 2.25e and [474]), especially if the pore surface is not treated by a surfactant with high molecular mobility. Figure 2.25e corresponds to an old rule that more structure results in higher T_g, i.e., lower mobility, because more structure diminishes the free volume (Sect. 5.2).

An isochoric glass transition in the bulk also has a WLF type trace in the Arrhenius diagram (Sect. 2.1.3, Fig. 2.3f). Note that an isochoric ($V =$ const.) α process in bulk, i.e., in large sample volumes $V \gg V_\alpha$, is quite different from a hindered α process in a given small volume of order the CRR size, $V_D = V_\alpha$. In the large volume V, we have many independent CRRs whose volumes can fluctuate independently [126] and can increase. This always results in WLF traces, one for $p =$ const. and another for $V =$ const. The isochoric WLF trace, however, is not as steep as the isobaric trace (Fig. 2.25f). This was first shown by Koppelmann [124] using dielectric investigations in polyvinylchloride PVC. As mentioned in Sect. 2.1.3, isobaric cooling involves more possibilities for diminishing the free volume than isochoric cooling. This means that the isochoric glass transition seems 'less fragile' [125]. This could also explain a $\Delta T_g < 0$ depression in small pores.

The first experimental evidence for an overall behavior similar to the hindered glass transition (Fig. 2.25c) was obtained by F. Kremer et al. for salol in porous glasses with silanized surfaces [464]. A careful data evaluation [475] also reveals the T_D shift to higher temperatures for smaller pores.

This discussion shows that the problem has not been completely settled as yet. Our goal with regard to cooperativity is to separate the three effects (i)–(iii) and compare the remainder, possibly a hindered glass transition, with the bulk glass transition.

▶

Fig. 2.25. Glass transition in confining geometries. (**a**) Comparison of a given spherical pore volume V_D with increasing volume $V_\alpha(T)$ of a cooperatively rearranging region CRR. T_D is explained by (2.124). (**b**) Comparison of traces of a hindered glass transition in a pore and a non-confined glass transition in bulk. (**c**) Hindered glass transition in a series of pores. The *arrow* is directed towards smaller diameters and C is the crossover region. (**d**) Dependence of calorimetrically determined characteristic length ξ_α on the layer thickness D of amorphous mobile phases in semicrystalline PET. The various *open symbols* correspond to different crystallization regimes. *Crosses* (✠) are obtained from an evaluation with a fluctuation formula from the Gibbs treatment [143, 456, 581]. (**e**) Example of the relationship between the hindered glass transition and the α process modified by the structure near the surface for strong interaction between surface and glass former. (**f**) Comparison of glass transitions along isobaric ($P =$ const.) and isochoric ($V =$ const.) paths in volumes much larger than a CRR. *Left-hand side* matched at T_g, *right-hand side* matched at high temperature

2.4.3 Tunnel States at Low Temperatures

We come now to the second group of long-running issues announced in the introduction to Sect. 2.4: the role of dynamic heterogeneity in the liquid above T_g for the glass structure below T_g in comparison to an amorphous structure generated without glass transition. The main problem seems to be the disengagement of the continuous evolution of fluctuational pattern for dynamic heterogeneity, with smooth increase of characteristic length $\xi_\alpha(T)$ for decreasing temperature, from more or less discrete structure elements such as the first coordination shell of molecules. This was also called the incommensurability problem in Sect. 2.2.6.

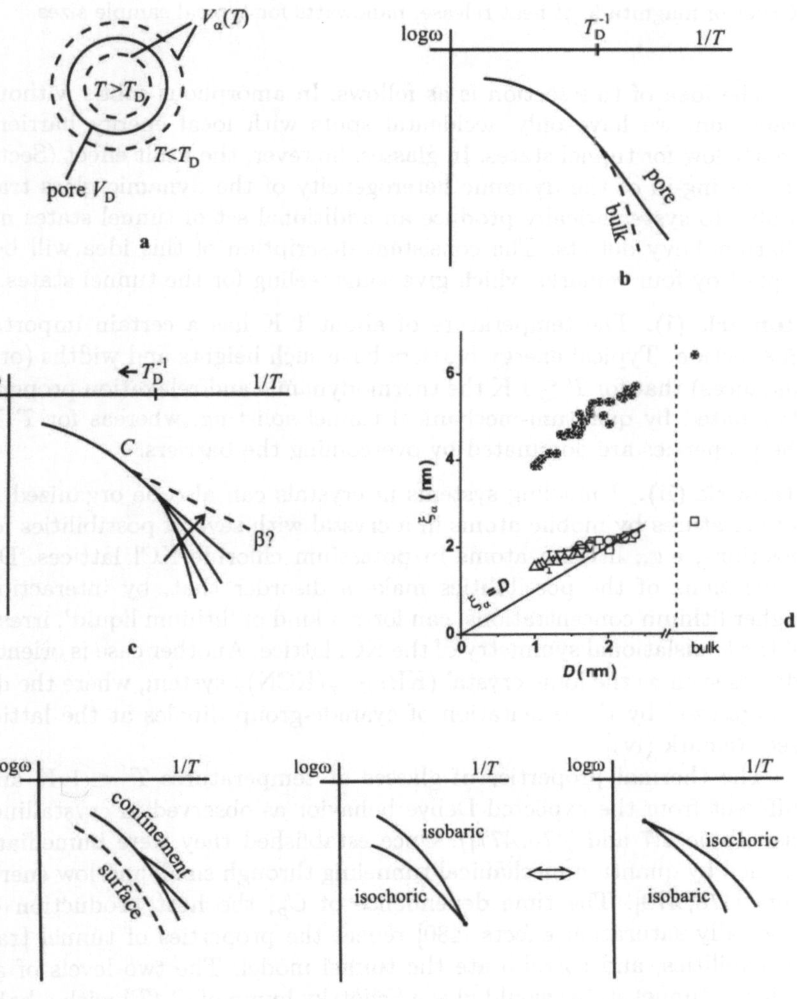

Table 2.7. Thermal properties of glasses and crystals at temperatures below 1 K (as known in 1980)

	Glass	Crystal
Heat capacity[a]	$C_p \sim n_0 T$	$C_p \sim T^3$
Time dependence[b]	$C_p \sim n_0 T_0 \ln(t/t_{\min})$	none
Heat production[c]	$\dot{Q} \sim n_0 (T_i^2 - T_0^2)\, t^{-1}$	none
Thermal conductivity	$\kappa \sim (1/n_0) T^2$	$\kappa \sim T^3$

[a]The number density of tunnel states is denoted by n_0 (number per unit volume).
[b]After quenching from T_i to T_0. The minimal time $t_{\min}$ is determined by phonon–tunnel interaction.
[c]Order of magnitude of heat release: nanowatts for typical sample sizes.

The idea of this section is as follows. In amorphous solids without glass transition, we have only accidental spots with local energy barriers sufficiently low for tunnel states. In glasses, however, the vault effect (Sect. 2.2.6) at freezing-in of the dynamic heterogeneity of the dynamic glass transition is able to systematically produce an additional set of tunnel states near the Glarum–Levy defects. The consistent description of this idea will be interrupted by four remarks which give some feeling for the tunnel states.

Remark (i). The temperature of about 1 K has a certain importance in this section. Typical energy barriers have such heights and widths (or tunnel distances) that for $T \lesssim 1$ K the thermodynamic and relaxation properties are dominated by quantum-mechanical tunnel splitting, whereas for $T \gtrsim 10$ K the properties are dominated by overcoming the barriers.

Remark (ii). Tunneling systems in crystals can also be organized at high temperatures by mobile atoms in a crystal with several possibilities for their positions, e.g., lithium atoms in potassium chloride KCl lattices. Different realizations of the possibilities make a disorder that, by interactions and higher lithium concentrations, can form a kind of 'lithium liquid', irrespective of the translational symmetry of the KCl lattice. Another case is orientational glasses such as the ionic crystal $(KBr)_{1-x}(KCN)_x$ system, where the disorder is organized by the orientation of cyanide-group dipoles at the lattice sites [see Remark (iv)].

The thermal properties of glasses at temperatures $T < 1$ K are quite different from the expected Debye behavior as observed in crystalline solids (see Table 2.7 and [476, 477]). Once established they were immediately explained by quantum mechanical tunneling through small and low energy barriers [478, 479]. The time dependence of C_p, the heat production $\dot{Q}$, and especially saturation effects [480] reflect the properties of tunnel transition probabilities, and corroborate the tunnel model. The two levels of a set of uniform tunnel states would give a Schottky hump of $C_p(T)$ with a half width of about 0.7 temperature decades. This is quite different from the observed

$C_p \sim n_0 T$. To explain the latter, a very broad distribution $P(\lambda, \Delta)$ of tunnel parameters and asymmetries is introduced. This hides all details. The glass term $c_1 T$ begins to dominate the Debye term $c_3 T^3$ below $T < 1$ K. The c_1 constant in the total heat capacity $c_p = c_1 T + c_3 T^3$ is then [481]

$$c_1 = \frac{\pi^2 k_{\mathrm{B}}^2 \bar{P}}{12\,\rho} \ln \frac{4\,t}{t_{\min}} , \tag{2.125}$$

where $\bar{P}$ is a constant average value of the P distribution, called 'the tunnel density', ρ is the mass density, and $t_{\min}$ is the tunnel–phonon coupling time. Experiments over huge temperature regions and for sufficiently different activities (different 'couplings' of tunnels to phonons and photons, for example; mechanical and dielectrical properties) would be necessary to find the details of the $P(\lambda, \Delta)$ distribution beyond the average $\bar{P}$ value. The orders of magnitude observed are, in general [482],

$$\bar{P} = 0.3\text{–}3.0 \times 10^{45}/\mathrm{Jm}^3 , \quad n_0 \approx 10^{17}/\mathrm{cm}^3 . \tag{2.126}$$

The number density n_0 corresponds to a tunnel fraction of order $n_{\mathrm{TLS}} \approx 10^{-5}$ per particle. In spite of many ingenious experiments [477,480,483], it has not been possible to give a final answer to the following questions: what are the tunnel entities and why are the tunnel parameter distributions so wide?

In the light of experimental details obtained in the last decade [484], together with recently detected collective effects for tunnel states in the millikelvin region [485], we shall first discuss the possibility of an approximate treatment for single tunnel states as independent objects at about $T = 1$ K. If the collective effects or small interactions between the entities are included from the beginning [484, 486], the term 'tunneling systems' has been established. If not, that is, if the entities can be considered as effectively isolated, we use the old terms 'tunnel states' or 'two-level states' TLS, as for the shear-angle fluctuation–dissipation theorem FDT (2.80). I think that such a treatment is possible.

Remark (iii). From a tunnel fraction $n_{\mathrm{TLS}} = 10^{-5}$ per particle we would find mean distances of 10 nm between neighboring tunnel states. Consider the energy hypersurface in the configuration space $\{q\}$. It is characterized by long valleys with low unevenness. Assume, as a worst case, that the wave function $\psi(q)$ from an otherwise localized tunnel state finds its way into the valleys. Then $\psi(q)$ is spread into the valleys. Their spatial range is mainly determined by kinetic energy, $(\hbar^2/2m)\Delta\psi$, and can therefore be estimated by a de Broglie wavelength λ_{B},

$$\lambda_{\mathrm{B}} = \frac{2\pi\hbar}{\sqrt{2m\,k_{\mathrm{B}}T}} . \tag{2.127}$$

Take the non-overlap of wave functions from different tunnel states as a criterion for their effective isolation. This avoids the need to consider finely tuned elastic and dipolar couplings with atomic energy parameters of order 1 eV = 11 604 K for estimates at low temperatures.

Take (2.127) as an estimate of the temperature range in which tunnel states can be considered as effectively isolated. For a mass m of 10 mass units (M = 10 g/mol, 1 m_u = 1.66 × 10^{-27} kg), we get λ_B = 0.97 nm for T = 1 K and λ_B = 31 nm for T = 1 mK. Compared with the 10 nm from the tunnel fraction, there is no overlap at T = 1 K, but a large overlap at T = 1 mK. As we are interested in the temperature interval between 0.1 K and 10 K, we can start our estimations from independent, isolated tunnel states.

As an aside, including many states, renormalization methods can scale the whole range between spurious overlappings at 1 K to heavy overlappings in the mK region.

We shall now pursue our approach to the glass structure (Sect. 2.2.6). We investigate an accidental part of the tunnel fraction, determined by accidental configurations of any amorphous structure, and a systematic part organized by partially vitrified dynamic heterogeneities in glasses (Fig. 2.16c). This approach aims to sharpen the present opinion of the low-temperature community [477]: "Tunnel states are not restricted to glasses. It may well be that variations in local environment resulting from any kind of disorder are sufficient to give the behavior ... ." Before the two fractions are estimated, some details concerning the tunnels will be presented.

The isolated tunnel model and its parameters are defined in Fig. 2.26a. For symmetric double well potentials, the tunneling splitting Δ_0 would be

▶

Fig. 2.26. Tunnel states at low temperatures. (**a**) $\varphi(q)$ double well potential of molecular energy, V barrier height, Δ asymmetry energy, d distance (to be calculated from canonical coordinate q), E total energy splitting. (**b**) Lattice model of cubic $(KBr)_{1-x}(KCN)_x$ orientational glass [489] (© 2001 American Physical Society). (**c**) Extrapolated dielectric loss around millihertz frequencies of $(KBr)_{1-x}(KCN)_x$ for different cyanide content x [488]. (**d**) Heat capacity in a C_p/T vs. T^2 diagram for different x [489]. (**e**) Excess heat capacity, $C_p^{ex} = C_p(T_0, t)-$ Debye term, for T_0 = 0.1 K as a function of x (log–log plot [490]). For the glass, the upper end is for t = 10 s, the lower for t = 1 ms. (**f**) Tunnel state fraction n_{TLS} as a function of square root of cooperativity $N_\alpha^{1/2}$ (schematic). Such behavior is expected from the vault model around Glarum–Levy defects. (**g**) Linear specific heat constant c_1 in $C_p = c_1T + c_3T$ (c_1 proportional to n_{TLS}) for a series of poly(n-alkyl methacrylates) (•), including some random copolymers (▪), as a function of the (average) number of C atoms in the n-alkyl remainder of the side group. The square root of cooperativity $N_\alpha^{1/2}$ at T_g (○ homo, □ copolymers) decreases continuously with increasing C number [319]. The c_1 maximum corresponds to a T_g cooperativity of about $N_\alpha^0 \approx 15$ monomeric units

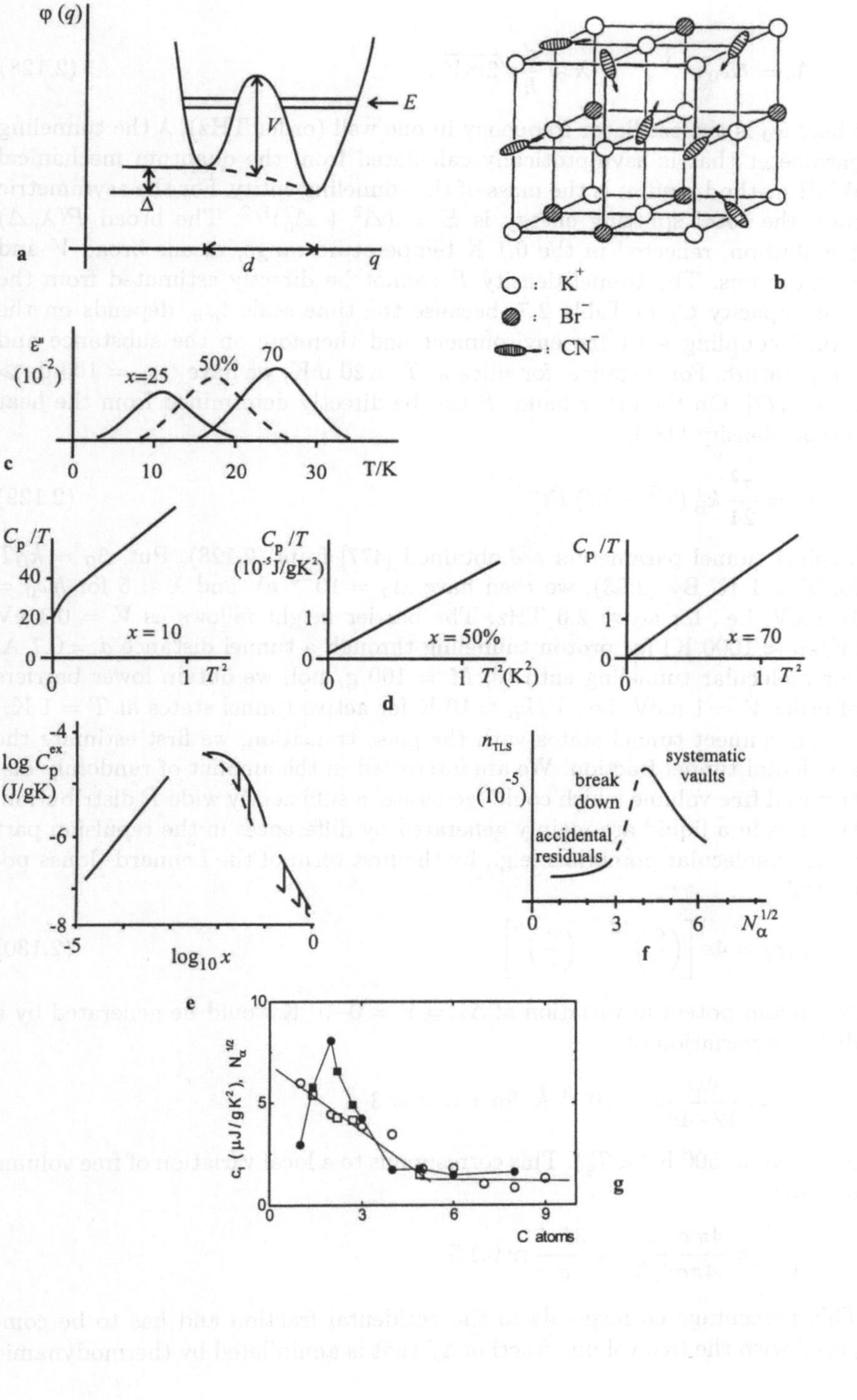
φ (q)
E
V
Δ
a
d
q
K+
Br-
CN-
b
ε"
(10-2)
x=25
50%
70
c
0
10
20
30
T/K
Cp /T
40
20
0
x = 10
T2
Cp /T
(10-5J/gK2)
1
x = 50%
T2(K2)
Cp /T
x = 70
d
log Cpex
(J/gK)
-4
-6
-8
-5
0
log10 x
e
nTLS
(10-5)
systematic
vaults
break
down
accidental
residuals
3
6
Nα1/2
f
cr (μJ/gK2), Nα1/2
10
5
9
C atoms
g

$$\Delta_0 = \hbar\omega_0 \, e^{-\lambda} , \qquad \lambda \approx \frac{d}{\hbar}\sqrt{2mV} , \tag{2.128}$$

where ω_0 is the oscillator frequency in one well (order THz), λ the tunneling parameter that is asymptotically calculated from the quantum mechanical WKB method, and m is the mass of the tunneling entity. For the asymmetric case, the total splitting energy is $E = (\Delta^2 + \Delta_0^2)^{1/2}$. The broad $P(\lambda, \Delta)$ distribution, reflected in the 0.1 K temperature range, means broad V and Δ variations. The tunnel density $\bar{P}$ cannot be directly estimated from the heat capacity C_p of Table 2.7, because the time scale $t_{\min}$ depends on the tunnel coupling with the environment and therefore on the substance and temperature. For instance, for silica at $T = 20$ mK, we have $t_{\min} = 100\ \mu s \gg 1/\omega_0$ [477]. On the other hand, $\bar{P}$ can be directly determined from the heat release density [487],

$$\dot{Q} = \frac{\pi^2}{24} \, k_B^2 \, (T_i^2 - T_0^2) \, \bar{P} t^{-1} . \tag{2.129}$$

Typical tunnel parameters are obtained [477] from (2.128). Put $\Delta_0 = k_B T$ for $T = 1$ K. By (2.93), we then have $\Delta_0 = 10^{-4}$ eV and $\lambda \approx 5$ for $\hbar\omega_0 = 10$ meV, i.e., for $\omega_0 \approx 2.6$ THz. The barrier height follows as $V = 0.1$ eV ($V/k_B \approx 1000$ K) for proton tunneling through a tunnel distance $d = 0.7$ Å. For molecular tunneling entities, $M = 100$ g/mol, we obtain lower barriers of order $V = 1$ meV, i.e., $V/k_B \approx 10$ K for active tunnel states at $T = 1$ K.

To connect tunnel states with the glass transition, we first estimate the accidental tunnel fraction. We are interested in the amount of randomly distributed free volume which could 'generate' a sufficiently wide P distribution. Barriers in a liquid are mainly generated by differences in the repulsion part of intermolecular potentials, e.g., by the first term of the Lennard–Jones potential

$$\varphi(r) = 4\varepsilon \left[\left(\frac{\sigma}{r}\right)^{12} - \left(\frac{\sigma}{r}\right)^{6} \right] . \tag{2.130}$$

A random potential variation of $\Delta\varphi = V = 0$–10 K would be generated by a distance variation of

$$\Delta r' = \frac{\Delta\varphi}{12 \cdot 4\varepsilon}\sigma \approx 10^{-3} \text{ Å} \quad \text{for } r = \sigma = 3 \text{ Å} ,$$

and $\varepsilon/k_B = 500$ K ($\approx T_g$). This corresponds to a local variation of free volume of order

$$\frac{\Delta V'}{V'} \approx \frac{4\pi\,\sigma^2 \Delta r'}{4\pi\sigma^3/3} = \frac{3\Delta r'}{\sigma} \approx 0.1\% .$$

This percentage corresponds to the accidental fraction and has to be compared with the free volume fraction Δf that is annihilated by thermodynamic optimization when the equilibrium liquid is cooled. The latter corresponds to the systematic tunnel fraction. Our estimate of $\Delta f \approx 1\%$ in Sect. 2.2.6 is about one order of magnitude larger than the 0.1% just estimated for the accidental fraction from the Lennard–Jones potential.

The $\Delta V'/V' = \Delta f = 1\%$ value is too large to be overlooked by optimization from self-organization during cooling. Such an amount, if at first regularly distributed over the sample, must be redistributed by optimization in the cold liquid. This indicates that the systematic part is important to get the usual tunnel densities and the broad parameter distribution observed in glasses.

As mentioned above, we include dynamic heterogeneity. The broad variation of tunnel parameters will be connected with a local concentration of free volume, in our approach near the Glarum–Levy defect of dynamic heterogeneity. Without the corresponding vaults (Sect. 2.2.6), we expect only the accidental tunnels corresponding to small tunnel fractions.

The number density n_0 of tunnels can be estimated from the experimental $\bar{P}$ values and from the estimated $V/k_B \approx 10$ K barrier height:

$$n_0 \approx V\bar{P} \approx k_B \times 10\ \mathrm{K} \times 10^{45}/\mathrm{Jm}^3 = 1.4 \times 10^{23}/\mathrm{m}^3 ,$$

corresponding to the n_0 value of (2.126) and therefore to a tunnel fraction of $n_{TLS} \approx 10^{-5}$ per molecule. From the shear FDT for independent or isolated tunnels, expressed by (2.80), we know that the tunnel distance is then of order $d \approx 1$Å. The tunneling entities are single molecules or small groups of molecules. The question is therefore: can the vault picture of glass structure (Sect. 2.2.6) explain a systematic tunnel fraction $n_{TLS} \approx 10^{-5}$ and the broad distribution P of tunnel parameters?

Remark (iv). To connect the general discussion with certain molecular pictures, let us briefly discuss, in this and the next three paragraphs, a concrete series of orientational glasses each with a given chemical disorder. Consider the ionic crystal $(KBr)_{1-x}(CKN)_x$ system (see [488,489,491] and Fig. 2.26b). The cyanide group is highly nonspherical and has a large dipole moment. Two kinds of disorder could in principle be responsible for the glass transition: the chemical disorder (Sect. 2.2.8) and the randomness of the CKN dipole orientations. For a cyanide content $x \lesssim 0.01\%$, the CKN events (180° flips) are isolated, for $x \approx 1\%$ the CKN flips are cooperative at least above the glass temperature T_g, and for $x \approx 50\%$, the glass is 'quasi-ordinary'. There is no antiferromagnetic order below $x \approx 75\%$.

For $x \lesssim 0.01\%$, we know the single, isolated tunnel process in this orientational glass exactly [492]. The flip coordinate is an angle. A Schottky-like hump in heat capacity is actually observed. Such knowledge does not exist for particle movements in glass-forming liquids of other types, e.g., in structural glasses. It was hoped that this knowledge would help procure a better understanding of the glass transition and glass properties. Could it be that there are aspects of the phenomena that can be explained by a separation of flips from the broad P distribution? This hope was not fulfilled. For large x values the covering of dynamic details by the broad P distribution was just as complete as for all other glasses.

The glass temperature in the KBrCN system is too low to allow the participation of chemical rearrangements; T_g changes from $T_g \approx 10$ K for $x = 25\%$ to $T_g \approx 20$ K for $x = 75\%$. The dynamic glass transition as detected by dielectrics in the millihertz range (Fig. 2.26c) shows loss curves with wide spectral distributions. The relative temperature dispersion is of order $\delta_\varepsilon T/T \approx 0.2$ for $x = 50\%$, much wider than expected from the usual estimate $\delta T/T \approx 0.65/\mathrm{m}$ [see (2.139) below]. This widening indicates a certain influence of static chemical disorder. The fragility m as determined by shear compliance [77] is about $m = 60$, i.e., this orientational glass is not as strong as several others ($m \lesssim 20$). The tunnel density $\bar{P}$ decreases dramatically for larger cyanide content x, from $\bar{P} \approx 4.0 \times 10^{45}/\mathrm{Jm}^3$ for $x = 0.25$ to $\bar{P} \approx 0.27 \times 10^{45}$ for $x = 0.75$. The $T < 1$ K heat capacity reflects the last remainders of the Schottky hump only for $x = 0.10$ (Fig. 2.26d). For higher cyanide content, the tunneling details are completely covered by the broad P distribution, as can be seen from the glass behavior of heat capacity $C_p/T = c_1 + c_3T^2$ for $x = 0.50$ and $x = 0.70$ (Fig. 2.26d). The small heat capacity values there indicate explicitly the small tunnel n_{TLS} fraction for the glass (see Fig. 2.26e and [490]). The specific excess heat capacity at $T_0 = 0.1$ K for the large $x = 0.70$ value is of the same order as that for the extremely small $x \approx 10^{-5}$ CN^- fraction!

Let us return to the general discussion. The first approach to tunnel states from finite molecular motions in islands of mobility was due to Johari [493]. He discussed a parallel behavior of tunnel density and β process intensity, e.g., after physical aging or from regions created in crystals by neutron irradiation. We connect the systematic part of tunnel states with the Glarum–Levy defects surviving vitrification at T_g by means of vaults in the glass structure (Sect. 2.2.6, Fig. 2.16c and d). We thus assume that enough free volume to form sufficiently small barriers is concentrated near the defects. This would be a systematic reason for tunnel states; outside defects, we expect only accidental possibilities for tunnel states. Combining this picture with the angstrom scale of tunnel entities from (2.80), we can give a qualitative answer to our above questions about the reason for the typical tunnel fraction n_{TLS}, the role of disengagement of the characteristic length from structural lengths, and the broad P distribution of tunnel parameters.

To begin with, assume a uniform distribution of barrier heights V/k_B between 0 and T_g. It is well known that this gives a factor $1\,\mathrm{K}/T_g$ for the part relevant to tunnel states. Recall that there is only one defect per cooperatively rearranging region CRR, with a cooperativity $N_\alpha(T_g)$ frozen in at T_g. This gives the factor $1/N_\alpha(T_g)$. Combining the two, we get

$$n_{\mathrm{TLS}} \approx \frac{1\,\mathrm{K}/T_g}{N_\alpha(T_g)} \tag{2.131}$$

for the systematic part of the tunnel fraction. Since $N_\alpha(T_g)$ is of order 100 for most glasses (Fig. 2.15g) [299], we actually get the order $n_{TLS} \approx 10^{-5}$. As the assumptions behind (2.131) are rough when used to compare different substances, this equation should only be applied to estimates within a particular series of chemically similar substances.

In the second place, (2.131) implies the existence of vaults ensuring that defects survive. In a series of similar substances with a large $N_\alpha(T_g)$ variation (disengaged from structure), this systematic behavior is expected for large cooperativities $N_\alpha(T_g)$ beyond the first coordination shell. On the other hand, we expect only a few accidental tunnel residuals to survive optimization during cooling if no vaults can be formed for small $N_\alpha(T_g)$, e.g., below the first coordination shell. This means that we expect a sharp drop in tunnel density near a certain 'vault-breakdown cooperativity' N_α^0 (see [319] and Fig. 2.26f). First experiments in a series of poly(n-alkyl methacrylates) including some random copolymers [319] indicate $N_\alpha^0 \approx 15$ monomeric units (Fig. 2.26g). The inclusion of copolymers is a further indication of the disengagement of dynamic heterogeneity from the liquid structure. The drop is a strong argument for a systematic reason (vaults) for the typical tunnel density. Vaults are not expected for the opposite dynamic heterogeneity: islands of immobility.

If the cooperativity N_α were calculated from the formula [143, 456, 581] of the Gibbs treatment (Sect. 3.2.1, and the formula was quoted in Sect. 2.4.2) rather than from (2.76), which is from the von Laue treatment of thermodynamics (Sect. 3.2.2), we obtain the drop in tunnel density at about 500 monomeric units. This comes from the large ratio of $1/\Delta C_p$, as used for Gibbs, to $\Delta(1/C_p) \approx \Delta C_p/\bar{C}_p^2$, as used for von Laue, when $(\Delta C_p/\bar{C}_p)^2$ becomes small. It would be hard to explain why the drop should occur at 500 monomeric units. The experimental finding of Fig. 2.26f and g thus favours the von Laue treatment of thermodynamics (Sect. 3.2.3).

Finally, since tunnel entities are small, a few angstroms in size, they can be localized at or near the center of defects. Since the defect is a consequence of the Levy distribution, with a preponderant component for Levy exponents $\alpha < 1$, it cannot be annihilated by optimization. It turns out that Glarum–Levy defects provide a large distribution of free volume as well as enough total free volume to generate wide tunnel parameter distributions. The Kohlrausch function (2.17) corresponds to Levy spectral densities with long high-frequency tails. For the spectral density of free volume fluctuations, the FDT implies

$$\Delta V'^2(\omega) \sim \omega^{-1-\alpha}, \qquad \alpha < 1, \tag{2.132}$$

where α is the Levy exponent. Large frequencies correspond to large local concentrations of free volume near the defect. Since such (2.132) distributions have expectation values neither for ω nor for $\delta\omega$, i.e., no average for a frequency or a dispersion of a frequency, we have a correspondingly wide distribution of free volumes $\Delta V'$. Since the position of the tunnel state near the defect is to a certain extent accidental, because of the disengagement mentioned above, the wide $\Delta V'$ distribution results in a wide P distribution of tunnel parameters.

In summary, it is suggested that the smallness of the tunnel fraction n_{TLS} is caused by the small probability of accidental tunnel states and, although about one order of magnitude larger, the smallness of the systematic tunnel fraction. The latter remains small because cooperativities that are too small, below the vault-breakdown cooperativity $N_\alpha(T_g) \lesssim N_\alpha^0$ (2.131), cannot build the vaults required for the systematic fraction. The broad tunnel distributions P can be explained by the divergence properties of the Glarum–Levy defects assumed necessary for the systematic fraction.

2.4.4 Mixed Alkali Effect

The mixed alkali effect [17,19,20,494,495] may be the most thought-provoking topic in the whole of glass science. The effect is observed in silicate (SiO_2), orthophosphate (P_2O_3), and metaphosphate (P_2O_5) [496] glasses containing two sorts (A, B) of alkali ions such as lithium and sodium (from Li_2O, Na_2O), and in some other systems, such as the BaO–CaO–SrO [497] and the $AgNO_3$–$NaNO_3$ [498] system. Let us mention four general features.

(i) The ionic d.c. conductivity has a deep minimum as a function of the mixed (A,B) alkali composition. This corresponds to a huge maximum in resistivity (Fig. 2.27a). The departure from any ad hoc additivity assumption is dramatic, up to a factor between 10^3 and 10^6 in the examples, although the conductivities of the pure alkali glasses are of the same order of magnitude.

(ii) The self-diffusivity of an impurity alkali ion is almost always lower than that of the majority alkali ion, regardless of the alkali size relationship. The alkali ion diameters range from 0.16 nm for lithium to 0.33 nm for cesium. The effect of minority concentration seems amplified (Fig. 2.27b). The diffusivity of the majority ion decreases by a factor of two, say, for a few percent increase in the impurity ion. We have a large amplification factor of order $2/0.02 = 10^2$.

(iii) A 'second' or 'additional glass transition' or 'mixed alkali peak' evolves between the majority-ion migration peak and the general glass transition (Fig. 2.27c). Since the migration peak is a decoupled glass transition (Sect.2.2.3), we have three glass transitions in total. The difference between the two low-T_g peaks is indicated by the difference in the ion diffusivities between A and B.

(iv) The mixed alkali effect becomes less pronounced as the temperature is raised towards the general glass temperature T_g, and at low total alkali content f_{alkali}.

A relationship between the general glass transition and the mixed alkali effect would be evident if the effect were missing in corresponding amorphous solids that are definitely produced without any glass transition. The present experimental situation seems unclear about this point.

The main problem is the two large factors mentioned above, viz., the conductivity minimum and the concentration amplification. Such large effects cannot be explained by conventional structure models (Fig. 2.27d) for silicate glasses (Sect. 2.2.3, Fig. 2.12e). Naively, the SiO_4 tetrahedral network could give way slightly in the vicinity of the larger or smaller alkali ions without any dramatic consequences. Instead, we try to connect the effects with partly frozen cooperativity and the correspondingly decreased possibilities for changing the structure (Sect. 2.2.6, Figs. 2.16b and e).

Concerning the conventional aspects, we partly follow the representation due to Ingram and Funke [268]. The ion cooperativity and the Coulomb interaction for the decoupled glass transition (the lowest of the three) is accessible to models [48, 499, 500] in which disorder in the glass host is based on the presence of different structural energy landscapes for the different ions. In our approach, disorder is a consequence of the general thermal glass transition at the conventional glass temperature T_g, e.g., at 470°C. We have A defects that are different from B defects [501], with mismatch and site memory effects. An explanation of these effects should be based on the cooperativity of the general dynamic glass transition above T_g, and the consequences of partial freezing of its dynamic heterogeneity at and below T_g (Sect. 2.2.6). As demonstrated by the two lower glass transitions, the molecular situation is not completely static. An important question is the possibility for molecular rearranging of A to B defects and vice versa.

An explanation of the large factors via exponentialization of different ion activation energies near the defects seems problematic. Neither computer simulations nor molecular theories of cooperativity can at present model the details of dynamic heterogeneity, not to mention its temperature dependence for $T > T_g$ and $T < T_g$. Exact knowledge of the temperature dependence would be necessary for gauging the different activation energies at high temperature, where the α process of the general dynamic glass transition becomes local and thus possibly easier to calculate. Gauging means an exact calculation of relaxation times from activation energies. We therefore seek an 'isothermal' treatment.

Let us stress again that it is important to know which of the three glass transitions the specific effects are related to, since the properties of the three glass transitions, e.g., their fragility or cooperativity, are different.

Above the general T_g, dynamic heterogeneity is self-organized in the cold liquid. The Glarum defect diffusion model has disconnected islands of high mobility, the defects, and connected cell walls of low mobility (Fig.2.27e). The cell size corresponds to the size of a cooperatively rearranging region CRR of the general glass transition. The mobility difference $\delta \log \omega$ between the islands and the cell walls varies between one and two decades in the liquid (Table 2.3). The characteristic length of our glasses is about $\xi_\alpha \approx 1.1$ nm at

T_g [226] corresponding to a cooperativity of about $N_\alpha = 35$ average molecular units. Thermograms of mixed alkali systems [502] show that these numbers do not dramatically change with the alkali composition. As mentioned several times above, the local density difference across a cell is small, well below 1%, since the free volume regulating the mobility is always a small quantity.

The cooperative self-organization is specific to the different sizes of A and B ions because this organization is a thermodynamic optimization that must operate with the low free volume. Although spherical, the ions are represented by different letters in Fig. 2.27e, viz., A and B. This is to indicate the specific environment as a result of the optimization, causing the A and B mismatch from the different SiO_4 tetrahedral configurations there. Below the general T_g, the mobility difference across the cell increases dramatically since the cell walls become partly immobile during vitrification, and because the mobility is differently accelerated due to the nonequilibrium (Fig. 2.4b). The vault effect (Sect. 2.2.6) seems important for explaining the remaining mobility needed for the two lower glass transitions.

A total alkali content of 30 percent, $f_{\text{alkali}} = 0.3$, for example, means that there will be about 10 alkali ions per CRR cell and at least one alkali ion at or near the defect (Fig. 2.27e). Self-diffusivity and ionic conductivity in $T < T_g$ glasses (i.e., the lower glass transitions) are carried by the mobile, i.e., near-defect ions. Exchange of equal near-defect ions A–A does not involve large rearrangements because there is no site mismatch for the equal alkali ions. One-component ionic conductance can thus largely decouple (Sect. 2.2.3) from the much slower structural relaxation including the cell walls. The lowest and the general glass transition remain decoupled. For mixed A–B exchange, however, there is a mismatch that cannot easily be surmounted. Cooperativity means that the whole CRR is involved in the optimization and hence that the whole CRR cell including its walls, the cooperativity shells, now partly immobilized, must be rearranged for mixed exchange. This is time-consuming,

▶

Fig. 2.27. Mixed alkali effect. (**a**) Resistivity maximum (logarithmic!) for mixed-alkali silicate glasses containing $f_{\text{alkali}} = 33$ mol % alkali oxide [19]. (**b**) Diffusivity data in the Na_2O–Cs_2O–SiO_2 system at $T = 480°C$ [19]. (**c**) Additional (middle) glass transition near $T \approx 100°C$ in lithium-sodium-trisilicate glasses at $\nu = 0.4$ Hz [495]. The 'general' glass transition is above 400°C, $T_g > 400°C$, the 'decoupled' glass transition of the majority ion transport (migration) is below 0°C. (**d**) Two-dimensional schematic representation of structure of alkali glasses. *Upper part*: only Na^+ ions. *Lower part*: mixture of Na^+ and Li^+ ions. For some oxygen atoms, the van der Waals radii are indicated. (**e**) Glarum defect diffusion model for a total alkali ion fraction $f_{\text{alkali}} = 0.3$ and a cooperativity size of $N_\alpha \approx 35$ units, i.e., a characteristic length of $\xi_\alpha \approx 1.1$ nm. *Top*: pure A ion glass, $f_B = 0$. *Middle*: ten percent B impurity ions, $f_B = 0.1$. *Bottom*: 50/50 AB glass, $f_B = 0.5 = f_A$. *Gray areas* are the connected CRR cell walls of low mobility, and *white spots* are the disconnected islands of high mobility, that is, the defects

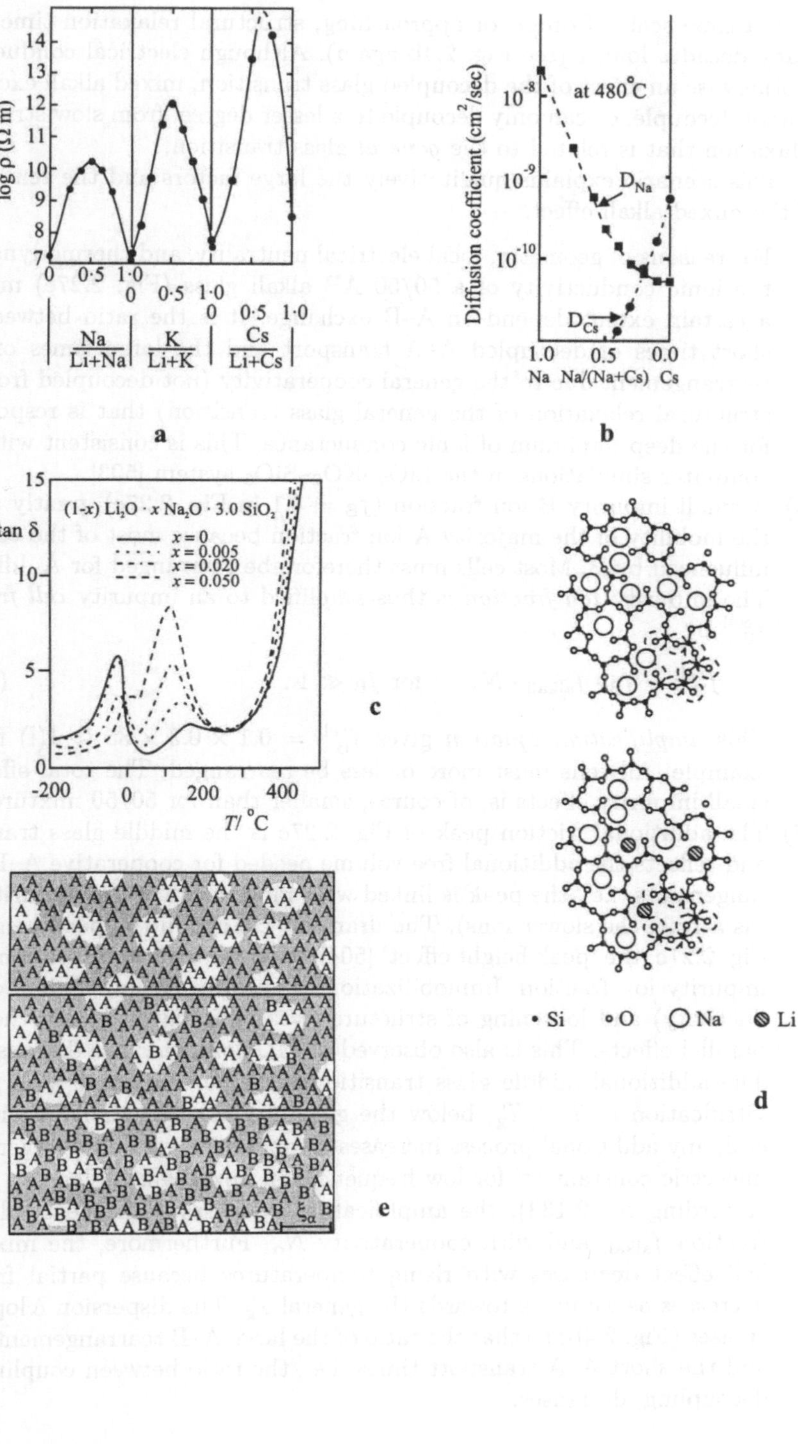

log ρ (Ω m)
14
13
12
11
10
9
8
7
0 0·5 1·0
0 0·5 1·0
0 0·5 1·0
Na / Li+Na
K / Li+K
Cs / Li+Cs
a
Diffusion coefficient (cm²/sec)
10⁻⁸
10⁻⁹
10⁻¹⁰
at 480°C
D_Na
D_Cs
1.0 0.5 0.0
Na Na/(Na+Cs) Cs
b
10³tan δ
(1-x) Li₂O · x Na₂O · 3.0 SiO₂
x = 0
x = 0.005
x = 0.020
x = 0.050
15
10
5
0
-200 0 200 400
T/ °C
c
• Si ∘ O ◯ Na ⊘ Li
d
ξ_α
e

with a time scale of order, or approaching, structural relaxation times, i.e., many decades longer (see Fig. 2.4b again). Although electrical conductivity is otherwise an effect of the decoupled glass transition, mixed alkali exchange cannot decouple, or can only decouple to a lesser degree, from slow structural relaxation that is related to the *general* glass transition.

This scenario explains qualitatively the large factors and the tendencies of the mixed alkali effect.

(i) For reasons of geometry, local electrical neutrality, and thermodynamics, the ionic conductivity of a 50/50 AB alkali glass (Fig. 2.27e) must to a certain extent depend on A–B exchange. It is the ratio between the short times of decoupled A–A transport and the large times of A–B rearrangement due to the general cooperativity (not decoupled from the structural relaxation of the general glass transition) that is responsible for the deep minimum of ionic conductance. This is consistent with MD computer simulations in the LiO_2–KO_2–SiO_3 system [503].

(ii) A small impurity B ion fraction ($f_B = 0.1$ in Fig. 2.27e) greatly affects the mobility of the majority A ion fraction because most of the cells are influenced by B. Most cells must therefore be rearranged for A diffusion. The impurity *ion fraction* is thus amplified to an impurity *cell fraction* f_B^{cell} by

$$f_B^{cell} = f_B \cdot f_{alkali} \cdot N_\alpha \,, \quad \text{for } f_B \ll 1 \,. \tag{2.133}$$

This *amplification equation* gives $f_B^{cell} = 0.1 \times 0.3 \times 35 \approx 1(!)$ for the example. All cells must more or less be rearranged. The total effect for small-impurity effects is, of course, smaller than for 50/50 mixtures.

(iii) The additional friction peak of Fig. 2.27c is the middle glass transition and reflects the additional free volume needed for cooperative A–B rearrangements, i.e., the peak is linked with movements of the impurity ions (as a rule, the slower ions). The dramatic change in the peak heights in Fig. 2.27c, the 'peak height effect' [504], expresses the amplification of the impurity ion fraction. Immobilization of ion mix (= low ionic d.c. conductivity) and loosening of structure (= additional peak) are therefore parallel effects. This is also observed in MD computer simulations [505]. The additional middle glass transition must be connected with partial vitrification at $T < T_g$, below the general glass transition. As, in general, any additional process increases compliances, we find a larger static dielectric constant (ε' for low frequency) in mixed alkali glasses.

(iv) According to (2.133), the amplification increases with the total alkali fraction f_{alkali} and with cooperativity N_α. Furthermore, the mixed alkali effect decreases with rising temperatures because partial freezing decreases as we move towards the general T_g. The dispersion $\Delta \log \omega$ decreases (Fig. 2.4b) so that the ratio of the large A–B rearrangement times and the short A–A transport times, i.e., the ratio between coupling and decoupling, decreases.

The opposite dynamic heterogeneity – disconnected islands of immobility instead of islands of mobility – would result in less dramatic effects, because the connected and voluminous cell walls would then be the carriers of ion transport. Little or no mixed alkali effect is expected from our scenario for small cooperativities, e.g., $N_\alpha \lesssim 15$. Such small cooperativities are obviously unable to form connected molecular cell walls. This situation is similar to the tunneling situation of Fig. 2.26f. Small cooperativities would be indicated by wide transformation intervals at T_g in homogeneous substances [see (2.77)].

Structure research into the ion environment (e.g., by EXAFS [506]), although it delivers a lot of detail, does not seem to have produced quantitative results for the cooperativity size so far, because the free volume is small for large cooperativities N_α. The more distinct dynamic heterogeneity is linked to very small structural variations (Sects. 2.5.2 and 2.5.3).

2.4.5 Relaxation Tails. No Deserts Between Dispersion Zones. Logarithmic Gaussian Distribution

Interest in the loss susceptibility minimum between the terahertz peak (vibration, boson peak, cage rattling), on the one hand, and the high-temperature *a* relaxation peak at lower frequencies, on the other hand, is a special case induced by mode-coupling theory MCT. We are usually interested in loss susceptibility maxima or shoulders and try to find a corresponding molecular process for them. We may ask whether the wings of the maxima (Figs. 2.2e and 2.15c) belong to the same process as the maxima or, adopting a more extreme position, whether the maxima are only 'crossovers' between two distinct wing processes. Since the loss susceptibilities between the dispersion zones are different from zero, $\chi'' > 0$, possibly larger than the wing extrapolations from the neighboring χ'' peaks, we may also ask if the processes between the peaks are distinct and independent. Finally, we should find the relations between all neighboring processes in the Arrhenius diagram for cold liquids (Sect. 2.5.2).

Consider for example the shear compliance $\tilde{J}(t) = J(t) - t/\eta$ or the more pronounced shear retardation spectrum $L(\log t)$ on the short-time side of the α relaxation (Fig. 2.28a) [106, 507]. A power law is usually observed between the cooperative α and the local Johari–Goldstein β process,

$$\tilde{J}(t) \propto t^s \ , \quad s = 0.3\text{–}0.4 \ . \tag{2.134}$$

This zone is called the Andrade zone (A1), reflecting the so called Andrade law $J(t) \propto t^{1/3}$ [508, 509] observed in many physical situations. The standard picture for small-molecule glass formers is on the left-hand side of Fig. 2.28a: the A1 zone is followed by a distinct maximum for the α process. But in some cases, as for orthoterphenyl OTP or tricresylphosphate TCP, we find a simple crossover from increasing A1 to a decreasing $\tilde{J}(t)$ or $L(\log t)$ function in the flow zone. We may then ask what the particularities of such an α process are. For entangled linear polymers such as polyvinylacetate PVAC, we observe two Andrade zones: A1 before the main relaxation, and A2 before the flow transition (right-hand side of Fig. 2.28a). It seems that, in the time domain, the Andrade zones reflect some isothermal precursors for the cooperative α or flow transition ft processes. The molecules may test some local elements for cooperativity. According to general scaling, they are faster than the cooperative modes since their modes are shorter. If local β modes are a prerequisite for operating the α precursor, the A1 Andrade zone is restricted to the region between α and β traces in an Arrhenius diagram (Fig. 2.28b). In the Glarum–Levy pattern for dynamic heterogeneity, we may consider local trials for cage door opening with local assistance of slow cell-wall parts. Similarly, if α is a prerequisite for the flow transition, the A2 Andrade zone is restricted to the region between the main and flow transitions. Note that A1 is important for physical aging below T_g; it seems difficult to separate it from the vault effect (Fig. 2.16f) there.

Let us now discuss the role played by *percolation* in the glass transition. In a Lindemann-type criterion for shear flow (Sect. 2.1.2), we interpreted the steady state shear compliance J_s^0 via the fluctuation–dissipation theorem FDT (2.13) as a minimal shear fluctuation necessary for flow. This means that, in the time domain, after the dynamic glass transition, all local shear precursors must be percolated. Does it also mean that, in the cold liquid, the α glass transition is itself a percolation threshold [510]? I think not, since no diverging length scale near the *dynamic* glass transition has so far been observed. Instead the cooperativity of the α process is of order $N_\alpha = 100$ particles, corresponding to a characteristic length of a few nanometers. We find a flow zone on the long-time side of the α process dispersion zone for all temperatures in the equilibrium above the glass temperature T_g. It seems that

▶

Fig. 2.28. Relaxation between dispersion zones. (**a**) Shear retardation spectrum L as a function of time (schematic). *Left*: Andrade zone A1 before the dynamic glass transition α. *Center*. the same, but without a distinct maximum for α. *Right*: two Andrade zones as isothermal precursors for the main transition α (A1) and flow transition ft (A2) for high-mass polymers. (**b**) Supposed position of the Andrade zones between the α and Johari–Goldstein β processes (*left*) and, for high-mass polymers, between α, β, and ft in an Arrhenius diagram. (**c**) Nagel wing on the high-frequency side of the α process in a log–log plot for dielectric loss ε'' as a function of angular frequency ω. $s' = -\mathrm{d}\log\varepsilon''/\mathrm{d}\log\omega\,|_T=$ Nagel wing exponent. (**d**) Temperature dependence of the wing exponent s'. Example glycerol, frequency $f = \omega/2\pi$. (**e**) Dynamic susceptibility χ'' for diffusion as a function of $\log\omega$ and $\log Q$, with scattering vector Q. (**f**) Definition of symbols for a phenomenological discussion of complicated loss curves. See text for details. (**g**) Interpretation of a logarithmic Gauss function for thermodynamic fluctuation transformed to a loss susceptibility via FDT and time–temperature equivalence. See text for details. (**h**) Derivation of (2.138) for time–temperature equivalence

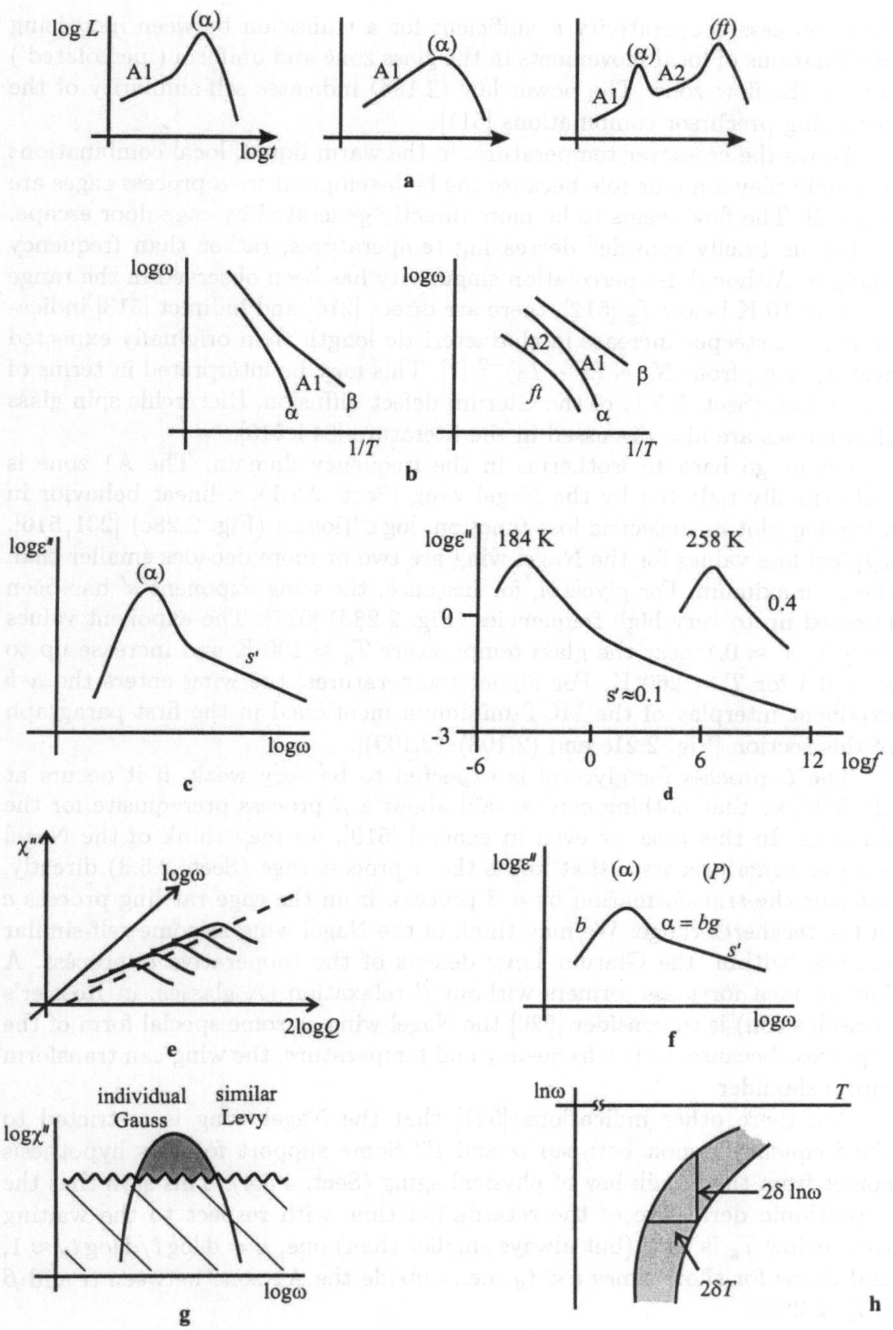
log L
(α)
A1
log t
(α)
A1
(ft)
(α)
A2
A1
a
logω
A1
β
α
1/T
logω
A2
A1
β
ft
α
1/T
b
logε″
(α)
s'
logω
c
logε″
184 K
258 K
0.4
0
s'≈0.1
-3
-6
0
6
12
logf
d
χ″
logω
2logQ
e
logε″
(α)
(P)
b
α = bg
s'
logω
f
individual
Gauss
similar
Levy
logχ″
logω
g
lnω
T
2δ lnω
2δT
h

the α process cooperativity is sufficient for a transition between increasing combinations of local movements in the glass zone and uniform ('percolated') flow in the flow zone. The power law (2.134) indicates self-similarity of the increasing precursor combinations [511].

Above the crossover temperature, in the warm liquid, local combinations may only play a minor role because the high-temperature a process cages are so small. The flow seems to be more directly generated by cage-door escape.

Let us briefly consider decreasing temperatures, rather than frequency changes. Although no percolation singularity has been observed in the range of about 10 K below T_g [512], there are direct [216] and indirect [513] indications for a steeper increase in characteristic length than originally expected near T_g, e.g., from $N_\alpha \sim (T - T_0)^{-2}$ [7]. This may be interpreted in terms of exhaustion (Sect. 2.5.1) of the Glarum defect diffusion. Hierarchic spin glass alternatives are also discussed in the literature [514, 515].

Let us go back to isotherms in the frequency domain. The A1 zone is dielectrically reflected by the Nagel wing (Sect. 2.2.1), a linear behavior in a log–log plot of dielectric loss function, $\log \varepsilon''(\log \omega)$ (Fig. 2.28c) [231, 516]. Typical loss values for the Nagel wing are two or more decades smaller than the ε'' maximum. For glycerol, for instance, the wing exponent s' has been pursued up to very high frequencies (Fig. 2.28d) [517]. The exponent values start at $s' \approx 0.1$ near the glass temperature $T_g \approx 190$ K and increase up to $s' \approx 0.4$ for $T = 260$ K. For higher temperatures, the wing enters the a–b exponent interplay of the MCT minimum mentioned in the first paragraph of this section [Fig. 2.21e and (2.108)–(2.109)].

The β process for glycerol is expected to be very weak, if it occurs at all [518], so that nothing can be said about a β process prerequisite for the A1 zone. In this case, or even in general [519], we may think of the Nagel wing as some precursor that forms the α process cage (Sect. 2.5.3) directly, without the transformation by a β process, from the cage rattling process c in the terahertz range. We may think of the Nagel wing as some self-similar process 'within' the Glarum–Levy defects of the cooperative α process. A further idea for glass formers without β relaxation (A glasses, in Rössler's classification) is to consider [520] the Nagel wing as some special form of the β process because, at low frequency and temperature, the wing can transform into a shoulder.

Are there other indications [521] that the Nagel wing is restricted to the frequency region between α and β? Some support for this hypothesis comes from the Struik law of physical aging (Sect. 4.5.4). This says that the logarithmic derivative of the retardation time with respect to the waiting time below T_g is near (but always smaller than) one, $\mu = \mathrm{d}\log t/\mathrm{d}\log t_e \approx 1$, and drops for short times $t < t_\beta$, i.e., outside the A1 zone between α and β (Fig. 2.28b).

The term 'constant loss' (reviews can be found in [52, 55]) may be used as a synonym for 'no deserts' as in the section heading, in the present case between the main process and the cage rattling process c. For an analysis of dielectric loss $\varepsilon''(\omega)$ between the high-frequency maximum for the c process and the low frequency maxima for the a process above the crossover, or the α and β processes below the crossover, we can look for a residual $\Delta\varepsilon''$ after subtracting all known maximum wings on both sides, e.g., [55],

$$\varepsilon''(\omega) = c_\beta\omega^{-\beta} + c_b\omega^{-b} + \Delta\varepsilon'' + c_3\nu^{0.3} + c_m\nu^m \, .$$

This residual is different from zero, $\Delta\varepsilon''(T) > 0$, and increases with temperature T. The $\Delta\varepsilon''(T)$ term is called *constant loss*. It seems difficult, after substraction of all wings, to associate $\Delta\varepsilon''(T)$ directly with one of the neighboring processes. Nevertheless, Ngai has shown that its temperature slope correlates with the nonexponentiality of the α process [54]. The slope $\mathrm{d}\log\Delta\varepsilon''(T)/\mathrm{d}\log T$ varies between about 5 (for $\beta_{\mathrm{KWW}} = 0.8$) and about 20 (for $\beta_{\mathrm{KWW}} = 0.5$). This also suggests looking for a property of the Glarum–Levy defect singularity as the origin for constant loss. Accepting general scaling (Sect. 2.2.5) for the frequency range and looking for a spatial scale, we have no alternative to the defect. The slope increases with the effect of the preponderant Levy component, $1 - \alpha$, $\alpha = \beta_{\mathrm{KWW}}$ (Sect. 3.6). Furthermore, let us recall that the a process can also be connected with a Levy distribution (Sects. 2.3.1 and 2.3.2).

Consider now the long-time side of the α process. The region between α and the ultraslow Fischer modes ϕ (Fig. 2.18b) is filled by some kind of spin diffusion modes, probably of preponderant CRRs (Sect. 2.2.7). This ϕ diffusion is identified [306] by a definite dispersion law,

$$\omega = Dq^2 \, , \tag{2.135}$$

where D is the interdiffusion coefficient. The intermediate scattering function for the diffusion equation is

$$S(Q,t) = S(Q)\exp\left(-DQ^2t\right) \, , \tag{2.136}$$

and the corresponding loss susceptibility in the frequency domain is a Lorentz line,

$$\chi'' = \frac{2\pi\chi x}{1+x^2} \, , \quad x = \frac{DQ^2}{\omega} \, . \tag{2.137}$$

This is a wall on the $\log Q$–$\log\omega$ plane (Fig. 2.28e). The $\chi = \chi(T)$ coefficient is defined by the FDT as

$$\chi \equiv \lim_{Q\to 0} \frac{1}{k_{\mathrm{B}}T} S(Q, t \to 0) \, ,$$

with the usual definition of the limit by means of small wave vectors Q and *times* t *in the* remaining structure-factor scattering plateau between the α

and ϕ processes [Fig. 2.17b and (2.82)]. The discussion of diffusion will be continued in Sect. 3.1.

Let us conclude this small collection of examples with two general remarks regarding phenomenology.

(i) Phenomenology needs an object that can be defined macroscopically. In the Nagel wing picture (Fig. 2.28f), we have a maximum (α) and three exponents for the wings $\{b,\ \alpha = -bg,\ s'\}$, where b and $\alpha = -bg$ are from the Havriliak–Negami HN function (2.14) with two exponent parameters b and g. The Nagel wing exponent s' is not described by (2.14). I see two variants for the interpretation. Does this curve represent a relaxation process (α) with a tightly connected precursor (P), or does it represent a long-time (low-frequency) region b where molecular individuality shines through, preceded at shorter times (higher frequencies) by a similarity region for a Levy process with Levy exponent α, itself preceded by a new, unknown, fast process with an exponent s'? Then we would have three processes and two crossovers between them. The phenomenology depends on the theoretical choice between such variants. Our present situation is difficult as the theory itself seems to be in a transition state between scaling laws and mastering of crossovers.

(ii) As a second, more radical remark let us discuss the *logarithmic Gauss function* [(2.22), Figs. 2.2e and 2.28g]. Assume that we are only interested in the χ'' maximum as such. Then the low-χ'' regions are not interesting for us (hatched in Fig. 2.28g). It may then also be of no interest that the Gaussian has no power law wings: the $\log \chi''(\log \omega)$ curve is a simple parabola. Such a wingless behavior has never been observed at the dynamic glass transition. Moreover, according to the Kramers–Kronig dispersion relation, this parabola would give overshots in the real-part curve, $\log \chi'(\log \omega)$, for $\delta \ln \omega \lesssim 2$ (cf. Table 2.2). This has also never been observed at the dynamic glass transition. That is, a Gaussian may only be interesting above the hatching (gray areas in Figs. 2.28g and h).

Two serious problems with this construction have not yet been finally solved. The first is more experimental. It is difficult to determine whether such data as heat capacity spectroscopy data for $C_p''(\log \omega, T)$, important for the determination of cooperativity, show power wings or a logarithmic Gaussian distribution with no power wings. Being some kind of diffusion process, the wings need not be entropy active. The experimental C_p'' curves, $C_p''(T - T_\omega)$ for $\omega = \text{const.}$, can so far be well approximated by Gauss [145]. The experimental sensitivity must be increased by at least one order of magnitude in order to decide this problem. The second is more theoretical. What is the relationship between our logarithmic Gaussian construction (Fig. 2.28g) and the Levy approach to Glarum defects? Roughly, can the logarithmic Gaussian construction represent the individual cooperativity shell, equal to the total cooperatively rearranging region CRR minus the Glarum–Levy defect? Or, more precisely, can our construction be related to a general scaling interpretation of the time or frequency succession discussed in connection with Fig. 2.28f? According to general scaling, shorter times (higher frequencies) correspond to shorter modes (Sect. 2.2.5, Fig. 2.15c). Then the α wing in Fig. 2.28f corresponds to the short, similar modes at the Glarum–Levy defect, and the b wing corresponds to the long, individual modes in the cooperativity shell (Sect. 2.2.5), responsible for the characteristic length. The Gaussian and Levy treatments become consistent when they are related to different length and frequency ranges. This means that the logarithmic Gaussian construction does not contradict our Glarum–Levy approach if the construction is primarily related to the low frequency b wing, possibly inclusive of the central part, i.e., if it is assumed to be representative of the cooperativity shell. The experimental finding [215, 522] that in most cases the isochronous $C_p''(T)$ curve has a symmetric peak formally allows the whole gray Fig. 2.28g peak to be included in the Gaussian evaluation.

Accepting this interpretation, we get a bridge to the usual fluctuation formulas of thermodynamics (Sect. 3.2), crucial for determining the cooperativity or characteristic length from the fluctuation formulas (2.76)–(2.78). A map from an intensive variable, such as temperature T, to the mobility $\log\omega$ would be mediated by time–temperature equivalence (Sect. 2.1.3). Since the dispersion of a Gaussian curve is given by the mean fluctuation, we see from Fig. 2.3b that, e.g.,

$$\frac{\delta\ln\omega}{\delta T} = \left.\frac{\mathrm{d}\ln\omega}{\mathrm{d}T}\right|_{\text{along}} \tag{2.138}$$

(see Fig. 2.28h). The subscript 'along' means along the dispersion zone. The symbol δT is the average temperature fluctuation of the representative functional subsystem, whilst $\delta\ln\omega$ is the $\ln\omega$ dispersion of the dispersion zone. This also means that the power-law wings are not described by large, seldom 'thermodynamic' fluctuations of the subsystems, but by the other processes, e.g., the preponderant components of Levy distributions in the high-frequency wing and the Fischer ϕ diffusion in the low-frequency wing. These have their own fluctuations relevant for the all-embracing FDT. The discussion concerning $\delta\ln\omega$ as a mobility fluctuation will be continued in Sects. 3.3 and 4.3.

One practical application of (2.138) is a rule of thumb for the relative transformation interval $\Delta T/T$ at the thermal glass transition $T = T_{\mathrm{g}}$. Note that the variation of $\delta\ln\omega \approx 1/\beta_{\mathrm{KWW}}$, $\beta_{\mathrm{KWW}} = 0.5$–$1.0$, is much smaller than the variation of the fragility, $m = 20$–200. Take $\delta\ln\omega = 3/2$ for a reasonable ad hoc estimate (Table 2.3). Use the definition (2.3) of fragility as a slope along the dynamic glass transition, $m = \mathrm{d}\log\omega/d\ln T$ at $T = T_{\mathrm{g}}$. Then, with $\ln = 2.3\log$, we have

$$\frac{\delta T}{T_{\mathrm{g}}} \approx \frac{0.65}{m}\,. \tag{2.139}$$

For a DSC heating curve, take the transformation interval $\Delta_h T$ as the temperature interval in which $C_p(T)$ increases from $C_p^{\mathrm{glass}} + 0.16\Delta C_p$ to $C_p^{\mathrm{liquid}} - 0.16\Delta C_p$. We find $\Delta_\mathrm{h} T \approx 2.0\delta T$ [317,523] and, from (2.139),

$$\frac{\Delta_\mathrm{h} T}{T_\mathrm{g}} \approx \frac{1.3}{m}, \quad \frac{\Delta_\mathrm{c} T}{T_\mathrm{g}} \approx \frac{2.9}{m}, \tag{2.140}$$

a useful relation between the transformation interval $\Delta_\mathrm{h} T$ at the glass temperature and the fragility m. The second equation in (2.140) follows from an analogous treatment for cooling using $\Delta_\mathrm{c} T \approx 4.5\delta T$ and (2.139). Shorter glasses also have smaller transformation intervals. Much wider transformation intervals than those given in (2.140) indicate additional reasons, e.g., from some structural nanoheterogeneity, as for folded proteins.

2.4.6 Glass Formation

Naturally, glass formation is important for glass technology and sample preparation in laboratories or computers. Consider a given substance or a given intermolecular potential. Can a glass be made by cooling, by densification, or by solvent extraction, etc., and if so, what rates must be used, and so on?

Since the dynamic glass transition is a typical relaxation for all liquids, glass formation means avoiding the intervention of a phase transition, usually crystallization. Several substances cannot crystallize because they cannot form a small unit cell, e.g., random copolymers or atactic homopolymers. Otherwise, glass formation often means avoiding crystal growth by applying moderate or high cooling rates and, as a rule, avoiding crystal nucleation, or avoiding a phase separation of mixtures into phases with good crystallization [17,19,20,524]. It must be checked that dynamic heterogeneities obtained from computer simulation are free of partial or aggregated ordering in the direction of a new phase.

The aim of this section is to bring some glass transition ideas to the fore with a future research objective in mind.

The rate I (nuclei/m^{3}s) for homogeneous nucleation usually has its maximum at lower temperatures than the maximum for the crystal growth rate G (m/s) (see Fig. 2.29a). This can be used for specific temperature–time programs to suppress or promote crystallization. In general, both maxima can be explained by two common tendencies. Firstly, at high temperatures, but of course below the crystal melting temperature $T < T_\mathrm{m}$, we have a thermodynamic potential difference that causes the system to crystallize or nucleate. The chemical potential for the crystal is lower than that for the liquid. This is the gain of crystallization. On the other hand, some activation is needed to overcome a barrier since the molecular order is different in both phases. The transition from melt to crystal can, for example, be macroscopically described by some interface energy. This is the cost of crystallization.

The thermodynamic gain is proportional to the volume, whilst the cost is, in the example, proportional to the transition interface. The gain/cost ratio is higher for crystal growth than for crystal nucleation, which explains the higher growth-maximum temperature. Secondly, at low temperatures, the increasing viscosity or decreasing diffusivity of the melt causes both rates to decrease. The transport and reorientation of molecules into the transition state needs more time at lower temperatures.

The peak shape can be described approximately by a product of two exponential factors, with different temperature dependencies for the thermodynamic (*a*) and the transport (*b*) activation. In the case of polymers, for example, we have the Turnbull–Fisher equation [26, 525]

$$I \approx I_0 \exp\left[-\frac{A}{T\Delta T} - \frac{B}{T - T_0}\right] . \tag{2.141}$$

The first term of the example expresses a specific nucleation regime where A is a constant and $\Delta T = T_\mathrm{m}^0 - T$ is the undercooling, with an ideal melting temperature T_m^0. In general, this term is different for different nucleation, and hence also for different crystallization regimes, and subsumes the thermodynamic cost/gain barrier. The second term is a VFT equation for low transportability at low temperature, with T_0 the Vogel temperature. The low temperature wings (*b*) of the nucleation or growth curves are therefore determined by the dynamic glass transition.

The overall situation can be explained by means of the *ttt diagram* (time, temperature, transition, Fig. 2.29b and Fig. 1.1 of the Introduction). For example, we may consider the transition from the melt. The time needed for crystallization of 1 ppt or 1 ppm of the volume, at the temperature indicated and for the technologies considered (in total, the transition criterion), is described by a nose. In a way, this nose is a map of the technological mix of growth and nucleation from the growth rate and nucleation rate curves into the transition criterion used. The upper part of the nose mainly represents the thermodynamic factor (*a*), and the lower part the transport factor (*b*) in (2.141). 'In the nose', crystallization is larger than the above 1 ppt or 1 ppm. The dynamic glass transition is defined by its trace in Fig. 2.29b, for example, a relaxation (or retardation) time, or a reciprocal fluctuation frequency as a function of temperature. The low temperature curves are nearly parallel because all curves are described by the same or similar VFT equations, for the (*b*) part branches and for the glass transition. For simplicity, in 2.29b, the glass transition and the (*b*) part of crystallization are matched by mutual scaling of the transition criterion and the glass transition trace.

Consider such a *ttt* diagram with respect to a given technological time, i.e., a preparation time τ_prep (↓ arrow in Fig. 2.29b). Note, that the admitted technologies are restricted by the transition criterion. Case I means no crystallization. We can easily find a cooling rate, for example, a vertical line near the technological time τ_prep in the *ttt* diagram, which is always

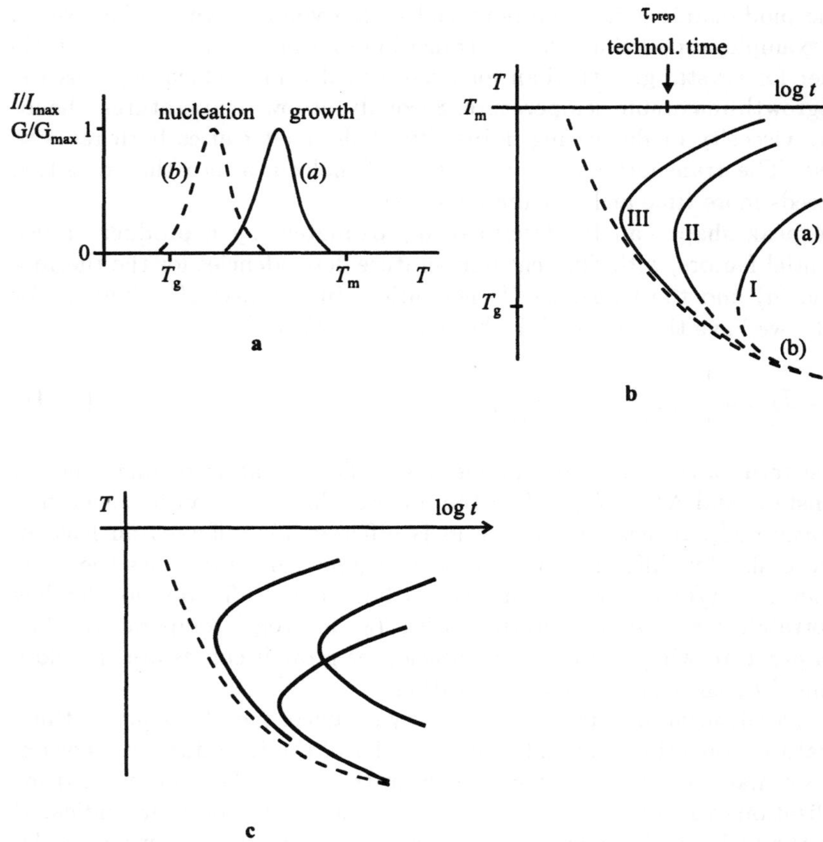

Fig. 2.29. Glass formation. (**a**) Schematic curves for crystal nucleation I and crystal growth G as a function of temperature. T_m = melting temperature, T_g = glass temperature. (**b**) ttt diagram for different crystallization noses with a thermodynamic part (a) and a transport part (b). The latter is matched here with a 'given' trace of the dynamic glass transition (*broken line*). The three variants I, II, III are explained in the text. Briefly, I corresponds to a 'low' and III to a 'high' crystallization rate. ttt means time ($\log t$), temperature (T), and transition. (**c**) The same, but without matching, to use the freedom for discussion of empirical trends

($T_g < T < T_m$) outside the nose in the liquid. Since the crystallization rate is small (nose at large time), there is not enough time for crystallization. Case III means that crystallization cannot be avoided since all accessible cooling rates intersect the nose. Case II means that glass formation requires a fast technology, for example, using a temperature–time program – a technological line in the ttt diagram (not shown) – that goes around the nose. The details require knowledge of mutual scaling, i.e., how the glass transition and crystallization kinetics are compared [526]. An example is the reduction of both transitions to rates dT/dt, for glass transition by means of (2.79), or empirical control of the matching.

Consider now *ttt* diagrams with practical definitions for the crystallization nose and the glass transition, i.e., with no matching between the (*b*) branch and the glass transition (Fig. 2.29c). Although puzzling in this treatment, we can more easily follow general and practical rules [526, 527]. Empirical factors that favor glass formation are:

- high viscosity at the nose of the *ttt* curve,
- absence of heterogeneous nucleation sites,
- large crystal–liquid interfacial energy,
- in multicomponent systems, a large concentration change between liquid and crystal, and a deep eutectic.

Experience also shows that low entropy of fusion also favors glass formation.

The long-running issue here is the calculation of absolute nucleation and growth rates for crystallization. Apart from certain substance series, where experience facilitates formula selection, the main problems seem to be that the crystal nucleus is not a small system in the thermodynamic sense [528], and that the specific conditions for molecular transport cannot a priori be represented by bulk diffusivity $D(T)$ or bulk viscosity $\eta(T)$ in the melt. They are, below the crossover region, different transport activities (Sect. 2.2.3) with individual divergences up to two time decades near T_g. For instance [529], we could put for the growth rate either $G(T) = D(T) \cdot f(T)$ or $[1/\eta(T)] \cdot f(T)$, where $f(T)$ is the same thermodynamic factor from free energy difference between melt and crystal. Detailed analysis [529] shows that the particularities of transport properties near T_c and T_A (Fig. 2.1c) are reflected in the $G(t)$ growth curve.

The particle transition from the melt to the nucleus or to the crystal requires a specific 'delivery' of molecules for their insertion into the new phase, with respect to position and orientation. This delivery tends to fluctuate around the transition state and its optimum may be different from the possibilities favored by the nanometer cooperative movements at the dynamic glass transition in the melt. Let us remember that the steady state compliance J_s^0 in the melt corresponds to an average angle fluctuation of only 10° per particle.

2.5 A Preliminary Physical Picture in 1999

The theoretical question of what underlies the dynamic glass transition may be formulated as a dichotomy: is it dynamic or thermodynamic? The *dynamic approach* tries to find a free volume explanation either without or with negligible (or at least avoidable) recourse to the J.W. Gibbs equilibrium distribution. It appeals to the generality of the dynamic scenario on p. IX in contrast to the multifariousness of structures and may require some change in the fundamentals of statistical physics and thermodynamics of liquids. A certain fraction of Chap. 3 will be devoted to a corresponding nonconventional codification. The *thermodynamic approach* tries to set off from a spurious reflection of the glass transition, and especially the free volume pattern, in the Gibbs distribution. It is based on the inevitability of reflecting in the Gibbs distribution the fluctuating density pattern which underlies dynamic heterogeneity. Formally, the thermodynamic approach reduces to trying to explain the glass transition via subtleties of the free energy.

The thermodynamic approach is conventional and began in 1948 with the Kauzmann paradox [24]. It continued in 1956 with the role played by configurational entropy in the glass transition, as described in the J.H. Gibbs–DiMarzio model [59,133], and in 1965 with the role played by configurational entropy in glass transition dynamics in the Adam–Gibbs paper [122]. The word 'preliminary' in the section heading indicates that no final decision between the alternatives was possible in and before 1999.

The thermodynamic approach is related to the following, seemingly rigorous problem: find a microscopic model for the dynamic glass transition with a structure optimized by the Gibbs distribution as the 'thermodynamic' boundary condition and with a rational Mori–Zwanzig projection for the dynamics. It may be that this problem could not be resolved as yet because the glass transition seems mainly related to dynamic heterogeneity, which is at best collaterally connected with liquid structure. This heterogeneity is monitored by extremely small variations in structure, expressed for example by extremely small values for the free volume. The thermodynamic approach will be pursued in Sect. 2.5.1, using phase transition concepts which perhaps express the generality needed for an explanation of the dynamic scenario on p. IX.

The dynamic approach must overcome the *inevitability of the thermodynamic variant.* Assume that we can find a molecular structure with intermolecular potentials that can model the dynamic scenario of the glass transition upon the following condition: the molecular structure is constant and the intermolecular potential V depends only on the configuration $\{q\}$, not the momenta $\{p\}$, i.e., $E(p,q) = E_{\text{kin}}(p) + V(q)$. Then the classical Gibbs distribution $\rho\{p,q\} \propto \exp[-E(p,q)/k_{\text{B}}T]$ separates into the Maxwell factor ρ_p and the Boltzmann factor ρ_q, $\rho\{p,q\} = \rho_p \cdot \rho_q$. The structure is thus obtained from an optimization of the configuration integral $Q\{q\} \propto \int \mathrm{d}q \exp[-V(q)/k_{\text{B}}T]$ alone. If dynamic heterogeneity is reflected in a concomitant free volume or density pattern, then this pattern must be calculable from $Q\{q\}$, i.e., from the Gibbs distribution. This would mean that, in the end, the glass transition has a thermodynamic explanation.

Pursuing the dynamic approach, we should ask: where could we change the fundamentals of statistical mechanics? The alternative to the Gibbs treatment to thermodynamics is the von Laue treatment [126, 530]. This will be described in Sect. 3.2. The following difference is important for the glass transition: in the von Laue treatment, entropy and temperature are conjugate partners of linear response, analogously to volume and pressure. Heat and work are treated equally with respect to their conjugate variables. This is not valid if the Gibbs approach is used for linear response. It seems possible that a new fundamental 'microscopic von Laue distribution' will be invented that is not restricted by the Maxwell–Boltzmann separation of the Gibbs distribution. Then the inevitability of the thermodynamic approach is bypassed and the glass transition may perhaps be explained solely on the basis of dynamics (starting possibly from Levy distributions for the spectral densities).

The von Laue treatment would also influence the map of MD simulations on linear response. The temperature fluctuation would not then follow from the Maxwell factor ρ_p alone, but rather from both subspaces of the phase space. That is, both $\{p\}$ and $\{q\}$ would participate in generating the temperature fluctuation. To find the map from MD simulations into thermodynamics, a special measuring process must also be modelled (see Fig. 3.6a and introduction to Sect. 3.4). Briefly, the computer experiment must model the way an input of minimal work into a freely fluctuating subsystem affects entropy and temperature as conjugate variables of heat, where heat is the internal energy minus all relevant kinds of work. I think that it may be possible to deduce Glarum–Levy defects from computer simulations.

Where else could the fundamentals be modified? The phenomenological description suggested (Sects. 2.2.5–7, 2.4.3–4) that the Levy treatment of Glarum defects should be built into the statistical physics of liquids. The compressibility paradox (Sect. 2.2.7) indicated that the Levy distribution (e.g., for the spectral density of free volume fluctuations) is not a priori well suited to conventional statistical mechanics. The problem is the consistency between a Gaussian distribution, an inevitable attribute of conventional thermodynamic subsystems [345], and a Levy distribution for Levy exponents $\alpha \leq 1$ as a widespread attribute for relaxation. A possibility for the future consistency of the two distributions was suggested, via general scaling, in Sect. 2.4.6. The manageability of this consistency will be pinned down to non-infinitesimal fluctuations in small representative subsystems (Sect. 3.2.3).

The hope for a non-contradictory insertion of both modifications, von Laue and Levy, into the otherwise well-established parts of statistical physics is based on the possibility that the fluctuation–dissipation theorem FDT remains unchanged if it is considered as an equation for a general quantum mechanical experiment. This suggests considering the FDT as the basis for thermodynamics (Sect. 3.4).

Is there an experimentum crucis that can decide between the alternatives? I think there is. The most striking result from a von Laue evaluation of heat capacity spectroscopy (HCS) data is the disengagement of dynamic heterogeneity from structure (Sect. 2.2.5). For moderate liquids, the characteristic length of the glass transition (cooperativity) tends smoothly [216] from a few nanometers near the glass temperature T_g to a few parts of a nanometer near the crossover temperature T_c [Fig. 2.23a (6)]. In and above the crossover region, $T \gtrsim T_c$, cooperativity is obviously [209] restricted to the a-process cage door, $N_a \approx 1$. I cannot see any way of modelling such a temperature dependence of characteristic length via the Gibbs distribution. In addition, the Gibbs evaluation of the HCS data [$1/\Delta C_p$ instead of $\Delta(1/C_p)$, see Sect. 2.4.2] would give much larger lengths. The confirmation of the small lengths by dynamic neutron or X-ray scattering would be strong evidence for the von Laue approach to thermodynamics [457, 615] and, to a certain extent, for the dynamic approach to the dynamic glass transition. Since the scattering contrast in pure substances is expected to be small, it seems unhelpful to concentrate on the structure factor $S(Q)$. Instead, the traces of the a, α, and β processes in the Arrhenius diagram should be followed by dielectric or calorimetric experiments and compared with the iso-Q lines from dynamic scattering. This Kahle approach for finding the characteristic length in dynamic scattering experiments was described in Sect. 2.3.1, Fig. 2.19b. A consistent evaluation method for dielectric data will be substantiated in Sect. 3.1. The best way to decide the glass transition issue would be a direct comparison of characteristic length from periodic calorimetry (e.g., by photopyroelectric methods [531]) with information about density fluctuation patterns from dynamic scattering (e.g., by the Kahle approach) in the crossover region of the same substance. This should be feasible within about 5 years.

We shall now continue the general discussion about the relation between dynamics and structure that started with the density contrast problem and the disengagement of characteristic length from structure lengths in Sects. 2.2.5 and 2.2.6. Let us make five remarks before announcing the contents of the present section.

Remark (i): Radical Definitions. The present discussion of dynamic heterogeneity in the literature often results from the strong definition by the Mainz school of glass transition [120, 168, 290, 300], see also [260]. Radical definitions (Sect. 2.2.5) that even eliminate the spatial aspects of this heterogeneity are certainly helpful for planning decisive experiments, but are not so helpful when integrating in other facets of the glass transition [139]. Thinking in terms of structure–property relations, spatial aspects of dynamic heterogeneity must be monitored by extremely small changes [see (2.150)] in the liquid structure, as mentioned above.

Remark (ii). Extreme Methods. In the separation model mentioned above for the thermodynamic approach, the structure is monitored by the potential energy $E(q)$ alone, well separated from the kinetic Maxwell factor. As also mentioned above, the Gibbs distribution is then self-consistently optimized by the configuration integral without any help from dynamics. Extremely fine and specific variational methods for finding the minimum of free energy would be required to obtain the extremely small free volume changes responsible for the thermodynamic boundary condition of glass transition dynamics.

Remark (iii). 'Too Small' Characteristic Lengths. Assuming that dynamic heterogeneity is really characterized by a fluctuating Glarum–Levy defect pattern, then in the thermodynamic approach, its spatial free volume pattern must inevitably be reflected in the structure factor $S(Q)$ for scattering. Hence, in the above $E(q)$ example, the 'intermolecular' potential energy must generate a characteristic length $\xi_\alpha(T)$ that more or less continuously increases from about 0.5 nm above and near T_c up to about 3 nm at T_g. This increase is (Sects. 2.2.5 and 2.2.6) disengaged from all characteristics of structure, e.g., the first coordination shell(s) or a correlation length as estimated from the width of the first sharp peak in $S(q)$. I do not think that it is possible to find a molecular $E(q)$ function for a Gibbs distribution that can reflect such a dynamic heterogeneity. It seems important that linear response is solely determined by fluctuations (FDT). The fluctuations are restricted in the Gibbs distribution for the thermodynamic approach and are 'free' in the von Laue treatment for the dynamic approach.

Remark (iv). Phase Transition at the Crossover Temperature T_c? The experimental quasistatic heat capacity $C_p(T)$ for glass formers displays no special features [532] or only weakly defined features [340] (bends, steps) at T_c (Fig. 2.30a) [533]. The thermodynamic approach would mean that all structural effects which accompany our dynamic pictures, such as cages for the a process at $T > T_c$ and Glarum–Levy defects for the α process at $T < T_c$, must be reflected by the $C_p(T)$ curve for the equilibrium liquid. The experimental entropy effect $\delta\tilde{S}$ linked with the crossover is small. This effect can be estimated from the $C_p(T)$ peculiarity, $\Delta\tilde{C}_p$, partly with the aid of a 'transformation interval' $\delta\tilde{T}$ of order $\delta\tilde{T} = 10$ K. Such a $\delta\tilde{T}$ corresponds to a length scale of about 1 nm. The quasistatic crossover effect $\delta\tilde{S} = \Delta\tilde{C}_p \cdot \delta\tilde{T}/T_c$, for the nine examples of Fig. 2.30a, including at least one substance with crossover scenario II (BIBE), is always small, $\delta\tilde{S} < 0.01R$, with R the molar gas constant. This quasistatic effect amounts to at most a few per cent of the analogously calculated dynamic effect at the glass temperature T_g. The small crossover effect corresponds to the restriction to quasistatic processes for thermodynamic paths in the equilibrium equation of state. On the other hand, the dynamic heat capacity $C_p^* = C_p' - \mathrm{i}C_p''$ spreads the picture. Figure 2.30b is a schematic sketch of C_p' $(T; \log\omega)$ curves for scenario I. Spreading is characterized by the relaxation intensity ΔC_p which increases for good glass formers at decreasing temperatures, up to several R at T_g. The dynamic heat capacity reflects not only the average of entropy fluctuations, $C_p(T) = \overline{\Delta S^2}/k_B$, but also their spectral density [FDT, (3.109)]. Through the dynamics, many large details become visible that are hidden in thermodynamics behind a possible transition and its transition interval in the equilibrium $C_p(T)$ curves.

Remark (v). What Other Than Phase Transition? As long as one adopts the thermodynamic approach, it seems a reasonable method to make

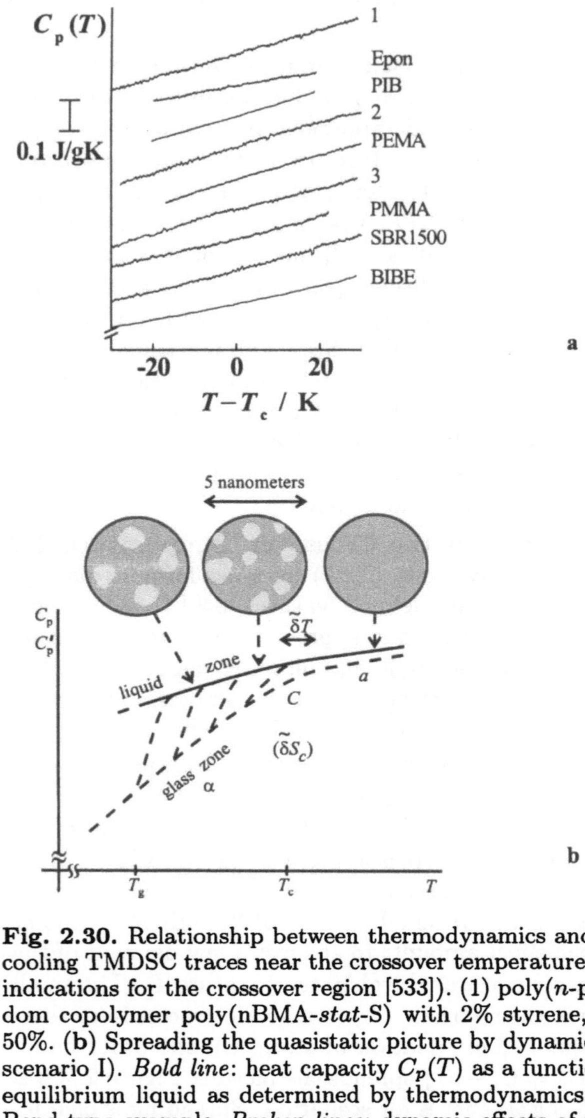

Fig. 2.30. Relationship between thermodynamics and dynamics. (**a**) Examples of cooling TMDSC traces near the crossover temperature T_c (bend type and step type indications for the crossover region [533]). (1) poly(n-pentylmethacrylate), (2) random copolymer poly(nBMA-*stat*-S) with 2% styrene, (3) poly(MMA-*stat*-EMA), 50%. (**b**) Spreading the quasistatic picture by dynamic calorimetry (schematic, for scenario I). *Bold line*: heat capacity $C_p(T)$ as a function of the temperature of the equilibrium liquid as determined by thermodynamics, $C_p = (\delta Q/\delta T)$ quasistatic. Bend-type example. *Broken lines*: dynamic effects of the dynamic glass transition from heat capacity spectroscopy. Real part $C'_p(T)$ isochrons from left to right for increasing frequencies. T_g is the glass temperature, T_c the crossover temperature, $\tilde{\delta T}$ crossover transition interval (if any), α the cooperative process in the cold liquid, C the crossover region, and a the high-temperature process in the warm liquid. *Circles* give a microscopic view of the fluctuating pattern for dynamic heterogeneity within the framework of the Glarum–Levy defect approach

a fibration (as for manifolds) of the statistical problem by introducing new and hidden (order) parameters that can induce large effects on fluctuations or dynamic heterogeneity by extremely small variations of the parameter values or functions. It then also seems reasonable to check known methods for getting general thermodynamic phenomena, e.g., phase transitions of every shade and color.

In the following sections, we shall first discuss some attempts of the thermodynamic approach to connect the glass transition with phase transitions at the crossover temperature T_c, at the Vogel or Kauzmann temperature T_0, at both temperatures, or along the pressure dependence of T_g, viz., $T_g(P)$, with some reference to Keesom–Ehrenfest relations (Sect. 2.5.1). Then, for the dynamic approach, we shall once again discuss the scenario of the dynamic glass transition in the Arrhenius diagram based on the Glarum–Levy defect model. It is much more general than might be expected from the multifarious possibilities for structural realization of its molecular origin. The generality, or even similarity, of dynamic heterogeneity is perhaps the clue to developing a final microscopic theory of the liquid state (Sect. 2.5.2). Thirdly, to be precise, simple-minded physical pictures with circles for molecules and arrows for rearrangements, with relation to a schematic representation of dynamic heterogeneity, will be presented as an example for a possible but hidden structural background underlying dynamic heterogeneity (Sect. 2.5.3). Section 2.5 and Chap. 2 close with some tables for parameters of the α process and the crossover region in different substances often investigated in glass transition research (Sect. 2.5.4).

2.5.1 Thermodynamic Approach. No Ideal Glass Transition?

The most famous homomorphisms between individual molecular variations and thermodynamic generality, similarity, or even universality are phase transitions. We shall use two characteristics of phase transitions for our discussion.

- A sharp transition temperature T_u can be generated either by homogeneity of large phase regions (as for phase transitions of first and second order) or by increasing and diverging correlation lengths ξ (as for critical states). For short-range intermolecular potentials, the latter correspond to a decrease in temperature fluctuations $\overline{\Delta T^2} = k_B T^2/C_V$, leading to a singular heat capacity at constant volume, $C_V \to \infty$. The existence of a finite transformation interval and related finite temperature fluctuations is a hard barrier to connect glass transition with sharp phase transitions. A temperature fluctuation of $\delta T = 20$ K would mean that states at 20 K above and below the transition temperature T_u also contribute to the actual transition phenomenon.
- A new phase has a new order parameter, absent in the old phase [313], or the complementary statement. In the classical example, the order parameter is the density difference $\Delta\rho$ between liquid and vapor or, including density fluctuations, the spatiotemporal function $\Delta\rho(\boldsymbol{r}, t)$.

The representation of specific methods of modern statistical mechanics [534] or dynamical density functional theories [535] lies outside the scope of our book.

Consider the pattern of dynamic heterogeneity for Glarum–Levy defects (Fig. 2.30b) from the thermodynamic standpoint. What about a corresponding order parameter? The new order parameter of the $T < T_c$ 'phase', the cold liquid, corresponds to the appearance of dynamic heterogeneity with the defects, the islands of mobility. The symmetry of the $T > T_c$ 'phase', the warm liquid, is the spatial homogeneity formed by the equivalence of all cage molecules. This symmetry is broken in the cold liquid by the pattern with a characteristic length ξ_α. On the α time scale, molecules near the defects are not equivalent to molecules in the assisting cooperativity shells around the cages.

The application of the term 'universality class' does not seem apt for the crossover region. The characteristic length becomes small at T_c. There is no sharpening at T_c because of the large, concomitant ξ_α dispersion near T_c (Fig. 2.34a, Sect. 4.3). As a rule, the Levy exponents α for the high-frequency wings of dynamic susceptibilities (Sect. 2.4.5) depend on temperature. This tends to indicate similarity (dependence on one parameter) rather than universality (fixed parameter inside a class).

The following contains a discussion of some examples of the thermodynamic approach from the literature and, sometimes, a comparison with (our) dynamic approach: crossover region near T_c, 'ideal glass transition' at Vogel temperature T_0, suggestions for order parameters, frustration, and the Prigogine–Defay ratio. Note 'two schools' in thermodynamics: some prefer functions such as enthalpy $H(T,p)$ or configurational entropy $\overline{S_{\text{conf}}(T,p)}$, e.g., Kauzmann [24], whilst the others like fluctuations such as $\overline{\Delta H^2}$ or $\overline{\Delta S^2}$, e.g., Jones [11]. The first start from $\Delta S = (C_p/T)\Delta T$, the second from $\overline{\Delta S^2} = k_B C_p$, for instance.

Crossover Region. In terms of dynamic parameters, the idealized mode-coupling theory (MCT) for the a process in the warm liquid, $T > T_c$, has a critical state of the A_2 fold type at T_c (Fig. 2.21f). This was sometimes called a 'kinetic' or 'dynamic' phase transition. A sharp transition temperature $T_u = T_c$ is obtained by extrapolating with (2.113) on the basis of dynamic homogeneity for the underlying cage picture. In other words, a diverging length scale is not necessary above T_c, since the phase is homogeneous and hence large. The new parameter of the idealized MCT is nonergodicity. This concept is of limited utility for the thermodynamic approach because the crossover is usually in the mega-to-giga hertz frequency range, not suited to any parameter on the time scale of quasistatic processes for the thermodynamic approach. Other ideas, such as hopping (Sect. 2.3.4) [158], were introduced for modified MCT versions to avoid nonergodicity and to pass the critical state continuously. This state then becomes unattainable, and such a phase transition in the off-side does not seem interesting.

In the end, an underlying conventional phase transition at the crossover temperature should be based on chemical potentials. For the thermodynamic approach, the free volume should be implied by these potentials. Since chemical potential and free volume are different concepts, the optimization of free energy could mean that some discontinuity or singularity occurs in the free volume and the corresponding mobility as a function of temperature at T_c.

The few examples of dynamic calorimetry experiments in the frequency–temperature range of the crossover region (Fig. 2.31a) [244] do not indicate any singularity (infinity) outside the admittedly large experimental uncertainty. However, a discontinuity is observed, at least for scenario I of Fig. 2.11a: the relatively weak calorimetric activity for the a process in the warm liquid dies off, and the cooperative α process in the cold liquid (2.49) sets in separately with steeply increasing calorimetric activity (Fig. 2.31a again). The contour map models a saddle between a and α for the imaginary part C_p'' of the heat capacity as a function of $\log \omega$ and T. The onset is assigned to increasing dynamic heterogeneity with small but also increasing characteristic lengths for cooperativity (2.49).

The relation between quasistatic thermodynamics and dynamic response near T_c was described in the context of Fig. 2.30b. Continuing this discussion for the dynamic approach, the right questions seems to be: why are the possibilities for the a cage process exhausted there and which new properties are to be connected with the α process onset? A dynamic answer seems possible because of the disengagement of dynamic heterogeneity from real structure. This is attempted in Sect. 2.5.3. The preliminary answers to the two questions are: exhaustion of the a process by cage closure, and α process onset by formation of a cooperativity shell of non-equivalent molecules that assists cage opening.

Vogel Temperature T_0. VFT or WLF equations for the α process indicate a finite Vogel temperature $T_0 > 0$ in the extrapolation to low temperatures. The Vogel temperature was defined dynamically by $\log \omega \to -\infty$ for $T \to T_0$. The interpretation of $T_0 = T_u$ as a certain phase-transition temperature to solve the entropy crisis (Kauzmann paradox) was supported by (2.36), $T_2 \approx T_0$, i.e., an expected identity of Kauzmann and Vogel temperatures, and by Adam–Gibbs relations like (2.28), connecting large relaxation times at low temperatures with small configurational entropy S_c there. Such a hypothetical phase transition, explained by the Kauzmann entropy crisis in generalization of the Gibbs–DiMarzio approach and seemingly unattainable only because of unattainably long experimental times, is sometimes called the *ideal glass transition* [133]. It should be repeated (Sect. 2.3.4) here that the existence of an ideal glass transition does not follow from an exact calculation of the Gibbs–DiMarzio model.

There is growing evidence that the characteristic length ξ_α for the free volume pattern increases with falling temperatures [215], i.e., with greater distance from the crossover, $T_c - T$ [215,216,536]. A fluctuation approach to dynamic heterogeneity (Sect. 4.3, (4.98), [298]) gives an extrapolated singularity at the Vogel temperature T_0, viz.,

$$\xi_\alpha \sim (T - T_0)^{-2/d} , \qquad N_\alpha \sim (T - T_0)^{-2} , \tag{2.142}$$

where d is the spatial dimension, $d = 3$ for the normal situation. The same temperature dependence was obtained by a generalized spin glass approach [537]. The fluctuation exponent corresponds to a critical exponent for the correlation length, $\nu = 2/d$ [7]. The $N_\alpha \to \infty$ limit corresponds to a sharpening of the transformation interval, $\delta T \to 0$ for $T \to T_0 = T_u$.

The length singularity (2.142) means that the liquid at T_0 is also 'ideal', in a sense: counting one Glarum–Levy defect per cooperatively rearranging region CRR of size ξ_α, most 'thermodynamic' Landau–Lifshitz subsystems have no defect for $\xi_\alpha \to \infty$. They thus belong to the cooperativity shell. We expect a certain homogeneity for the 'ideal liquid'. Transition of such an ideal liquid at $T = T_0$ into an 'ideal glass' at $T = T_0$ means doing without partly frozen dynamic heterogeneity of the glass state, that is, without the vault effect, for example (Sects. 2.1.5–6, 2.4.3–4). It seems that any thermodynamic approach to the ideal glass transition fails to recognize this effect which is all important for understanding structure and mobility in most real glasses. The

▶

Fig. 2.31. Glass transition and phase transition. (**a**) Contour map of the imaginary part of the heat capacity, $C_p''(\log\omega, T)$, as a function of log frequency and temperature for poly(n-hexyl methacrylate) in the crossover region (splitting scenario I) [244]. Observe the saddle between the a process at large frequencies and the α process at low frequencies. (**b**) Comparison of real relaxation of a quantity f_Q (*curve 1*), with a tendency to nonergodicity because of too few possibilities for molecular movements (*curve 2*), and with acceleration because of partial freezing-in (*curve 3*). (**c**) Mezard–Parisi scenario. *Bold lines* indicate the main interests of many contemporary theorists. (**d**) Schematic pictures for illustration of the frustration concept in disordered materials. In low dimensions, $d = 2$ or $d = 3$, a hard core accompanied by a specific interaction cannot be realized between many particles in the same way as between a few particles (paradise problem). (**e**) VFT or WLF extrapolated Vogel temperatures (Table 2.8) as compared with other marked temperatures for orthoterphenyl OTP (T_2 Kauzmann temperature, T_c crossover temperature, T_m melting temperature). (**f**) Heat capacity of OTP for crystalline (c), glass (g), and liquid (l) states [575]. (**g**) Bestul's calculation and extrapolation of configurational entropy S_c (Fig. 2.5d) for OTP [575]. (**h**) Experimental curves from which susceptibility steps are determined (ΔC_p here). *Continuous curve* marked by an *arrow*: from DSC curves at given cooling rate $\dot{T}$ near T_g between liquid and glass state. *Dashed curve*: from heat capacity spectroscopy at given frequency (e.g., $\nu = 100$ Hz) between the flow and glass zones of the equilibrium liquid state. *Dotted curves*: 'quasistatic' extrapolations of the liquid state or flow zone below T_g, and the glass zone above T_g

approach to statistical mechanics of the ideal glass seems easier than that of the real glass with vaults or frozen-in cages. The general difficulties of a statistical description of glasses will be mentioned briefly below (under the heading Order Parameters).

Two alternatives to (2.142) should be mentioned. An alternative approach to spin glasses [538] yields a stronger singularity at a critical temperature far

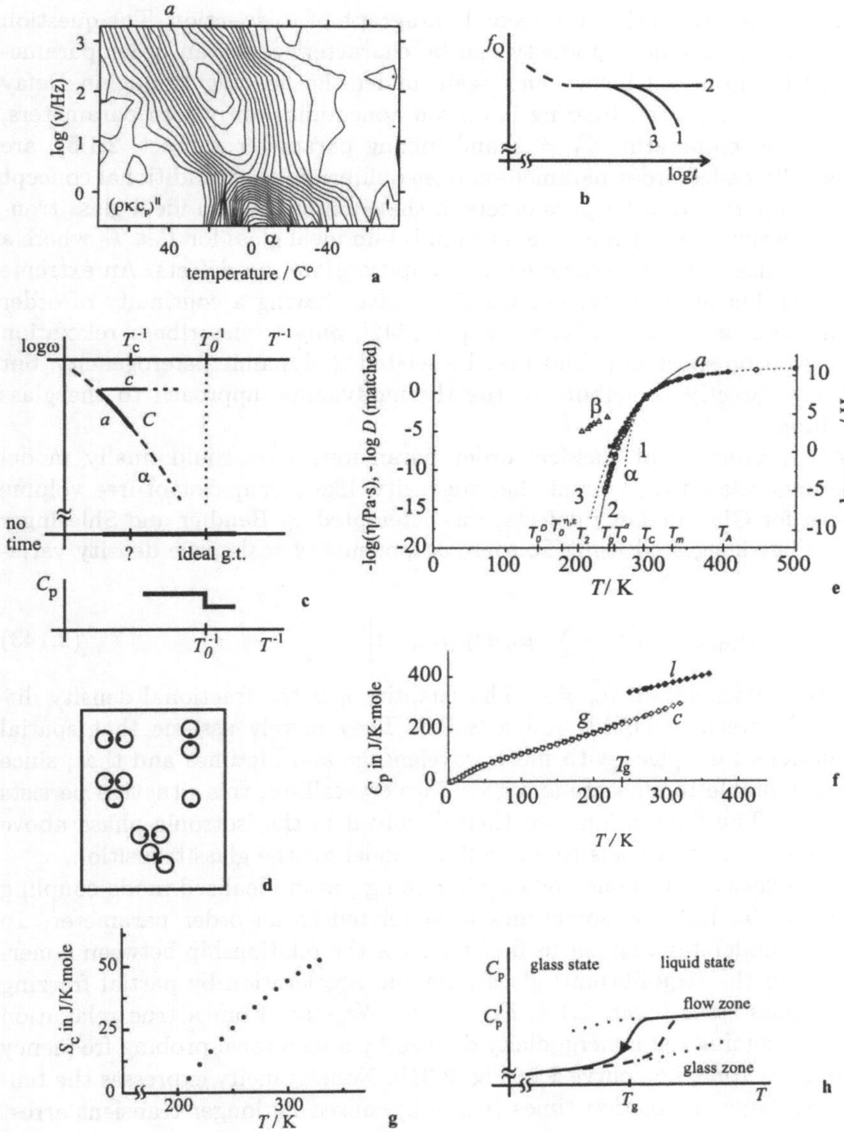

above T_0, near but below the glass temperature. On the other hand, there are indications that the real dynamic possibilities of the cooperative α process are exhausted far above the Vogel temperature (2.150). This alternative will also be discussed below.

Order Parameters. Only a few examples of this growing field of research will be presented. Attempts to apply order parameters to the glass transition constitute a long and involved story [11,23,343,539,540]. How order parameters could describe dynamic heterogeneity, such as a Glarum–Levy defect, has already been discussed in the second paragraph of this section. The question of whether or not nonergodicity can be characterized by an order parameter will be discussed below, and again under the heading Prigogine–Defay Ratio (below). Partial freezing implies a nonequilibrium whose parameters, e.g., fictive temperature $T_\mathrm{f} \neq T$ and mixing parameter x (Sect. 2.1.5), are not usually called order parameters. Nonequilibrium is an additional concept when compared to order parameters in the equilibrium. An ideal glass transition, however, would generate an equilibrium ideal glass for $T < T_0$ where a complementary order parameter seems appropriate: no defects. An extreme interpretation of order parameters [7] involves having a continuity of order parameters, or at least a large set [541, 542], only to describe a relaxation or retardation spectrum. This may be related to dynamic heterogeneity, but does not directly contribute to the thermodynamic approach to the glass transition.

A construction of classical order parameters that could finally model structures related to dynamic heterogeneity, like a snapshot of free volume pattern for Glarum–Levy defects, was attempted by Bendler and Shlesinger [343]. They suggested using Fourier components of real-space density variations,

$$\rho(\boldsymbol{r}) = \rho_{\text{liquid}}\left[1+\eta+\sum \mu_n \exp(\mathrm{i}\boldsymbol{k}_n\cdot\boldsymbol{r})\right] , \tag{2.143}$$

as order parameters if $\mu_n \neq 0$. The quantity η is the fractional density difference between a 'liquid' and a 'solid'. They merely assume that spatial fluctuations take place with finite wavelengths and lifetimes and that, since they are unable to nucleate and grow into crystallites, this situation persists below T_u. The fluctuations are then dissolved in the isotropic phase above T_u. For $\mu_n \neq 0$, this leads to a two-fluid model for the glass transition.

The transition to nonergodicity 'jamming', as in idealized mode-coupling theory MCT, has also sometimes been related to an order parameter. To avoid misunderstanding, let us first reiterate the relationship between nonergodicity in the (equilibrium) glass zone and acceleration by partial freezing in the glass state (Sect. 2.1.4, Fig. 2.4b). We start from a true relaxation after overcoming the nonergodicity defined by a reciprocal probing frequency in the glass zone, i.e., curve 1 in Fig. 2.31b. Nonergodicity expresses the tendency to larger relaxation times (curve 2) caused by longer transient arrest due to decreasing possibilities for molecular movements. Acceleration caused by additional driving forces of the partial nonequilibrium due to freezing-in means shorter relaxation times (curve 3).

The *Edwards–Anderson order parameter* [543],

$$q_{\mathrm{EA}} = \frac{1}{N} \sum_i \langle S_i(0)\, S_i(t) \rangle \ \text{ for } t \to \infty \,, \tag{2.144}$$

is directly related to nonergodicity. Applied to a nonzero $q_{\mathrm{EA}} \neq 0$ in the glass zone of an equilibrium liquid (Sect. 2.1.4), the $t \to \infty$ limit in (2.144) means a 'large' probing time $t = \tau_{\mathrm{probe}} \approx 1/\omega$. It is, however, smaller than the relaxation time for the α process, $t \ll \tau_\alpha$. The sum is over the ensemble average of a spin system $\{S_i\}$, where i counts all spin sites. A nonzero order parameter $q_{\mathrm{EA}} \neq 0$ means that the final state ($t \to \infty$) remembers, in a way, the initial state ($t = 0$). In the discussion of the Prigogine–Defay ratio below, it will be shown that nonergodicity of the glass zone cannot be described by a Landau order parameter, i.e., $q_{\mathrm{EA}} \neq \eta_{\mathrm{L}}$.

The sum over all i sites does not allow us to discuss spatial aspects of dynamic heterogeneity from the order parameter (2.144) alone. Detailed discussion of an underlying Hamiltonian H would be necessary for this purpose, e.g.,

$$H = -J \sum_{\text{neighbors}} S_i S_j + V \sum (S_i \times S_j) \{\boldsymbol{R}_i - \boldsymbol{R}_j\} + \dots \,, \tag{2.145}$$

where the second term on the right-hand side is a symbolic notation for an interaction functional depending on 'spin' properties and spatial relations between them. The actual form of H varies for the many approaches suggested: replica methods [544], spin glasses [537], frustration [545], more conventional approaches [343], and others. A further example are p-spin models [546] where the spins interact via p-spin interactions, e.g., for $p = 3$,

$$H = \sum_{ijk} J_{ijk}\, S_i\, S_j\, S_k, \,, \tag{2.146}$$

with interactions J_{ijk} that are randomly quenched and distributed over positive and negative values with zero average, $\langle J_{ijk} \rangle = 0$.

Some recent ideas for overcoming models with quenched disorder for application to the structural glass transition with self-induced disorder are discussed for example by Mezard and Parisi [547] and Wolynes [548]. The first define an order parameter as the inverse radius of a cage seen by each particle. This order parameter jumps discontinuously from 0 in the 'liquid phase' to a finite value in the 'glass phase'. In a first principles thermodynamic approach, the transition obtained is continuous (a second order transition) with a continuous free energy and a jump in the heat capacity. Such properties are indeed observed in generalized spin glasses described by Kirkpatrick et al. [549] and Mezard et al. [550], although they have quenched disorder.

The discovery of some generalized spin glass systems without quenched disorder [551] strongly suggests that this similarity is not fortuitous. The replica technique (see below) is applicable to such atomic fluid systems. This and other techniques indicate a sharp downwards jump in specific heat for the ideal glass transition [552].

The other paper [548] discussed the thermodynamic approach via a random first order transition theory at T_0. Many exactly solvable models of disordered magnetic systems have been shown to exhibit freezing into many structures [553–556]. The largest class of these also show a first order jump in a locally defined order parameter without any latent heat. This defines a 'random first order' transition. Unlike Ising spin glasses, these models possess no symmetry between local states but have long range, quenched random interactions, like *p*-spin glasses, for example [555], and exhibit a Kauzmann entropy crisis. Further, in exactly solvable statistical mechanical models [553–555], also with finite-range interactions, a dynamic transition occurs at high temperatures (T_c), coincident with mode-coupling and stability analysis, but the thermodynamic transition (ideal glass transition) does not occur until a lower temperature (T_0), where the configurational entropy of different frozen solutions vanishes [557].

The *replica method* corresponds to a 'thermodynamic fibration' of the Edwards–Anderson approach, to find hidden order parameters of the models. The two times in (2.144) are replaced by two replicas. These are different copies of the same system that are connected by a virtual interaction. The *Parisi order parameter* Q_P is based on overlaps of two copies S, S':

$$Q_\mathrm{P}(S,S') = \frac{1}{N}\sum_i S_i S_i' \,. \qquad (2.147)$$

For high temperatures, $Q_P = 0$, and for low temperatures, $Q_\mathrm{P} \neq 0$. Q_P is not based on nonergodicity.

The replica method is an idea for solving the difficult problem of giving a statistical mechanical description of the various local structures in a glass phase by using several copies of the same system. The correlation between the copies is intended to be translated into the physical properties of the glass phase. To give some details, we partly follow [558]. The need for coupling two copies comes from the need to break the symmetry to find the order parameter. For example, in an Ising ferromagnet, the symmetry broken in the ordered phase is the simple up-down symmetry. We can couple the system to a local magnetic field, and then let the field go to zero. In application to the glass transition, we do not know a priori which symmetries are broken in the glass zone or glass state. The approach therefore couples the system to a copy of itself. Each of the copies (S or S') plays the role of a staggered field acting on the corresponding sites of the other copy. For $\varepsilon > 0$, in the overlap Hamiltonian $H' = -\varepsilon \sum_i S_i S_i'$, for instance, the two copies are constrained in the same valley of the (free) energy landscape. This property is also retained for $\varepsilon \to 0^+$. The symmetry will then be explicitly broken and we may perhaps find a useful order parameter $Q_\mathrm{P} \neq 0$.

In binary mixtures with repulsive interaction, a certain transition temperature T_c [559] corresponds to the edge of a metastability region. These authors, in their thermodynamic approach, tried to show how the glass transition can be described as an ordinary phase transition within the framework of coupled replicas. Enlarging the space of parameters to include the (hidden) coupling of replicas, they find a first order transition line terminating in a critical point. A similar 'phase diagram' was also obtained [560] with an average disorder strength parameter $\bar{S}$.

In an 80/20 Lennard–Jones mixture, T_c was estimated as $T_\mathrm{c} = 0.435$ [561] and $T_0 = 0.32$ [552, 562] Lennard–Jones units.

In summary, I sometimes think that many theorists are satisfied with a theoretical picture of the glass transition that I will call a *Mezard–Parisi scenario* (Fig. 2.31c) [534]. The dynamics is applied to the Williams–Götze process and cage rattling (a and c processes) with the main ideas of idealized mode-coupling theory MCT. The rest is explained by application of the Gibbs distribution in equilibrium (no time seems better than $t \to \infty$), often concentrated to find the ideal glass transition as the final cause for all that comes between. It seems that, in this scenario, the α process is only some kind of an intermezzo between the MCT critical point C at the crossover temperature T_c and the ideal glass transition at the Kauzmann temperature $T_2 = T_0$. I have doubts about this view. I think that certain features of the α process such as the disengaged increase in cooperativity with decreasing temperature, a possible exhaustion effect due to too small a free volume far above T_0, and the vault effect below freezing-in at the conventional glass temperature T_g, making the real glass different from the ideal glass, can probably not be understood from the two limits. I believe that they justify a study of the α process as an independent and distinct dynamic phenomenon.

As a pedagogical appendix, we repeat some properties of real *spin glasses* as a popular reference system [563]. Magnetic components with a true spin and a true magnetic moment, in a large dilution of order several percent, are randomly distributed in a solid non-magnetic matrix. We are interested in the magnetic interaction upon the condition of a given, fixed constant spatial disorder of the magnetic components. Cooling the system, the spin degrees of freedom can freeze. We observe some kind of glass transition in the spin glass. Compare it with the structural glass transition (Sect. 2.2.8). In the spin glass we have a given disorder in the interaction, e.g., we have exchange integrals like

$$J_{ij}(r) \sim r^{-3} \cos(\boldsymbol{k}_\mathrm{F} \cdot \boldsymbol{r}) , \tag{2.148}$$

where $\boldsymbol{r} = \boldsymbol{r}_i - \boldsymbol{r}_j$ represent the given site disorder and $\boldsymbol{k}_\mathrm{F}$ some spatial periodicity damped by increasing distance as r^{-3}. Only the spin orientations are subject to optimization. In the structural glass, on the other hand, disorder is self-induced. The intermolecular interaction does not have any given disorder (apart from statistical copolymers, for example). Attractive and repulsive parts of the molecular potentials, mixing of different particles, or the conformation of flexible polymer chains depend on mutual distances and orientations which are optimized by the free energy of subsystems. This leads to optimized, self-induced disorder in which interactions between the particles are actually different.

In real spin glasses the frustration of spins (see below) may be dissolved by local preferences to ferromagnetic or antiferromagnetic phases, and phase transitions are *not* excluded. This may be a motive for seeking hidden phase transitions as a reason for the glass transition.

It was not my intention to discuss the many variants of generalized spin glasses in detail. We may mention once again the relationship with orientational glasses (Sect. 2.2.8). One question was (Sect. 2.4.3) whether the initial separation between given spatial disorder and free orientational dependencies facilitates the understanding of self-induced disorder in structural glasses. It seemed in the 1990s that it did not.

Frustration. *Frustration* means that a molecular orientational situation that can be saturated inside isolated groups of two or a few molecules cannot be saturated in the same way for all partners of disordered dense states (Fig. 2.31d). In the example, an optimal saturation of a (weak) chemical bonding or an optimal orientation between dipolar molecules may be realized for any isolated pair. Adding a third molecule we find frustration: the third molecule cannot find the same bonding simultaneously with each of the other two. The conflict involved in selecting the partner cannot be resolved in a stable and complete manner in disordered states.

Consider, for instance, the densification of a real gas [564]. Binary saturation of specific interaction has a dramatic effect on the second virial coefficient, a moderate effect on the third virial coefficient, and only a relatively small effect on the reduced thermodynamic equation of the dense state (like Pitzer's acentric factor [565]). Applied to the dynamic glass transition, we may perhaps connect the unresolved conflicts with free volume, increasing cooperativity with refined multiparticle resolutions, and the rare defects with localized few-particle frustrations.

Consider some newer examples of frustration recently applied directly to the glass transition or glass structure. An interesting one is a topological variant called geometrical frustration [66, 67, 566–569]. It is assumed that a glass-forming substance has a preferred local structure into which it could crystallize, for example, an isohedral structure or a structure for an appropriately curved space. In the Euclidean three-dimensional (3-d) space, however, the strain associated with such an 'ideal structure' would diverge with the size of the 3-d system. Put differently, the inability to extend the locally preferred configuration [341, 570] throughout the phase is a kind of frustration. Another formulation is that frustration is the inability to tile the common space. Such models usually favor the formation of less frustrated grains with

Table 2.8. Marked temperatures for OTP

Name			(Extrapolated) property
Vogel	T_0^{D}	145 ± 10 K	Translational diffusion [259] from α process, decoupled from η and ε
Vogel	$T_0^{\eta,\varepsilon}$	178 ± 10 K	Viscosity [79,106] and dielectrics [84] from α process
Kauzmann	T_2	200 ± 10 K	Calorimetry [575], Fig. 2.31g
Glass	T_g	243 ± 1 K	Standard DSC
Vogel	$T_0^{a,\varepsilon}$	250 ± 10 K	Dielectrics [84] from a process
Crossover	T_c	290 ± 10 K	β extrapolation to the $a\alpha$ trace
Melting	T_m	329.35 K	–
Arrhenius	T_A	≈ 400 K	Viscosity [79, 106]

more frustrated grain boundaries. Connecting more frustration with more free volume, we would get islands of immobility, instead of islands of mobility, and the mobility would be higher in the grain walls. The thermodynamic approach yields a domain model with a correlation length.

Kivelson et al. constructed such a frustration-limited domain theory for many aspects of the glass transition [545, 571–574]. The basis is a concrete effective Hamiltonian of the type in (2.145), viz.,

$$H = -J \sum_{\text{neighbors}} Q_i Q_j + \frac{1}{2} V_{ao} \sum_{i \neq j} \frac{Q_i - Q_j}{|r_i - r_j|}, \tag{2.149}$$

where Q_i is an 'order variable' that represents a coarse-grained measure of local ordering whose projections onto experimental, time-dependent quantities must be separately defined.

The Experimental Viewpoint Revisited. From an experimental point of view, the extrapolations from glass temperature T_g to Vogel or Kauzmann temperature T_0 are not unequivocal for the example orthoterphenyl OTP (Figs. 2.31e–g and Table 2.8). In general, T_0 is usually about 50 K below T_g. This temperature difference increases with decreasing fragility. The extrapolation is wide, and the assumption that the α process possibilities cannot be exhausted between T_0 and T_g is questionable. Furthermore, the value of T_0 depends on the selected activity, e.g., viscosity or translational diffusivity (Table 2.8). If the high temperature a process trace is also of WLF type, we can ask where its Vogel temperature is (T_0^a). We expect a higher one from Fig. 2.24f, $T_0^a > T_0$, if the a process is more local and the α process is more cooperative, with a larger characteristic length. For OTP, T_0^a for the a process is really higher than both values of T_0 for the α process (Fig. 2.31e). T_0^a for OTP is even near the glass temperature. However, no indication of any caloric peculiarity has so far been observed near T_0^a (Fig. 2.31f).

The extrapolated configurational entropy S_c for OTP gives a Kauzmann temperature T_2 that differs from all Vogel temperatures (Fig. 2.31g, Table 2.8). The extrapolation of data from Bestul et al. [575] is based on the definition (2.33) that the full difference between liquid and crystal is configurational. Other contributions, for example, from different vibrations in the phases [89], should be included in a more precise extrapolation. To get the Vogel temperature $T_0^{\eta,\varepsilon}$ from viscosity and dielectrics by means of the calorimetric Kauzmann construction would require the liquid heat capacities C_p^{l} to become different from Bestul's [575] extrapolation below T_{g}. Dramatic C_p^{l} effects in the direction of lower values would be needed to reach the small Vogel temperature from the translational diffusivity, $T_2 = T_0^{\mathrm{D}}$, or even to completely avoid the Kauzmann paradox.

Let us discuss some further consequences of the dynamic approach for the thermodynamic approach. Consider the increasing cooperativity of the α process. The characteristic length of dynamic heterogeneity increases with falling temperatures (Sect. 2.2.5). Counting one defect per island in the mobility pattern, the defects become scarce objects at low temperature. From the standpoint of a phase transition, if the regions of different mobility are identified with different phases, the amount of the mobile phase would become very small and this phase would be highly disconnected. The natural question that arises from this situation is whether the cooperative α process is exhausted or otherwise further modified above its Vogel temperature, due to a lack of free volume, for instance (*exhaustion* of the α process, [576] and the second paper of [202]). An experimental verification [135] from a drop in high-frequency indications for the α relaxations – also called the ideal glass transition by the authors – needs a separation from the vault effect caused by the frozen cell walls of the islands.

Another point is the estimation of phase property differences using dynamic arguments. An analysis of the Glarum defect diffusion model by means of Levy distributions leads to a 'fluctuating free volume' $v_{\mathrm{f}}(T)$ that becomes extremely small for large cooperativities $N_\alpha(T)$ and even moderate Levy exponents $\alpha(T)$ (Sect. 4.3),

$$\frac{v_{\mathrm{f}}(T)}{v_{\mathrm{f}}(T_{\mathrm{c}})} \approx [N_\alpha(T)]^{-1/\alpha(T)} . \tag{2.150}$$

(Examples: $N_\alpha = 500$, $\alpha = 1/2$, $500^{-2} = 4 \times 10^{-6}$ or $N_\alpha = 100$, $\alpha = 1/2$, $100^{-2} = 1 \times 10^{-4}$.) This small ratio may be used as a measure for the small density contrast of dynamic heterogeneity or of thermodynamic phases. A larger contrast seems possible in mixtures if the dynamic heterogeneity or the phase separation is accompanied by a redistribution of mixture components.

Prigogine–Defay Ratio. A relation between the glass transition and the second order phase transition is suggested by the steps in the heat capacity, ΔC_p, thermal expansion, $\Delta\alpha_p$, and isothermal compressibility, $\Delta\kappa_T$, occurring in both transitions. For the glass transition, we consider the rounded steps across the dispersion zone (Fig. 2.2a–b), i.e., the relaxation intensities (2.35) between the glass zone and the liquid (flow) zone in the equilibrium liquid. The difference in using the glass state instead of the glass zone is small and unimportant here (Sect. 2.1.4, Fig. 2.11c). The use of the glass zone, however, has the advantage of remaining entirely within equilibrium. Nonequilibrium caused by freezing-in is not important for the discussion of our relation. For a second order phase transition, the steps are sharp, i.e., we consider a sharp discontinuity in the caloric and mechanical coefficients $\{\Delta C_p, \Delta\alpha_p, \Delta\kappa_T\}$ between two equilibrium phases along a line $T(p)$ in the temperature–pressure (pT) plane. The quantity of interest is the *Prigogine–Defay ratio* Π. This ratio stems from a division of the caloric Keesom–Ehrenfest relation by the mechanical one [313,577]; $\Pi = 1$ for the phase transition. Excluding the glass state concentrates the discussion on the difference between nonergodicity (for the dynamic glass transition) and the Landau order parameter (for a phase transition).

Consider first the thermodynamic derivation of the Keesom–Ehrenfest relations. The assumptions are:

- A sharp line $T(p)$ in the Tp diagram separating the states 0 ($\eta_\mathrm{L} = 0$) and + ($\eta_\mathrm{L} > 0$), where η_L is the Landau order parameter.
- The existence of a common free energy function $F(V, T; \eta_\mathrm{L})$ for comparing both states via a mutual equilibrium and for using the Maxwell relation $(\partial S/\partial V)_T = (\partial p/\partial T)_V$ $\left[= \overline{\Delta T \Delta p}/\overline{\Delta T^2}\right]$ in each state. The attribute 'common' refers to equilibrium with respect to the same degrees of freedom on both sides, independently of whether or not $\eta_\mathrm{L} \neq 0$.
- We may put $\Delta V = 0$ and $\Delta S = 0$ across the line to define the second order of the thermodynamic phase transition, viz., continuity of volume V and entropy S.

To derive the Keesom–Ehrenfest equations, we consider an isobaric section ($p =$const., action of $\partial/\partial T$) and an isothermal section ($T =$ const., action of $\partial/\partial p$) of the $T(p)$ line in the pT plane. For the isobar,

$$\frac{\mathrm{d}T}{\mathrm{d}p} = \lim_{\Delta\to 0}\left(\frac{V^+ - V^0}{S^+ - S^0}\right) = \frac{\partial/\partial T(\Delta V)}{\partial/\partial T(\Delta S)} = \frac{\Delta(\partial V/\partial T)_p}{\Delta C_p/T}\,, \tag{2.151}$$

and for the isotherm,

$$\frac{\mathrm{d}T}{\mathrm{d}p} = \lim_{\Delta\to 0}\left(\frac{V^+ - V^0}{S^+ - S^0}\right) = \frac{\partial/\partial p(\Delta V)}{\partial/\partial p(\Delta S)} = -\frac{\Delta(\partial V/\partial p)_T}{\Delta(\partial V/\partial T)_p}\,, \tag{2.152}$$

whence

$$\Pi = \frac{\Delta C_p \Delta\kappa_T}{VT(\Delta\alpha_p)^2} = 1\,. \tag{2.153}$$

Since any equilibrium must allow actual fluctuational exchange of internal energy between the phases, the isobaric equation (2.151) is interpreted as heat exchange, and the isothermal equation (2.152) as exchange of work. The equilibrium is complete, with thermal and mechanical contact between the phases. It is not supposed that an extra energy related to η_L is exchanged.

Consider now the α dispersion zone in the equilibrium liquid, i.e., the dynamic glass transition in equilibrium. Steps in susceptibilities are not sharp here. We always observe a finite transformation interval. Nevertheless, take the steps of corresponding susceptibilities across the α dispersion zone, i.e., between the flow zone and the glass zone of the liquid state (Fig. 2.31g, see also Figs. 2.11c and 2.15d), and put them formally into the Π expression:

$$\Pi \equiv \frac{\Delta C_p \Delta \kappa_T}{VT\,(\Delta \alpha_p)^2} = \frac{\overline{(\Delta S)^2} \cdot \overline{(\Delta V)^2}}{(\overline{\Delta S \Delta V})^2} \,. \qquad (2.154)$$

The second equation follows from the fluctuation–dissipation theorem. The fluctuations of entropy S and volume V in (2.154) are functional to the α relaxation in the sense of Fig. 2.15d, i.e., they are the contributions of the α process fluctuations to the total fluctuations. The fluctuations are directly related to the dispersion zone. This zone has a functional independence (Sect. 3.5). It describes the transition between the glass zone in nonergodic equilibrium and the liquid zone in ergodic equilibrium. The fluctuations are for freely fluctuating subsystems of the von Laue treatment in a dynamic state characterized by the frequency and temperature (ω, T) of the dispersion zone. More precisely, they are integrals of the corresponding spectral densities over the dispersion zone.

Equation (2.154) has nothing to do with a phase transition. From the Schwarz inequality of mathematics, it follows that $\Pi \geq 1$. The Π ratio increases with the decoupling of entropy and volume fluctuations or, generalizing, with the difference between heat and work conversion in the dispersion zone for the α process.

The $\Pi > 1$ tendency of the dynamic equation (2.154) is prescribed by the Third Law of thermodynamics. The volume is a continuous parameter of the energy eigenvalues from the Schrödinger equation so that ΔV reflects the continuity aspect of fluctuations. The entropy is a fluctuation in the logarithm of a population number $\Delta\Gamma$ of states, $S = k_B \ln \Delta\Gamma$, so that ΔS reflects the discreteness aspect of fluctuations. The two fluctuations must therefore decouple for $k_B T \lesssim \Delta E$ where ΔE is the energy splitting in the relevant representative subsystem. From experience with low temperature physics, we know that the first indications for the Third Law do occur at much higher temperatures than $\Delta E / k_B$.

Experimental values for Π at T_g usually lie between 1 and 5 (Table 2.11, Sect. 2.5.4). Note once again that there is no significant difference between the glass zone and glass state susceptibilities at T_g, C_p(glass state) $\approx C_p'$(glass zone).

Historically, the route to (2.154) was long. Davies and Jones [578] described Π with a similar fluctuation formula but using enthalpy fluctuations

ΔH instead of ΔS. This follows from the Gibbs distribution not allowing temperature fluctuations (whose inclusion leads to ΔS in the von Laue treatment Sect. 3.2). Lesikar and Moynihan [579] used the entropy but with TIP gradients ∇S instead of fluctuations (TIP = Thermodynamics of Irreversible Processes). Each gradient needs some index indicating the thermodynamic path. Fluctuations of freely fluctuating subsystems, however, do not require such indices (Sects. 3.2–3.4) [18, 313]. Jäckle [580] clarified the relationship with dynamics in the dispersion zone, but he used ΔH again, instead of ΔS. Formally, (2.154) could also be used for the phase transition. Then $\overline{(\Delta S)^2}$ would mean $\Delta\overline{(\Delta S)^2}$, where the first Δ expresses the difference in $\overline{(\Delta S)^2}$ between the two phases, and so on.

The pressure dependence of glass temperature T_g – or, remaining within the zones of equilibrium states, of some dynamic glass temperature T_ω above T_g – is sometimes compared with expressions for dT_ω/dP calculated in analogy to the Keesom–Ehrenfest relations. We have a mechanical and a calorimetric comparison,

$$\frac{dT_\omega}{dP} \quad \text{vs.} \quad \frac{\Delta\kappa_T}{\Delta\alpha_p}, \tag{2.155}$$

$$\frac{dT_\omega}{dP} \quad \text{vs.} \quad \frac{TV\Delta\alpha_p}{\Delta C_p}, \tag{2.156}$$

respectively ($dT_\omega/dp \approx dT_g/dp$ for $T_\omega \approx T_g$). As mentioned above, Π is the ratio between the two right hand sides. For a second order phase transition, both sides of both equations are the same. For the glass transition, as a rule, the two sides of the mechanical comparison (2.155) are different, but those of the caloric comparison (2.156) are similar (Table 2.11 in Sect. 2.5.4).

The interpretation of this result is difficult [539, 581, 582]. Comparing the thermodynamic approach to (2.151) and (2.152) with the dynamic approach to (2.154), we see that the difference can be reduced to the question of whether or not nonergodicity can be described by a Landau order parameter. The experimental answer via (2.155)–(2.156) is no. In thermodynamics, both equilibrium phases, 0 and +, are described by a single state function $F(V, T; \eta)$ embracing the same degrees of freedom in both phases. Both phases are ergodic and the possibly slow relaxation of $\eta_L \to \eta_L$(equilibrium) can be understood by small driving forces, on the basis of this ergodicity. The difference between the two phases is defined by the Landau stability construction, $\eta_L = 0$ and $\eta_L \neq 0$. In dynamics, both the liquid zone and the glass zone belong to the same equilibrium liquid state, embracing the small frequencies below and the high frequencies above the dispersion zone for the α process. The dispersion zone, separating the two other zones, is defined by peaks in loss susceptibilities via the ω identity of the measuring equation, the fluctuation–dissipation theorem. The difference between the two zones is defined by the restrictiveness of the ω identity (no exchange between the zones, Sect. 3.5), and by the nonergodicity of one of them, the glass zone. Note (Sect. 2.1.5) that the time needed for the ergodicity concept was the probing time $\tau_{\text{prob}} \approx 1/\omega$. If a certain order parameter $\eta_{\text{L}} \neq 0$ is connected with the liquid state, it is connected with all three zones.

Nevertheless, something must be indicated by the differing degrees of fulfillment of the mechanical comparison (2.155) (usually bad) and the caloric comparison (2.156) (not usually bad). In (2.154), we reduced the issue to properties of the α process in the dispersion zone. The Keesom–Ehrenfest analogy tells us to compare it with neighboring zones. Take a simplified hopping picture of the α process in terms of failed and successful attempts to overcome barriers. The glass zone corresponds to the frequent attempts on their own being nonergodic, by comparison with the ergodic liquid zone which contains both attempts and a small number of successes. The existence of mechanical and caloric comparisons points to work and heat conversion during α process hopping. Good fulfillment of the caloric equation (2.156) means that heat conversion could be successfully estimated from a model using a Landau order parameter (replacing zones by phases), whilst the poor fulfillment of the mechanical comparison (2.155) means that work conversion might be poorly estimated from such a model. Historically, this may have been the reason for changing from free volume arguments to configurational entropy arguments for the dynamic glass transition, as initiated in the famous paper of Adam and Gibbs in 1965 [122]. This does not mean, however, that space–time aspects should be given up. It is not only because Flory has had a large influence up to now that free volume arguments are so important [39], and dynamic heterogeneity should not be reduced to non-spatial types of independence [128].

Let us discuss some examples where phenomena that appear to have volumetric relevance can still be reduced to free volume arguments. The experimental finding of $\Delta\alpha_p < 0$ and, correspondingly (2.155)–(2.156), $\text{d}T_{\text{g}}/\text{d}p < 0$ in the exceptional glass-forming alkali acetate + water system $\text{LiOAc}\cdot 10\text{H}_2\text{O}$ [711] can be explained by the structural particularities of water. Less volume for higher temperature does not exclude (Fig. 2.16b) more free room for mobility, i.e., more *free volume* there. Note that $\Delta\alpha_p < 0$ means $\alpha_p(\text{glass}) > \alpha_p(\text{liquid})$, i.e., for cross fluctuations of volume and entropy (3.38), $\overline{\Delta S \Delta V}(\text{glass}) > \overline{\Delta S \Delta V}(\text{liquid})$. The coupling of volume and entropy ('disorder') fluctuation is larger in the glass state. A similar example is B_2O_3 [712]. Adding a few percent of water increases both density and mobility. Let us further recall the reasons for a WLF equation in PVC under isochoric conditions (Fig. 2.3f) [124]. More free volume is generated by the smaller effective cores of the particles at higher temperatures.

2.5.2 Dynamic Approach: Strategy

Prominent theorists have claimed that a general molecular theory of liquids will be difficult or even impossible, for two reasons.

- There is no small parameter that can be used in a Taylor series for the strong interaction, for example, of 15 particles inside a dense disordered state.
- The intermolecular interaction in different substance classes is so different that no general picture can be achieved which includes both structure and dynamics.

Glass transition research in the last 15 years has proved, however, that a general and partly similar dynamic scenario can be observed in the Arrhenius diagram (Fig. 2.32a), at least for glass formers of moderate complexity. The scenario does not vary essentially between different substance classes, but is at most completed by additional traces and zones, as in the case of polymers (Fig. 2.32b) or liquid crystals (not shown). Occasionally, in special substances, a particular relaxation trace is missing. Examples are the Nagel wing in several orientational glasses [583], although it is observed in cyclohexanol [584]. Otherwise the β relaxation is missing in a few substances, as mentioned in Sect. 2.2.1. I think that this is mainly an activity or intensity problem for these substances, not a fundamental one. Furthermore, different scenarios are observed in the crossover region.

We may make some rather trivial remarks.

(a) There is no conceptual difference between 'normal' dynamics of a liquid and the dynamic glass transition in a cold liquid [585]. Both belong to the general scenario, and the frontier between them, the frequency and reduced temperature of the crossover region, can vary. Bearing this in mind, the dynamics of liquids is completely described by the traces in the Arrhenius diagram and their molecular dynamic models.

(b) The scenario is general although, apart from thermal frequencies in the terahertz region, there is no a priori given 'typical' time or frequency scale in the liquid. The only characteristic frequency–temperature 'point' of the Arrhenius diagram is defined by the crossover region, but this point is also not given a priori.

(c) Consider the main dispersion zone of the Arrhenius diagram: the high-temperature Williams–Götze a process and the cooperative α process. Both are usually nonexponential, i.e., they have a large mobility width, expressed for example by the stretched exponential of (2.17) with KWW exponents $\beta_{\mathrm{KWW}} < 1$ or by the mobility dispersion $\delta \ln \omega$ (Table 2.3). Nevertheless, it seems that there exist small parameters for both a and α processes: free volume, whose decrease with falling temperatures regulates cage-closing for the a process and the development of a cooperativity shell around the Glarum defect for the α process. However, the nonlinearity of mode-coupling theory for the a process (Sect. 4.1), as well as the Levy distribution approach to the Glarum defects (Sect. 4.3), both show that a Taylor series development with respect to free volume is not appropriate.

(d) The multiplicity of the scenario increases with molecular complexity, and a certain *polyamorphism* cannot be excluded for more complex liquids, e.g., for the exceptional glass former water [586, 587]. Even first order phase transitions between amorphous phases are observed. Obviously, self-organization in cold liquids can yield sufficiently different structures with different thermodynamic properties, relaxation times, etc.

(e) Several important features of the dynamic glass transition can also be obtained from a landscape discussion for a Lennard–Jones system [588], such as nonexponentiality (stretching of relaxation dynamics), temperature dependence of the stretching parameter, and breakdown of the Stokes–Einstein relation at a critical temperature. All this is possible without using the concept of dynamic heterogeneity. The stretching exponent follows there from a fit of the off-diagonal stress tensor autocorrelation function without reference to a Levy distribution with a Levy exponent $\alpha < 1$. It seems important to reiterate the quantitative experimental indications we have in favor of our characteristic length for dynamic heterogeneity, independently from the fluctuation formula (2.76): confinement effects (Sect. 2.4.2, Fig. 2.25d) and vault effect in tunnel density (Sect. 2.4.3, Fig. 2.26f).

Formally, the dynamics of liquids is a mathematical map from multifarious molecular situations to rather general dynamics characterized by a reduction in the number of parameters. Such a map was called a homomorphism (Fig. 2.1f), and the appropriate terms for a discussion are individuality (as for the molecules and structures) on the one side and similarity (as for the dynamic scenario) on the other side, with some graduation of generality between. These concepts were introduced in Sect. 2.1.1. This differs from the universality concepts suited to phase transitions. The long route followed by glass transition research seems to indicate that it is the *dynamics* of liquids which is subject to a general treatment through a homorphism map.

In this sense, the concept of structure–property relations does not seem the best starting point for understanding the glass transition.

Let us restate the question in other words. The mystery of liquids seems to be that large general effects are restricted to the dynamics. The individuality of molecular situations, i.e., specific aspects of molecular disorder in different substance classes does not prevent generality in the dynamics. The free volume generating the mobility is so small that its development with increasing temperature has only small, non-specific back-effects on the individual molecular structures and on general thermodynamic properties. It seems as if it is not the structure and not the thermodynamic properties which regulate the dynamics, but only their 'negligible' variations of the order of the free volume, variations which can hardly be detected by structure research and thermodynamic experiments, e.g., through the quasistatic heat capacity $C_p(T)$. Put in yet another way, the disengagement of dynamic heterogeneity from structure (Sect. 2.2.6) or, in other words [589], the intermittent release

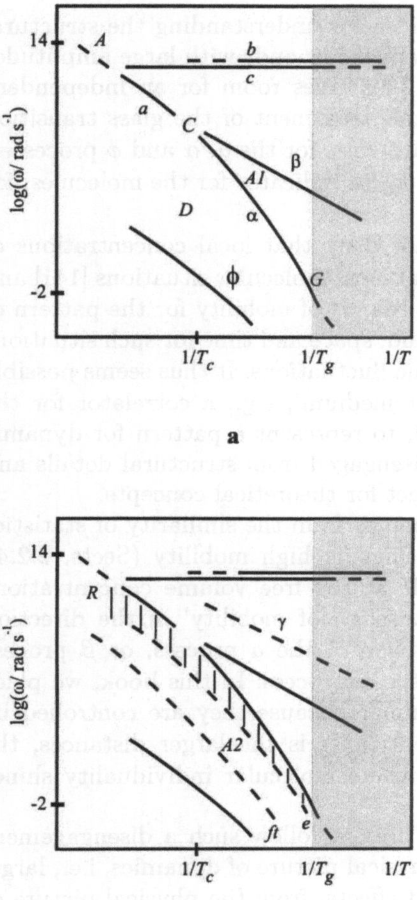

Fig. 2.32. Arrhenius diagram. General scenario for dynamics in the equilibrium liquid state (*white*) and glass state (*gray*). The considerable width of the dispersion zones is not drawn. **(a)** Standard for small-molecule glass formers. A transient to molecular thermal frequencies (Sects. 2.3.1 and 2.3.2). (b) Boson peak (Sect. 2.3.1). c cage rattling (Sects. 2.3.1 and 2.3.2). a Williams–Götze high-temperature process, i.e., escape from the cage formed by neighboring molecules (Sects. 2.3.1 and 2.3.2). C crossover region with perhaps two scenarios (Sect. 2.2.1). β Johari–Goldstein local process (Sect. 2.2.1). α cooperative process (Sect. 2.1.3). Decoupling of some transport processes (Sect. 2.2.3) is neglected. $A1$ Andrade precursor for α (Sect. 2.4.5). ϕ ultraslow Fischer modes (Sect. 2.2.7). D diffusive precursor for Fischer modes. G thermal glass transition (Sect. 2.1.4). T_c crossover temperature, T_g glass temperature. **(b)** Additional traces and zones for the example of amorphous polymers (Sect. 2.4.1). R Rouse modes (modified at low temperatures). γ side group relaxation or local relaxation from additives. e entanglement relaxation. ft flow transition. $A2$ Andrade precursor for flow transition in the plateau zone

from constraints (Fig. 2.16b), is a key idea for understanding the structural dynamics of complex systems. The dynamics responds with large amplitudes to minute variations in the structure. This gives room for an independent though, at present, non-rigorous dynamic treatment of the glass transition in three sections of Chap. 4. These treatments, for the a, α and ϕ processes, do not contain the Boltzmann constant k_B as indicator for the molecules, for instance.

To avoid misunderstanding, I do not deny that local concentrations of free volume are connected with particular local molecular situations [141] and that these situations are the carriers of islands of mobility for the pattern of dynamic heterogeneity. On the other hand, space and time for such situations change by molecular and thermodynamic fluctuations. It thus seems possible and appropriate to look for a 'transfer medium', e.g., a correlator for the a process or a mobility field $\log \omega(\boldsymbol{r}, t)$, to represent a pattern for dynamic heterogeneity whose fluctuations are disengaged from structural details and which can serve as an independent object for theoretical concepts.

Similarity of dynamics probably emerges from the similarity of statistics governing the many possibilities for achieving high mobility (Sects. 2.2.4–5). Such possibilities become accessible at the free volume concentrations on small scales. We observe a 'breakthrough of mobility' in the direction of thermal frequencies for the cage escape of the a process, or β process frequencies at the Glarum defects for the α process. In this book, we place on Glarum–Levy defects, which are similar because they are controlled by only one parameter, the Levy exponent α. It is the larger distances, the cooperativity shell for the α process, where molecular individuality shines through for the dynamics.

In summary, I think we should at present follow such a disengagement strategy (Sect. 2.2.6), separating the physical picture of dynamics, i.e., large, similar or general, and clearly arranged effects, from the physical picture of their background in structure or the Gibbs boundary, i.e., tiny, multifarious, individual, and unknown effects.

2.5.3 Molecular Pictures

We are relatively free to choose a molecular model to illustrate the a, α, and β processes since the dynamic scenario of liquids is rather general, irrespective of the multifarious molecular individualities involved in their structure and interaction (Sect. 2.5.2). The disengagement strategy does not allow solution of the inverse problem, that is, determination of liquid structure from dynamics. We shall select circles for the molecules and arrows for typical movements (Fig. 2.33). We should bear in mind two things, however, that cannot easily be prevented or realized, respectively, in such a model.

- The dynamic glass transition in such a circle model would be interrupted by fast crystallization.

- In reality, it is only small, mainly orientational or compositional variations that actually create the small free volume necessary for dynamic heterogeneity (Fig. 2.16b).

To reflect these subtleties in our pictures, dynamic heterogeneity will also be indicated by different characteristics of the circles (Fig. 2.33) or by an underlying pattern like Fig. 2.12d (Fig. 2.34a). The length scale will always be related to characteristic lengths ξ_α from entropy or density fluctuation patterns [Fig. 2.23a (6)]. For pedagogical reasons, we first refer only to the crossover scenario I of Fig. 2.11a, i.e., to the splitting scenario; the possible modifications caused by scenario II will be described later in this section. The ultraslow ϕ process (Fischer modes) does not allow a molecular picture because of its large spatial scale (Fig. 2.17a in Sect. 2.2.7, Fig. 4.3a in Sect. 4.4).

Williams–Götze or High Temperature *a* Process. In the warm liquid, approaching the crossover from above with decreasing temperature or increasing density, each molecule becomes progressively more trapped in the transient cage formed by its neighbors [376]. Many attempts within the cage are necessary before the molecules find a way out by escaping the cage, i.e., cage diffusion. Successive cage diffusion events provide the long-range diffusion which slows down drastically as the cage effect strengthens.

Each cage is composed of molecules that are themselves surrounded by cages of their own neighbors. All partial systems of cage size, and therefore all molecules, are equivalent on the *a* time scale. The cage undergoes collective distortions that may open the way for the cage diffusion event defining the *a* time scale. The *cage door* (the environment of the arrow tip in the *a* process pictures of Fig. 2.33) is a collective effect and corresponds to a temporary concentration of the ample free volume which is otherwise regularly distributed. Back flow [376] prevents holes of particle size from surviving and precludes binary changes of place. The *a* process is thus not determined by simple one-dimensional barriers, as for activated processes which are perhaps typical above Barlow's T_A temperature far above the crossover. Hopping does not seem typical for the *a* process.

As the temperature continues to decrease, or the density to increase, the opening frequency of escape paths in a cage may become small. The *a* process cages have a tendency to become closed. Hence the *a* process goes down, seems to become exhausted, and we expect an *a* process death at a 'critical', or better, a crossover temperature T_c.

Diffusion means that the given molecule does change neighbors. Since all experiments show that diffusion is not stopped in the crossover region we must find an alternative for the main relaxation below T_c.

Götze's mode-coupling theory MCT [33] describes the cage effect strengthening by a Langevin equation with a braking memory caused by coupling of increasing mode amplitudes (Sect. 2.3.2). This leads to a 'kinetic' critical temperature T_c for the *a* death. The extended MCT tries to use an alternative suggested by Goldstein [158]: a molecule can still escape from its cage by activated hopping (Sect. 2.3.4) that carries it over a local potential minimum. A corresponding extension of MCT by time derivatives leads to a continuous path in kinetic coordinates avoiding the critical point. The continuity, however, contradicts the separateness of a death and α onset observed at least for scenario I (Fig. 2.31a).

Cooperative or Low Temperature α Process. In the cold liquid, we look for an α process picture that, conceptually, also starts from homogeneity of the equilibrium liquid, i.e., from equivalence of all its partial systems. The alternative, new element is cooperativity. A cooperatively rearranging region CRR was defined by statistical independence from the environment. According to Adam and Gibbs, this is a subsystem which, upon a sufficient thermodynamic fluctuation, can rearrange into another configuration independently of its environment [122]. This subsystem is functional, i.e., related to the Fourier components of the slow α dynamics. The components shift along the α dispersion zone when temperature is changed. Furthermore, this subsystem is representative for the susceptibilities in the α dispersion zone and is freely fluctuating (no boundary conditions). One average CRR is presented in Fig. 2.33 for α and β.

Phenomenologically, a CRR is also considered as the representative region for the balance of fluctuational redistribution of free volume (Sect. 4.3). To this end, any CRR is partioned into a number of equivalent partial systems. We assume first that the balance is the only characteristic of cooperativity between partial systems, which are otherwise independent, and second, that the mobility of each partial system is controlled by its own local share of free volume. This democratic model implies that the spectral density of thermodynamic fluctuation of a random extensive variable is in the domain of attraction towards a Levy distribution (Sect. 3.6) with a Levy exponent $\alpha \leq 1$, at least for the high frequency part of isothermal loss susceptibilities. The $\alpha \leq 1$ relation follows from the relaxational character of mobility for frequencies much lower than the thermal frequencies of order terahertz (Example 1 of Sect. 3.1). This exponent depends on the activity and, especially near the crossover, the temperature.

A Levy distribution with $\alpha < 1$ implies that the random variable of one partial system inside each CRR is preponderant. This partial system has an extraordinarily high mobility and represents, in a way, a localized breakthrough to higher mobilities. Referring to the presumed locality of free volume, this means that this one partial system in any CRR has an extraordinarily high concentration of free volume. The partial systems become dynamically nonequivalent although, conceptually, we started from the equivalence of all partial systems. The preponderant partial system was called a Glarum–Levy defect (Sect. 2.2.4). To reiterate, we do not assume, but in fact find the defect, from a democratic starting point. The Glarum–Levy defects are self-organized in the cold liquid. The spatiotemporal fluctuation of the

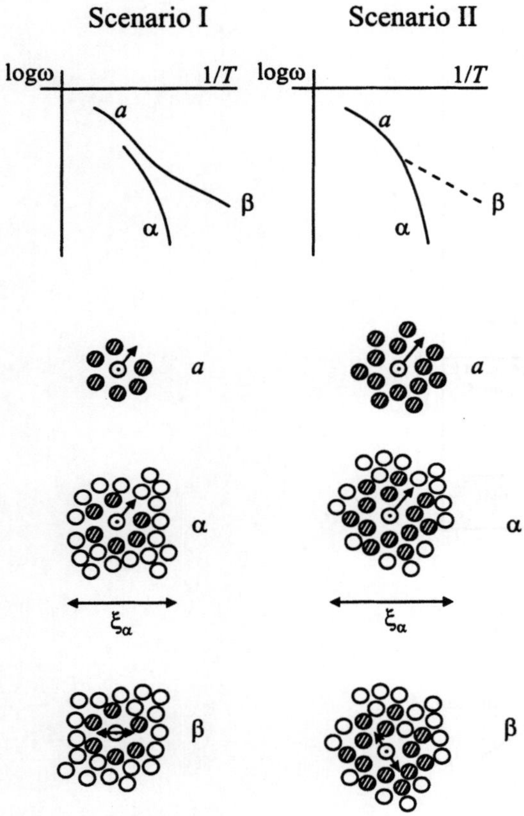

Fig. 2.33. Molecular pictures for the high-temperature a process, the cooperative α process, and the local Johari–Goldstein β process, for the two crossover scenarios of Sect. 2.2.1 (schematic). The dynamic heterogeneity is separated (disengaged) from the structure and is indicated by different symbols, as follows. ⊙ is a particle considered at its naming movement, with high mobility near a defect. ⊘ are cage molecules which mediate mobility (as related to the relevant zone). ○ are cooperativity shell molecules with small mobilities

preponderant partial system corresponds to diffusion of the defect inside the CRR. This defines the α time scale.

This approach automatically generates a dynamic heterogeneity in the sample, viz., a fluctuating pattern with islands of mobility (Fig. 2.34a). Assume, at least for scenario I, that the α process diffusion is also generated microscopically by escaping the cage, and that escape is only possible near defects. For the a process, all molecules were considered as equivalent. For the α process, however, escaping the cage is assisted by the cooperativity of the whole CRR (Fig. 2.33). In the α regime we have two dynamically distinct sets of molecules: faster ones near α *cages*, i.e., near the Glarum–Levy de-

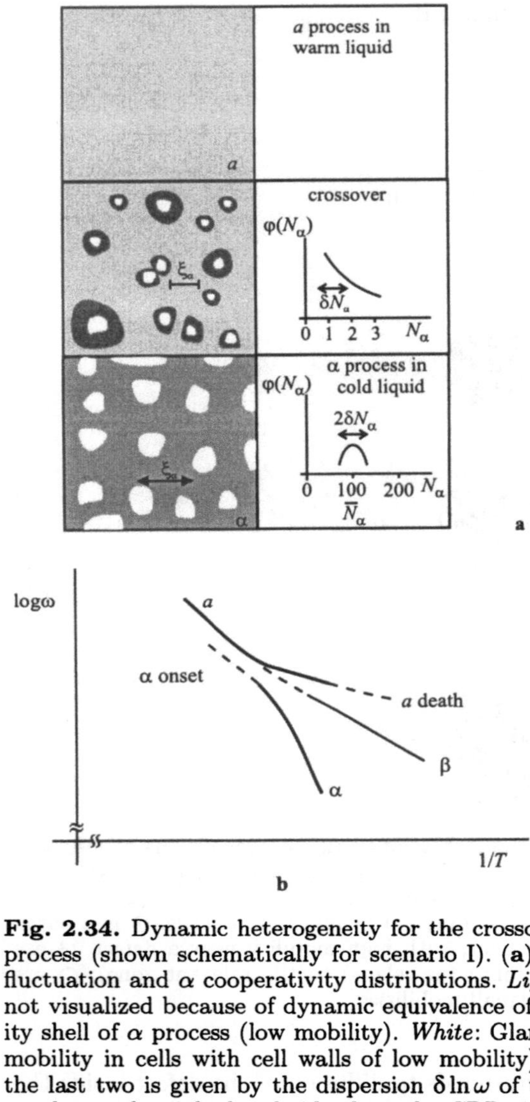

Fig. 2.34. Dynamic heterogeneity for the crossover region and the cooperative α process (shown schematically for scenario I). (**a**) Pattern from entropy or density fluctuation and α cooperativity distributions. *Light gray*: a process cages that are not visualized because of dynamic equivalence of particles. *Dark gray*: cooperativity shell of α process (low mobility). *White*: Glarum–Levy defects (islands of high mobility in cells with cell walls of low mobility). The mobility contrast between the last two is given by the dispersion $\delta \ln \omega$ of the α relaxation (Table 2.3). The graphs on the right hand side show the CRR size distribution $\varphi(N_\alpha)$. (**b**) Hypothetical Arrhenius plot with three processes near the crossover (a: Williams Götze, β: Johari–Goldstein, α: cooperative)

fects, and slower ones in the environment of α cages but also inside the CRR. On the α process time scale, the molecules are not dynamically equivalent.

The α cages are different from the a cages. The former are built from dynamically equivalent molecules, the latter from dynamically nonequivalent molecules. The cooperativity-assisting molecules form a 'cooperativity shell' for the α cages. Escaping the α cage is a molecular picture for Glarum defect diffusion.

The indications for Goldstein hopping from computer experiments (see Fig. 2.23b, Sect. 2.3.4) can be inserted in this picture by the assumption that a part of escaping the α cage is concentrated on the escaping particle. The other part is attributed to the Levy distribution for the dynamic breakthrough near the defect.

A characteristic length is induced for the α process by the average size of CRRs or, equivalently, by the average distance between Glarum–Levy defects. We have one defect per CRR.

Crossover Region. Let us recall the pedagogical restriction to scenario I. In the crossover region, both a and α processes operate in parallel. Their mobility is separated by about one frequency decade (Fig. 2.31a). This can be understood from the dynamic heterogeneity pattern (Fig. 2.34a), in particular from the CRR size distribution that may be derived [296] from our α approach (Sect. 4.3). Let the CRR size be characterized by the cooperativity N_α, the number of particles in an average CRR. In the crossover region, the approach results in an N_α distribution with both average and dispersion of order 1, $\bar{N}_\alpha \approx 1$ and $\delta N_\alpha \approx 1$. The a and α cages operate simultaneously. The α cages are slower because α cage escape requires the assistance of its cooperativity shell. This corresponds to general scaling (Sect. 2.2.5). The α mode lengths are larger than those for the a cage escape without cooperativity. This explains the separate α onset below the $a\beta$ trace in the Arrhenius diagram.

The middle pictures of Fig. 2.34 suggest a second length scale in the crossover region: the average distance between α process defects. However, this distance is not active with regard to entropy or density fluctuations as measured by calorimetry or compressibility [see (2.76)–(2.78)]; changing the distance does not change entropy or density. In scattering, however, the distance may be detected. For the α process picture in Fig. 2.34a, the distance and CRR size coincide.

Small cooperativities of order one, $N_a \approx 1$ and $N_\alpha \approx 1$, were first observed in a few examples for the crossover region in scenario I of moderate liquids by Beiner et al. [209, 399]. How can we understand such small CRR sizes in our particle pictures? Using the many attempts and the rare successes in a landscape-barrier picture and the spatiotemporal thermal fluctuation in the natural 3-d space, a quasi-continuous (Table 3.2 in Sect. 3.5) pattern for a mobility field $\log \omega(\boldsymbol{r}, t)$ can be constructed. This is dense enough in space, on the a and α time scales, to define small CRR sizes independently of the particle structure (Sect. 3.4). $N_a = 1$ and $N_\alpha = 1$ correspond to the size of cage doors in scenario I. Let us recall that the calorimetric fluctuation formula (2.76) primarily defines a volume. Irrespective of how many molecules collectively (for the a process) or cooperatively (for the α process) help to open the door, this volume is assigned to the actual a or α process: escaping the a cage or the α cage. To ensure diffusion, it is one particle, and not a part or a bead of the molecule, that must move through a door of the appropriate size. $N_\alpha \approx 1$ and $N_\alpha \approx 1$ describe the sizes of collectivity and cooperativity, respectively, for the actual a and α process being a success after the many attempts needed to escape the cage through the cage door in scenario I.

The average cooperativity $\bar{N}_\alpha$ values increase with decreasing temperature (2.142), initially as $\bar{N}_\alpha \sim (T_{\mathrm{on}} - T)^2$, $T < T_{\mathrm{on}}$. Two points seem important. Firstly, the increase is smooth and does not show [216, 590] any particularities at N_α that could have some relation to structure, e.g., $N_\alpha \approx$ 10–15 for the first coordination shell. This is consistent with the disengagement concept. Secondly, the CRR size distribution becomes centered at $\bar{N}_\alpha$ of order 100 at the glass temperature T_g for many glass formers (Sect. 4.3). The relative dispersion $\delta N_\alpha/\bar{N}_\alpha$ becomes small there, of order 10%. This also means that the parallel a cages have become vanishingly rare compared to α cages, which results in the a death in the crossover region. Discussion of the crossover region will be continued after discussion of the β process.

The first to mention that the a process is a distinct and separate process from α (and β) was G. Williams [206].

Johari–Goldstein β Process. Besides the diffusive, cooperativity-assisted α cage escape, the concentration of free volume at and near the isolated Glarum–Levy defects may allow less cooperative [154] finite molecular motions with relatively fixed mechanisms, which do not seem possible elsewhere because of insufficient free volume there. An example would be some libration simulating the first stage of escaping the cage for the splitting scenario I (Fig. 2.33). This would explain the continuity of the $a\beta$ trace in the Arrhenius diagram. As the a cage and the α cage have different properties, we expect a bend in the $a\beta$ trace in the crossover region. No such separate β process is possible in the a regime region as long as all cage molecules are equivalent. Instead, the many attempts before success in a cage escape generate the cage rattling process c of Fig. 2.21a. In MCT, the corresponding non-activated dispersion zone at terahertz frequencies is usually labelled by β_{fast} (see Table 2.6).

From a mechanical point of view, it does not seem to be excluded that the molecular mechanisms for the β process and a and α cage escape have common components which are differently and gradually modified for the various processes. A long-standing question concerns the number of particles that take part in the β process [591]. If the finite motion of the central particle(s) is modified by cooperativity, then all particles take part, the central ones more and the assisting-shell ones less. NMR experiments are consistent with this view [592]. Counting only the central particles with their larger amplitudes, we arrive at a black-and-white depiction of the β process: the central particles 'yes', the cooperatively helping particles from the shell 'no' (Sects. 2.2.1 and 4.2). This depiction should not tempt us to forget that the cooperativity shell, i.e., the α process, is a necessary condition for the β process. NMR experiments [592] also show the rich properties of the Glarum–Levy defects and differences between α and a cages.

To my knowledge, the first to describe a similar picture for the β process was Johari [593].

Let us continue the crossover discussion. Figure 2.31a in Sect. 2.5.1 from heat capacity spectroscopy [209] and results from a solid-state NMR exchange method [713] seem to indicate that, in a small temperature interval of the crossover region for scenario I, all three processes a, α and β can be simultaneously observed in an isothermal section (Fig. 2.34b). This would be consistent with an insertion of the pictures for a, α, and β on the left-hand side of Fig. 2.33 into the middle pattern for dynamic heterogeneity of Fig. 2.34a. The task for a corresponding theory would be to find, for a given state at a temperature near T_c, three different processes in a frequency region of a few decades. Large temperature fluctuations δT would be helpful for this task since contributions from states are organized when they have a large temperature distance from the crossover temperature T_c ($\delta T = \pm 20$ K for instance). At such temperatures all three processes (a for $T_c + \delta T$, β and α for $T_c - \delta T$) are well developed, distinct and independent.

Speculation about the Difference Between the Two Crossover Scenarios I and II. The difference between the two scenarios is defined in Fig. 2.11a. Let as assume that the a cage formed in scenario II substances is more complex and includes the second coordination shell, or at least a larger symmetrical part of it (right-hand side of Fig. 2.33). A diffusion event for the a process is then more complex than in scenario I. The collectivity for the a process may be larger than that for scenario I. A larger a cage door would imply $N_a > 1$ and $N_\alpha > 1$ in the crossover region for scenario II. More internal possibilities inside the larger cage may include larger polarization fluctuations ($\Delta\varepsilon$) or larger entropy fluctuations (ΔC_p). Near the crossover, the difference between the a and α processes is therefore not so large and we expect a tendency towards greater continuity between a and α; $\Delta\varepsilon^{(\alpha)}$ and $\Delta C_p^{(\alpha)}$ need not tend to zero for $T \to T_c$. Cage closing seems more radical because cage opening for escape from the larger a cage includes longer distances across the thicker 'cage walls'. Finite simulation of the total escape then seems less probable below the crossover. We may consider finite simulation, for example, of the first step or another part of escape. This enlarges the difference between the a and β processes – the β activation energy may be rather different from the apparent a process activation energy defined by the slope of the a trace in the Arrhenius plot near the crossover – and the β intensity near the crossover is expected to be small when the finite attempts concern only a small and less active part of the larger α cage. Another reason for small β intensities may be larger CRRs growing from the larger cages. One defect per CRR implies a smaller defect density for larger CRRs. A certain amplification of the β intensity after quenching below T_g [594] may be generated by the additional free volume from the vault effect of glass structure (Sect. 2.2.6).

Relation of the α Onset to Crystallization/Melting. In principle, the dynamic glass transition and crystallization could hardly be more different. The dynamic glass transition is an equilibrium phase phenomenon that also occurs in stable phases [595]. A glass state near T_g is unstable against its liquid state, whereas a liquid, or more generally, a glass former in the temperature range of the dynamic glass transition or even at the thermal glass transition, is not necessarily metastable against a crystal. Some substances with a thermal glass transition in a stable phase were listed in Table 2.5 (Sect. 2.1.5). The term 'undercooled' or 'supercooled', as thought to be essential for the glass transition, is therefore misleading and has been replaced by 'cold' in our book. Crystallization, on the other hand, is a first order phase transition with a nonzero melting entropy between the crystalline and liquid states, accompanied by a new symmetry or an order parameter.

The α onset has no definite attributes corresponding to such a prominent phase transition (Sect. 2.5.1). The new property (cooperativity) that emerges in the crossover seems to be mainly of a dynamic nature.

Nevertheless, in moderate liquids (i.e., liquids without overcomplicated glass-transition situations, Sect. 2.2.8), the crossover from the α to the a process is reminiscent of some properties of melting. For the α process, Glarum–Levy defects are isolated, in a similar way to crystal defects. In a sense, the a process situation in the warm liquid can be considered to be generated from the α process situation in the cold liquid by overlarge fluctuations of the cooperativity shells, just as in the application of the Lindemann criterion for melting caused by overlarge fluctuations in the crystal. This relationship is supported by the fact that, for moderate liquids, the melting temperature T_m and crossover temperature T_c generally have the same order of magnitude, with a difference of the order of a few tens of kelvins [see Table 2.8 for orthoterphenyl; further examples for (T_c/T_m) are salol (280 K/317 K), sorbitol (330 K/373 K), and propylene carbonate (215 K/218 K)].

More important for such ideas is the flow criterion (2.13) from the steady state compliance J_s^0. The mean shear fluctuation necessary for flow, related to one particle, should be larger than about 10%.

2.5.4 Tables

Our description of the glass transition phenomenon will be concluded with three tables of data. The substances are mainly chosen by availability of

Table 2.9. Properties of the α process [299]. The quantities $\log \Omega_\alpha$, B, and T_0 are parameters of the WLF or VFT equation, $\log_{10}(\Omega_\alpha/\omega) = B/(T - T_0)$. T_g is the conventional glass temperature, Δc_p the step in the specific heat at T_g. N_α is the cooperativity at T_g, i.e., the number of particles (molecules, monomeric units) in a cooperatively rearranging region CRR calculated from calorimetry by (2.77). The accuracy for the absolute values of ΔC_p is $\pm 10\%$, and for N_α, $+100\%$, -50%

Substance	$\log \Omega_\alpha$ [rad s^{-1}]	B [K]	T_0 [K]	T_g [K]	Δc_p [J/gK]	N_α –	Ref.
OTP	22.6	1820	168	249	0.46	83	[684]
BMPC [a]	31.8	4250	115	242	0.41	108	[684]
Salol	15.9	823	175	219	0.53	173	[685]
B_2O_3 [b]	(23.7)	(20120)	(0)	547	0.50	52	[686]
PEMA	8.3	366	304	343	0.26	33	[687]
PnBMA	7.6	361	256	298	0.19	8	[687]
Copolymer [c]	7.2	318	266	303	0.22	16	[688]
Glycerol	15.1	1065	126	189	1.02	350	[684]
Propylenglycol	14.4	930	110	167	0.86	280	[684]
PVAC	12.3	650	265	313	0.44	258	[689]
Sorbitol	12.5[d]	460[d]	229[d]	261	1.04	195	[690]
CKN	14.6	782	288	337	0.55	339	[691]
BIBE	14.4	683	180	221	0.49	69	[692]
Standard glass I [e]	12.1	4317	521.8	823	0.25	67	[693]

[a] BMPC = bis methoxy phenyl cyclohexane.
[b] The B_2O_3 trace in an Arrhenius plot has only a weak curvature. The WLF equation seems inappropriate. The $\log \Omega_\alpha$, B, and T_0 figures for B_2O_3 are of pedagogical interest to warn the reader against formal fittings without subsequent testing.
[c] Random copolymer of n-butylmethacrylate with 2% styrene.
[d] Both a and α parts of the dynamic glass transition are formally included.
[e] Standard glass I of the DGG (Deutsche Glastechnische Gesellschaft, i.e., the German Society of Glass Technology), mainly a CaO–Na_2O–SiO_2 system.

data. Strong glass formers (small fragility), substances which tend to crystallize or have small α activities, and substances with T_g beyond easily manageable temperatures are underrepresented. Instead, several favorites of the glass transition community are included. The general possibility of finding 'typical glass formers' is limited. Substances that are too simple tend towards crystallization or other phase transitions (e.g., spin glasses), whilst substances that are too complex tend to exhibit additional phenomena, such as polyamorphism with a phase transition (e.g., water), nanophase separation (e.g., proteins), and additional traces in the Arrhenius plots (e.g., polymers, liquid crystals).

Table 2.9 lists parameters for the dynamic α process. This table is intended to give a short overview concerning the variation of parameters for good glass formers of moderate complexity.

Table 2.10 [596] lists some parameters connected with the crossover region. It seems that the crossover region parameters $\{T_c, \log \omega_c\}$ are rather insensitive to the activity used for their determination. This is consistent with the hypothesis that crossover phenomenology can be reduced to one physical cause, viz., the onset of α process cooperativity and its steep increase as we move towards lower temperatures.

The symbols used in Table 2.10 are as follows: T_c are crossover temperatures, m is the fragility parameter at T_g (2.3); x_g is the reduced (conventional) glass temperature between the Vogel and crossover temperatures, $x_g = (T_g - T_0)/(T_c - T_0)$; $\log \Omega_\alpha$, B_α and $T_{0\alpha}$ are parameters of the WLF or VFT equation for the α process, $\log_{10}(\Omega_\alpha/\omega) = B_\alpha/(T - T_{0\alpha})$; T_g is the conventional glass temperature; $\log_{10} \Omega_\beta$ and $E_{A\beta}$ are parameters of the β process Arrhenius equation, $\log_{10}(\Omega_\beta/\omega) = E_{A\beta}/RT$; R is the gas constant; and n.a. means that the information is not available.

Substance acronyms used in Table 2.10 are as follows: PFBSE is poly(2-propenoicacid 2-[[(nonaflourobutyl) sulfonyl] methylamino] ethylester), $(C_{11}H_{12}F_9NO_4S)_n$, ($\overline{M}_W$ = 6 kg/mol); P(nBMA-*stat*-S) x%S are random copolymers of n-butylmethacrylate and styrene containing x% styrene; BMPC is bis-methoxy-phenyl-cyclohexane; BMMPC is bis-methyl-methoxy-phenyl-cyclohexane; PDE is phenyl-phthalate-dimethylether; CDE is cresol-phthalate-dimethylether; FDE is flourene-9,9-bisphenol-dimethylether; ODE is 4,4-(octahydro-4,7-methano-5H-indene-5-ylidene)-bisphenol-dimethylether $(C_{24}H_{28}O_2)$; Epon 828 is an epoxy resin, viz., diglycidyl ether of bisphenol-A; CKN is a melt containing 62 mol% KNO_3 and 38 mol% $Ca(NO_3)_2$.

Table 2.11 lists some mechanical and calorimetric coefficients of glass and liquid states near T_g for a few favorites. The problems involved in a precise determination of the Prigogine–Defay ratio Π are discussed in [714]. A value of $\Pi = 1.3 \pm 0.1$ is obtained for atactic polystyrene with $T_g = 353$ K. The main uncertainties in the ΔC_p and $\Delta\kappa_T$ steps result from uncertainties in the tangents for the liquid and glass states. It seems sure, however, that the mechanical Keesom–Ehrenfest relation I is more seriously violated than the calorimetric one and that, as a rule, $\Pi > 1$. Note that the experimental value of the step ΔC_V in heat capacity at constant volume is different from zero. The evaluation of a larger data set shows that $\Delta(1/C_V)$ – as needed for the calculation of cooperativity from calorimetric data according to (2.77) – is about 74% of $\Delta(1/C_p)$, with a variation between 25% and 120%. The existence of $\Delta(1/C_V) \neq 0$ or $\Delta C_V \neq 0$ implies that C_V has a dispersion $C_V(\log\omega)$ across the dispersion zone of the dynamic glass transition (α process), against a tough theoretical prejudice.

The symbols used in Table 2.11 are as follows: V specific volume; C_p specific heat capacity at constant pressure; α_p cubic thermal expansion coefficient at constant pressure, $\alpha_p = (1/V)(\partial V/\partial T)_p$; κ_T isothermal compressibility (reduced bulk compliance) $\kappa_T = -(1/V)(\partial V/\partial p)_T$; Π Prigogine–Defay ratio (2.153); I, II left-hand sides of the Keesom–Ehrenfest comparisons (2.155) and (2.156), $\Pi = I/II$; $\mathrm{d}T_\mathrm{g}/\mathrm{d}p$ experimental pressure dependence of glass temperature; C_V calculated specific heat capacity at constant volume, $C_V = C_p - VT\alpha_p^2/\kappa_T$.

The small glass temperature T_g differences between the tables indicate methodological differences between different laboratories.

Table 2.10. Crossover parameters and properties of α and β processes [596]. See text for explanation

Label	Substance	T_c [K]	$\log \omega_c$ [rad s^{-1}]	m	x_g	$\log \Omega_\alpha$ [rad s^{-1}]	B_α [K]	$T_{0\alpha}$ [K]	T_g [K]	$\log \Omega_\beta$ [rad s^{-1}]	$E_{A\beta}$ [kJ mol^{-1}]
	Polymers										
1	Poly(methyl methacrylate)	450^a ($\approx 455^d$)	6.2	136	0.42	10.8	366	371	404	15.5	79
2	Poly(ethyl methacrylate)	380^a ($\approx 395^d$)	3.6	68	0.57	8.3	361	304	347	14.8	73.1
3	Poly(*n*-butyl methacrylate)	323^a ($\approx 340^d$)	2.3	48	0.72	7.6	361	256	304	15.0	76.2
4	Poly(*n*-hexyl methacrylate)	282^a ($\approx 295^d$)	1.1	≈35	n.a.	n.a.	n.a.	n.a.	254	(20.5)	(98.1)
5	P(S-*stat*-*n*BMA) 2%S	351^a ($\approx 355^d$)	3.0	70	0.44	7.2	318	266	303	14.5	74
6	P(S-*stat*-*n*BMA) 8%S	356^a ($\approx 360^d$)	3.5	71	0.43	8.0	375	264	304	16.9	86
7	P(S-*stat*-*n*BMA) 19%S	373^a ($\approx 360^d$)	4.7	72	0.36	8.0	323	269	306	17.2	86
8	1,4 polybutadiene	210^d ($\approx 230^a$, 216^e)	6.5	59	0.54	13.3	456	141	178	16.5	39.6
9	PFBSE	374^a ($\approx 370^d$)	3.2	86	0.47	7.5	296	301	335	12.9	69.2
10	Poly (vinyl acetate)	395^d ($\approx 390^b$)^m	7.3	96	0.35	12.3	650	265	311	11.5	39
	Small Molecules										
11	BMPC	270^b (270^d)	4.4	63	0.83	31.8	4250	115	243	14.0	50.1
12	Orthoterphenyl	290^b ($290^{c,d}$, 289^e)	7.3	74	0.64	22.6^f	1820^f	168^f	246	16.0	48.9
13	Epon 828	350^a (354^b, $\approx 375^d$)	9.2	153	0.20	11.9	315	234	257^g	13.2	23.9
14	Cresyl glycidyl ether	265^b (360^a, 265^d)	9.6	135	0.27	13.9	350	181	204^g	13.2	18.4
15	Propylene carbonate	189^b (187^e)	7.1	91	0.46	13.9	389	132	158	n.a.[h]	n.a.[h]
16	Salol	265^b (260^e)	6.8	89	0.50	15.9	823	175	220	n.a.	n.a.
17	*n*-propanol	140^b (138^d)	6.8	36	0.52	n.a.	824	50	97^g	13.7	19.7
18	Methyltetrahydrofurane	107^b	6.5	93	0.52	≈15.3	≈295	≈74	91	n.a.[h]	n.a.[h]
19	Glycerol	285^b (262^e)	8.5	49	0.40	15.1	1065	126	190	n.a.[h,i]	n.a.[h,i]

Label	Substance	T_c [K]	$\log \omega_c$ [rad s^{-1}]	m	x_g	$\log \Omega_\alpha$ [rad s^{-1}]	B_α [K]	$T_{0\alpha}$ [K]	T_g [K]	$\log \Omega_\beta$ [rad s^{-1}]	$E_{A\beta}$ [kJ mol^{-1}]
20	B_2O_3	≈620[b,c] (≈800[e])	≈1.0	32	0.87	(23.7)[j]	(20120)	(0)	(537)	n.a.	n.a.
21	Benzoin isobutylether	262[a] (270[b],256[d])	6.2	94	0.49	14.4	683	180	220	16.3	55
22	3-bromopentane	≈140[d] (≈137[b])[n]	≈6.5	56	0.53	16.6	671	72	108	14.8	23.1
23	Propylene glycol	280[b] (251[e])	8.9	48	0.34	14.4	930	110	167	n.a.[h,i]	n.a.[h,i]
24	BMMPC	330[b]	7.5	59	0.60	20.8	2206	164	263	n.a.	n.a.
25	PDE	325[b]	4.2	78	0.75	23.1[f]	2348[f]	200[f]	294	13.2	17.4
26	CDE	390[b]	7.3	75	0.52	17.8[f]	1700[f]	229[f]	313	n.a.	n.a.
27	FDE	365[b]	7.0	84	0.63	21.9	2023	228	315	10.4	4.7
28	ODE	325[b]	5.8	71	0.76	31.8	4600	148	283	n.a.	n.a.
29	CKN	378[e] (390[b])	5.9	129	0.50	14.6	782	288	333	n.a.	n.a.
30	*m*-tricresyl phosphate	245[b] (260[e], ≈270[c])	5.8	63	0.58	14.4	716	161	210	n.a.[h]	n.a.[h]
31	Triphenyl phosphite[k]	215[b]	3.5	97	0.80	22.4	1152	154	203	n.a.	n.a.
32	Toluene	140[d] (143[e])	6.6	98	0.41	12.5[l]	215[l]	101[l]	117	16.1	25.3
33	1-butyl bromide	129[b]	7.0	44	0.59	18.9	913	53	98	n.a.	n.a.
34	2-butyl bromide	118[b]	5.8	52	0.74	25.4	1350	49	100	n.a.	n.a.
35	Isobutyl bromide	117[b]	5.3	49	0.80	31.5	2200	33	100	n.a.	n.a
36	Isopropyl benzene	171[b] (150[e],≈150[d])	≈6.0	189	0.40	(8.2)[j]	1455	94	125	15.0	25.9
37	Butyl benzene	172[b] (≈181[b],160[e])	7.5	172	0.29	(5.3)[j]	436	110	128	n.a.	n.a.
38	Propyl benzene	170[b] (≈176[b])	8.0	70	0.38	14.0	440	97	125	n.a.	n.a.

[a]Onset temperature for the α process. [b]Crossover temperature from change of WLF or VFT parameters. [c]Starting temperature for divergence of different transport activities after matching at high temperatures $T > T_c$. [d]Extrapolated (formal) crossing of α and β traces in Arrhenius plot from far below. [e]'Critical temperature' from evaluation by use of mode-coupling theory. [f]Bad fit of VFT or WLF equation. [g]Dynamic glass temperature for $\tau_\alpha = 100$ s. [h]β process not yet detected. [i]β process 'behind' α process? [j]As estimated from viscosity $[-\log(\eta/\mathrm{Poise})]$. [k]Glacial phase below $T = 214$ K. [l]common evaluation of α and a process. [m]Using temperature dependence of $\beta_{KWW}(T)$ and electron spin resonance. [n]Using Adam–Gibbs approach.

Table 2.11. Thermodynamic data for liquid (l) and glass (g) states near T_g [72]

		B_2O_3	CKN	PVAC	Glycerol
T_g	[K]	550[a]	340[a]	303.8[a]	183[b]
V	[cm^3/g]	0.56[a]	0.46[a]	0.84[a]	0.76[b]
C_p^l	[J/gK]	1.93[a]	1.49[a]	1.80[a]	1.87[e]
C_p^g	[J/gK]	1.30[a]	0.95[a]	1.30[a]	0.99[e]
ΔC_p	[J/gK]	0.63[a]	0.54[a]	0.50[a]	0.88[b]
α_p^l	[$10^{-4}K^{-1}$]	4.0[a]	3.5[a]	7.1[a]	4.8[b]
α_p^g	[$10^{-4}K^{-1}$]	0.5[a]	1.2[a]	2.8[a]	2.4[b]
$\Delta\alpha_p$	[$10^{-4}K^{-1}$]	3.5[a]	2.3[a]	4.3[a]	2.4
κ_T^l	[$10^{-10}Pa^{-1}$]	4.0[a]	1.3[a]	5.0[a]	1.8[b]
κ_T^g	[$10^{-10}Pa^{-1}$]	1.2[a]	0.6[a]	2.9[a]	0.9[b]
$\Delta\kappa_T$	[$10^{-10}Pa^{-1}$]	2.8[a]	0.7[a]	2.1[a]	0.9
Π	–	4.7[a]	4.5[a]	2.2[a]	9.4
$I = \Delta\kappa_T/\Delta\alpha_p$	[10^{-7}K/Pa]	8.0	3.0	4.8	0.36
$II = TV\Delta\alpha/\Delta C_p$	[10^{-7}K/Pa]	1.7	0.7	2.2	0.38
dT_g/dP (exper.)	[10^{-7}K/Pa]	1.97[j]	n.a.	2.17[j]	0.39[j]
C_V^l	[J/gK]	1.81	1.35	1.54	1.69
C_V^g	[J/gK]	1.29	0.91	1.23	0.90
ΔC_V	[J/gK]	0.51	0.43	0.31	0.79[b]
$\Delta(1/C_p)$	[10^{-1}gK/J]	2.51	3.81	2.14	4.75
$\Delta(1/C_V)$	[10^{-1}gK/J]	2.20	3.51	1.63	5.14

[a] [694], [b] [695], [c] [696], [d] [697], [e] [698], [f] [699], [g] [700], [h] [701], [i] [702], [j] [703]

Table 2.11. (Continued.) See text for explanation of symbols

n-propanol	Selenium	Polyisobutene	PVC	Metallic glass Zr Ti Cu Ni Be
95[b]	304[b]	198[c]	350[c]	610[d]
0.91[b]	0.24[b]	1.05[c]	0.73[c]	–
1.75[f]	0.51[g]	1.53[h]	1.46[i]	–
1.08[f]	0.33[g]	1.13[h]	1.16[i]	–
0.67[b]	0.19[b]	0.40[c]	0.30[c]	0.70[d]
8.0[b]	4.2[b]	6.2[c]	5.7[c]	–
4.0[b]	1.7[b]	1.5[c]	2.0[c]	–
4.0	2.5	4.7	3.7	0.19[d]
0.8[b]	3.0[b]	4.0[c]	4.4[c]	–
0.4[b]	2.4[b]	3.0[c]	2.4[c]	0.097[d]
0.4	0.6	1.0	2.0	0.000785[d]
1.9	2.4	0.9	1.7	2.4[d]
1.0	2.3	2.1	5.4	0.04
0.52	0.98	2.4	3.1	
0.69[j]	1.28[j]	2.36[j]	1.57[j]	0.036[d]
1.06	0.47	1.33	1.27	
0.74	0.32	1.11	1.12	
0.32[b]	0.15[b]	0.22[c]	0.15[c]	
3.5	11.2	2.3	1.77	
4.1	10.4	0.65	1.08	

3. Theoretical Framework

The aim of this chapter is to describe some theoretical methods which seem necessary for a better understanding of the text and formulas in Chap. 2. Moreover, Chap. 3 will prepare tools for understanding selected theories of the dramatic slowing down mechanisms which occur (Chap. 4).

This chapter was called 'Codes' in a preliminary draft of the manuscript. It seems typical for glass transition research that the results of experiments or computer simulations contain messages that are not understood even in terms that are widely used, such as cooperativity. Furthermore, I think that, not only the molecular situations, but also their transfer to measured quantities is unclear. Chapter 3 is devoted to this 'phenomenological channel' and to the unravelling of the codes.

We continue here the discussion of the thermodynamic vs. dynamic approach to the glass transition started in the introduction to Sect. 2.5. All linear response and all dynamic scattering (reduced by the structure factor) is determined by thermal fluctuation. We should therefore look carefully to the phenomenological fundamentals: character of relevant subsystems, subsystem thermodynamics, linear response, and fluctuation–dissipation theorem FDT (Sects. 3.1–3.4). In particular, we are interested in the limits of Landau–Lifshitz subsystems with Gaussian statistics as used in the conventional theory of liquids. Section 3.5 analyses preconditions for the definition of a mobility field $\log(\omega(\boldsymbol{r}, t))$ used to describe a fluctuating pattern of dynamic heterogeneity for the α process (Sect. 2.2.5, Figs. 2.12d, 2.15b). The pattern was related to density or entropy fluctuations and was used to define a dynamic kind of subsystem, the cooperatively rearranging region CRR. This pattern, with islands of mobility, was explained by Levy distributions of probability theory. Section 3.6 describes the mathematical facts for these distributions insofar as it seems necessary to understand the pattern.

The topics of Chap. 3 are the representativeness theorem as a consequence of the Sect. 2.2.5 gedankenexperiment (Sect. 3.1), the von Laue alternative for freely fluctuating subsystems with a suggestion for a 'microscopic von Laue distribution' (Sect. 3.2), the interpretation of FDT as the equation for linear response experiments to measure thermal fluctuation (Sect. 3.4), and the evidence for a sufficient spatiotemporal event density to justify a quasi-continuous definition of the mobility field to describe α process cooperativities smaller than the first coordination shell, $N_\alpha < 10$ (Sect. 3.5).

Remark. In the past, hydrodynamic methods were sometimes used on the nanometer scale, e.g., monomeric friction coefficients, or Rouse modes in polymer physics. I think that such methods are not applicable to α process cooperativity. In the linear region for nanometer processes, we have no alternative but to calculate or measure the desired fluctuations of the relevant subsystems. The assumption of prescriptive molecular modes and the discussion of their different 'couplings' to external fields or to mean field transport coefficients does not seem appropriate for the molecular α process.

3.1 Representative Freely Fluctuating Subsystems

Many experimental methods for the dynamic glass transition investigate fluctuations of subsystem variables or their partition in the frequency–time domain or wave vector–length domain, e.g., thermodynamics, linear response, and dynamic scattering. Only a few methods seek out underlying mechanical correlations, e.g., two- or higher-dimensional exchange NMR. This means that subsystem fluctuations must be understood as completely as possible. For heat capacity spectroscopy, it seems interesting if we have a conjugate pair, e.g., both entropy and temperature fluctuation, or if the results can be reduced to enthalpy fluctuation only.

Let us first describe some well known aspects of conventional liquid theory from the fluctuation standpoint. The text will seem lengthy, but it is my experience that serious misunderstandings can arise from a careless application of conventional theory to the glass transition. The final point will be that, in some subtle places, the underlying Gaussian distribution should be replaced by the Levy distribution to explain the desired heterogeneities. The word 'subtle' means that the Levy distribution is related to the fact that free volume fluctuation is only a fine shadow over the particle picture (Figs. 2.16e and 2.27e).

Partitions of fluctuations. Through measurement by FDT, linear response is the partition of thermodynamic fluctuation in the frequency or time domain. Consider the dynamic heat capacity at constant pressure, for example, measured by the 3ω method or temperature modulated DSC of heat capacity spectroscopy.

$$k_B \Delta C_p = \Delta(\overline{\Delta S^2}) , \tag{3.1}$$

$$\Delta(\overline{\Delta S^2}) = 2 \int_{\mathrm{DZ}} \Delta S^2(\omega)\, \mathrm{d}\omega , \tag{3.2}$$

$$\Delta S^2(\omega) = k_B T \, J_S''(\omega) \, / \, \pi\omega \, , \tag{3.3}$$

$$J_S''(\omega) = C_p''(\omega) \, / T \, . \tag{3.4}$$

The C_p step between liquid zone and glass zone, denoted ΔC_p, is determined by the contribution of the dispersion zone DZ to the time averaged entropy fluctuation (3.1). This contribution is given by the integral over the spectral density of the entropy fluctuation $\Delta S^2(\omega)$ (3.2). The FDT relates the spectral density to the imaginary part of the dynamic entropy compliance $J_S''(\omega)$ by (3.3), and this in turn is simply related to the imaginary part of the dynamic heat capacity $C_p''(\omega)$ by (3.4).

Scattering is, in a way, the fluctuation partition in the Q domain, where Q is the wave vector of the scattering experiment. The wave vector reflects the momentum change of the scattered particle and corresponds to a (de Broglie) wavelength $\lambda = (4\pi/Q)\sin(\theta/2)$ (2.87) that feels for length scales of the structure. The Q variation may be simply managed by varying the scattering angle θ. Scattering detects density fluctuations of subsystems. For liquids made of structureless particles, we have

$$S(Q) = \frac{1}{N}\left\langle \sum_{jk} \exp\left(\mathrm{i}\boldsymbol{Q}(\boldsymbol{r}_j - \boldsymbol{r}_k)\right) \right\rangle , \tag{3.5}$$

$$S(Q \ll Q_{\mathrm{max}}) = \langle \Delta N^2 \rangle = V \, \bar{n} \, k_B T \left(\frac{\partial \bar{n}}{\partial p} \right)_T , \tag{3.6}$$

$$c(r) = \frac{1}{2\pi^2 \bar{n} r} \int_0^\infty Q \left[1 - \frac{1}{S(Q)} \right] \sin(Qr) \, \mathrm{d}Q \, , \tag{3.7}$$

$$k_B T \left(\frac{\partial \bar{n}}{\partial p} \right)_T = 1 - 4\pi\bar{n} \int_0^\infty c(r) \, r^2 \mathrm{d}r \, . \tag{3.8}$$

The structure factor is the ensemble average of the Huygens wave interferences from particle scattering (3.5). The conventional relation to the time average of the thermodynamic compressibility, $\kappa_T = (1/\bar{n})(\partial\bar{n}/\partial p)_T$, is realized at small Q in the (corresponding) length range of 5 nm (3.6). This range is also related to the particle number fluctuation $\langle \Delta N^2 \rangle \equiv \langle (\Delta N)^2 \rangle$ considered for subsystems with particle exchange rather than work exchange. The partition of fluctuation in the distance domain is expressed as a spatial Fourier transformation of the structure factor (3.7) that leads to the conventional compressibility equation (3.8). As the spatial weight of the density correlations, we chose the direct correlation function $c(r)$, with $c(Q)$ from the Ornstein–Zernike equation in the Q domain, $h(Q) = c(Q) + \bar{n}c(Q)h(Q)$, where $h(Q) = g(Q) - 1$ and $g(Q)$ is the radial distribution function in the Q domain.

The relation of the dynamic glass transition to the ω and to the Q partition could hardly be more different. Across the dispersion zone, ΔC_p is a large and significant indication (10%) of the glass transition, whereas the structure factor $S(Q)$ remains unchanged. Combine both partitions in the dynamic scattering function $S(Q,t)$. Crossing the dispersion zone, this function vanishes completely with no reflections in the $S(Q)$ shape (Fig. 2.5a). The main result from dynamic scattering is that the time scales for vanishing are not the same for different Q values. After reduction by $S(Q)$, we get some kind of diffusion with a dispersion law $\omega(Q)$.

An example of the difference between spatial and temporal Fourier analysis of fluctuations is the FDT applied to Q components of susceptibilities, $\chi^*(Q,\omega)=\chi'(Q,\omega) \pm i\chi''(Q,\omega)$:

$$\tilde{S}(Q,\omega) \equiv k_B T \frac{\chi''(Q,\omega)}{\pi\omega} . \tag{3.9}$$

In principle, any susceptibility can be partitioned in the Q domain. The ω identity of the FDT is separately valid for any given Q component. The wave vector Q is more like an index than a variable, $\chi''(Q,\omega) = \chi''_Q(\omega)$.

Let us recapitulate some details of dynamic scattering (Sect. 2.3.1). This is a combination of the ω and Q partition of density fluctuation and results in the structure function $S(Q,\omega)$ or the intermediate scattering function, $S(Q,t)$ = the temporal Fourier transform of $S(Q,\omega)$. The ω dependence reflects the energy loss of the scattered particles, $\Delta E = \hbar\omega$ (2.90). The transfer from the wave vector Q to 3-d space distances r is the connection between coherent scattering and the distant part of the van Hove correlation function, and between incoherent scattering and the self-correlation part of this function (Sect. 3.4.2).

Dynamics vs. Structure. The first paragraph summarizes the strategy of the dynamic approach outlined in Sect. 2.5.2. Attempts are often made to compare the large dynamic effects observed in crossing the α dispersion zone with presumed structural effects. The α dispersion is wide (Table 2.3), the susceptibility steps are relatively large (Table 2.11, $\Delta C_p/\bar{C}_p$, $\Delta\kappa_T/\bar{\kappa}_T$ and $\Delta\alpha_p/\bar{\alpha}_p$ are of order 10% for many glass formers), and the reduced dynamic structure function decays to zero like a correlation function (Fig. 2.5a). These large effects are obviously regulated by free volumes that are small fractions of the total volume. The expected structural effects are correspondingly small. The density contrast of an underlying fluctuating density pattern for dynamic heterogeneity is estimated to be extremely weak, of order 0.01% at T_g. The pattern is therefore hardly reflected in the structure factor. Understanding the glass transition means understanding how large 'dynamic' fluctuations are compatible with no structure effects or only spurious structure effects. We consider two examples.

- *Diffusion* is based on the evolution of an order parameter $\psi(\boldsymbol{r},t)$, e.g., a concentration. Let us repeat (2.135)–(2.137) in the context of structure. The experimental variable for the stationary situation is the correlation function $S(\boldsymbol{r},t) \equiv \overline{\psi(\boldsymbol{r},t)\psi(\boldsymbol{0},0)}$.

$$\left(\frac{\partial}{\partial t} - D\Delta\right) S(r,t) = 0 \,, \tag{3.10}$$

$$S\,(\text{small } Q, t = 0) = k_{\mathrm{B}}T\chi \,, \tag{3.11}$$

$$S(Q,\omega) = 2k_{\mathrm{B}}T\chi \frac{DQ^2}{\omega^2 + (DQ^2)^2} \,. \tag{3.12}$$

The Q–t boundary condition for the diffusion equation (3.10) with diffusion coefficient D is conventionally defined by a susceptibility χ (3.11). The structure function, multiplied by ω, is then a $\chi''(x)$ loss function representing a Lorentz line on the $\log\omega$–$2\log Q$ diagram of the dispersion zone around $x \equiv DQ^2/\omega = 1$ (3.12) (see Fig. 2.28e). The structure factor $S(Q)$ of the liquid is not reflected in this first example.
- The step in compressibility, $\Delta\kappa_T = \kappa_T^{\text{liquid}} - \kappa_T^{\text{glass}}$, across the dispersion zone, i.e., the compressibility difference between the liquid and glass zones of the equilibrium liquid (see the analogous calorimetric Figs. 2.11c and 2.31g), is not reflected in the compressibility equations (3.6) and (3.8). It is certainly a misunderstanding to say that the 'static' structure factor $S(Q)$ belongs to the short-time limit of the glass zone, whereas the compressibility κ_T belongs to the long-time limit of the liquid zone. The misunderstanding is resolved by the ergodicity of the liquid zone: equivalence of time average and ensemble average is required to get a compressibility of the liquid state that is equivalent to the structure factor. Both averages are sampled in the same liquid state, and the nonergodic glass zone has the same structure as the ergodic liquid zone (Sect. 2.1.4).

The above property is used, for example, in the evaluation of dynamic scattering for the Williams–Götze process (Sect. 2.3.1). The intermediate structure function $S(Q,t)$ is reduced by a 'structure factor fraction' to visualize the correlation function behavior across the dispersion zone of the dynamic glass transition, including $S(Q,t) \to 0$ for large times in the liquid zone (Figs. 2.5a and 2.19d). Although the correlation time depends on the wave vector Q, the peak structure of $S(Q)$ is not reflected in the reduced dynamic scattering. As mentioned above, the disappearance of $S(Q,t)$ is not accompanied by a change in $S(Q)$. If the dynamic heterogeneity corresponds to a spatiotemporal density pattern, this must be reflected in the structure factor $S(Q)$. The 'decay' of the pattern across the dispersion zone, however, is reflected neither in the structure factor $S(Q)$ nor in the correlation function $c(r)$.

Let us now turn to subsystems.

Subsystems for Fluctuations. Statistical Independence. Assume that the α process really has a characteristic length $\xi_\alpha(T)$ of order 3 nm at the glass temperature T_g, with a strong decrease down to 0.5 nm in the crossover region. Such a length is largely disengaged from the liquid structure (Sect. 2.2.5). We connect this length with dynamic heterogeneity, i.e., exclusively with fluctuations. This length is defined as the size of the smallest subsystems (cooperatively rearranging subsystems CRRs) whose fluctuations are statistically independent from the environment. The analysis must be based on a sound definition of *subsystems* that can be representative for a large sample (Fig. 3.1a). We continue here the discussion of the gedankenexperiment in Sect. 2.2.5.

A statistically independent subsystem inside a large total system, sufficiently distant from the walls of the total system, will be called a *freely fluctuating subsystem*. It is not defined by such boundaries as could restrict its fluctuations. It is surrounded on all sides by equivalent subsystems that themselves fluctuate freely. Its size is defined, for example, by a small scattering volume inside a large sample cell, or a number of smallest independent subsystems. In a homogeneous liquid, i.e., without dynamic heterogeneity, the desired additivity of such subsystems is defined via the nonoverlapping covering induced by the Riesz representation theorem of probability theory (see below). In a liquid with a pattern of dynamic heterogeneity, additivity can be defined via this physical pattern (Fig. 2.12d).

Representativeness was defined with reference to the gedankenexperiment of Sect. 2.2.5. Each representative subsystem can represent the dynamic susceptibilities and the thermal fluctuations, apart from scaling, of the total system (sample) in a certain dispersion zone, e.g., for the α process. All subsystems are representative if they are larger than a typical cell size of the pattern. Since the pattern is related to the α process dispersion zone, the representativeness is functional to this dispersion zone (see below). If such subsystems are much smaller than the sample volume then they are freely fluctuating, in particular, their fluctuations do not depend on the boundaries of the total system (e.g., isobaric, isochoric, adiabatic, isothermal, etc.). The smallest representative subsystem functional to the α process was called a cooperatively rearranging region or CRR (Sect. 2.2.5).

The scaling properties of representative subsystems yield the representativeness theorem important for the input of Levy distribution into thermodynamics before any distribution of statistical mechanics can be constructed (see below).

A linear response experiment or a dynamic scattering experiment in the linear region does not create new states that would not be covered by fluctu-

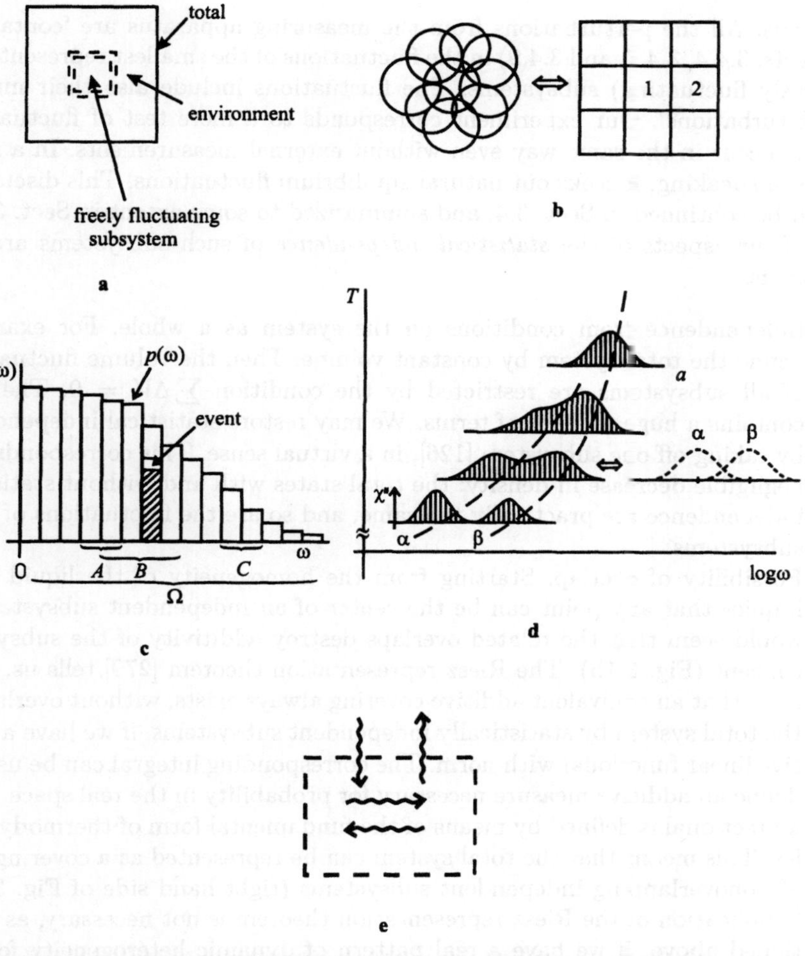

Fig. 3.1. Subsystems defined by independent fluctuations inside the sample. (a) Thermodynamic situation for freely fluctuating subsystems. (b) The Riesz representation theorem allows us to consider a covering of the sample by nonoverlapping subsystems. (c) Illustration of the sample space for a probability histogram. (d) Overlap of two dispersion zones (α, β) in the temperature–mobility plot. (e) Measurement of susceptibilities by exchange of quanta inside and between freely fluctuating subsystems

ations. All the perturbations from the measuring apparatus are 'contained' (Sects. 3.4.4, 3.4.7, and 3.4.9) in the fluctuations of the smallest representative (freely fluctuating) subsystems. The fluctuations include also their mutual 'perturbations'. Our experiment corresponds to a mere test of fluctuations that occur in the same way even without external measurements. In a manner of speaking, it seeks out natural equilibrium fluctuations. This discussion will be continued in Sect. 3.4, and summarized to some extent in Sect. 3.4.4.

Four aspects of the *statistical independence* of such subsystems are important.

- Independence from conditions on the system as a whole. For example, define the total system by constant volume. Then the volume fluctuations of all subsystems are restricted by the condition $\sum \Delta V = 0$. The sum contains a huge number of terms. We may restore statistical independence by taking off one subsystem [126], in a virtual sense. This corresponds to a negligible decrease in density: the total states with and without statistical independence are practically the same, and so are the fluctuations of their subsystems.
- Possibility of overlap. Starting from the homogeneity of the liquid state implies that any point can be the center of an independent subsystem. It would seem that the related overlaps destroy additivity of the subsystem concept (Fig. 3.1b). The Riesz representation theorem [279] tells us, however, that an equivalent additive covering always exists, without overlaps of the total system by statistically independent subsystems, if we have a positive linear functional with norm. The corresponding integral can be used to define an additive measure necessary for probability in the real space. Such a functional is defined by means of the fundamental form of thermodynamics. This means that the total system can be represented as a covering sum of nonoverlapping independent subsystems (right-hand side of Fig. 3.1b). [Application of the Riesz representation theorem is not necessary, as mentioned above, if we have a real pattern of dynamic heterogeneity for the α or ϕ process. However, this theorem is required to understand such a pattern if we start out from a homogeneous liquid concept or situation (Fig. 3.1b).]
- Functionality of CRRs was defined as relevance to Fourier components in a certain dispersion zone (Sect. 2.2.5, Fig. 2.15d). For the α process, for instance, only the Fourier components of fluctuations in the frequency region of the α dispersion zone are relevant. It is therefore statistical independence related to these slow components that is important for the α components (Fig. 3.1c). The CRRs and corresponding dynamic heterogeneity are exclusively related to these components. The large and slow temperature fluctuation of CRRs is therefore also functional to α, e.g., it cannot be quenched by thermal conductivity from fast phonons. The functional independence of different dispersion zones will be further discussed in Sect. 3.5.

- Independence from activity. As mentioned in Sect. 2.2.5, the size of CRRs, i.e., the characteristic length determined from calorimetry (2.76) or compressibility (2.78) should be the same because of the coupling of entropy and volume fluctuations via thermal expansivity (3.38). This will be discussed again in Sect. 4.3.

Statistical independence of such subsystems is disengaged from statistical independence in the sense of the Gibbs distribution. The latter is based on energetic independence, i.e., energetic additivity:

$$w(1,2) = \exp\left[-\frac{E(1,2)}{k_B T}\right] = \exp\left[-\frac{E(1)}{k_B T}\right] \cdot \exp\left[-\frac{E(2)}{k_B T}\right] = w(1) \cdot w(2)\,, \tag{3.13}$$

for $E(1,2) = E(1) + E(2)$.

Note that the separation of the Gibbs distribution into a dynamic and a configurational factor, viz., $w(p,q) = \rho_p \cdot \rho_q$ (Sects. 2.5 and 3.2), is another aspect. Equation (3.13) refers to neighboring subsystems (1, 2), whereas separability is based on $E = E_{\text{pot}}(q) + E_{\text{kin}}(p)$.

The intermolecular potentials are switched off neither for the overlap picture nor for the covering picture of Fig. 3.1b. Statistical independence directly derived from (3.13) would prevent large variations in the CRR size V_α with temperature and, in particular, would forbid any V_α smaller than the volume of the first coordination shell.

In general, energetic independence caused by spatial separation within the framework of the Gibbs distribution implies statistical independence, leading to the definition of Landau–Lifshitz subsystems. The converse statement, however, is not always true. An old example for statistical independence in systems whose parts are strongly coupled by energy is provided by playing cards. An ace of hearts couples physically the independent events of getting aces and hearts. The independence and functionality of CRRs corresponds to the game, not to the deck of cards.

Statistical Independence and the Representativeness Theorem. The general concept of *statistical independence* [279, 597, 598] will be described through the example of spectral density $x^2(\omega)$ for a stationary fluctuation of a continuous stochastic variable $x(t)$. The probability field (sample space) comprises three parts (Fig. 3.1c).

(i) The probability space Ω is the ω axis in the relevant dispersion zones for the α, β, a, ... processes.
(ii) The events are observations of contributions to the spectral density ('power spectrum') $x^2(\omega)$ in certain frequency ω intervals, e.g., A, B, C of Fig. 3.1c.
(iii) The probability density $p(\omega)$ for events is the normalized spectral density $\tilde{x}^2(\omega) \geq 0$ with

$$\int_{\mathrm{DZ}} \tilde{x}^2(\omega)\,\mathrm{d}\omega = 1\,, \tag{3.14}$$

related to the dispersion zone(s) DZ under consideration.

We must carefully distinguish between statistical independence and mutual exclusiveness. Two events (A, B) are said to be statistically independent if, for their probabilities P,

$$P(AB) = P(A) \cdot P(B) \tag{3.15}$$

holds in the overlap region of A and B, $AB = A \cap B$. Two events (A, C) are mutually exclusive if

$$A \cap C = 0 \quad \Rightarrow \quad P(A + C) = P(A) + P(C)\,. \tag{3.16}$$

The following notions should not be confused. Two stochastic functions $\Delta x_1(t)$ and $\Delta x_2(t)$ with zero expectations, $\overline{\Delta x_1} = 0$ and $\overline{\Delta x_2} = 0$, are statistically independent for $\overline{\Delta x_1(t)\Delta x_2(t)} = 0$. The product equation for mean square (ms) fluctuations, $\overline{(\Delta x_1 \Delta x_2)^2} = \overline{\Delta x_1^2} \cdot \overline{\Delta x_2^2}$, the 'Schwarz equality', indicates maximal *interdependence*, whereas the product in (3.15) or the product of spectral densities, $\Delta x_1^2(\omega) \cdot \Delta x_2^2(\omega)$ in (3.17) below, means statistical *independence*.

We shall now discuss three important examples for the dynamic glass transition.

Example 1. Let us continue the discussion of freely fluctuating subsystems. This includes the representativeness theorem which could, I think, help to replace the conventional aspects of liquid theory. Consider representative functional subsystems for the α process and two equivalent nonoverlapping neighboring subsystems 1 and 2 (Fig. 3.1b). Consider the conditional probability $P(1 \mid 2)$, where the condition 2 means that the second subsystem is a neighbor, in contrast to the trivial case that the second subsystem is so far away that independence is evident. The two subsystems are statistically independent if $P(1 \mid 2) = P(1)$. The same holds for $P(2 \mid 1) = P(2)$. This means that the spectral densities of the two equivalent subsystems are the same and that both are proportional to the spectral density of any sum of nonoverlapping subsystems, i.e., of any subsystem. The spectral density of any, including the smallest such subsystem (CRR), is representative of the whole sample, as long as the influence of the boundary of the large total system can be neglected (as for small CRRs in a large bulk sample).

The representation condition severely restricts the shape of the spectral density. Assume that we could distinguish the probability spaces of the two subsystems, Ω_1 with the frequency variable ω_1, Ω_2 with the frequency variable ω_2. Consider additive stochastic variables in these spaces, $\mathbf{X} = \mathbf{X}_1 + \mathbf{X}_2$ (thermodynamically extensive variables). Then the probability for observing a spectral density in a region A_{12} of the $\omega_1\omega_2$ plane for independent fluctuations reads

$$P(A_{12}) = \int_{A_{12}} \tilde{x}_1^2(\omega_1) \cdot \tilde{x}_2^2(\omega_2)\, \mathrm{d}\omega_1\, \mathrm{d}\omega_2 \,. \tag{3.17}$$

Measuring the relevant susceptibility by the FDT with its ω identity means mapping $\{\omega_1, \omega_2\} \mapsto \{\omega\}$, where $\omega_1 = \omega_2 = \omega$, i.e., the reduction to one common variable ω. Then, from (3.17), we get a convolution [279] for $\mathbf{X} = \mathbf{X}_1 + \mathbf{X}_2$ on the dispersion zone DZ,

$$\tilde{x}^2(\omega) = \int_{\mathrm{DZ}} \tilde{x}_1^2(\omega - \omega')\, \tilde{x}_2^2(\omega')\, \mathrm{d}\omega' = \tilde{x}_1^2 * \tilde{x}_2^2(\omega) \,. \tag{3.18}$$

The representation condition requires $\{\tilde{x}^2(\omega), \tilde{x}_1^2(\omega), \tilde{x}_2^2(\omega)\}$ to be equivalent. The ω identity narrows this equivalence condition to a proportionality, and the normalization condition (3.14) yields the identity

$$\tilde{x}^2(\omega) = \tilde{x}_1^2(\omega) = \tilde{x}_2^2(\omega) \,, \tag{3.19}$$

for $\mathbf{X} = \mathbf{X}_1 + \mathbf{X}_2$. Remember that statistical independence yields (3.17), the ω identity yields (3.18), and normalization yields (3.19).

Equation (3.18) and its enlargement to all binary combinations of representative subsystems, also of arbitrary sizes,

$$c_1\mathbf{X}_1 + c_2\mathbf{X}_2 = c\mathbf{X} \,, \tag{3.20}$$

with arbitrary scaling constants, is a very restrictive condition for the spectral density. This 'stability condition' for probability densities restricts their shape to stable distributions, i.e., Levy distributions with Levy exponent $\alpha \leq 2$ (Sect. 3.6). Special cases are the Gauss function of ω ($\alpha = 2$, not a priori suitable for relaxation without resonances), the Debye retardation which corresponds to the Cauchy distribution ($\alpha = 1$), and those Levy distributions which correspond to Kohlrausch correlation functions for retardation of compliances ($\alpha = \beta_{\mathrm{KWW}} < 1$). As we seek a dynamic approach to the a, α, and ϕ processes, we also need a dynamic reason for the restriction of the α range. It is therefore the relaxation character of dynamics that restricts the Levy exponents to $0 < \alpha \leq 1$. The three conditions for (3.20) and this relaxation character thus lead to the *representativeness theorem*. Consider the classical spectral density for stationary fluctuation of free volume or corresponding 'free entropy' in a retardation regime of a freely fluctuating subsystem that is a representative for the total system. This spectral density is a symmetric Levy limit distribution $p(x)\mathrm{d}x$ with measure $\mathrm{d}x = \mathrm{d}\omega$ and Levy exponent $\alpha \leq 1$.

This theorem is a physical expression of the freedom connected with the partition procedure of the Sect. 2.2.5 gedankenexperiment. The freedom is reflected in the arbitrariness of the scaling constants $\{c_1, c_2, c\}$ of (3.20). Freed from the 'equipartition' of the Gaussian distribution, the free volume can in principle be more or less concentrated in one or a few preponderant partial systems or subsystems. The equivalence (3.20) of arbitrary partial systems, with their scalings, regulates the concentration and 'stabilizes' the limit distribution to Levy distributions. Although the Levy distribution for $\alpha \leq 1$ clearly lies outside any conventional Gaussian statistics, the smallness of free volume guarantees a certain subtlety in the action of the Levy distribution.

A physical illustration for the dynamic character of the $\alpha \leq 1$ condition may be the idea that overdamping of molecular motion is necessary for the Glarum defects or the speckles. They need correspondingly slow, sublinear diffusion for their existence in the long τ_α or τ_ϕ time scales; otherwise, for underdamping, they would be crushed by the oscillation.

The restriction to symmetrical spectral densities [no centering constant in (3.20)] follows from the classical context (no Planck constant for the usual glass transition, Sect. 3.4.2). The representativeness theorem will be applied to the cooperativity situation inside CRRs (Sect. 4.3, 'internal application') and to the large-scale situation with many CRRs, as for the Fischer modes (Sect. 4.4, 'external application', see also Fig. 2.17a in Sect. 2.2.7).

The representativeness theorem may be the reason for the common occurrence of Kohlrausch correlation functions for compliances at the glass transition.

Other familiar functions such as the Havriliak–Negami (HN) function (2.14) with exponents $b < 1$ and $g < 1$ are not stable in the sense of (3.20). This means that, for a sample in which an HN function is measured, representative subsystems for the whole dispersion zone, including the wings, do not exist. The way out to save the CRR concept is again indicated by the ω identity of the FDT. This allows a separate consideration of any limited frequency interval of the dispersion zone.

We now continue the discussion of Sect. 2.4.5 concerning the consistency of different retardation functions. A logarithmically ($\log \varepsilon''$–$\log \omega$) linear wing at the low-frequency side of the loss peak corresponds, for slope $b < 1$, more to self-similarity than to limit stability with $b = 1$ (white noise without physical scaling, Fig. 2.2e). According to general scaling (Sect. 2.2.5), for this wing of the HN function (Figs. 2.14d and 2.15c), we expect a power-like length scaling for the low frequency region, with the larger length for the smaller frequency. This prevents the definition of independent subsystems: such a length dependence must lead to correlations between neighboring subsystems as long as such modes do not establish a linear hydrodynamic regime. On the other hand, the high-frequency wing of the HN function may correspond to a Levy exponent $\alpha = bg$. The limitations of this wing far above the loss maximum frequency are given by the widespread occurrence of the Nagel wing (Fig. 2.28f).

The functionality of CRRs with their characteristic length ξ_α is thus restricted to the central part of the compliance loss peaks for the α retardation. An approximation of the Levy part by a logarithmic Gaussian function (Gaussian bell curve on $\log\omega$) facilitates the input of thermodynamic variables via the local time–temperature equivalence, $\mathrm{d}\log\omega/\mathrm{d}t = \delta\log\omega/\delta T$ [(2.138), Fig. 2.28g], since fluctuations of thermodynamic variables are conventionally distributed in Gaussian manner to a good approximation [313]. The subtlety needed for consistency between the dynamic Levy and thermodynamic Gaussian distributions is guaranteed by the smallness of the free volume resulting from the Levy situation. This will be further discussed in Sects. 3.5 and 4.3.

Example 2. Let us now discuss $\alpha\beta$ *separation* in the crossover region. Consider two neighboring dispersion zones, e.g., α and β in Fig. 3.1d, where overlapping spectral densities $x_\alpha^2(\omega)$ and $x_\beta^2(\omega)$ of additive variables, $x = x_\alpha + x_\beta$, are not excluded. The problem is to separate $x_\alpha^2(\omega)$ from $x_\beta^2(\omega)$ with a minimum of preconditions. Let us assume that we can distinguish the frequencies ω_α and ω_β (Fig. 2.10e) and that we can a priori define the function type for α and β retardation, e.g., each being a Havriliak–Negami HN function with adjustable parameter sets $\{b, g, \omega_{\max}, \Delta\varepsilon\}_\alpha$ and $\{b, g, \omega_{\max}, \Delta\varepsilon\}_\beta$ as for the dielectric response. Then separation can be managed by (3.16) assuming only that α and β signals are mutually exclusive, referring to different probability spaces, ω_α and ω_β. The ω identity of FDT then operates only on the ω_α and ω_β axes, not in the $\omega_\alpha\omega_\beta$ plane, and we get additivity of the spectral densities,

$$x^2(\omega) = x_\alpha^2(\omega) + x_\beta^2(\omega) \,, \tag{3.21}$$

which means $\varepsilon^*(\omega) = \varepsilon_\alpha^*(\omega) + \varepsilon_\beta^*(\omega)$ for the dielectric compliance, for example. This procedure has weaker preconditions than the assumption that α and β are statistically independent fluctuations, $x^2(\omega) \sim x_\alpha^2(\omega) \cdot x_\beta^2(\omega)$, leading to the Williams convolution formula for their correlation functions (Sect. 3.4.1) [238]. Statistical independence is here an additional restriction and contradicts the picture in Fig. 2.33. The cooperativity shell influences both the local β process and the cooperative α process. Their retardations are not therefore a priori independent, especially not in the overlap region of Fig. 3.1d where the retardation times are comparable. This discussion will be continued in Sect. 3.4.1, Table 3.1.

Example 3. We shall now comment on the correlation function. The Fourier transform of a probability density $p(x)$ is called a characteristic function $f(t)$ in probability theory,

$$f(t) = \int p(x)\,\mathrm{e}^{\mathrm{i}xt}\mathrm{d}x \,. \tag{3.22}$$

Since the Fourier transform of a convolution is the product of the two components, the characteristic function for two independent distributions (3.18) is the product of their characteristic functions,

$$f(t) = f_1(t) \cdot f_2(t) \,. \tag{3.23}$$

Consider the spectral density $x^2(\omega)$ as the probability density. Its characteristic function is proportional to the correlation function $x^2(t)$ (2.72). For $x^2(\omega) \sim x_\alpha^2(\omega) \cdot x_\beta^2(\omega)$, however, we must use the inverse Fourier transformation and would get a convolution of their correlation functions, as mentioned above. The characteristic function, and therefore the correlation function, is not in general a probability density, because $f(t)$ can be negative. The condition $\left|x^2(\tau)\right| < x^2(0) = \overline{x^2}$ alone does not exclude negative values of $x^2(\tau)$. More details will also be described in Sect. 3.4.1.

Freely fluctuating subsystems as the subject of experiments involving internal and external exchange of quanta (Fig. 3.1e) will be discussed on the basis of the FDT in Sect. 3.4.

3.2 Gibbs Fluctuations and Thermodynamic Fluctuations

The aim of this section is to compare thermodynamic fluctuations for freely fluctuating subsystems with statistical fluctuations that can be calculated from the Gibbs distribution. We shall see that the Gibbs distribution cannot model genuine temperature fluctuations. If the complete collection of all possible fluctuations is really important for understanding the dynamic glass transition, in particular for the α process, its statistical treatment with the aid of the Gibbs distribution must be considered with caution.

Referring to freely fluctuating subsystems (Fig. 3.1a), we use the following symbols for the number of states Γ, the energy E, and the other variables such as temperature:

$$\begin{array}{ll} \text{subsystems without indices} & \Delta E,\, \Delta\Gamma,\, T,\, \dots\,, \\ \text{environment with an upper prime} & \mathrm{d}E',\, \mathrm{d}\Gamma',\, T',\, \dots\,, \\ \text{total system with an upper zero} & E^0,\, V^0,\, T^0,\, \dots\,. \end{array} \tag{3.24}$$

The fluctuations of extensive variables such as ΔE and $\Delta(\ln \Delta\Gamma) \sim \Delta S$ increase with subsystem size, for Gaussian statistics, in proportion to $\sqrt{N}$. Their relative fluctuations, however, decrease, e.g., $\Delta E/\bar{E} \sim 1/\sqrt{N}$. This means that, for small subsystems like CRRs ($N_\alpha \approx 100$), the relative fluctuations may be large, e.g., of order 10%. This justifies the notation ΔE, and so on, indicating that they are not considered to be infinitesimally small. A Taylor series expansion up to the second term and perhaps beyond seems meaningful. For larger subsystems, however, all relative fluctuations are small (of order 10^{-10} for $N \approx 10^{20}$). This sharpening of fluctuations will be indicated by the infinitesimal notation $\mathrm{d}E'$, etc.

This difference is also important for frequency ω and wave vector Q partitions of fluctuations (Sect. 3.1). Neither linear response nor dynamic scattering leads to any sharpening of distributions in the $\log \omega$ and Q domain. Its amplitude, however, is determined by the fluctuation of the smallest representative subsystems, the CRRs for the α process, often denoted by δS, δT, and so on, in this book. This fluctuation is by no means infinitesimally small for the dynamic glass transition.

It is well known that energy plays a fundamental role in thermodynamics [313] because it is conjugate to the homogeneity of time as represented, for example, by the equilibrium concept (no change after removing insulating walls) or by the stationary nature of equilibrium fluctuations. In the following, we discuss two variants (Gibbs vs. von Laue) with regard to:

- how energy is introduced,
- how the subsystem–environment relationship is managed,
- how the temperature is put in.

3.2.1 Gibbs Treatment of Statistical Mechanics

Here, energy E comes from *mechanics*. The degeneracy of eigenvalues in many-particle groups is counted by statistical weights $\mathrm{d}\Gamma = \Delta\Gamma\,\mathrm{d}\Gamma'$ in the probability distribution of the microcanonical ensemble,

$$\mathrm{d}w = \text{const.} \times \delta\,(E + E' - E^0)\,\Delta\Gamma\mathrm{d}\Gamma' . \tag{3.25}$$

Consider several statistically independent subsystems with $w^{(1)}, w^{(2)}, w^{(3)}, \ldots$ [313]. Then from statistical independence, $w = w^{(1)} \cdot w^{(2)} \ldots$, from additivity of energy, $E = E^{(1)} + E^{(2)} + \ldots$, and from the property that the energy is a mechanical integral, we obtain

$$\log w_n = \alpha + \beta E_n \tag{3.26}$$

for the probability w_n of the nth state of any subsystem. *Thermodynamics* is subsequently introduced by the integral over the environment. The $S'(E')$ function is the environment entropy as a 'sharp' function of its energy, sharp because of the infinitesimality of $\mathrm{d}\Gamma'$ and the corresponding $\mathrm{d}E'$. Let us forget the small subsystem temporarily and put $T^0 = T' = \mathrm{d}E'/\mathrm{d}S'$ for the system temperature. The question is: what is the temperature of the subsystem?

Strictly speaking, the subsystem temperature is determined in the Gibbs approach by the Zeroth Law. The mutual equilibrium between any pair of systems (here subsystem and environment) is an equivalence relation. This implies that any set of systems is partitioned into equivalence classes with the equivalence index T for equilibration by heat exchange. Equilibrium means then $T = T'$ (sharp). Such a subsystem temperature cannot fluctuate,

$$\overline{\Delta T^2} = 0 , \tag{3.27}$$

and the Gibbs distribution

$$w_n = A \exp(-E_n/k_{\mathrm{B}}T) \tag{3.28}$$

is obtained from (3.26).

This procedure for finding T is described in [313] under the questionable assumption that there should exist 'by definition' a sharp function $\Delta\Gamma(\bar{E})$, and therefore a sharp function $S \equiv k_B \ln \Delta\Gamma = S(\bar{E})$ for the subsystem itself, with $\bar{E}$ the mean value of the subsystem energy.

This assumption does not correspond to the ability of small subsystems far from the boundaries of the large total system (Fig. 3.2a) to fluctuate freely (Fig. 3.1a). Formally, such a fluctuation ability is still expressed in the Dirac delta of the microcanonical distribution by $E = E^0 - E'$ being a small difference between huge numbers (3.25). The possibility of different paths across subsystem fluctuation regions (Fig. 3.2a) corresponds to different subsystem temperatures, i.e., to *temperature fluctuations* within each subsystem.

Let us repeat this important issue in other words. The statistical weight $\Delta\Gamma$ is defined in [313] as "the number of quantum states corresponding to the interval ΔE" of order the mean energy fluctuation. The use of $\Delta\Gamma$ for entropy $S = k_B \ln \Delta\Gamma$, however, is factual as always in science: $\Delta\Gamma$ is based on factual occupation numbers that do not define, as pure numbers, any exact E value inside the ΔE interval. Different E values can be realized by different occupation-number situations, i.e., by different populations for a given $\Delta\Gamma$ number. Both $\Delta\Gamma$ and ΔE are real fluctuations with no one-to-one dependence $\Delta\Gamma(\Delta E)$ inside the small system. For subsystems, the $\Delta\Gamma/\Delta E$ ratio does not define a function in a mathematical sense. The textbook equation of [313], viz., $\Delta\Gamma \stackrel{?}{=} (\mathrm{d}\Gamma(E)/\mathrm{d}E)\Delta E$, is therefore wrong for our subsystems. It is this equation that finally eliminates temperature fluctuations from the Gibbs distribution.

▶

Fig. 3.2. Gibbs and von Laue treatment of fluctuations. (**a**) No Zeroth Law temperature can be defined for freely fluctuating subsystems, because different paths (1, 2) can be defined in the fluctuation region. (**b**) Input of minimal work to the subsystem. $\Delta W_{\min}$ is considered to be generating the fluctuation. (**c**) Schematic sequence showing genesis of (3.41) for the von Laue energy fluctuation $\overline{\Delta U^2}$ from a Gibbs $\overline{\Delta E^2}$ fluctuation with no temperature fluctuation. The von Laue fluctuation is automatically separated into a 'work part' $\sim \overline{\Delta V^2}$ and a 'heat part' $\sim \overline{\Delta T^2}$ because volume and temperature fluctuations are not correlated but are statistically independent, $\overline{\Delta V \Delta T} = 0$ (3.40). Note, however, that $\overline{\Delta S \Delta V} \neq 0$ (3.38). The slope of the level-occupation number distribution (the population, *hatched*) is the subsystem temperature $T = (\partial E/\partial S)_V$. ΔU and ΔS are shifts representing the prefactors on the right-hand side of (3.41), $(\partial U/\partial V)_T = T(\partial p/\partial T)_V - p$ and $(\partial U/\partial T)_V = T(\partial S/\partial T)_V = C_V$. The symbol Fl stands for fluctuation. It is an interesting exercise to check all differences in this diagram, both internal fluctuations or external shifts by heat or work. (**d**) Illustration of (3.29) for canonical Gibbs energy fluctuation $\langle \Delta E^2 \rangle_V$ restricted by the isochoric (V = const., no work) boundary condition and additionally restricted by the requirement of no temperature fluctuation $\overline{\Delta T^2} = 0$. The latter means no change in the population slope

Since pressure p and chemical potential μ are also equivalence class indices (for equilibration by volume work exchange and by particle exchange, respectively), one could ask why pressure fluctuation $\overline{\Delta p^2}$, for example, can be calculated from the Gibbs distribution. This is possible because work can be externally parametrized as an energy part, and heat cannot. Heat can only be reduced to equilibrium change of internal energy by the entropy, $\tilde{d}Q = T\mathrm{d}S$, leading in the end to the heat numerator $k_{\mathrm{B}}T$ in the Gibbs

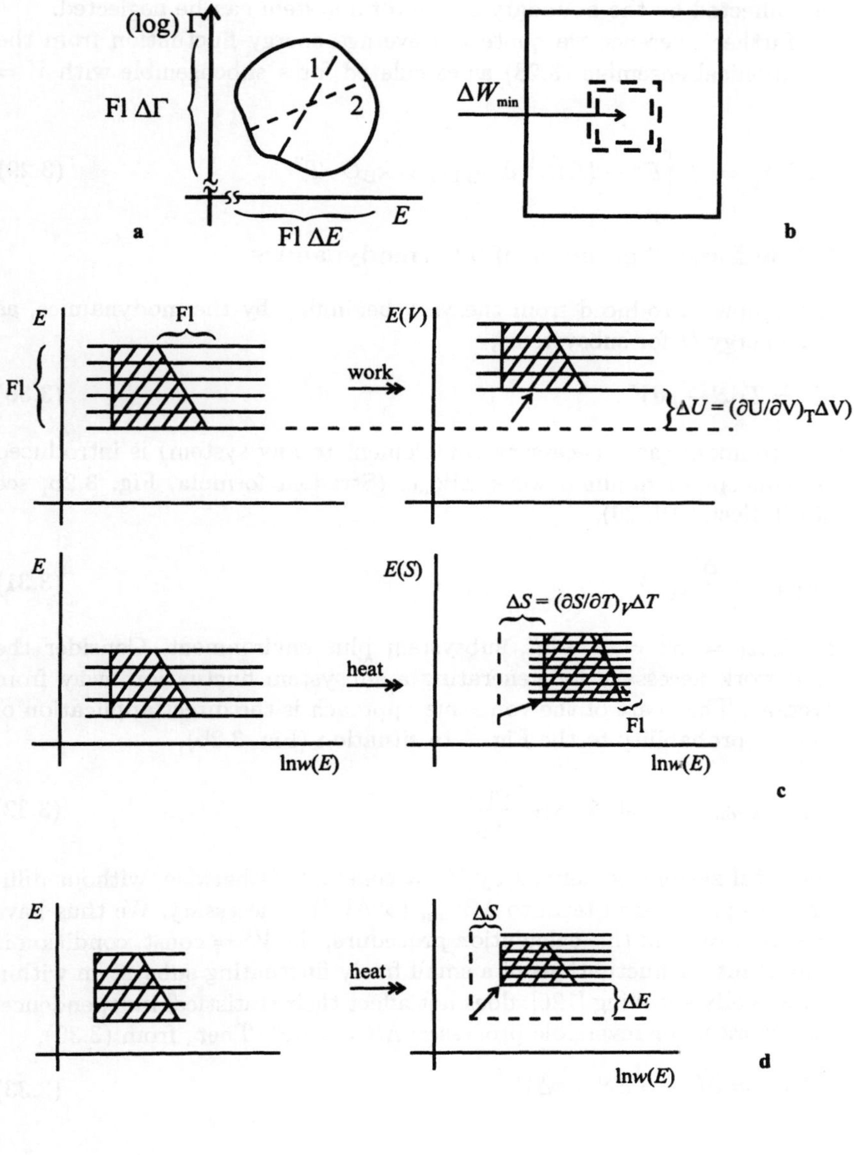

distribution. There is no corresponding work numerator $k_B'p$. Consequently, a pressure fluctuation can be calculated from the canonical ensemble (see equation (19.9) in [715]), but it depends on boundary conditions needed for the application of any ensemble calculation to fluctuations. Application of boundary conditions does not correspond, for instance, to a small scattering volume inside a sample cell (Fig. 3.1a). Neither does it correspond to calorimetry of samples with small CRRs, because any CRR is representative for the linear response of the total sample, and the overwhelming number of CRRs correspond to the freely fluctuating Fig. 3.1a situation. The few CRRs that are affected by the boundary of the total system can be neglected.

For further reference we quote the average energy fluctuation from the Gibbs canonical ensemble (3.28) as calculated for a subensemble with $V =$ const.,

$$\langle \Delta E^2 \rangle_V = \int \left[E^2 - \langle E \rangle_V^2 \right] \, \mathrm{d}w_{\mathrm{Gibbs}} = k_B C_V T^2 \, . \tag{3.29}$$

3.2.2 Von Laue Treatment of Thermodynamics

Energy is now introduced from the very beginning by thermodynamics, as internal energy U for subsystems,

$$\mathrm{d}U = T\mathrm{d}S - p\mathrm{d}V \, . \tag{3.30}$$

The environment (as a necessary complement to any system) is introduced by the concept of minimal work ΔW_{min} (Stradola formula, Fig. 3.2b, see also [313], Sects. 19, 20),

$$\Delta S_{\mathrm{total}} = \frac{\Delta W_{\mathrm{min}}}{T^0} \, , \tag{3.31}$$

where $S_{\mathrm{total}} = S^0 = S + S'$, subsystem plus environment. Consider the minimal work necessary for generating a subsystem fluctuation away from the average. The heart of the von Laue approach is the direct application of Boltzmann probability to the Fig. 3.1a situation (Fig. 3.2b),

$$w_{\mathrm{fluctuation}} = \mathrm{const.} \times \exp \frac{\Delta S_{\mathrm{total}}}{k_B} \, . \tag{3.32}$$

Let the total system be defined by $V^0 =$ constant. Otherwise, without difficulties arising, a second term to ΔW_{min} ($\propto \Delta V^0$) is necessary. We thus have $\Delta V = -\Delta V'$. Beyond this calculation procedure, the $V^0 =$ const. condition is not important for fluctuations of a small freely fluctuating subsystem within and, practically speaking [126], does not affect their statistical independence. Minimal work is for reversible processes: $\Delta S = -\Delta S'$. Then, from (3.30),

$$\Delta W_{\mathrm{min}} = \Delta U - T\Delta S + p\Delta V \, , \tag{3.33}$$

where once again the Δ for the subsystem need not be infinitesimally small. This opens the way to calculation of subsystem fluctuations. The formal treatment of $T\Delta S$ and $p\Delta V$ is the same, so that temperature fluctuations are not excluded. From a second order series expansion in Δ, we obtain

$$w_{\text{fluctuation}} = \text{const.} \times \exp \frac{\Delta p \Delta V - \Delta T \Delta S}{2\, k_{\mathrm{B}} T} . \tag{3.34}$$

We then put $T^0 = T$, since the treatment of fluctuations is separated from the average temperature of equilibrium. Equation (3.34) implies, in contrast to (3.27) of the Gibbs approach, a nonzero temperature fluctuation,

$$\overline{\Delta T^2} = \frac{k_{\mathrm{B}} T^2}{C_V} , \tag{3.35}$$

and yields, for comparison with (3.29), another formula for fluctuations in the internal subsystem energy:

$$\overline{\Delta U^2} = -k_{\mathrm{B}} \left[T \left(\frac{\partial p}{\partial T} \right)_V - p \right]^2 T \left(\frac{\partial V}{\partial p} \right)_T + k_{\mathrm{B}} C_V T^2 . \tag{3.36}$$

Using the notation $\Delta x^2 = (\Delta x)^2$ and $\Delta f^2 = (\Delta f)^2$, fluctuations in the thermodynamic variables read [313]:

$$\overline{\Delta T^2} = \frac{k_{\mathrm{B}} T^2}{C_V} , \quad \overline{\Delta p^2} = -k_{\mathrm{B}} T \left(\frac{\partial p}{\partial V} \right)_S , \quad \overline{\Delta p \Delta T} = \frac{k_{\mathrm{B}} T^2}{C_V} \left(\frac{\partial p}{\partial T} \right)_V , \tag{3.37}$$

$$\overline{\Delta S^2} = k_{\mathrm{B}} C_p , \quad \overline{\Delta V^2} = -k_{\mathrm{B}} T \left(\frac{\partial V}{\partial p} \right)_T , \quad \overline{\Delta S \Delta V} = k_{\mathrm{B}} T \left(\frac{\partial V}{\partial T} \right)_p , \tag{3.38}$$

$$\overline{\Delta V \Delta p} = -k_{\mathrm{B}} T , \quad \overline{\Delta S \Delta T} = k_{\mathrm{B}} T , \tag{3.39}$$

$$\overline{\Delta V \Delta T} = 0 , \quad \overline{\Delta S \Delta p} = 0 , \tag{3.40}$$

$$\overline{\Delta U^2} = \left[T \left(\frac{\partial p}{\partial T} \right)_V - p \right]^2 \overline{\Delta V^2} + C_V^2 \overline{\Delta T^2} , \quad \overline{\Delta H^2} = V^2 \overline{\Delta p^2} + T^2 \overline{\Delta S^2} , \tag{3.41}$$

where $H = U + pV$ is the enthalpy and, from (3.38),

$$\overline{\Delta N^2} = k_{\mathrm{B}} T \left(\frac{\partial N}{\partial \mu} \right)_{T,V} = -\frac{k_{\mathrm{B}} T N^2}{V^2} \left(\frac{\partial V}{\partial P} \right)_T . \tag{3.42}$$

Note that all symbols x standing for extensive variables refer to values for the whole subsystem, $x \sim N$, not specific or molar quantities. We see that $\Delta x \sim \sqrt{N}$ for extensive variables and $\Delta f \sim 1/\sqrt{N}$ for intensive variables, corresponding to the Gaussian aspect of the Gibbs distribution.

A consequent treatment of freely fluctuating subsystems with three or more thermodynamic dimensions does not yet exist. The problems come from additional mutual dependence beyond the simple independence equation (3.40) for the two-dimensional case. A harmless quasi two-dimensional example is the relation between volume fluctuation (3.38) and particle fluctuation (3.42). For freely fluctuating subsystems, we may also think about particle number fluctuation for given volume or volume fluctuation for given particle number. Not so trivial (and unsolved) is the three-dimensional example with simultaneous fluctuation of entropy, concentration, and volume in a binary mixture [599] or the three-dimensional example with simultaneous fluctuation of entropy, volume, and shear angle [600]. The correction to our two-dimensional fluctuation formula for the CRR size (2.76) can amount to 10% when shear angle fluctuation is also considered [601]. A complete listing of thermodynamic cross-relations for three dimensions, the Maxwell relations, is given in [602].

Consider now two examples in two thermodynamic dimensions.

(i) The Prigogine–Defay ratio (2.154) describes the decoupling of entropy and volume fluctuations ΔS and ΔV at 'low' temperatures on the basis of (3.38). The ΔS and ΔV fluctuations there are functional to the α process.

(ii) The thermodynamic expression for the Landau–Plazek ratio Π_{LP} describes the limited decoupling of ΔS and ΔT fluctuations, or equivalently, of ΔV and Δp fluctuations on the basis of (3.37)–(3.39). This ratio is, among factors of lesser interest here, the intensity ratio of the central component to the shifted components in Brillouin light scattering. Because of the high frequency of light (of order 500 THz), the fluctuations are related to the equilibrium liquid structure. We may therefore take either ensemble averages $\langle\ \rangle$ or time averages: Π_{LP} is not functional to the α process.

$$\begin{aligned}\Pi_{\mathrm{LP}} &= \frac{C_p}{C_V} - 1 = \frac{\langle\Delta S^2\rangle\langle\Delta T^2\rangle - \langle\Delta S\Delta T\rangle^2}{\langle\Delta S\Delta T\rangle^2} \\ &= \frac{(\partial p/\partial V)_T}{(\partial p/\partial V)_S} - 1 = \frac{\langle\Delta V^2\rangle\langle\Delta p^2\rangle - \langle\Delta V\Delta p\rangle^2}{\langle\Delta V\Delta p\rangle^2}\,. \qquad (3.43)\end{aligned}$$

The Π_{LP} ratio is always positive for stable materials. Negative values are prevented by the Schwarz inequality, $\langle\Delta S\Delta T\rangle^2 \le \langle\Delta S^2\rangle\langle\Delta T^2\rangle$. Values are limited by (3.39), $\langle\Delta S\Delta T\rangle = k_{\mathrm{B}}T > 0$ or $|\langle\Delta V\Delta p\rangle| = k_{\mathrm{B}}T > 0$.

Finally, let us describe the difficulties involved in comparing the two energy fluctuations, for Gibbs and von Laue. The energy fluctuation $\overline{\Delta U^2}$ of the von Laue treatment, given by (3.41), is automatically separated into a contribution from volume fluctuation $\overline{\Delta V^2}$ and a contribution from temperature fluctuation $\overline{\Delta T^2}$, since volume and temperature fluctuations are statistically independent, $\overline{\Delta V \Delta T} = 0$ [see (3.40) and Fig. 3.2c]. An additivity of $\overline{\Delta S^2}$ and $\overline{\Delta V^2}$ terms, as might be expected from the potential $U(S,V)$ as a function of natural variables S and V, is not possible because we have $\overline{\Delta S \Delta V} \neq 0$ from (3.38). The first contribution to (3.41) is defined by the internal pressure $p_{\text{int}} = (\partial U/\partial V)_T = [T(\partial p/\partial T)_V - p]$, and the second contribution by the heat capacity at constant volume $C_V = (\partial U/\partial T)_V$. The temperature fluctuation is indicated by a fluctuation in the slope for the population in the lower right part of Fig. 3.2c. This change corresponds to the two paths $(1,2)$ in Fig. 3.2a. On the other hand, the $\overline{\Delta T^2}$ contribution is suppressed in the canonical Gibbs energy fluctuation (3.29) for $V = \text{constant}$. Although the second von Laue contribution is formally identical with the total isochoric Gibbs fluctuation, $k_B C_V T^2$, the latter can never be interpreted as the remaining temperature fluctuation for the canonical boundary condition $V = \text{const.}$, since $\overline{\Delta T^2} = 0$ for Gibbs [see (3.27)].

This situation has caused much confusion. The solution lies in the difference between subsystems: the von Laue treatment uses freely fluctuating subsystems with two thermodynamic dimensions $T\mathrm{d}S$ and $p\mathrm{d}V$, whilst the canonical Gibbs fluctuation treatment uses isochoric subensembles, with $\Delta V = 0$, in one thermodynamic dimension $T\mathrm{d}S$, and additionally, restricts the remaining 'entropy fluctuations' [ΔS in $(\partial E/\partial S)_V \Delta S$, heat] by the Zeroth Law temperature with no fluctuation, $\overline{\Delta T^2} = 0$ (Fig. 3.2d). The two shift components $(\Delta E, \Delta S)$ in Fig. 3.2d are determined by one single quantity for the Gibbs fluctuation, $C_V = (\partial E/\partial T)_V = (\partial S/\partial \ln T)_V$.

3.2.3 Differences and Relationships Between the Two Treatments

We begin with the differences. The Gibbs treatment starts from mechanics, the subsystem fluctuations are restricted by the Zeroth Law, which excludes temperature fluctuations, and the other fluctuations are affected by boundary conditions. On the other hand, the von Laue treatment starts from thermodynamics (i.e., the mechanics must be introduced afterwards by means of minimal work, as described below), the subsystem fluctuations are not restricted because the Zeroth Law is only applied to averages of the total system [T^0 in (3.31)], and the fluctuations of freely fluctuating subsystems are not affected by boundary conditions.

All temperatures are nevertheless well defined in the von Laue treatment, since the temperature fluctuations ΔT are separated before the Zeroth Law is applied to $T = T^0$ in the denominator of (3.31). This T^0 corresponds to the time average of the fluctuating temperature ($\overline{\Delta T} = 0$), and ΔT is gauged by the same equation as the average temperature. The starting equation for both is the same fundamental form (3.30).

Let us repeat the definition of *temperature* for freely fluctuating subsystems. For particles, energy E and momentum $\boldsymbol{p}$ are defined by mechanics, whereas for subsystems the energy $\mathrm{d}U$ is defined by the internal energy. The energy terms which can be parametrized by external variables (e.g., in $-p\mathrm{d}V$ with pressure p) are called work, whilst the residual term which cannot is called heat ($\tilde{\mathrm{d}}Q$). Thermodynamic equilibrium by heat exchange between subsystems defines a Zeroth Law equivalence class index called average temperature T (or T^0). The heat term is then separated into a product by $\tilde{\mathrm{d}}Q = T\mathrm{d}S$, hence also defining the entropy S. Temperature fluctuation is no problem because their gauging and calibration uses exactly the same fundamental form as the average temperature, e.g., $\mathrm{d}U = T\mathrm{d}S - p\mathrm{d}V$.

In this two-dimensional example, according to the von Laue approach, we get $\overline{\Delta T^2} = k_\mathrm{B}T^2/C_V$, etc. The restriction of $\Delta(1/C_V)$ to a certain dispersion zone in the log frequency domain is made by experiment: the fluctuation–dissipation theorem FDT with its ω identity (Sect. 3.4). The connection with mechanics is more complicated than for the Gibbs approach (see below). Even for the classical (non-quantum mechanical) case, the separation of a Maxwell factor seems insufficient for illustration of temperature. However, a new microscopic von Laue distribution has not yet been invented.

The existence of a temperature fluctuation ΔT on the nanometer scale of CRRs is not related to a hydrodynamic treatment. The volume elements needed for the definition of space $\boldsymbol{r}$ in a spatiotemporal temperature field $T(\boldsymbol{r}, t)$ for hydrodynamics seem to be larger than those needed for application inside the cooperatively rearranging regions (Sect. 4.3). In large volumes, the temperature can of course be gauged by the Zeroth Law independently from any decision between the Gibbs and von Laue approach, and the temperature fluctuation can subsequently be calibrated by $\overline{\Delta T^2} = k_\mathrm{B}T^2/C_V$. The nanometer pattern of dynamic heterogeneities with Glarum–Levy defects does not seem accessible to a hydrodynamic treatment [276].

Let us now consider the relationships between the two approaches. The von Laue treatment implies the Gibbs treatment,

$$\text{von Laue} \Longrightarrow \text{Gibbs}\,, \tag{3.44}$$

but, of course, the converse [(3.27) and (3.35)] is not true.

The starting point for understanding the implication (3.44) may be Fermi's golden rule of quantum mechanics (QM), where heat exchange between subsystem and environment is mediated by quanta,

$$\hbar\omega_\mathrm{fi} = E_\mathrm{f} - E_\mathrm{i}\,. \tag{3.45}$$

Absorption and emission are microscopically equivalent. This implies that the (reversible per definitionem) minimal work is equivalent to a change of mechanical energy E in the microcanonical QM ensemble, (3.25),

$$\Delta W_\mathrm{min} = \Delta E_\mathrm{QM}\,. \tag{3.46}$$

In the Gibbs treatment, the (3.46) equivalence is reduced to the implication (3.44) by the restriction originating from application of the Zeroth Law to the subsystem before the fluctuations are separated. If fluctuations are not important, e.g., for quantities related to translational symmetry, then the application of the Gibbs distribution is sufficient.

Notwithstanding, the von Laue treatment is by no means a purely phenomenological one. Minimal work can in principle be fed in as a mechanical energy change (3.46). We have no experience of this approach because no statistical distribution has yet been invented that can produce all fluctuations (3.36)–(3.42). The Gibbs distribution yields only exact average C_V values from which a $\overline{\Delta T^2}$ can be formally calculated a posteriori, although this fluctuation is not contained in it.

Let us finish by discussing some problems and suggestions for a microscopic von Laue distribution, as is obviously needed for a comprehensive microscopic theory of minimal representative freely fluctuating subsystems of nanometer size in liquids and disordered materials.

The largest conceptual disadvantage of the Gibbs distribution is its *separability*. If the energy is a sum of kinetic and potential energy,

$$E = E_{\text{kin}}(p) + E_{\text{pot}}(q) \, , \tag{3.47}$$

where E_{kin} depends only on the momenta p and E_{pot} depends only on the coordinates q, then the quasiclassical Gibbs distribution separates into a kinetic and a potential factor,

$$\mathrm{d}w_{\text{Gibbs}} = \rho_p \mathrm{d}p \cdot \rho_q \mathrm{d}q \tag{3.48}$$

(Sects. 2.5 and 3.1), and the quasiclassical sum of states for the Gibbs distribution separates into a kinetic factor depending only on temperature, $M_p(T)$, and a configurational integral, $Q(V,T)$, viz.,

$$Z = M_p(T) \cdot Q(V,T) \, , \tag{3.49}$$

where $M_p(T)$ is the Maxwell factor, and

$$Q(V,T) = \frac{1}{N!} \int\limits_{(V)} \mathrm{d}q \exp\left[-\frac{E_{\text{pot}}(q)}{k_{\text{B}}T} \right] . \tag{3.50}$$

Since the dynamics is generated by the molecular momenta p, the fluctuational dynamic heterogeneity is bound to temperature and therefore forbidden in the Gibbs distribution, because temperature cannot fluctuate there (3.27).

To find a fluctuating pattern of dynamic heterogeneity, we need a *microscopic von Laue distribution* with at least the following three possibilities. [An instant $p \to (1+\varepsilon)p$ scaling [603, 604] in the microcanonical ensemble would not appear adequate to this task.]

(i) **Introduction of Dynamic Levy Distribution for Spectral Density of Free-Volume Fluctuation.** Small representative subsystems have relatively large fluctuations that can easily go beyond the Gaussian distribution motivated [345] by the extensiveness of entropy in the 'Boltzmann exponent' of (3.32). This is implied by the non-infinitesimality symbol Δ in the phenomenological von Laue distribution (3.32)–(3.34) that can give, for small systems, significant contributions beyond the quadratic Gaussian term. Moreover, free volume is a small part of total volume [small contrast equation (2.150)] so that third order terms can enter the Δ without seriously disturbing the second order Gaussian terms for calculating average fluctuations.

(ii) **Giving up the Quasi-Classical Separability of the Gibbs Distribution.** This would allow us to construct patterns for dynamic heterogeneity from statistical mechanics, perhaps without the need to base them on the configurational integral. In an ad hoc approximation, we could divide the system into the cell walls (cooperativity shells) accessible to Gaussian statistics and the Glarum–Levy defects not accessible to Gaussian statistics, e.g.,

$$E \overset{?}{\approx} E_{\mathrm{pot}}(q, \mathrm{Gaussian}) + E_{\mathrm{kin}}(p, \mathrm{Gaussian}) + E_{\mathrm{defect}}(p, q, \mathrm{Levy}) \,. \quad (3.51)$$

I believe that statistical mechanics must be preceded by an analysis of conditions concerning the way a subsystem can be defined. Statistical independence seems the most important aspect for such a definition, and neither dynamics nor a Levy distribution should be excluded a priori from the discussion, by mere prejudice.

(iii) **Free Fluctuation of Subsystems.** To avoid artificial boundary conditions for the pattern of dynamic heterogeneity, the properties of representative freely fluctuating subsystems should be reflected by the probability measure of the microscopic von Laue distribution whose invention may be crucial for a full theoretical understanding of the dynamic glass transition.

3.3 Linear Response

Linear response is the time or frequency partition of thermodynamic variables in the linear region, $\mathrm{d}x = (\partial x/\partial f)\mathrm{d}f$ or $\mathrm{d}f = (\partial f/\partial x)\mathrm{d}x$. Some basic facts were described in Sect. 2.1.2. The aim in this section is to give details. We confine ourselves to one thermodynamic dimension in the fundamental form $\mathrm{d}U = \ldots \pm f\mathrm{d}x \pm \ldots$, i.e., one extensive variable x, the conjugate intensive variable f, the corresponding compliance $J \sim (\partial x/\partial f)$, and the modulus $G \sim (\partial f/\partial x)$ (see Table 2.1). Equation (3.52) lists the symbols generally used for the (heat, volume work, shear, dielectric, diffusion) set of energy contributions (see Table 2.1):

$$\begin{aligned} \mathrm{d}x &= \mathrm{d}S\,,\ \mathrm{d}V\,,\ V\mathrm{d}\gamma\,,\ V\mathrm{d}P\,,\ \mathrm{d}N_i\,,\\ f &= T\,,\quad p\,,\quad \sigma\,,\quad E\,,\quad \mu_i\,,\\ J &= J_S\,,\ B\,,\ J\,,\quad \alpha\,,\quad -\,,\\ G &= K_T\,,K\,,\ G\,,\quad M\,,\quad -\,. \end{aligned} \tag{3.52}$$

The treatment of cross response (e.g., $\mathrm{d}x = \mathrm{d}V$, $f = T$) is similar but lies outside the scope of this book.

For small enough perturbations (Sect. 2.1.5), the response of the subsystem is linear and causal although the mechanical molecular motion inside the subsystem is highly nonlinear and necessarily chaotic. The term 'linear' means that the response of two temporally overlapping arbitrary perturbation programs is a linear superposition of the individual responses (Boltzmann's superposition principle), and the term 'causal' means that the current response (the observable at the present time t) is only influenced by the perturbation program in the past, $t' \leq t$.

Comprehensive representations of linear response are to be found in the books by Ferry [3] and Tschoegl [96], and there is a useful review in [29]. A short and clear mathematical account is given in [605]. A more physical representation including scattering is [606]. One of the first groups to use linear response methods for glass transition was, to my knowledge, Bennewitz and Rötger's group in Jena [607].

Step Experiments and Periodic Experiments. The situation for step programs is shown in Figs. 3.3a, b and c for the example Δx = shear angle γ and Δf = shear stress σ. The *relaxation experiment* gives the intensive variable $\sigma(t)$ after a step in the extensive variable $\gamma(t)$, with step height γ_0. The ratio

$$G(t) \equiv \frac{\sigma(t)}{\gamma_0} \tag{3.53}$$

is called the (shear) relaxation *modulus*. Conversely, the *creep* or *retardation experiment* gives the extensive variable $\gamma(t)$ after a step in the intensive variable $\sigma(t)$, with step height σ_0. The ratio

$$J(t) \equiv \frac{\gamma(t)}{\sigma_0} \tag{3.54}$$

is called the (shear) retardation *compliance*. Typical time scales are called relaxation and retardation times (τ_{R}, $\tau_{\mathrm{R}'}$), respectively. Linearity and causality properties of thermodynamic experiments for arbitrary programs lead in a straightforward manner to convolutions,

$$\sigma(t) = \int_{-\infty}^{t} G(t-t')\,\dot{\gamma}(t')\,\mathrm{d}t' \equiv G * \dot{\gamma}(t)\,, \tag{3.55}$$

$$\gamma(t) = \int_{-\infty}^{t} J(t-t')\,\dot{\sigma}\,(t')\,\mathrm{d}t' \equiv J * \dot{\sigma}(t) \,. \tag{3.56}$$

Linearity is expressed by the additive properties of integrals. Causality is expressed through the integration limits: t' covers the whole history up to the present time t. The 'memory' functions $G(\tau)$ and $J(\tau)$, $\tau = t - t'$, evaluate the shear history given by the rates $\dot{\sigma}(t') = \mathrm{d}\sigma/\mathrm{d}t(t')$ and $\dot{\gamma}(t') = \mathrm{d}\gamma/\mathrm{d}t(t')$, respectively. The time situation of a convolution is indicated in Fig. 3.3d. It is a useful exercise, using $\gamma(-\infty) = 0$ and the relation

$$\int_{-\infty}^{t} \mathrm{d}t' \quad \Longleftrightarrow \quad \int_{0}^{\infty} \mathrm{d}\tau \,, \tag{3.57}$$

to prove the following equations:

$$\begin{aligned} G * \dot{\gamma} &= \int_{0}^{\infty} G(\tau)\,\dot{\gamma}\,(t-\tau)\,\mathrm{d}\tau = \frac{\mathrm{d}}{\mathrm{d}t}\int_{-\infty}^{t} G(t-t')\,\gamma\,(t')\,\mathrm{d}t' \\ &= G_{\mathrm{g}}\gamma(t) + \int_{-\infty}^{t} \dot{G}(t-t')\,\gamma\,(t')\,\mathrm{d}t' \,, \end{aligned} \tag{3.58}$$

where the 'glass modulus' is the short time limit of $G(\tau)$, $G_{\mathrm{g}} = G(\tau \to 0)$. The long time limit is called the equilibrium modulus, $G_{\mathrm{e}} = G(\tau \to \infty)$. From a theoretical point of view, $\tau \to 0$ and $\tau \to \infty$ mean times much shorter and much longer than typical for the relevant dispersion zone. From an experimental point of view, $\tau \to 0$ and $\tau \to \infty$ often mean times shorter and longer than the time window of the experiment. In the latter case, a check of the formulas is recommended.

Using the simultaneous substitution $\sigma \leftrightarrow \gamma$ and $G \leftrightarrow J$, (3.55)–(3.56) and (3.58) are also true for compliances. In general, relations between $\sigma(t)$ and $\gamma(t)$ are called *material equations* or *constitutive equations*. The convolutions (3.55)–(3.56) are their linear forms.

Complex variables are used for a stationary *periodic experiment*, sometimes called a *dynamic experiment* (Fig. 3.3e). This is indicated by an asterisk (not to be confused with complex conjugate). Then

$$\gamma^* = \gamma_0 \exp(\mathrm{i}\omega t) \,, \quad \sigma^* = \sigma_0 \exp\left[\mathrm{i}(\omega t + \delta)\right] \,. \tag{3.59}$$

The corresponding susceptibilities,

$$G^*(\omega) = \frac{\sigma^*}{\gamma^*} = G'(\omega) + \mathrm{i}G''(\omega) \,, \tag{3.60}$$

$$J^*(\omega) = \frac{\gamma^*}{\sigma^*} = J'(\omega) - \mathrm{i}J''(\omega) \,, \tag{3.61}$$

are called the *dynamic modulus* and *dynamic compliance*, respectively. For a given thermodynamic state, the frequency region where G^* and J^* depend nontrivially on the frequency (ω, or $\nu = \omega/2\pi$) was called the dispersion zone (Sect. 2.1.2).

The tangent of the loss angle, $\tan\delta$, is called the *loss factor*. From (3.60)–(3.61), we see that

$$\tan\delta(\omega) = G''/G' = J''/J' \,. \tag{3.62}$$

Modulus and compliance of energetically conjugate intensive and extensive variables are connected by

$$G^*(\omega) \cdot J^*(\omega) = 1 \text{ (real)} \,, \tag{3.63}$$

as can be seen from the one-dimensional definitions (3.60)–(3.61).

The real parts (G', J') are called elastic or storage parts, and the imaginary parts (G'', J'') loss parts. The latter are related to heat production in the periodic experiment (energy dissipation for shear). The relation

$$G''(\omega) \geq 0 \,, \quad J''(\omega) \geq 0 \,, \tag{3.64}$$

is sometimes called the dissipation theorem.

If resonances are excluded [consider only responses that can be built up additively from positive exponential decays, $\exp(-t/\tau_\mathrm{R})$, or exponential growths, $1 - \exp(-t/\tau_{\mathrm{R}'})$], then $G' \geq 0$, $J' \geq 0$, and from (3.62) and (3.64), we obtain $\tan\delta \geq 0$. The γ sine peak 'comes earlier' than the σ sine peak in Fig. 3.3e. This does not reflect any causality between σ and γ – both (3.55) and (3.56) are entirely equivalent as far as causality is concerned – but the $\tan\delta \geq 0$ equation, in connection with the different signs in (3.60) and (3.61) and the dissipation theorem (3.64), reflects the increase in the extensive variable $\gamma(t)$ and the decrease in the intensive variable $\sigma(t)$ in Figs. 3.3b and c. Briefly, excluding resonances means that $J(\tau)$ increases and $G(\tau)$ decreases with τ, for a given state.

Debye Relaxation. The basic facts and problems of Debye relaxation were discussed in Sects. 2.1.2 and 2.2.4. We shall now discuss further details.

The Debye relaxation describes the exponential decay in the time domain between a positive and constant modulus and for short times, the glass, $G_\mathrm{g} = J_\mathrm{g}^{-1} = \text{const.} > G_\mathrm{e}$ for long times, the equilibrium, $G_\mathrm{e} = J_\mathrm{e}^{-1} = \text{const.} > 0$. The *relaxation intensity* and the *retardation intensity* are, as for any type of relaxation function, denoted by

$$\Delta G = G_\mathrm{g} - G_\mathrm{e} \,, \quad \Delta J = J_\mathrm{e} - J_\mathrm{g} = G_\mathrm{e}^{-1} - G_\mathrm{g}^{-1} \,. \tag{3.65}$$

The *Debye formulas* for the modulus of an intensive variable f are

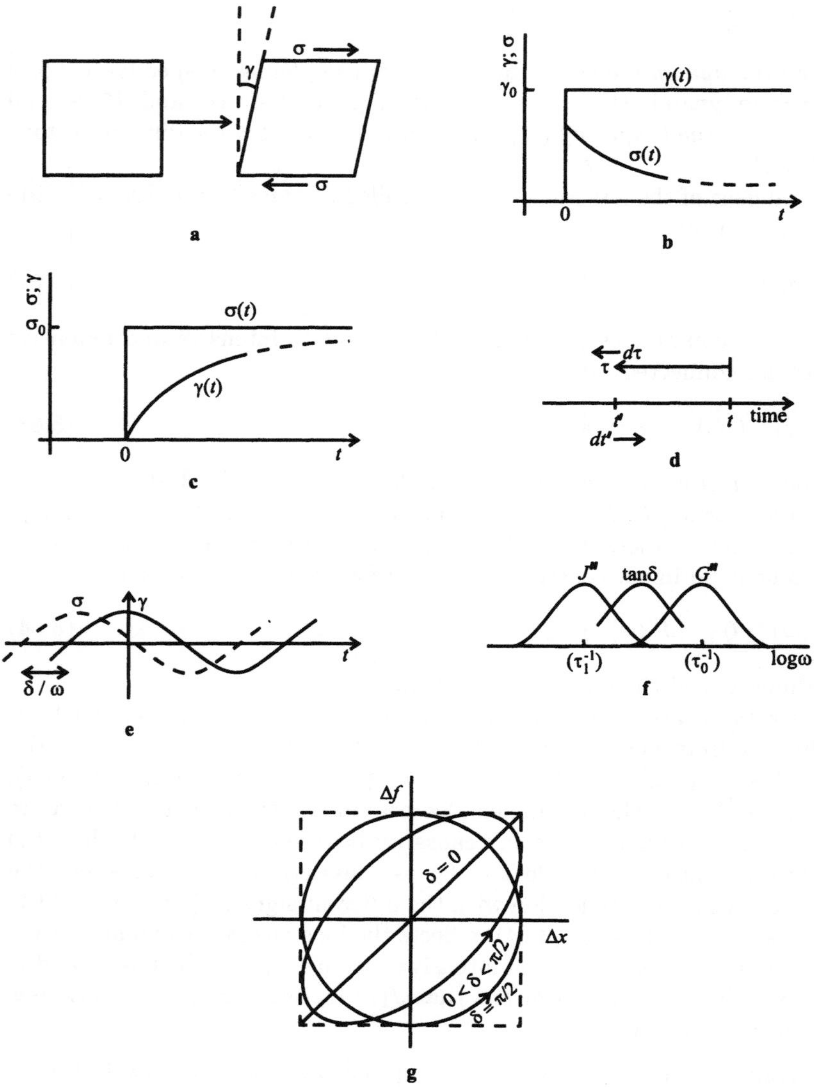

Fig. 3.3. Linear response for shear. (**a**) Shear experiment, γ shear angle, σ shear stress. (**b**) Shear stress relaxation. (**c**) Shear deformation retardation (creep experiment). (**d**) Times: t present time, t' past time, $\tau = t - t'$. (**e**) Stationary periodic experiment. δ is the (shear) loss angle. (**f**) Relative positions of the (semi-logarithmic and normalized) Lorentz lines for loss compliance J'', loss factor $\tan\delta$, and loss modulus G'', for a Debye relaxation with large relaxation intensity, $G_g/G_e = \tau_1/\tau_0 \approx 100$. (**g**) Linear response as left thermodynamic cycle. *Axes* are formally scaled to equal amplitudes Δf, Δx

$$G(\tau) = G_{\mathrm{g}} - \Delta G \exp(-\tau/\tau_0) , \quad G^*(\omega) = G_{\mathrm{e}} + \Delta G \frac{\mathrm{i}\omega\tau_0}{1 + \mathrm{i}\omega\tau_0} , \tag{3.66}$$

and for the compliance of an extensive variable x,

$$J(\tau) = J_{\mathrm{e}} - \Delta J \exp(-\tau/\tau_1) , \quad J^*(\omega) = J_{\mathrm{e}} - \Delta J \frac{\mathrm{i}\omega\tau_1}{1 + \mathrm{i}\omega\tau_1} . \tag{3.67}$$

Note that a Debye relaxation has two characteristic times. The retardation time τ_1 is larger than the relaxation time τ_0,

$$\tau_1 = \frac{G_{\mathrm{g}}}{G_{\mathrm{e}}}\tau_0 = \frac{J_{\mathrm{e}}}{J_{\mathrm{g}}}\tau_0 > \tau_0 . \tag{3.68}$$

Not only the loss parts $G''(\log\omega)$ and $J''(\log\omega)$ are Lorentz lines [see (2.68) and Fig. 2.14a], but the loss factor $\tan\delta = G''/G' = J''/J'$ is also such a curve. The maximal values are

$$\Delta G/2 , \quad \Delta J/2 , \quad (\tan\delta)_{\max} = \frac{\Delta G}{2\sqrt{G_{\mathrm{g}}G_{\mathrm{e}}}} , \tag{3.69}$$

at the maximum frequencies

$$\omega_{\max}\tau_0 = 1 \text{ for } G'' , \quad \omega_{\max}\tau_1 = 1 \text{ for } J'' , \quad \omega_{\max}\sqrt{\tau_0\tau_1} = 1 \text{ for } \tan\delta . \tag{3.70}$$

For a Debye relaxation, the Lorentz line of the loss factor, i.e., $\tan\delta$ as a function of $\log\omega$, lies exactly midway between those for $J''(\log\omega)$ and $G''(\log\omega)$ (Fig. 3.3f).

Thermodynamic Cycles. A periodic linear response experiment can be considered as a stationary anticlockwise (left) thermodynamic cycle on a subsystem (Fig. 3.3g). The curves are ellipses that degenerate to the diagonal for zero loss angle, $\delta = 0$. The parametric representation is (3.59),

$$\Delta f(t) = \Delta f \cdot \sin\omega t , \quad \Delta x(t) = \Delta x \cdot \sin(\omega t - \delta) . \tag{3.71}$$

For the extensive variable, for instance, the compliance $J^* = J' - \mathrm{i}J''$ is introduced by (3.61). From (3.71),

$$\Delta x(t) = \left[\frac{\Delta x}{\Delta f}\cos\delta\sin\omega t - \frac{\Delta x}{\Delta f}\sin\delta\cos\omega t\right]\Delta f , \tag{3.72}$$

where Δx and Δf (without the t variable) are the amplitudes. The real part J' is the term in phase with the intensive variable $\Delta f(t)$, whilst the imaginary part J'' is $\pi/2$ out of phase with $\Delta f(t)$:

$$\frac{\Delta x(t)}{\Delta f} = J'\sin\omega t - J''\cos\omega t , \quad J' = \frac{\Delta x}{\Delta f}\cos\delta , \quad J'' = \frac{\Delta x}{\Delta f}\sin\delta . \tag{3.73}$$

The energy turnover per cycle is

$$\Delta U = \oint f \, \mathrm{d}x = \text{ energy storage} + \text{heat production} . \tag{3.74}$$

The first part is stored and released two times per cycle,

$$(\Delta f)^2 \, J' \omega \oint \sin(\omega t) \cos(\omega t) \, \mathrm{d}t = 0 . \tag{3.75}$$

The real part of the compliance, J', is therefore called the *storage compliance*. The second part of (3.74) comes from the internal entropy production times the average temperature,

$$Q = T \Delta S_i = (\Delta f)^2 J'' \omega \oint \sin^2(\omega t) \, \mathrm{d}t = \pi (\Delta f)^2 J'' = \pi \Delta f \Delta x \sin \delta . \tag{3.76}$$

The imaginary part, J'', is therefore called the *loss compliance*. The integral in (3.76) is equal to the ellipse area of the thermodynamic cycle in Fig. 3.3g.

The same calculation can be made starting from the intensive variable $\Delta f(t)$, instead of $\Delta x(t)$ as in (3.72). We have to substitute

$$x \longleftrightarrow f \quad \text{and} \quad J^* = J' - \mathrm{i} J'' \longleftrightarrow G^* = G' + \mathrm{i} G'' . \tag{3.77}$$

The average heat production is Q/τ_p with $\tau_p = 2\pi/\omega$, i.e.,

$$\dot{Q}_{\mathrm{av}} = \frac{1}{2} \omega \, \Delta f \, \Delta x \sin \delta = \frac{1}{2} \omega \, (\Delta f)^2 J'' = \frac{1}{2} \omega \, (\Delta x)^2 G'' . \tag{3.78}$$

Otherwise, for the linear response, $\omega \tau = 1$ is the more appropriate relation between angular frequency ω and time intervals τ. Note that the heat pair of (3.52) ($x =$ entropy S, $f =$ temperature T) is not excluded from consideration, as long as we start from $\mathrm{d}U = T\mathrm{d}S - p\mathrm{d}V \pm \ldots$. This means that, in periodic calorimetry, additional heat is produced:

$$\dot{Q}_{\mathrm{av}} = \frac{1}{2} \omega \, \Delta T \, \Delta S \sin \delta = \frac{1}{2} \omega \, (\Delta T)^2 J_S'' = \frac{1}{2} \omega \, (\Delta S)^2 K_T'' . \tag{3.79}$$

Using the loss part of dynamic heat capacity, $C_p''(\omega) = T \, J_S''(\omega)$, we obtain

$$\dot{Q}_{\mathrm{av}} = \frac{1}{2} \omega \, (\Delta T)^2 \frac{C_p''(\omega)}{T} , \tag{3.80}$$

with T the thermodynamic temperature (in kelvin). The imaginary part C_p'' of the dynamic heat capacity thus describes the average additional entropy production of stationary periodic calorimetry [91],

$$\dot{S}_{i,\mathrm{av}} = \frac{1}{2} \omega \left(\frac{\Delta T}{T} \right)^2 C_p''(\omega, T) . \tag{3.81}$$

Note that, in practical periodic calorimetry, $C_p''(\omega, T)$ is obtained from the phase angle δ, and not from direct measurement of the additional entropy production.

For the work variables, e.g., (x = volume V, f = pressure p), the left cycle corresponds to a heat pump working between fluctuating temperature differences from temperature fluctuations of the subsystems. For a left Carnot cycle operating between T_1 and $T_2 = T_1 - \Delta\tilde{T} < T_1$, we have an efficiency $\eta = T_1/(T_1 - T_2) \approx T/\Delta\tilde{T}$. Since $\overline{\Delta\tilde{T}} = 0$ for fluctuations, we get an infinite pump efficiency corresponding to the thermal chaos with no first order heat loss.

Inside the dispersion zone, however, the phase shifts δ (different for different activities) give a second order heat effect proportional to $\sin\delta$ (3.79). Since, for freely fluctuating subsystems in two thermodynamic dimensions, the cycle area in the pV diagram is always the same as in the TS diagram ($|\partial(S,T)/\partial(V,p)| = 1$, First Law) and since, for example, temperature and pressure fluctuations are coupled (3.37), we always get a second order effect for both work and heat input, viz., (3.78) for work variables and (3.79)–(3.81) for heat variables, where $(\Delta T/T)^2$ in the latter reflects the reciprocal Carnot efficiency to second order.

General Considerations. Fourier–Laplace Transform. Kramers–Kronig Relations. Consider again an extensive observable (a 'displacement' x) and an intensive observable (a 'force' f) associated with a subsystem, energetically conjugated by the contribution to the First Law fundamental form ($f\mathrm{d}x$ or $x\mathrm{d}f$). The compliance is defined by a convolution for x,

$$\overset{\longrightarrow}{f(t')} \quad \boxed{J(\tau)} \quad \overset{\longrightarrow}{x(t)}\,, \quad x(t) = J * \dot{f}(t)\,, \tag{3.82}$$

and the general behavior at a dispersion zone is shown in Fig. 3.4a. The dynamic compliance is obtained from (3.82) by a *Fourier–Laplace transformation* of $\dot{J}(\tau) = \mathrm{d}J/\mathrm{d}\tau(\tau)$ covering the time domain of the whole dispersion zone,

$$J^*(\omega) = J_\mathrm{g} + \int\limits_0^\infty \dot{J}(\tau)\,\mathrm{e}^{-\mathrm{i}\omega\tau}\mathrm{d}\tau = J'(\omega) - \mathrm{i}J''(\omega)\,, \tag{3.83}$$

where the glass limit compliance J_g is extracted to ensure convergence. The 'area' used by the integration is hatched in Fig. 3.4b.

In a complementary way, the modulus is defined by a convolution for the intensive f variable (see Figs. 3.4c, d)

$$\overset{\longleftarrow}{f(t)} \quad \boxed{G(\tau)} \quad \overset{\longleftarrow}{x(t')}\,, \quad f(t) = G * \dot{x}(t)\,, \tag{3.84}$$

$$G^*(\omega) = G_g + \int_0^\infty \dot{G}(\tau)\,\mathrm{e}^{-\mathrm{i}\omega\tau}\mathrm{d}\tau = G'(\omega) + \mathrm{i}G''(\omega)\,. \tag{3.85}$$

Note that, inside the same dispersion zone, the extrema of the loss compliance are at longer times ($1/\omega$) and lower frequencies than those for the loss modulus. The differences on logarithmic scales ($\log\tau$, $\log\omega$) increase with the ratio of glass and equilibrium values, $\log(G_g/G_e)$ and $\log(J_e/J_g)$. For shear, for example, at both the main and flow transition of polymers, this difference amounts to 3 or 4 orders in τ or ω corresponding to 3 or 4 logarithmic decades. Generally, the modulus accentuates the transition to the solid, and the compliance accentuates the flow onset. These two phenomena are not the same in broad viscoelastic dispersion zones.

Analyzing shear for a true flow transition ($G_e = 0$), we must take care over the convergence of the integrals, for example, by two subtractions,

$$J^*(\omega) - J_s^0 - \mathrm{i}/\omega\eta = \mathrm{i}\omega F\left[J(\tau) - J_s^0 - \tau/\eta\right]\,, \tag{3.86}$$

where F stands for the Fourier–Laplace transform. The symbol $J_s^0 = \lim J'(\omega)$ for $\omega \to 0$ is called the steady state compliance or recoverable compliance of a liquid (Table 2.1, Sect. 2.1.1). It can also be calculated from the shear modulus by an integration over the relevant dispersion zone,

$$J_s^0 = \frac{1}{\eta^2}\int_0^\infty \mathrm{d}\tau\,\tau G(\tau)\,, \quad G_e = 0\,, \tag{3.87}$$

where the viscosity is given by

$$\eta = \int_0^\infty \mathrm{d}\tau\,G(\tau)\,. \tag{3.88}$$

The corresponding formulas in the frequency domain are (2.11). From Fig. 3.4e, we see that the steady state compliance J_s^0 is defined formally by the $\tau \to \infty$ limit,

$$J_s^0 = \lim_{\tau\to\infty}\left[J(\tau) - \tau/\eta\right]\,, \tag{3.89}$$

but we should remember that J_s^0 is related to the whole dispersion zone.

Modulus and compliance are connected by (2.8), $J^*(\omega)G^*(\omega) = 1$. Transformation into the time domain gives

$$\int_{-\infty}^t G(t-t')\,\dot{J}(t')\,\mathrm{d}t' = 1\,, \qquad \int_0^t J(t-t')\,G(t')\,\mathrm{d}t' = t\,. \tag{3.90}$$

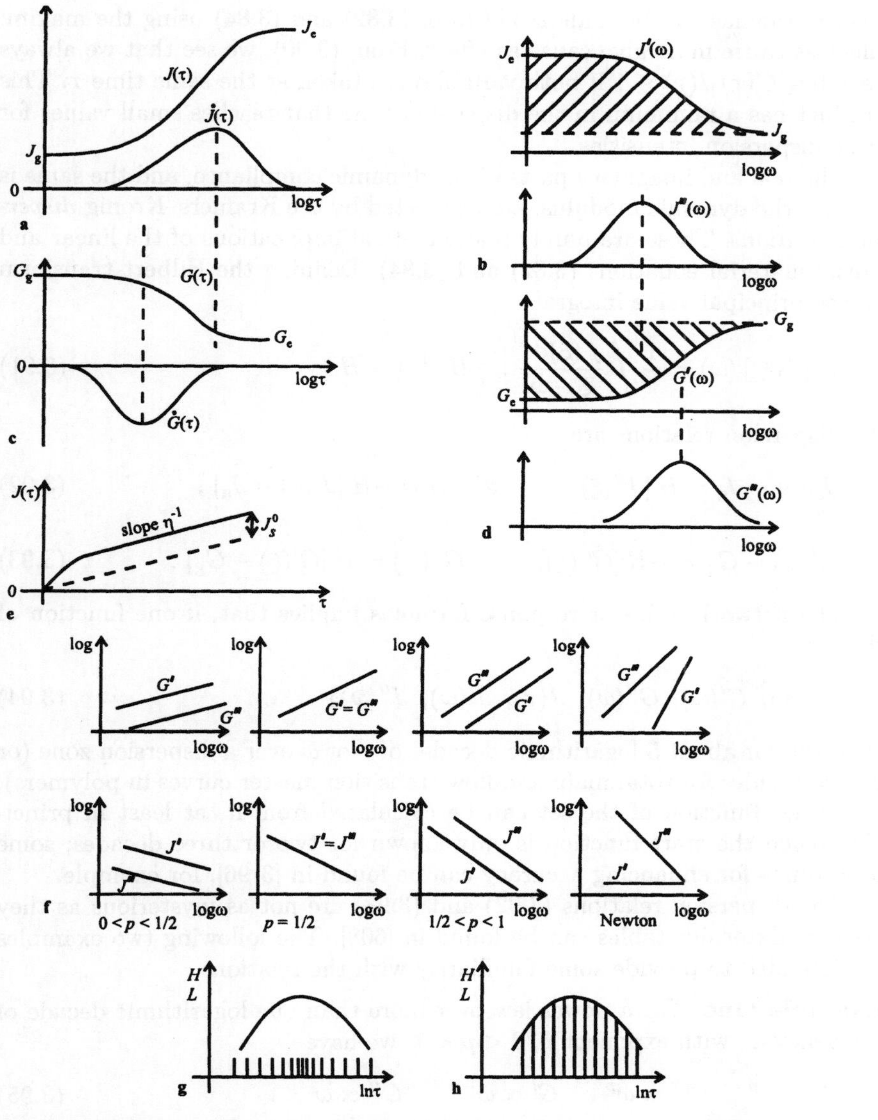

Fig. 3.4. Linear response (continued). **(a)** Compliance $J(\tau)$ for extensive observables. **(b)** Dynamic compliance $J^*(\omega) = J'(\omega) - iJ''(\omega)$. **(c)** Modulus $G(\tau)$ for intensive observables. **(d)** Dynamic modulus $G^*(\tau) = G'(\omega) + iG''(\omega)$. **(e)** Illustration of steady state compliance J_s^0 for flow transition. **(f)** Dispersion relation for power laws. $0 < p < 1/2$ solid-like behavior. $p = 1/2$ Rouse modes. $1/2 < p < 1$ liquid-like behavior. For the Newtonian liquid, all four exponents are different. **(g)** Realization of a spectrum by a distribution of elements with constant intensities (e.g., corresponding to one k_BT) over the $\ln\tau$ axis. **(h)** Fitting of element intensities at equidistant $\ln\tau$ values

These formulas can be understood from (3.82) and (3.84) using the maxim: effect as cause must give cause as effect. From (3.90), we see that we always have $0 \leq G(\tau)\, J(\tau) \leq 1$ if both variables are taken at the same time τ. This product has a minimum in the dispersion zone that reaches small values for large dispersion intensities.

The real and imaginary parts of the dynamic compliance, and the same is true for the dynamic modulus, are connected by the Kramers–Kronig *dispersion relations.* These are purely mathematical implications of the linear and causal material equations (3.82) and (3.84). Defining the Hilbert transform by the principal value integral

$$H\left[f\left(\xi\right)\right]\left(\omega\right) = \frac{1}{\pi}\int \frac{f\left(\xi\right)\mathrm{d}\xi}{\xi - \omega}\,, \quad H^{-1} = -H\,, \tag{3.91}$$

the dispersion relations are

$$J'(\omega) - J_{\mathrm{g}} = H\left[J''(\xi)\right]\,, \qquad J''(\omega) = -H\left[J'(\xi) - J_{\mathrm{g}}\right]\,, \tag{3.92}$$

$$G'(\omega) - G_{\mathrm{g}} = -H\left[G''(\xi)\right]\,, \qquad G''(\omega) = H\left[G'(\xi) - G_{\mathrm{g}}\right]\,. \tag{3.93}$$

The network of linear response formulas implies that, if one function of the set

$$\{G(\tau),\ G'(\omega),\ G''(\omega),\ J(\tau),\ J'(\omega),\ J''(\omega)\} \tag{3.94}$$

is known for about 5 logarithmic decades of τ or ω over a dispersion zone (or 10–15 decades for total main and flow transition master curves in polymers), any other function of the set can be calculated from it, at least in principle. When the start function is only known for two or three decades, some procedures for enhancing accuracy can be found in [3, 96], for example.

The dispersion relations (3.92) and (3.93) are not as mysterious as they appear. Extensive tables can be found in [608]. The following two examples are intended to provide some familiarity with the relations.

Example One. For a power law over more than one logarithmic decade of frequency ω, with exponent p, $0 < p < 1$, we have

$$J' \propto \omega^{p}\,, \quad J'' \propto \omega^{p}\,, \quad G' \propto \omega^{-p}\,, \quad G'' \propto \omega^{-p}\,, \tag{3.95}$$

where $\tan\delta = G''/G' = J''/J' \approx \tan(p\pi/2)$. This means parallel straight lines for J' and J'', and for G' and G'', in a log–log plot (Fig. 3.4f). The vertical distance between storage and loss parts is

$$\Delta \log\tan\delta = \log\tan\frac{p\pi}{2}\,. \tag{3.96}$$

Measuring both real and imaginary parts, the exponent p can be determined more easily from this distance than from the slopes. Care must be taken for $p > 0.9$ because (3.95) for $p \to 1$ goes over to (2.12) for a Newtonian liquid.

Example Two. For logarithmically broad curves, with hardly any details within one decade, the Hilbert transformation acts rather like a logarithmic differentiation (Staverman–Schwarzl approximation). Written down for the moduli, for example, one finds

$$G''(\omega) \approx \frac{\pi}{2}\frac{\mathrm{d}G'(\xi)}{\mathrm{d}\ln\xi} \quad \text{for } \xi = \omega\,, \tag{3.97}$$

$$G'(\omega) - G_{\mathrm{g}} \approx -\frac{\omega\pi}{2}\frac{\mathrm{d}[G''(\xi)/\xi]}{\mathrm{d}\ln\xi} \quad \text{for } \xi = \omega\,, \tag{3.98}$$

$$\delta(\omega) \approx \frac{\pi}{2}\frac{\mathrm{d}[\ln|G^*|(\xi)]}{\mathrm{d}\ln\xi} \quad \text{for } \xi = \omega\,. \tag{3.99}$$

In case of emergency, of course, one has to look to computer programs with reasonable procedures for smoothing and extrapolation, for handling flow, and for making accessible measurements at different temperatures, e.g., by WLF scaling (i.e., master curve construction), and so on.

The general connection between linear response and linear thermodynamics of irreversible processes (TIP) is difficult. The former tries to describe broad time or frequency spectra directly, and the latter looks for a link between flux and force, including rate, acceleration, and perhaps higher derivatives. Restricting ourselves to the rate alone, the link is simply mediated by Debye relaxation (see [313], Chap. 12, for example). Small gradients can be modelled by a set of subsystems with smoothly varying state parameters shifting the spectra [609]. A general connection, using a spectrum of internal parameters λ for linear TIP, gives nonlinear, complicated formulas between the λ and the time spectra [610].

Relaxation and Retardation Spectra. A spectrum is a distribution of exponential decays or exponential growths over the logarithmic time scale, $\ln\tau$, to model the relaxation modulus or the retardation compliance in wide dispersion zones. In other words, the spectrum is a distribution of mechanical Maxwell elements for the modulus, or of Voigt–Kelvin elements for the compliance [3, 96]. These are models made from elastic springs and viscous dampers. To make the distribution definite, e.g., one-dimensional, two limiting cases are available. Either we think of a distribution of elements with the same intensity (Fig. 3.4g), or we think of elements at fixed, equidistant $\ln\tau$ values and fit the distribution of their intensities (Fig. 3.4h). Modern methods to get spectra from susceptibilities are based on Tikhonov regularization of 'ill proposed problems' to handle a few pieces of true information from large data sets with more or less known uncertainties [611], in particular, inverting noisy linear operator equations. The equilibrium modulus G_{e}, the glass compliance J_{g}, and the flow contribution τ/η are usually extracted. The reduction to exponential decays excludes resonances (weakly damped vibrations) from the description, although such observables can also be integrated into the scheme of linear responses, including the FDT.

The *relaxation spectrum* H is defined by

$$G(\ln\tau) = G_{\mathrm{e}} + \sum_i G_0\,\mathrm{e}^{-\tau/\tau_{0i}} \longrightarrow G_{\mathrm{e}} + \int_{-\infty}^{\infty} \mathrm{d}\ln\tau_0\, H(\ln\tau_0)\,\mathrm{e}^{-\tau/\tau_0}\,, \tag{3.100}$$

and the *retardation spectrum* L correspondingly, by

$$J(\ln\tau) = J_{\mathrm{g}} + \frac{\tau}{\eta} + \int_{-\infty}^{\infty} \mathrm{d}\ln\tau_1\, L(\ln\tau_1)\left[1 - \mathrm{e}^{-\tau/\tau_1}\right]. \tag{3.101}$$

Both spectra can be incorporated into the set (3.94).

In a similar way to the difference between modulus and compliance, the relaxation spectrum underlines the transition to solid-like behavior at shorter times, and the retardation spectrum underlines the flow onset at larger times. As a reminder, the dummy variables in (3.100) and (3.101) are chosen differently (τ_0, τ_1).

For dispersion zone activities with moderate intensities (ΔG, ΔJ) and nonzero equilibrium values (G_{e}, J_{e}), normalized spectra ($\tilde{H}$, $\tilde{L}$) can be used,

$$H = \Delta G \cdot \tilde{H}\,, \quad \int \tilde{H}\,\mathrm{d}\ln\tau_0 = 1\,, \tag{3.102}$$

$$L = \Delta J \cdot \tilde{L}\,, \quad \int \tilde{L}\,\mathrm{d}\ln\tau_1 = 1\,. \tag{3.103}$$

Using the Debye formulas (3.66) and (3.67), after some arithmetic, the distribution can be expressed by

$$\begin{aligned}
G'(\omega) &= G_{\mathrm{e}} + \Delta G \int \tilde{H}(\ln\tau_0)\,\frac{\omega^2\tau_0^2}{1+\omega^2\tau_0^2}\,\mathrm{d}\ln\tau_0\,,\\
G''(\omega) &= \Delta G \int \tilde{H}(\ln\tau_0)\,\frac{\omega\tau_0}{1+\omega^2\tau_0^2}\,\mathrm{d}\ln\tau_0\,,\\
J'(\omega) &= J_{\mathrm{g}} + \Delta J \int \frac{\tilde{L}(\ln\tau_1)}{1+\omega^2\tau_1^2}\,\mathrm{d}\ln\tau_1\,,\\
J''(\omega) &= \Delta J \int \tilde{L}(\ln\tau_1)\,\frac{\omega\tau_1}{1+\omega^2\tau_1^2}\,\mathrm{d}\ln\tau_1\,.
\end{aligned} \tag{3.104}$$

Equation (3.104) is also a convenient framework for conversion within the set (3.94). The responses are sometimes rationalized by an ansatz for $\tilde{H}$ or $1-\tilde{L}$, for instance, by the stretched exponential (2.17):

$$\tilde{H} \propto \exp\left[-(\tau/\tau_{\mathrm{KWW}})^{\beta_{\mathrm{KWW}}}\right], \quad 0 < \beta_{\mathrm{KWW}} \leq 1 . \tag{3.105}$$

Although spectra are in common use they should be handled with caution.

- Not all of the four quantities $\{G(\tau),\ H(\tau),\ 1 - J(\tau),\ L(\tau)\}$ can have the same functional shape, and this goes for the stretched exponential. The latter was derived quite generally for the α process compliance $J(\tau)$ in Sect. 3.1, with different $\beta_{\mathrm{KWW}} = \alpha$ values for the activities. Then $G(\tau)$, $H(\tau)$, and $L(\tau)$ cannot simultaneously be modelled by the stretched exponential for $\alpha < 1$. Some care should be taken to find the quantity (response or spectrum) that is actually fitted in a given paper.
- The spectra do not have a direct molecular interpretation. They originate from mechanical models such as the Maxwell or Voigt–Kelvin models [3,96] which fail for molecular movements. The latter are fluctuationally reversible and do not feel any viscosity. Phenomenologically, the spectra accentuate peculiarities of the response insofar as they are real pieces of information. Tikhonov regularization should be considered as some kind of logarithmic differentiation of correlation functions, as moduli $G(\tau)$ and compliances $J(\tau)$ really are.

We conclude this section on linear response with two nontrivial examples which anticipate aspects of the fluctuation–dissipation theorem FDT in the next section.

Example 1. We shall prove that the average temperature fluctuation δT of cooperatively rearranging regions (CRRs), needed for the determination of their size according to (2.76), $V_\alpha = k_{\mathrm{B}}T^2\Delta(1/c_V)/\rho(\delta T)^2$, or (2.77), $N_\alpha = RT^2\Delta(1/c_V)/M_0(\delta T)^2$, can be determined from Heat Capacity Spectroscopy (HCS). Since HCS measures the entropy compliance, we must ask, according to the fluctuation–dissipation theorem (FDT), whether or not, or in which approximation, the temperature fluctuation can be determined from the entropy fluctuation of CRRs.

The fluctuation formulas (2.76)–(2.77) for determination of characteristic length involve two problems. Firstly, does the temperature really fluctuate, i.e., does δT exist? This problem was discussed in Sect. 3.2. Secondly, can this δT be determined from the shape of (e.g., the imaginary part of) the dynamic heat capacity curve for given frequency, $C_p''(T)$? This problem will now be discussed in detail as our first example.

In general, from the functional fluctuation formulas $\overline{\Delta S^2} = k_{\mathrm{B}}\Delta C_p$ and $\overline{\Delta T^2} = k_{\mathrm{B}}T^2\Delta(1/C_V)$ of the von Laue treatment (Sect. 3.2.2), we see that the spectral density for entropy fluctuation $\Delta S^2(\omega)$ is extensive whilst for temperature fluctuation $\Delta T^2(\omega)$ it is intensive: $\Delta S^2(\omega) \propto N$ and $\Delta T^2(\omega) \propto 1/N$, because the heat capacities are extensive, $C_p \propto C_V \propto N$, where N is the number of particles in a subsystem. For macroscopic subsystems, e.g., $N = 10^{20}$, we have the orders of magnitude $(\overline{\Delta S^2})^{1/2}/S \sim (\Delta T^2)^{1/2}/T \sim 10^{-10}$.

It is important to distinguish between the probability density $p(\Delta T)$ of finding a certain temperature fluctuation amplitude ΔT and the spectral density $\Delta T^2(\omega)$ for finding an event ω of the power spectrum (Fig. 3.1c). The two distributions are distinct with regard to their dependence on the size (V or N) of the subsystem (Fig. 3.5a). Only the dispersion of the former becomes smaller for larger size, whilst the dispersion of the latter does not depend on size (ω identity of the FDT, Sect. 3.4.5).

From the representativeness gedankenexperiment of Sect. 2.2.5, we see, however, that the macroscopic compliances J_S^* and moduli G_T^* are representatives for the smallest independent subsystems, the CRRs functional to the α or a process (Fig. 3.5b). Since $J_S''(\omega)$ is connected with $\Delta S^2(\omega)$, and $G_T''(\omega)$ is connected with $\Delta T^2(\omega)$, via the FDT (2.75), and since both $\Delta T^2(\omega)$ and $\Delta S^2(\omega)$ contain information about the size of the subsystems, we may ask whether the information about the CRR size is also hidden in the shape of the responses. As HCS measures $C_p''(\omega)$, and hence, via the FDT, $\Delta S^2(\omega)$, and δT is needed for the CRR size, we arrive at the above question as to whether the measurable $C_p''(\omega, T)$ curves contain information about δT. The problem cannot be solved unambiguously, because we have two unknowns, the fluctuation δT and the size (N or V), e.g., $\delta T(N)$, that must be deduced from one known quantity, the width of the $C_p''(T)$ curve.

Intuitively, we get a second relation [Fig. 3.5b (1)] by relating the vertical and horizontal sections of the dispersion zone. We can determine δT from the local time–temperature equivalence (2.138), $\delta T/\delta \ln \omega = (\mathrm{d}T/\mathrm{d}\ln \omega)\,|_{\mathrm{along}}$, if $\delta \ln \omega$ is interpreted as the dispersion of a mobility fluctuation in the dispersion zone. The $C_p''(\log \omega, T)$ peak may then be considered as a consequence of an invariant mobility distribution, invariant against the size (V or N) because of the ω identity of FDT. Only one unknown is left, and δT may now be determined from an isochronal section of the $C_p''(\log \omega, T)$ function as a dispersion δT that belongs to one average CRR of size $N = N_\alpha$ [Fig. 3.5b (2)]. This was suggested in 1979 [7, 226, 298].

The local time–temperature equivalence (2.138) is thus related, not only to a path along the $\log \omega(T)$ trace, but also across the dispersion zone: both $\delta \ln \omega$ and δT are considered as dispersions of distributions. This idea solves the problem. However, it needs [612] a model that realizes $\delta \ln \omega$ as an average fluctuation.

The following is, in a manner of speaking, a check on this procedure, under the condition that there exists a temperature fluctuation conjugate to the entropy fluctuation. Let us continue with details.

The imaginary part of the complex heat capacity as a function of temperature T for given frequency ω can be approximately fitted by a Gaussian function [Fig. 3.5b (2)],

$$\tilde{C}_p''(T) \approx \frac{\sqrt{\pi}}{2\sqrt{2}}\frac{\Delta C_p}{\delta T} \exp\left[-(T - T_\omega)^2/2\delta T^2\right] \,, \quad \omega = \text{const.} \,, \qquad (3.106)$$

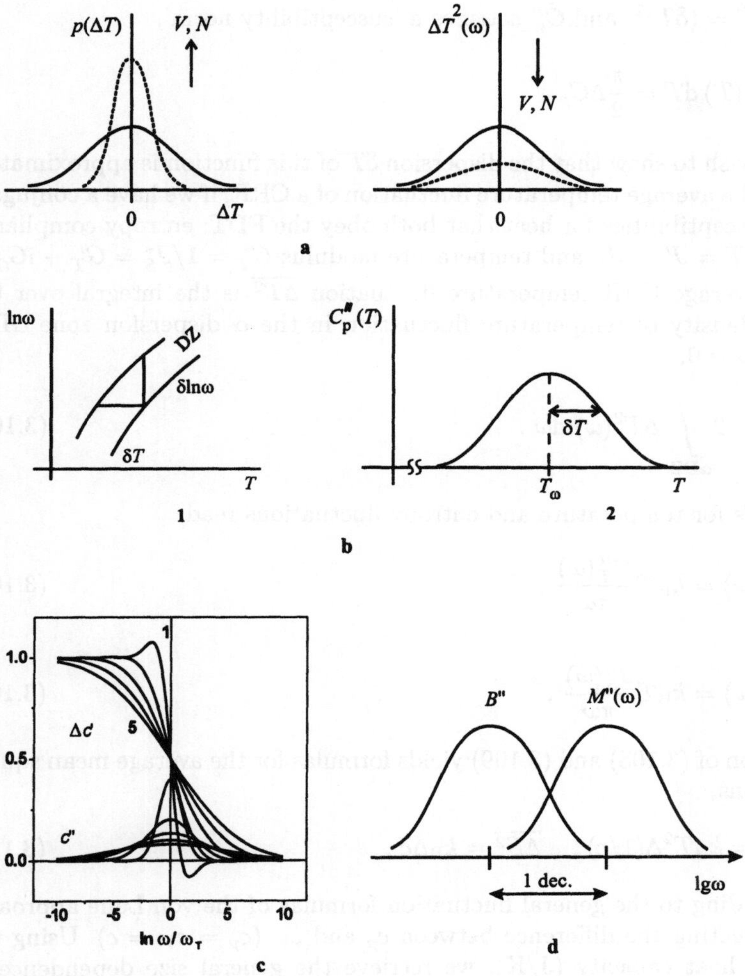

Fig. 3.5. Linear response (continued). (**a**) Distribution of temperature fluctuation, $p(\Delta T)$, and spectral density of temperature fluctuation, $\Delta T^2(\omega)$, for two different sizes (V or N) of representative subsystems. The size increases with the *arrow*. (**b**) (1) Local time–temperature (more precisely: log frequency–temperature) equivalence across the dispersion zone DZ. (2) Imaginary part of the dynamic heat capacity for the α process at the 'dynamic glass temperature' T_ω. The Gaussian dispersion δT is approximately equal to the average temperature fluctuation of a CRR. (**c**) The α process contribution to the real and imaginary parts of the heat capacity $\Delta c^*(\omega) = \Delta c'(\omega) - ic''(\omega)$ for logarithmic Gaussian functions. The parameter is $\delta \ln \omega = 1, 2, \ldots, 5$. The contributions of the other dispersion zones to the compliance are substracted. (**d**) Comparison of loss bulk compliance $B''(\omega)$ from dynamic dilatometry and loss part of longitudinal wave modulus $M''(\omega)$ from dynamic light scattering (photon correlation spectroscopy) for PVAC at about $T = 50°\text{C}$, normalized to the same height

where $\delta T^2 = (\delta T)^2$, and $\tilde{C}''_p$ satisfies a 'susceptibility norm',

$$\int \tilde{C}''_p(T)\,\mathrm{d}T = \frac{\pi}{2}\Delta C_p \,.$$

We thus wish to show that the dispersion δT of this function is approximately equal to the average temperature fluctuation of a CRR, if we have a conjugate pair of susceptibilities for heat that both obey the FDT: entropy compliance $J_S^* = C_p^*/T = J' - \mathrm{i}J''$ and temperature modulus $G_T^* = 1/J_S^* = G'_T + \mathrm{i}G''_T$.

The average CRR temperature fluctuation $\overline{\Delta T^2}$ is the integral over the spectral density of temperature fluctuation in the α dispersion zone αDZ, $\Delta T^2(\omega)$, $\omega \geq 0$,

$$\overline{\Delta T^2} = 2 \int\limits_{\alpha\mathrm{DZ}} \Delta T^2(\omega)\,\mathrm{d}\omega \,. \tag{3.107}$$

The FDTs for temperature and entropy fluctuations read

$$\Delta T^2(\omega) = k_\mathrm{B}T\frac{G''_T(\omega)}{\pi\omega} \,, \tag{3.108}$$

$$\Delta S^2(\omega) = k_\mathrm{B}T\frac{J''_S(\omega)}{\pi\omega} \,. \tag{3.109}$$

Integration of (3.108) and (3.109) yields formulas for the average mean square fluctuations,

$$\overline{\Delta T^2} = k_\mathrm{B}T^2\Delta(1/c) \,, \quad \overline{\Delta S^2} = k_\mathrm{B}\Delta c \,, \tag{3.110}$$

corresponding to the general fluctuation formulas of the von Laue approach, after neglecting the difference between c_p and c_V ($c_p = c_V = c$). Using the extensive heat capacity (J/K), we retrieve the general size dependence of fluctuations (Fig. 3.5a). Using some 'specific' heat capacity (J/gK), we hope to get representative formulas for the CRR. After so many details, our task in Example 1 is now to show the following.

Proposition. *δT in (3.106), i.e., the dispersion in the calorimetric peak of entropy retardation, is the same as ΔT in (3.110), i.e., the temperature fluctuation of an average CRR.*

Proof. To simplify the treatment, we consider the approximation of small c'' peaks, $c'' \ll c$. The main problem is of course the transition from entropy to temperature. There are two steps: firstly, the general transition from entropy compliance to temperature modulus, $J'' \to G''$, and secondly, the proof that $\delta T^2 = \overline{\Delta T^2}$ in analogy with this transition.

First Step. To calculate $\overline{\Delta T^2}$ from the spectral density $\Delta T^2(\omega)$, we need the imaginary part of the temperature modulus G''_T from the calorimetrically accessible spectral density for entropy fluctuation. For $c'' \ll c$, we obtain from $G^*_T = 1/J^*_S$, with $J^*_S(\omega) = c^*(\omega)/T$,

$$G''_T(\omega) \approx J''_S / J'^2_S \,, \tag{3.111}$$

and, using the FDT (3.108),

$$\Delta T^2(\omega) \approx k_B T^2 \frac{c''(\omega)}{c^2 \pi \omega} \,. \tag{3.112}$$

After integration we obtain, with $\Delta(1/c) \approx \Delta c/c^2 = \Delta c \cdot (1/c)^2$, the formula (3.110). The interesting point here is that the transition contains a separation between two factors:

(i) the Δc factor related to the dispersion zone,
(ii) the total heat capacity $c = T\mathrm{d}S/\mathrm{d}T$ from total entropy S and temperature T in the $(1/c)^2$ factor not related to the dispersion zone considered.

Second Step. Consider the explicit formula for δT from the entropy fluctuation formula (3.106) for $\tilde{C}''_p$, viz.,

$$\delta T^2 = \frac{2}{\pi \Delta c} \int\limits_{-\infty}^{\infty} (T - T_\omega)^2 \tilde{c}''(T)\, \mathrm{d}T \,. \tag{3.113}$$

We should once again find the entropy temperature transition with the (i)–(ii) factor separation. Using the mean value theorem of integral calculus, we get

$$\delta T^2 = (\widetilde{\Delta T})^2 \cdot \frac{2}{\pi \Delta c} \int\limits_{-\infty}^{\infty} \tilde{c}''(T)\, \mathrm{d}T = (\widetilde{\Delta T})^2 \,, \tag{3.114}$$

since the integral is $\pi \Delta c/2$. The last equality is based on the property that the function $c''(T)$ can be considered as a distribution. This follows from (3.111), since $J'^2_S(\omega) \approx$ const. for $c''(\omega) \ll c$, leading to (3.112) with $\Delta T^2(\omega)$ a distribution and $c^2 \approx$ const.

All this means that $(\widetilde{\Delta T})^2$ can be considered as the mean-squared temperature dispersion calculated from an entropy distribution. The (i)–(ii) factor separation is achieved by

$$(\widetilde{\Delta T})^2 = \left(\frac{\Delta T}{\Delta S}\right)^2_{\text{total}} \cdot \overline{\Delta S^2}_{\alpha \text{ process}} \,. \tag{3.115}$$

This leads to $(\widetilde{\Delta T})^2 = (T/c)^2 \cdot k_B \Delta c$, or with (3.110), to $\delta T^2 = \overline{\Delta T^2}$. Hence, the temperature fluctuation of a CRR can be obtained from the $C''_p(T)$ curve of heat capacity spectroscopy.

QED

The proof holds for small relaxation intensities, $c_p''/c_p \ll 1$, and for the approximations related to the consistency of logarithmic Gaussian functions with more realistic functions in the central part of the dispersion zone. A more formal proof without the $c'' \ll c$ restriction can be found in [299].

Remarks. Physically, the problem of whether or not $\delta \ln \omega$ can really be considered as an average fluctuation of mobility $\ln \omega$ in the dispersion zone has some depth. We touch upon nonconventional arguments that up to now have hindered general acceptance of the fluctuation formulas (2.76)–(2.77) for the size of an average CRR. We discuss this issue in the fluctuation picture with measure $\mathrm{d}\omega$, leaving the $\mathrm{d}\omega \to \mathrm{d}\ln\omega$ transition to a logarithmic frequency measure in the FDT, (2.75) and Sect. 3.4.

(a) Where do we get the frequencies ω for a distribution? This distribution comes from the model of minimal coupling in Sect. 4.3, suggested in 1978 [138]. This model is consistent with the disengagement concept (Sects. 2.2.5 and 2.2.8). The ω_i for partial systems V_i (Fig. 4.2a) get their independence from the quanta $\hbar\omega$ in Nyquist's transmission lines for explanation of the FDT (Fig. 3.6b). The possibility of a quasi-continuous treatment is discussed in Sect. 3.5 (Table 3.2).
(b) What is the connection between the ω distribution and a temperature distribution? We assume that different 'populations' $\{\omega_i\}$ of the model can be connected with different temperatures for the whole CRR. The definition of partial temperatures T_i for the partial systems V_i makes no sense, since their fluctuations would be much too large.
(c) How do we get the Gaussian distributions? The model unambiguously yields a Levy distribution with Levy exponent $\alpha \le 1$. The consistency of the two distributions is restricted to the central part of the dispersion zone (Sect. 2.4.5). This is also discussed in Sects. 2.5.2 and 3.2.3.

For the sake of completeness, we transform the $\tilde{C}_p''(T)$ function for $\omega =$ const. from the temperature domain into the frequency or mobility domain by using

$$\frac{\mathrm{d}T}{\mathrm{d}\ln\omega} = \frac{\delta T}{\delta \ln \omega} \,. \tag{3.116}$$

We obtain a logarithmic Gauss function explicitly

$$\tilde{C}_p''(\ln\omega) = \frac{\sqrt{\pi}}{2\sqrt{2}} \frac{\Delta C_p}{\delta \ln \omega} \exp\left[-\frac{\ln^2(\omega/\omega_T)}{2(\delta\ln\omega)^2}\right] , \quad T = \text{const.} , \tag{3.117}$$

where ω_T is the inverse of the maximum temperature T_ω, i.e., the dynamic glass temperature of Fig. 3.5b (2), and $\delta \ln \omega$ is the logarithmic frequency dispersion (Table 2.3). For $\delta \ln \omega \lesssim 2$, i.e., also for reasonable dispersions of Table 2.3, we find overshots in the real part curves $c'(\omega)$ (Fig. 3.5c) that do not fit the relaxation character of the a and α processes. We must then restrict the Gaussian fit to the central part of the dispersion zone (Sect. 2.4.5).

Example 2. We now consider how *composed activities* should be handled. There are no problems for addition of different susceptibilities, such as for the longitudinal wave modulus

$$M(\tau) = K(\tau) + \frac{4}{3}G(\tau) , \tag{3.118}$$

interesting for ultrasonic experiments, or for the linear creep compliance

$$D(\tau) = \frac{1}{3}J(\tau) + \frac{1}{9}B(\tau) , \tag{3.119}$$

where $D^* = 1/E^*$ and E^* is the Young modulus. As shown in Fig. 3.5d [613], one can also try to compare such things as the bulk compliance $B''(\omega)$ with the longitudinal modulus $M''(\omega)$ in the main transition.

In general, however, the composition is not linear, and the calculation would give quite different results depending on whether we start from the frequency or the time domain [96]. For example,

$$E^*(\omega) = \frac{9K^*(\omega)G^*(\omega)}{3K^*(\omega) + G^*(\omega)} \tag{3.120}$$

is equivalent to

$$\int_0^t E(t-u)[3K(u) + G(u)]\,\mathrm{d}u = 9\int_0^t K(t-u)G(u)\,\mathrm{d}u , \tag{3.121}$$

quite different from an equation that would follow from (3.120) when $X^*(\omega)$ is simply substituted by $X(\tau)$, $X = E$, K, or G. The rule is [614] always to start from the frequency picture. This corresponds to the ω identity of the FDT.

3.4 Fluctuation–Dissipation Theorem FDT

The greater the difference between the concepts underlying the left- and right-hand sides of an equation, the greater the significance of that equation. The *fluctuation–dissipation theorem* [94] therefore ranks as one of the great equations of physics, since such different concepts as reversible fluctuation and irreversible response are considered to be equivalent.

In this section, the FDT is considered as the equation for a quantum mechanical (QM) experiment where both the QM object and the measuring apparatus are identified with freely fluctuating subsystems (Fig. 3.6a) [615]. Such an FDT is based on the fundamental energy form of thermodynamics (Table 2.1). It therefore holds equally well for all its variables T, S, p, V, etc., and hence justifies the von Laue treatment of fluctuations (Sect. 3.2.2).

This view of the FDT is based on a *separation model* for quantum physics [615]. This model separates the QM object and the QM experiment of Bohr's pristine complementarity concept. I will only use plausible properties to describe the object and the experiment. Nevertheless, the separation model goes beyond conventional quantum theory, which claims to treat the two things in a single common frame from the very beginning. This issue is further discussed in Sect. 3.4.7.

The general idea for the FDT is based on Nyquist's derivation, which also separates the object from the measurement [94]. The object of the Nyquist gedankenexperiment is the thermal voltage noise $V'(t)$ of a 'Brownian' complex resistance $Z'(\omega)$ (Fig. 3.6b). This was called the emitter and is now identified with the QM object aspect of the subsystem. The apparatus of the Nyquist experiment is characterized by a resistance $R(\omega)$, a $V(t)$ generator, and (not shown in Fig. 3.6b) a voltmeter where the signal is transformed to a pointer position. This was called the absorber and is now identified with the apparatus aspect of the subsystem. All emitted voltage waves can be absorbed by adjustment of $R(\omega)$ and the apparatus $V(t)$ generator, resulting in the balance of transmission lines between emitter and absorber [280].

The circuit behavior is now identified with quantum jumps representing the QM collapse of the wave function when experiments are carried out (reduction map of Sect. 2.1.5). The transmission lines enforce detailed balance for a given frequency ω. They act as a kind of Planckian blackbody radiator, that is, a set of oscillators that delivers the oscillator energy factor of the FDT (Bose factor),

$$\frac{1}{2}\hbar\omega \coth\frac{\hbar\omega}{2k_\mathrm{B}T} \longrightarrow k_\mathrm{B}T \quad \text{for} \quad \hbar\omega \ll k_\mathrm{B}T\,. \tag{3.122}$$

This factor determines the ratio between fluctuation and response. Note that the Planck constant is not contained in the classical $k_\mathrm{B}T$ case of (3.122).

►

Fig. 3.6. Fluctuation–dissipation theorem FDT. (**a**) Identification of the apparatus and the object of a quantum mechanical experiment with the same thermodynamic subsystem. The collapses of a stationary succession of such 'self-experiments' yields a fluctuation–dissipation theorem (FDT) that includes temperature fluctuation. (**b**) Nyquist's scheme for the derivation of the FDT. E emitter, A absorber, TL transmission lines. (**c**) Typical behavior of a correlation function $\varphi_x(\tau) = \Delta x^2(\tau)$ in a simple dispersion zone. (**d**) Corresponding spectral density $\Delta x^2(\omega)$. (**e**) Dependence of the self-part (s) and distant part (d) of the van Hove correlation function [616] on distance r for three values of time t. The *solid line* corresponds to the average density ρ. The example is related to a correlation time τ_0 in the picosecond range, not to the dynamic glass transition at longer times. (**f**) and (**g**) Comparison of response and fluctuation. (**h**) ω identity of the FDT

The famous derivation by Callen and Welton [277, 313] uses conventional QM theory where the QM object and QM experiment are not separated. It is based on a mechanical Hamiltonian (instead of the thermodynamic fundamental form) and uses the Gibbs distribution without temperature fluctuation. The QM experiment is reflected by the quantum jumps $\hbar\omega_{\mathrm{fi}} = E_{\mathrm{f}} - E_{\mathrm{i}}$ fed into Fermi's golden rule.

In a stochastic treatment [167, 314, 615], each jump is connected with the temporal properties of a measurement: before the measurement, the unmeasured stochastic variable and the apparatus pointer are not correlated; after the measurement the pointer is fixed at the value of the stochastic variable

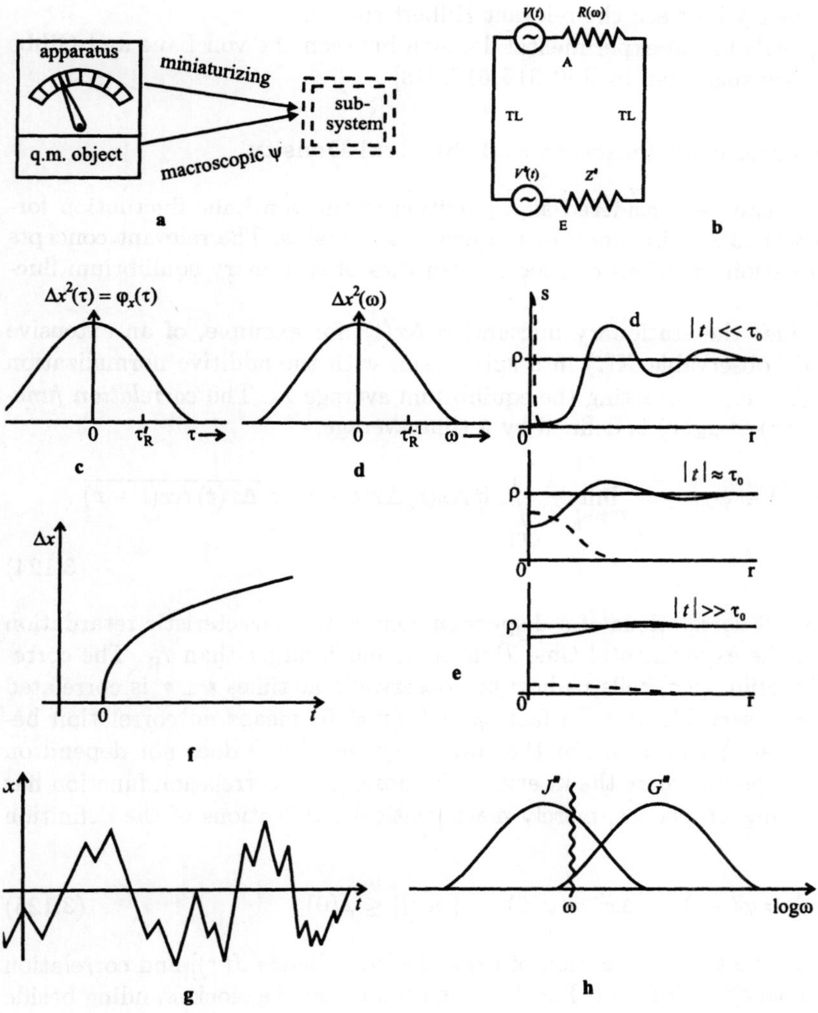

at measuring time t'. The FDT follows explicitly from a stationary succession of such measurement steps, $\{t'\}$, when both object and apparatus are identified with representative subsystems. The underlying mathematics is a Fourier series treatment, as represented for example in [279], Chap. XIX. The map between the stochastics of a thermodynamic variable, $\mathbf{X}_n$, and the associated process with uncorrelated increments, $\mathbf{Y}_t$, standing for the stationary succession of quantum jumps,

$$\mathbf{X}_n = \int_{-\pi}^{\pi} \mathrm{e}^{-int} \mathrm{d}\mathbf{Y}_t \,, \tag{3.123}$$

is an isometry between the relevant Hilbert spaces.

Proposals for an experimental decision between the von Laue and Gibbs variants are suggested in [140, 319, 617, 618].

3.4.1 Correlation Function and Spectral Density

The FDT can be considered as a partition of the von Laue fluctuation formulas (Sect. 3.2.2) in time t or frequency ω domains. The relevant concepts are correlation functions or spectral densities of stationary equilibrium fluctuations.

Consider the stationary fluctuation $\Delta x(t)$, for example, of an extensive stochastic observable $\mathbf{X}(t)$ in a subsystem, with the additive normalization $\overline{\Delta x(t)} = 0$, i.e., subtracting the equilibrium average x_e. The *correlation function* $\Delta x^2(\tau) \equiv \varphi_x(\tau)$ is defined by a time average,

$$\Delta x^2(\tau) \equiv \varphi_x(\tau) = \lim_{T \gg \tau'_\mathrm{R}} \frac{1}{T} \int_{(T)} \mathrm{d}t \, \Delta x(t) \, \Delta x(t+\tau) \equiv \overline{\Delta x(t) \, \Delta x(t+\tau)} \,. \tag{3.124}$$

To cover all correlations of a dispersion zone with characteristic retardation time τ'_R, the experimental time T must be much larger than τ'_R. The correlation function $\varphi(\tau)$ tells us how the observable at times $t \pm \tau$ is correlated with the observable at t. In fact, $\varphi = 0$ $(\tau \neq 0)$ means no correlation between $x(t+\tau)$ and $x(t)$. For the stationary case [$\varphi(\tau)$ does not depend on the time (epoch) where the interval T is chosen], the correlation function has the following properties, merely mathematical implications of the definition (3.124):

$$\varphi(\tau) = \varphi(-\tau) \,, \quad \overline{\Delta x^2} = \varphi(0) \,, \quad |\varphi(\tau)| \leq \varphi(0) \,. \tag{3.125}$$

Note that the time conception of response [compliance $J(\tau)$] and correlation function [$\varphi(\tau)$] is different. For $J(\tau)$, one can imagine a clock standing beside the creep experiment, and for $\varphi(\tau)$, the time τ is composed of many intervals

τ at t_1, t_2, and so on. This is a probing time in the terminology of Sect. 2.1.4. The correlation function of an intensive observable f is defined analogously,

$$\Delta f^2(\tau) \equiv \varphi_f(\tau) = \overline{\Delta f(t)\,\Delta f(t+\tau)}\,. \tag{3.126}$$

Let us now change the time domain into the frequency domain. The *spectral density* is defined by Fourier components of correlation functions,

$$\Delta x^2(\omega) = \frac{1}{2\pi}\int_{-\infty}^{\infty} \Delta x^2(\tau)\,\mathrm{e}^{\mathrm{i}\omega\tau}\,d\tau\,, \tag{3.127}$$

$$\Delta f^2(\omega) = \frac{1}{2\pi}\int_{-\infty}^{\infty} \Delta f^2(\tau)\,\mathrm{e}^{\mathrm{i}\omega\tau}\,d\tau\,. \tag{3.128}$$

The following properties are implications of this definition, again for the stationary case [619],

$$\overline{\Delta x^2} = \varphi(0) = \int_{-\infty}^{\infty} \Delta x^2(\omega)\,\mathrm{d}\omega\,, \quad \Delta x^2(\omega) = \Delta x^2(-\omega)\,, \quad \Delta x^2(\omega) \geq 0\,, \tag{3.129}$$

where negative ω are formal constructions based on the symmetry $\varphi(\tau) = \varphi(-\tau)$ of (3.125). Negative ω can be avoided by use of the Fourier cosine transformation,

$$\begin{aligned}\Delta x^2(\tau) &= \varphi(\tau) = 2\int_0^{\infty} \mathrm{d}\omega\cos(\omega\tau)\,\Delta x^2(\omega)\,,\\ \Delta x^2(\omega) &= \frac{1}{\pi}\int_0^{\infty} \mathrm{d}\tau\cos(\omega\tau)\,\Delta x^2(\tau)\,.\end{aligned} \tag{3.130}$$

The spectral density for intensive variable fluctuations is defined analogously. The interesting property $\Delta x^2(\omega) \geq 0$, $\Delta f^2(\omega) \geq 0$ comes from the positive definiteness concept (Sect. 3.4.6) underlying the correlation function and has led to the name 'density'. Note, however, that negative values $\varphi(\tau) < 0$ are allowed. Whilst the spectral density of fluctuations, $\Delta x^2(\omega)$, can be interpreted as a probability density with measure $\mathrm{d}\omega$, the correlation function $\Delta x^2(\tau) = \varphi(\tau)$ cannot in general. Typical graphs of correlation function and spectral density are shown in Figs. 3.6c, d.

Let us now examine how two processes $(1, 2)$ can be composed to form a common correlation function or spectral density. Examples are the $\alpha = 1$ and $\beta = 2$ processes in the crossover region, or the fluctuations of two independent representative subsystems (Sect. 3.1).

Table 3.1. Three possibilities for compositions (A, B, C) of spectral densities and correlation functions. See text for details

A	$p(x) \propto p_1(x)p_2(x)$	$f(t) \propto f_1 * f_2(t)$	$x_1^2(\omega)x_2^2(\omega)$	$x_1^2 * x_2^2(t)$
B	$p(x) = p_1 * p_2(x)$	$f(t) = f_1(t)f_2(t)$	$x_1^2 * x_2^2(\omega)$	$x_1^2(t)x_2^2(t)$
C	$p(x) = \tilde{p}_1(x) + \tilde{p}_2(x)$	$f(t) = \tilde{f}_1(t) + \tilde{f}_2(t)$	$\tilde{x}_1^2(\omega) + \tilde{x}_2^2(\omega)$	$\tilde{x}_1^2(t) + \tilde{x}_2^2(t)$

Three compositions of probability theory are presented in Table 3.1. The general symbols are the probability density $p(x)$ with $\int p(x)\,\mathrm{d}x = 1$ and the characteristic function $f(t) = \int_0^\infty \mathrm{e}^{\mathrm{i}tx}p(x)\,\mathrm{d}x$ (3.22). Their relation to the spectral density $x^2(\omega)$ and correlation function $x^2(t)$ is

$$x^2(\omega) \sim p(x)\,, \quad x^2(t) \sim f(t)\,. \tag{3.131}$$

Composition A is the simplest situation of statistical independence for two densities $p_1(x)$ and $p_2(x)$ on the same x axis (the same probability space) and corresponds to the 'and' composition of (3.15), $P = P_1 \cdot P_2$. The proportionality sign $\propto$ permits a separate normalization of $p_1(x)$ and $p_2(x)$.

Composition B is the additivity variant of 'probability of a function' [620], here the probability density for a sum

$$\eta = \xi_1 + \xi_2 \tag{3.132}$$

of two random variables (ξ_1, ξ_2) with densities $p_1(x_1)$ and $p_2(x_2)$ on $\{x_1, x_2\} \in \{x\}$, i.e., where x_1 and x_2 are in the same probability space (x axis). The convolution is defined by

$$p_1 * p_2(x) = \int p_1(z)p_2(x - z)\,\mathrm{d}z\,, \tag{3.133}$$

and follows from statistical independence of the random variables ξ_1 and ξ_2, $p(x_1, x_2) = p_1(x_1) \cdot p_2(x_2)$ (subsystem representativeness, Example 1 of Sect. 3.1).

Composition C is mutual exclusivity for nonoverlapping probability spaces (e.g., x_α and x_β of the *$\alpha\beta$ separation problem*, Example 2, Sect. 3.1) and corresponds to the 'or' composition of (3.16), $P = P_1 + P_2$. The idea for additivity is the primary nonidentity of $x_1 = \omega_\alpha$ and $x_2 = \omega_\beta$ with a subsequent identification by the measuring process (experiment), e.g., by the ω identity of FDT ($\omega_\alpha = \omega_\beta = \omega$). The tilde on the symbols, e.g., $\tilde{p}$, $\tilde{f}$, permits the separate normalization of p_1 and p_2. In detail, we have $p(x) = A_1p_1(x) + A_2p_2(x)$, etc., where in the dielectric case for Example 2, the A_i constants are related to the dielectric intensities, $A_1 = A_\alpha \propto \Delta\varepsilon_\alpha$ and $A_2 = A_\beta \propto \Delta\varepsilon_\beta$.

The Williams or Williams–Watts formula [238, 621] for evaluation of the reduced dielectric compliance in the $\alpha\beta$ crossover region,

$$\varphi(\tau) = f_\alpha\varphi_\alpha(\tau) + (1 - f_\alpha)\varphi_\alpha(\tau)\varphi_\beta(\tau)\,, \tag{3.134}$$

with temperature dependent α intensity $f_\alpha = f_\alpha(T)$, is an example of an entangled mixture of composition C (the sum) and composition B (the product). The product term uses the picture of a sum of two stochastic polarizations, α and β, that are assumed to be statistically independent.

The additivity alternative (3.21), $\varepsilon^* = \varepsilon^*_\alpha + \varepsilon^*_\beta$, (Example 2 of Sect. 3.1) is the pure composition C and does not need any restriction, such as statistical independence of α and β polarizations. Such a statistical independence is not consistent with the physical pictures Fig. 2.33. Interpretations of the C composition of the type "several dipoles make the β process and the other dipoles make the α process" are inappropriate. It is only necessary to distinguish the spectral densities (or correlation functions) of α and β, which includes a preliminary distinction between α and β frequencies, ω_α and ω_β. As mentioned above, the latter are eventually identified by the ω identity of FDT measurement.

On the other hand, the treatment of subsystem representativeness (Example 1 of Sect. 3.1), Levy distribution (Sect. 3.6), and Fischer modes (Sect. 2.2.7 and Sect. 4.4) uses the composition B and is therefore related to statistical independence.

3.4.2 Van Hove Correlation Functions for Dynamic Scattering

The aim in this section is to describe an example of the difference between spectral density (probability distribution) and correlation function (no probability distribution) and to sharpen the view with respect to certain experimental subtleties.

The basic concepts were introduced in Sect. 2.3.1. $S(Q,\omega)$ is the dynamic structure factor and $S(Q,t)$ the intermediate scattering function. The concepts of incoherent and coherent parts were introduced by (2.91) and (2.92).

As long as no complete picture for the crossover region of the dynamic glass transition is available, we need a detailed knowledge of scattering phenomenology for the evaluation of dynamic scattering experiments, expected in the next few years to resolve the crucial problems of the dynamic glass transition. The van Hove correlation functions are conceptually different from a probability density. Furthermore, dynamic scattering is interpreted as linear response, where a part of the fluctuating quanta is extracted for registration. The only difference with linear response is a partition in the scattering vector Q domain. The time t domain remains regulated by a correlation function, whilst only the frequency ω domain is regulated by a probability density.

We refer to the classic description in [334], p. 178 ff. In order to focus on the description, all scattering lengths (b_i for neutron scattering) and all form factors (for X-ray scattering) are dropped. Then dynamic scattering can be straightforwardly reduced to the *van Hove correlation function* [616]

$$G(\boldsymbol{r},t) = \frac{1}{N}\left\langle \sum_{ij} \int_V \mathrm{d}\boldsymbol{r}' \delta\left[\boldsymbol{r} - \boldsymbol{r}_i(0) - \boldsymbol{r}'\right] \delta\left[\boldsymbol{r}' - \boldsymbol{r}_j(t)\right] \right\rangle , \tag{3.135}$$

where $\boldsymbol{r}_i(0)$ and $\boldsymbol{r}_j(t)$ are Euclidean coordinates at $t = 0$ and $t \neq 0$ for the particles i and j, respectively. The structure of (3.135) and the reasons for its use [334] show how the Huygens wave interferences are reduced to particle correlations in space $\boldsymbol{r}$ and time t. Mathematical simplification yields

$$G(\boldsymbol{r},t) = \frac{1}{N}\left\langle \sum_{ij} \left[\delta(\boldsymbol{r} + \boldsymbol{r}_i(0) - \boldsymbol{r}_j(t)\right] \right\rangle . \tag{3.136}$$

At $t = 0$ and for isotropic liquids, we get

$$G(\boldsymbol{r},0) = \delta(r) + \rho g(r) = G_\mathrm{s}(\boldsymbol{r},0) + G_\mathrm{d}(\boldsymbol{r},0) , \tag{3.137}$$

where G_s is the self-part and G_d the distant part of the correlation function. The radial distribution function $g(r)$ is normalized by $g(r) \to 1$ [corresponding to $G(\boldsymbol{r},0) \to \rho$] for $r \to \infty$. Since $g(r) \geq 0$, one may find a formal probability interpretation for it if an appropriate convention is introduced for the normalization.

Defining the number density ρ by

$$\rho(\boldsymbol{r},t) = \sum_i \delta(\boldsymbol{r} - \boldsymbol{r}_i(t)) , \tag{3.138}$$

we get

$$G(\boldsymbol{r},t) = \frac{1}{N}\int \mathrm{d}\boldsymbol{r}' \left\langle \rho(\boldsymbol{r} - \boldsymbol{r}',0)\rho(\boldsymbol{r}',t)\right\rangle , \tag{3.139}$$

which shows how the correlation function is related to spatiotemporal correlations of thermally excited density fluctuations.

The dynamic structure factor is a spatial and temporal Fourier transform of the van Hove correlation function,

$$S(Q,\omega) = \frac{1}{2\pi}\int_V \int_{-\infty}^{\infty} \mathrm{d}\boldsymbol{r}\ \mathrm{d}t\ \left[G(\boldsymbol{r},t) - \rho\right] \exp[\mathrm{i}(\boldsymbol{Q}\cdot\boldsymbol{r} - \omega t)] . \tag{3.140}$$

The correlations are partitioned into the distant part and the self-part,

$$G_\mathrm{d}(\boldsymbol{r},t) - \rho \quad \text{and} \quad G_\mathrm{s}(\boldsymbol{r},t) . \tag{3.141}$$

The distant part $G_\mathrm{d}(\boldsymbol{r},t) - \rho$ can become negative, which excludes any direct probability interpretation (Fig. 3.6e).

Turning to the details, we get the intermediate scattering function as a spatial Fourier transform,

$$S(Q,t) = \int \mathrm{d}\boldsymbol{r} \exp(\mathrm{i}\boldsymbol{Q}\cdot\boldsymbol{r})\,[G(\boldsymbol{r},t) - \rho] \,. \tag{3.142}$$

Normalizing $S(Q,\omega)$ as a probability distribution, we obtain the structure factor $S(Q)$,

$$\int_{-\infty}^{\infty} \mathrm{d}\omega\, S(Q,\omega) = 1 + \rho \int \mathrm{d}\boldsymbol{r}\, \exp(\mathrm{i}\boldsymbol{Q}\cdot\boldsymbol{r})\,[g(r) - 1] = S(Q) \,. \tag{3.143}$$

The classical symmetry is obtained from detailed balance,

$$S(Q,-\omega) = S(Q,+\omega) \exp\left(-\frac{\hbar\omega}{k_{\mathrm{B}}T}\right) , \tag{3.144}$$

where the exponential stems from the ratio $(k_0 \to k)/(k \to k_0)$ of statistical weights for the neutron wave vectors k, k_0 before and after scattering. The symmetry is obtained in the quasi-classical case,

$$S(Q,-\omega) = S(Q,+\omega) \quad \text{for} \quad \frac{\hbar\omega}{k_{\mathrm{B}}T} \ll 1 , \tag{3.145}$$

usually fulfilled for dynamic glass transition frequencies ω. $G(\boldsymbol{r},t)$ becomes real (not complex) for symmetric $S(Q,\omega)$, with the boundary condition

$$G(\boldsymbol{r},t) \to \rho \quad \text{for} \quad r \to \infty \quad \text{or} \quad t \to \infty \,. \tag{3.146}$$

In summary, for each Q value,

$$\frac{S(Q,t)}{S(Q)} \text{ behaves as a correlation function ,} \tag{3.147}$$

and

$$\frac{S(Q,\omega)}{S(Q)} \text{ behaves as a spectral density ,} \tag{3.148}$$

similarly to the situation for the FDT. Only the spectral density has a rigorous probability interpretation, $x^2(\omega)\,\mathrm{d}\omega \Leftrightarrow p(x)\,\mathrm{d}x$.

Remark. The difference between dynamic neutron and dynamic X-ray scattering [622] is:

- the density for the first is given by scattering length distributions of different nuclei, and for the second by electron distributions,
- the dispersion law for neutrons is $E \sim k^2$, whilst for X-ray photons it is $E \sim k$.

It is speculated that dynamic X-ray scattering may sooner be successful in the crossover region of the dynamic glass transition than dynamic neutron scattering.

3.4.3 Differential and Integral Forms of the FDT

The differential form of FDT is, as mentioned above, the time (τ) or frequency (ω) partition of von Laue's fluctuation formulas [Sect. 3.2.2, (3.37)–(3.40)] for $\hbar\omega \ll kT$:

$$\Delta x^2(\tau) = -k_B T\,[J(\tau) - J_e]\,, \quad \Delta f^2(\tau) = k_B T\,[G(\tau) - G_e]\,, \tag{3.149}$$

$$\Delta x^2(\omega) = -k_B T\; J''(\omega)/\pi\omega\,, \quad \Delta f^2(\omega) = k_B T\; G''(\omega)/\pi\omega\,. \tag{3.150}$$

The susceptibilities on the right-hand side feel for the fluctuations on the left-hand side of (3.149)–(3.150). Their ratio is determined by the 'molecular quantum' $k_B T$ = "correlation divided by response" (von Laue–Nyquist theorem). The term 'dissipation' in the FDT comes from the loss parts of susceptibilities in (3.150).

The integral form of the FDT defines the relaxation intensities (ΔJ, ΔG) as integrals over the dispersion zone (DZ) under consideration:

$$\begin{aligned}\Delta J = J_e - J_g &= \frac{\overline{\Delta x^2}}{k_B T} = \frac{\varphi_x(0)}{k_B T}\\ &= \frac{2}{\pi}\int_{DZ} J''(\ln\omega)\,\mathrm{d}\ln\omega = \frac{2}{k_B T}\int_{DZ} \Delta x^2(\omega)\,\mathrm{d}\omega\,,\end{aligned} \tag{3.151}$$

$$\begin{aligned}\Delta G = G_g - G_e &= \frac{\overline{\Delta f^2}}{k_B T} = \frac{\varphi_f(0)}{k_B T}\\ &= \frac{2}{\pi}\int_{DZ} G''(\ln\omega)\,\mathrm{d}\ln\omega = \frac{2}{k_B T}\int_{DZ} \Delta f^2(\omega)\,\mathrm{d}\omega\,.\end{aligned} \tag{3.152}$$

The integrals in (3.151)–(3.152) show that the values of susceptibilities at the low frequency end of the dispersion zone (J_e, G_e) are the result of fast molecular motions in the dispersion zone(s) at higher frequencies.

3.4.4 Equilibrium or Nonequilibrium?

The interpretation of the FDT as an experiment that realizes susceptibilities only from equilibrium fluctuations in the linear region excludes the following picture. The external program would generate a new state that relaxes into equilibrium. The FDT shows instead that the external program makes observable only such correlations as also occur in the equilibrium, with no program; for subsystems, any possible 'external program' is already contained in the stochastic fluctuational 'internal disturbances' from equilibrium fluctuations of the environment or from within. The only difference concerns thermodynamics: the external program, i.e., measurement from outside, generates an entropy production (3.78)–(3.81) and is therefore irreversible no matter how small the program amplitude may be, whereas the internal, fluctuational disturbances are completely reversible. The response is thus characterized by a time arrow ('after' the step), by thermodynamic causality in the sense of linear material equations, and by entropy production, however small the perturbations may be. In contrast, the thermodynamic fluctuation is reversible and correlative, and is not connected with any entropy production (Figs. 3.6f, g). Otherwise the concepts of fluctuation, relaxation or retardation, dispersion, and molecular dynamics are equivalent in the framework of the linear FDT.

Nonlinearity means that the program is not contained in the internal disturbance from dynamic equilibrium, i.e., in the fluctuations. Decisive for the fluctuation amplitude is the size of the smallest representative subsystems (Sects. 2.1.2, 2.1.5, and 3.1).

3.4.5 ω Identity of the FDT

The simple fact that the frequency ω in the spectral density is exactly equal to the probing frequency in the linear dynamic susceptibility (3.150) is called the ω *identity* of the FDT. This identity ensures that Fourier components of microscopic processes (a, α, β, etc.) are transferred without any frequency change into the susceptibilities. This permits a corresponding labelling of the traces in the Arrhenius diagram (Fig. 2.32) and also means that their frequency dispersions $\delta \ln \omega$ (Table 2.3) can be discussed in terms of microscopic dynamic heterogeneity. The ω identity was also crucial for the first two examples of Sect. 3.1: the representativeness theorem for a basic introduction of Levy distributions in liquid dynamics and the additivity of α and β compliances in the crossover region.

The ω identity emerges from the convergence of many factors: the balance condition for the transmission lines in Nyquist's derivation of the FDT, energy conservation ($\Delta E = \hbar\omega$) in statistical mechanics, and the indistinguishability of $\hbar\omega$ quanta arising from different processes or different representative subsystems.

The ω identity holds for the following examples:

- No net quanta are taken away for the experiment, such as for linear response, because of the adjustment of the transmission lines in Fig. 3.6b.
- Some quanta are set aside for equilibration the system (violation of FDT because of partial freezing-in does not violate the ω identity, Sect. 4.5.2).
- The frequency shift of neutrons (2.90) or X rays in dynamic scattering exactly corresponds to Fourier components of microscopic processes.
- The quantum mechanical amplification of the FDT, for $\hbar\omega/k_{\mathrm{B}}T \gtrsim 1$, is in the frequency domain [(3.179), Sect. 3.4.8)], not in the time domain.

Two further examples will be discussed, with some details.

(A) Figure 3.6h is reminiscent of Fig. 2.14b for the discussion of the maximum frequency difference between compliance and modulus loss peaks of a Debye relaxation (*modulus–compliance displacement*). The ω identity of FDT is much sharper than the frequency dispersion of Lorentz lines for the loss peaks, so that the same physical frequency can simultaneously test the high-frequency wing of the J'' peak and the low-frequency wing of the G'' peak. The point here is not that the compliance can be converted into the modulus (or how this happens), and vice versa, but rather that the J'' and G'' peaks are really at different frequencies and that, consequently, compliance and modulus are sensitive in different ways to different frequencies, and that the various peak frequencies may be sensitive to in different ways to the length scales of molecular motion, and so on.

(B) Consider two situations for a logarithmically wide dispersion zone:
- an ensemble of mutually isolated, independent local mechanisms each corresponding to a Debye relaxation, where the spectrum comes from a wide distribution of τ_0 parameters (activation energy distribution);
- a cooperative process including, for example, 100 particles where the wide spectrum comes from cooperativity.

The FDT, and hence the thermodynamic experiment cannot distinguish between the two situations.

3.4.6 Mathematical Background

A unified understanding of the various aspects of FDT can be achieved by a stochastic treatment in the time domain and a map to the frequency domain. This includes such features as the equation for the linear response experiment, the QM experiment on freely fluctuating representative subsystems, and the consequences of a stationary succession of quantum jumps due to self-experiments, corresponding to running ruptures of the unitary state development by reduction (collapse) of the wave function, like the reduction map of Sect. 2.1.5. For mathematical rigor, we refer to the textbook by Feller [279] Vol. II, Chap. XIX. In the following, we aim to outline the basic mathematical facts without introducing physical arguments for the treatment with the nonconventional separation model (Sect. 3.4.7).

A complex-valued function φ of the real variable t is said to be *positive definite* if and only if

$$\sum_{j,k} \varphi(t_j - t_k)\; z_j\; \bar{z}_k \geq 0 \tag{3.153}$$

holds for every choice of finitely many real numbers $t_1, \ldots, t_n$ and complex numbers $z_1, \ldots, z_n$.

Bochner Theorem. A continuous function φ is the characteristic function, i.e., the Fourier transform (3.22) of a probability distribution if and only if it is positive definite and $\varphi(0) = 1$.

Bochner–Khintchin Theorem. Let $\{\mathbf{X}_t\}$ be a family of random variables such that the expectation

$$\rho(t) = \mathrm{E}\left[\mathbf{X}_{t+s}\mathbf{X}_s\right] \tag{3.154}$$

is a continuous function independent of s. Then ρ is positive definite, that is

$$\rho(t) = \int_{-\infty}^{\infty} \mathrm{e}^{\mathrm{i}\omega t} R\{\mathrm{d}\omega\} , \tag{3.155}$$

where R is a measure on the real line with total 'mass' $\rho(0)$. [The translation to our symbols is response $X = (x, f)$, spectral density $R = (\Delta x^2(\omega), \Delta f^2(\omega))$ and correlation function $\rho = (\Delta x^2(t), \Delta f^2(t))$. The independence of $\rho(t)$ from the time s means stationarity.]

We now consider the associated process with *uncorrelated increments*, needed for the underlying random succession of quantum jumps in the separation model. For each t with $-\pi < t < \pi$, define y_t to be the step function

$$y_t(x) = \begin{cases} 1 & \text{for} \quad x \le t \,, \\ 0 & \text{for} \quad x > t \,, \end{cases} \tag{3.156}$$

and let $\mathbf{Y}_t$ denote the corresponding random variable in a Hilbert space $\mathfrak{H}$. The increments $\mathbf{Y}_t - \mathbf{Y}_s$ for non-overlapping intervals have covariance 0. Furthermore, the variance is $\mathrm{Var}(\mathbf{Y}_t) = (2\pi)^{-1} R\{-\pi, t\}$. Thus $\{\mathbf{Y}_t\}$ is a process with uncorrelated increments and variances given by R. If the $\mathbf{X}_k$ are normal (see below), the increments of the $\mathbf{Y}_t$ process are actually independent.

We have two Hilbert spaces: the Hilbert space L^2_R consists of all functions on the circle with finite norm, and the Hilbert space $\mathfrak{H}$ is spanned by $\{\mathbf{X}_n\}$ with inner products given by the expectation $\mathrm{E}[\mathbf{U}\bar{\mathbf{V}}]$. L^2_R may serve as (a very robust) model for $\mathfrak{H}$: the mapping $\mathfrak{H} \longleftrightarrow L^2_R$, $\mathbf{X}_n \longleftrightarrow \mathrm{e}^{-\mathrm{i}kx}$ induces a one-to-one correspondence between the random variables in $\mathfrak{H}$ and the functions in L^2_R. This correspondence preserves inner products and norms, so the two spaces are isometric.

The function $\exp(-\mathrm{i}nt)$ corresponds to the variable $\mathbf{X}_n$. The basic spectral representation of an arbitrary sequence $\{\mathbf{X}_n\}$ in terms of the associated process with uncorrelated increments is the stochastic integral (3.123),

$$\mathbf{X}_n = \int_{-\pi}^{\pi} \mathrm{e}^{-\mathrm{i}nt} \mathrm{d}\mathbf{Y}_t \,. \tag{3.157}$$

Wiener–Khintchin Criterion. A necessary and sufficient condition for f to be the characteristic function of a probability density p is that there should exist a normalized function u, i.e., with $\|u^2\| = 1$, such that

$$f(\tau) = \int_{-\infty}^{\infty} u(t)\, \bar{u}(t+\tau)\, \mathrm{d}t \,. \tag{3.158}$$

In this case $p = |\hat{u}|^2$, with

$$\hat{u}(\omega) = \frac{1}{\sqrt{2\pi}} \int_{-\infty}^{\infty} u(t)\, \mathrm{e}^{\mathrm{i}\omega t} \mathrm{d}t \quad \text{and} \quad u(t) = \frac{1}{\sqrt{2\pi}} \int_{-\infty}^{\infty} \hat{u}(\omega)\, \mathrm{e}^{-\mathrm{i}\omega t} \mathrm{d}\omega\,. \tag{3.159}$$

The choice of u is not unique.

All this gives the possibility for random quantum jumps if u can be considered as a QM wave function ψ [623].

Remark. The usual 'physical' formulation of the *Wiener–Khintchin theorem* is as follows. Let $x(t)$ be a continuous stationary stochastic function of time t. Then its spectral density can be calculated directly from

$$x^2(\omega) \propto \lim_{T\to\infty} \frac{1}{T} \left| \int_{(T)} \mathrm{e}^{\mathrm{i}\omega t} x(t)\, \mathrm{d}t \right|^2 . \tag{3.160}$$

Our mathematical outline shows the relationship between fluctuation stochastics and jumps or ruptures (3.156)–(3.157), and also possible wave functions (3.159) behind this theorem.

For physical reference, we use three statements from the textbooks of Landau and Lifshitz [345, 716].

(i) Thermodynamic fluctuations of representative subsystems are Gaussian as long as their extensive entropy variation is much larger than k_B (and we may add, as long as a preponderant component from a Levy distribution with Levy exponent $\alpha < 1$ can be neglected in the smallest representative subsystem).

(ii) The matrix elements of QM tend to Fourier components in the quasi-classical limit,

$$f_{mn} \to f_{m-n} \quad \text{in} \quad \bar{f} = \sum_s f_s\, \mathrm{e}^{\mathrm{i}\omega s t}\,. \tag{3.161}$$

(iii) The new state after any QM experiment is independent of the previous QM state, before the experiment.

Bringing the mathematical and physical references together in our separation model, considering linear response as a succession of QM experiments, we take any QM experiment as a rupture in the description that changes the QM state vector (or wave function) abruptly and discontinuously with time in accordance with the laws of probability. A succession of such ruptures can be modelled mathematically by a process $\{\mathbf{Y}_t\}$ with uncorrelated and, for Gaussian, independent increments. It is its variance given by the probability measure R in (3.155) which realizes the link with the physical laws of QM probability, here, with the spectral density.

3.4.7 Stochastic Derivation of the FDT

As mentioned above, the phenomenology of the dynamic glass transition requires a thorough understanding of all possible thermal fluctuations, use of the $d \log \omega$ measure, and different time concepts, in particular, the preparation time τ_{prep} (Sect. 2.1.4). We try to get this understanding from a new analysis of the FDT on the basis of separate models for the QM object and the QM experiment (separation model). In other words, our approach is a synthesis of three classic papers by von Laue [126], Szilard [624], and Nyquist [94].

We know the following four properties of any QM experiment.

(a) Bohr wholeness. A QM experiment can only be understood if both aspects, the macroscopic apparatus with pointer ϕ_{B} (index B for observed) and the QM object with state ψ (without index B) are taken into consideration.
(b) The time coordinate t of the state $\psi(t)$ is defined by the apparatus clock.
(c) Each QM experiment (state collapse, quantum jump) is characterized by a time order. *Before*: a state ψ is reached by preparation, e.g., by the preceding experiment and the continuous evolution between, described by the Schrödinger equation, for example. *At*: verification of one arbitrary ψ_n from $\psi = \sum c_n \psi_n$ with probability $|c_n|^2$, where the index n characterizes the states, i.e., the eigenvalues of the experiment under consideration. The verified state n is indicated by the apparatus pointer $\phi_{n\text{B}}$, e.g., $x_{\text{B}} = \phi_{n\text{B}}$. *After*: the pointer is fastened at $\phi_{n\text{B}}$ and the new state develops independently, analogously as before.
(d) The verification step has a zero collapse time due to the spooky action at a distance ('EPR correlations', a lucid reference is [625]).

These four properties suffice to describe our model for the QM experiment [615].

The general idea was called the separation model above and arises from a dichotomy.

(i) The conventional option, in which quantum mechanics (QM) is considered to be in its final form. Extending its range of application then means making new physics (a new physical model) and applying the formal rules of QM to it. Then QM is some kind of theory of everything.
(ii) The nonconventional option, in which QM is considered as a holistic description of two things, separated by Bohr's original complementarity concept. We have a QM object beyond the common space–time, and an apparatus that gives observables in the common space–time. Although this measurement (experiment) is inherent to conventional QM, we could try to dissect QM with a view to separating the QM object from the apparatus, to describe the QM object and, after this description, to reinsert it and stitch it back to the apparatus. The hope is that we might learn

about previously hidden features, such as whether the temperature fluctuates (or about the parameters of standard model of elementary particle physics). Then QM is a restrictive theory only suited to our real world.

It seems necessary to distinguish three time aspects from the very beginning.

(A) A general, classical time as indicated by the apparatus clock.
(B) Time intervals or, more precisely, any two times used in the definition of correlation functions. After the Fourier transformation (3.127)–(3.128), these time intervals correspond to the probing time $1/\omega$ of dynamic experiments.
(C) A 'vital time', defined by the succession of quantum jumps.

The following derivation also shows how these three time aspects are identified by the FDT. The new derivation of the FDT is not restricted to glass transition relaxations.

One might think once again that the starting point for our aim must be mechanics, i.e., particles and fields. As repeatedly mentioned, however, the basic object of thermodynamics [126] is a freely fluctuating representative subsystem with an internal energy U described by the fundamental form $U = \sum f_i \mathrm{d}x_i$. This form ensures an equivalent treatment of heat variables ($T\mathrm{d}S$) and work variables ($-p\mathrm{d}V + \dots$), i.e., temperature fluctuation is not excluded.

We start, therefore, with the following identification (Fig. 3.6a). Let us assume that no external conscious observer is needed for the experiment. Then we can miniaturize the apparatus to the size of any representative subsystem with the two following properties: it remains a classical body in the sense that we can at least define thermodynamic variables, and it is statistically independent from the surroundings.

Concerning the dynamic glass transition, the following aspects seem important for such an apparatus.

(i) Representativeness allows an appropriate homogeneity concept of space and time for thermodynamics. This is important for a primary use of the fundamental form.
(ii) The ω identity allows functionality, i.e., the separate application to each dispersion zone. We may therefore refer to the pattern of dynamic heterogeneity for the α process, for example, formulated as a fluctuating mobility field $\log \omega(\boldsymbol{r}, t)$ relevant for entropy and density fluctuations. For spatial scales larger than the characteristic length ξ_α and for times longer than the retardation time τ_R, we can define homogeneity (similar to such a definition in a molecular liquid). For smaller scales, we can get experimental information about dynamic heterogeneity.
(iii) Since the pattern of dynamic heterogeneity is disengaged from molecular structure (Sect. 2.2.6), we can get, near the $\alpha\beta$ crossover region, small representative subsystems with size of order one molecule (Sect. 2.5.3). This is no problem since, according to Nyquist [94], we measure quanta

$\hbar\omega$ directly, and the definition of the apparatus only needs a sufficient event density of the mobility field (Sect. 3.5).

(iv) The relationship with the molecular nature of liquids is established subsequently, after the experiment, by the Boltzmann constant k_B in the FDT.

On the other hand, we extend the QM system, the object of the experiment, to the size of a representative subsystem and define a macroscopic wave function ψ and Fourier components (3.161).

We thus assume the identification

$$\text{QM object} = \text{subsystem} = \text{apparatus}\,. \tag{3.162}$$

Equation (3.162) means that we consider a spontaneous succession of *self-experiments* in any subsystem. We must now carefully distinguish $\{x(t), f(t)\}$ without index, the stochastic functions that are not yet observed, from $\{x_B, f_B\}$ with B index, the values observed (indicated by a pointer or being the pointer). Once again, detailed knowledge of the 'hidden' stochastics of the QM system prior to the experiment is not necessary, except that the stationary stochastics of the spontaneous fluctuation is modified by the succession of self-experiments. Since the $\{x, f\}$ variables of the fundamental form are the only important variables for our subsystems in a homogeneous environment in the linear region, we assume that the ψ collapses are reflected in certain jumps of these $\{x, f\}$ variables. This property suffices to describe the model for the QM object.

We show [167] that the two model properties give the fluctuation–dissipation theorem FDT. Consider first the pointer. The dynamics of a thermodynamic subsystem imply that two things act on the pointer x_B:

(i) the subsystem memory,
(ii) the succession of QM self-experiments.

Any experimental equation for x_B must contain the time t from the first time aspect (A), and must be causal. Referring to the fundamental form, the memory must be characterized by a phenomenological subsystem susceptibility J describing how the actual state is influenced by the proceeding random quantum jumps. Causality implies

$$\Delta x_B(t) = \int_{-\infty}^{t} J(t - t')\, \dot{f}\,(t')\, \mathrm{d}t'\,, \tag{3.163}$$

for the linear region. Equation (3.163) is merely a consequence of causality and linearity for a succession of self-experiments in systems with thermodynamic memory. The symbols of this equation get the following meaning in the present context: $\Delta x_B(t)$ is the pointer aspect of the experiment, $\dot{f}(t') = \mathrm{d}f/\mathrm{d}t(t')$ the QM object aspect, and $J(t - t')$ the subsystem memory aspect. Equation (3.163) is a *stochastic integral* similar to (3.157). There is no B index on f, and the integral defines $\Delta x_{\mathrm{B}}(t)$ as a continuous stochastic variable. The use of the time derivative $\dot{f}$ (instead of f) is permitted by the robustness of the $\mathfrak{H} \longleftrightarrow L^2_R$ mapping (3.156)–(3.157) and is dictated by the insight that it is the change of state, and not the state itself, that causes a response. There would be no response for $\dot{f}(t) \equiv 0$, as for example, in the case of zero response in the glassy state with no relevant fluctuation.

The selection of different activities, e.g., different f and x from a higher-dimensional form $\sum f_i \mathrm{d}x_i$, is made by different pieces of external apparatus. Referring to the general 'cloud of quanta' in and around any subsystems of Fig. 3.1e, it is the specific external apparatus which selects the specific population of quanta for the concrete modulus G or the concrete compliance J from the stock of (3.52), or for cross activities, or for different kinds of dynamic scattering.

Secondly, let us consider the succession of self-experiments. Let $\Delta f(t')$ be a stationary stochastic process with the expectation value $\mathrm{E}[\Delta f] = 0$ (and $\mathrm{E}[\Delta x] = 0$). To find the FDT, we need the correlation function for the fluctuating observable x_{B},

$$\Delta x^2(t,s) \equiv \varphi_{\mathrm{B}}(t,s) = \mathrm{E}\left[\Delta x_{\mathrm{B}}(t)\Delta x_{\mathrm{B}}(s)\right] \ . \tag{3.164}$$

Multiplying the experimental equation (3.163) by $\Delta x_{\mathrm{B}}(s)$, we obtain, after partial integration,

$$\varphi_{\mathrm{B}}(t,s) = -\int_{-\infty}^{t} \mathrm{d}t' \dot{J}(t-t')\,\mathrm{E}\left[\Delta f(t')\Delta x_{\mathrm{B}}(s)\right] \ . \tag{3.165}$$

We must calculate the expectation value of the product $\Delta f(t')\Delta x_{\mathrm{B}}(s)$, where Δf is without B index. Both t' and s are times that must be considered as the two times in a correlation function (the second time aspect B in the above list).

The main idea is to follow the way indicated first by (3.156) and later by (3.157). We show that a Heaviside step function ε is obtained for

$$\mathrm{E}\left[\Delta f\ (t')\Delta x_{\mathrm{B}}(s)\right] = \overline{E}\epsilon(s-t') \ , \quad \epsilon(\tau) = \begin{cases} 0 \text{ for } \tau < 0 \\ 1 \text{ for } \tau > 0 \end{cases} , \tag{3.166}$$

where $\overline{E}$ has energy dimensions.

We begin by proving the ϵ step. It is important that the Δx_{B} observable is the result of the QM experiment, whilst Δf is not. In fact, Δf and Δx_{B} are 'complementary'. Without any experiment we would have zero correlation between Δx_{B} and Δf.

Let t' be the time of one experiment in the succession of self-experiments and consider different s times. Before the experiment $s < t'$, the state

$\Delta f(t')$ is not 'measured', i.e., the state $\Delta f(t')$ has no influence on $\Delta x_{\mathrm{B}}(s)$, $\mathrm{E}\,[\Delta f(t')\Delta x_{\mathrm{B}}(s)] \equiv 0$ for $s < t'$. At the moment when the experiment takes place $s = t'$, the state $\Delta f(t')$ realizes its role with regard to the indication of the pointer $\Delta x_{\mathrm{B}}(s)$. After the experiment $s > t'$, the pointer is fastened [624] at $\Delta x_{\mathrm{B}}(s)$. We have a fixed relation between $\Delta x_{\mathrm{B}}(s)$ for all $s > t'$ and the realized state $\Delta f(t')$ at this one t', i.e., $\mathrm{E}[\Delta f(t')\Delta x_{\mathrm{B}}(s)] = \text{const.}$ for $s > t'$. Thus the ϵ step in (3.166) describes the quantum jump of the experiment at t', this being one event in the $\{t'\}$ succession of self-experiments (time aspect C on the above list). The subsystem memory $J(t - t')$ enters the stage when we integrate over this succession, (3.163) and (3.165).

Secondly, we prove that $\overline{E}$ is an energy. The conjugate of a homogeneous time, represented by the stationary stochastic process $\Delta f(t)$ and by reversible fluctuation $\Delta x_{\mathrm{B}}(s)$, must be an energy. In particular, if x is extensive, then f is intensive, and $x\mathrm{d}f$ or $f\mathrm{d}x$ must be a part of the fundamental energy form for the subsystem. This completely proves the statement of (3.166).

Furthermore, stationarity implies a relation between the first (A) and second time (B) aspects: stationarity identifies these two time aspects by a difference,

$$\varphi_{\mathrm{B}}(t, s) = \varphi_{\mathrm{B}}(t - s)\ . \tag{3.167}$$

Provided that $\overline{E}$ does not depend on time or frequency, we obtain the FDT in the time domain from (3.165)–(3.167),

$$\varphi_{\mathrm{B}}(\tau) = -\overline{E}\,[J_{\mathrm{B}}(\tau) - J_{\mathrm{B}}(\text{equil})]\ , \quad \tau \geq \tau_{\mathrm{c}} \geq 0\ . \tag{3.168}$$

Since the left-hand side of (3.168) is observable, we have to put the same for the right-hand side. In other words, (3.168) implies

$$J = J_{\mathrm{B}}\ , \tag{3.169}$$

which means that the subsystem memory J in (3.163), starting out as a susceptibility for the so far unobserved stochastics of $f(t')$, advances without any change in (3.168) to an observable compliance J_{B} with index B, because φ_{B} on the left-hand side is an observable correlation for $x_{\mathrm{B}}(t)$. In this sense, (3.169) represents the Onsager hypothesis [626], where the decay of an accidentally large internal, spontaneous fluctuation is identified with the linear response to a corresponding external disturbance. The collapse or jump time is denoted by τ_{c} in (3.168). The spooky action at a distance implies that $\tau_{\mathrm{c}} = +0$ is not, for instance, the time-of-flight of a photon. This identifies the vital time aspect (C) with the two others, (A) and (B). We thus conclude the stochastic derivation of the FDT as an equation for QM experiments applied to the thermodynamic situation of freely fluctuating subsystems.

The results of such experiments are susceptibilities: compliances or moduli. The minus sign in (3.168) is for compliances (because x was assumed to be extensive). For moduli G, we obtain analogously

$$\Delta f^2(\tau) = \varphi_{\mathrm{B}f} \equiv \mathrm{E}[\Delta f_{\mathrm{B}}(t)\Delta f_{\mathrm{B}}(t+\tau)]$$
$$= +\overline{E}\,[G_{\mathrm{B}}(\tau) - G_{\mathrm{B}}(\mathrm{equil})] \quad \text{for } \tau \geq 0 \,, \tag{3.170}$$

where f is intensive. Instead of (3.163), the experimental equation is now

$$\Delta f_{\mathrm{B}}(t) = \int_{-\infty}^{t} G(t-t')\,\dot{x}(t')\,\mathrm{d}t' \,. \tag{3.171}$$

Calculating $\Delta x_{\mathrm{B}}(s)$ from (3.163) by substituting $\dot{f}(t')$ as obtained from the experimental equation (3.171), we model a self-experiment for $\Delta x_{\mathrm{B}}(s)$ by:

$$\Delta x_{\mathrm{B}}(s) = \int_{-\infty}^{s} J(s-t)\frac{\mathrm{d}}{\mathrm{d}t}\left[\int_{-\infty}^{t} G(t-t')\,\dot{x}(t')\,\mathrm{d}t'\right]\mathrm{d}t \,. \tag{3.172}$$

The right-hand side of (3.172) gives $\Delta x_{\mathrm{B}}(s)$ only for $G * J(t) \sim t$, $t > 0$, i.e., after Fourier transformation [605], for

$$J_{\mathrm{B}}^{*}(\omega)\;G_{\mathrm{B}}^{*}(\omega) = 1 \,, \tag{3.173}$$

in thermodynamically one-dimensional systems. The consistency with external perturbation results from $J = J_{\mathrm{B}}$, $G = G_{\mathrm{B}}$, and (3.171):

$$\Delta x_{\mathrm{B}}^2(t,s) = \int\!\!\int \mathrm{d}t'\mathrm{d}s'\,\dot{J}_{\mathrm{B}}(t-t')\;\dot{J}_{\mathrm{B}}(s-s')\;\Delta f_{\mathrm{B}}^2(t,s) \,, \tag{3.174}$$

and an analogous equation for $\Delta f_{\mathrm{B}}^2(t,s)$ on the left-hand side with two $\dot{G}'_{\mathrm{B}}$s on the right-hand side.

3.4.8 Quantum Aspects

We must now calculate the $\bar{E}$ parameter in (3.166). The quantum jumps (now interpreted as part of the self-experiments) generate frequencies which are suggested to come from oscillators. We therefore identify $\overline{E}$ with the average energy of an oscillator with frequency ω_0. But this is not trivial [627, 628], and we need further details.

In the QM derivation of the FDT (e.g., [313], Sect. 124), the energy enters via the quantum jump condition

$$E_n - E_m = \hbar\omega_{nm} \,, \tag{3.175}$$

for example, as the energy of a photon, phonon, or oscillator. This problem was first described by Planck for the equilibration of black body radiation, and a modified Planck approach was later used by Nyquist for his transmission lines, as mentioned above.

Thus, bearing in mind that $\overline{E}$ in (3.168) and (3.170) is a thermodynamic average of energy, and reflecting that, from general QM arguments, energy enters via (3.175), we must substitute $\overline{E}$ by the mean energy of an oscillator $\overline{E}(\omega_0, T) = -(1/2)\hbar\omega_0 \coth(\hbar\omega_0/2k_BT)$. We therefore obtain for the moment

$$\varphi_B(\tau) = -\frac{1}{2}\left[J_B(\tau) - J_B(\text{equil})\right] \coth\frac{\hbar\omega_0}{2k_BT} \,. \qquad (3.176)$$

In the classical case $\hbar\omega_0/k_BT \ll 1$, we do indeed obtain the FDT, in the time domain,

$$\Delta x^2(\tau) = -k_BT\left[J_B(\tau) - J_B(\text{equil})\right] \,,$$
$$\Delta f^2(\tau) = +k_BT\left[G_B(\tau) - G_B(\text{equil})\right] \,, \qquad (3.177)$$

and in the frequency domain,

$$\Delta x^2(\omega) = k_BTJ''_B(\omega)/\pi\omega \,, \quad \Delta f^2(\omega) = k_BT\, G''_B(\omega)/\pi\omega \,. \qquad (3.178)$$

In the QM case $\hbar\omega/k_BT \gtrsim 1$, we must find the relation between ω_0 and ω in the frequency domain. Formally, the reduction of the many jump frequencies (3.175), $\{\omega_{mn}\} \mapsto \omega$, corresponds to energy conservation. Reflecting that the experiments are self-experiments without additional external parameters, and from consistency of internal and external perturbation, (3.169) and (3.174), we are forced to identify the frequency ω_0 of the internal oscillator with the external frequency ω of a dynamic experiment,

$$\omega = \omega_0 \,. \qquad (3.179)$$

This identification expresses a detailed balance connected with the $\{\omega_{mn}\} \mapsto \omega$ reduction and the ω identity of the FDT. The detailed balance can therefore be partitioned into dω intervals [280]. From (3.179), we finally get

$$\Delta x_B^2(\omega) = \overline{E}(\omega, T)\frac{J''_B(\omega, T)}{\pi\omega} \,, \quad \Delta f_B^2(\omega) = \overline{E}(\omega, T)\frac{G''_B(\omega, T)}{\pi\omega} \,, \qquad (3.180)$$

which correspond exactly to the formulas first obtained by Callen and Welton [277].

In a way, our derivation of (3.180) is an explicit example of the consistency of the separation model we used for *QM experiments* and *QM objects*, on the one hand, and the predictive powers of *quantum mechanics* plus the quantum jumps (3.175), on the other. Equation (3.179) again shows that the ω identity has a genuine physical background. One advantage of our derivation is that the FDT can also be applied to the temperature fluctuation (as conjugate to the entropy fluctuation). This is not possible within the framework of the Gibbs distribution. Another gain is the demonstration that the relevant representative subsystems (CRR for the α process here) can be unexpectedly small. A further gain is the explanation of the Second Law of thermodynamics (Sect. 3.4.9).

We conclude this section with two remarks.

(i) The correlation functions in the time domain for the $\hbar\omega/k_BT \gtrsim 1$ case must be calculated by Fourier transforms of the spectral densities (3.180) in the frequency domain. The result is, of course, different from the original findings in the time domain, (3.168) and (3.170). This means that, for matrix elements beyond the Fourier components (3.161), the QM experiment is not so simple in the time domain. Several arguments should be translated into the frequency domain if the apparatus pointer itself has QM aspects.

(ii) On the one hand, the question about the wave function ψ of a subsystem is obscured by the memory. On the other hand, both the unitary evolution of ψ and the response are linear. One might consider [623] using $\hat{u}(\omega)$ and $u(t)$ of the Wiener–Khintchin criterion, (3.158) and (3.159), as conjugate wave functions in frequency and time representations. Being mutual Fourier transforms, they obey an uncertainty relation [$\hbar$ is introduced by (3.175)]. The $|\psi|^2$ of the probability interpretation is substituted by spectral densities and correlation functions. The running rupture in the unitary evolution induces stochastic aspects into ψ as well as into u and $\hat{u}$.

3.4.9 Thermodynamic Consequences: Subtleties of the Second Law

Defining thermodynamics as phenomenological relations for state changes resulting from the FDT as the equation for QM self-experiments, we show that such a thermodynamics obeys the *Second Law* without the need for any additional input [167].

The transfer of the time order of a quantum mechanical (QM) experiment to a general time arrow is subtle. Thermodynamic equilibrium fluctuations, although containing the succession of self-experiments in vital time, remain reversible. A first description of the situation with demons making the experiments was published by Szilard [624].

To begin with, we show that, due to the self-experiments, the internal entropy increases or remains constant during or after external disturbances, as expected. The label of an external disturbance is again the B index, e.g., for f_B.

Consider first the frequency domain. In the vicinity of local equilibrium, the increase in internal entropy is equivalent to the non-negativity of all imaginary parts of dynamical compliances and moduli, e.g., $J''_B \geq 0$. The property $J''_B \geq 0$ and $G''_B \geq 0$ in the FDT equation (3.180) for any term in the fundamental form is a consequence of the purely mathematical property of the corresponding spectral densities, $\Delta x^2(\omega) \geq 0$, $\Delta f^2(\omega) \geq 0$ [Bochner–Khintchin theorem, (3.155)].

Now consider the time domain. The entropy production for external disturbances is calculated from an equation that has the property of positive definiteness, as defined in (3.153). Consider a 'noncomplementary' situation with two conjugate observables: (i) Δx_{B} from (3.163) with $J = J_{\mathrm{B}}$, and (ii) the conjugate observable f_{B} of a subsystem. Both have the index B, i.e., noncomplementarity means that both are observed by (possibly different) experiments.

Example. Consider a viscoelastic creep experiment that observes (measures) the shear angle $\gamma(t) = \Delta x_{\mathrm{B}}$ after or during an external shear stress program $\sigma(t') = \Delta f_{\mathrm{B}}$. This program is also 'observed' by the macroscopic program device. The B character of γ and σ is transferred across the sample to any subsystem.

The entropy production $\dot{S}_{\mathrm{int}} = \dot{Q}/T$ is then calculated from the expectation E

$$\dot{Q} = \mathrm{E}\int_{-\infty}^{t} J_{\mathrm{B}}(t-s)\dot{f}_{\mathrm{B}}(t)\dot{f}_{\mathrm{B}}(s)\,\mathrm{d}s \overset{\mathrm{FDT}}{=} \ldots \tag{3.181}$$

$$\ldots \overset{\mathrm{FDT}}{=} \frac{1}{E}\mathrm{E}\int_{-\infty}^{t} \varphi_{\mathrm{B}}(t-s)\dot{f}_{\mathrm{B}}(t)\dot{f}_{\mathrm{B}}(s)\,\mathrm{d}s \geq 0\,.$$

The inequality in (3.181) follows from positive definiteness, i.e., from a comparison with (3.153) using

$$\sum_{j,k} \sim \mathrm{E}\int\,, \quad z_j \sim \dot{f}_{\mathrm{B}}(t)\,, \quad \bar{z}_k \sim \dot{f}_{\mathrm{B}}(s)\,. \tag{3.182}$$

Since the FDT is a consequence of the QM experiment only we have proven the statement. *The time order (before, at, after) of the QM experiment is sufficient to explain the increase in internal entropy during and after an external disturbance, i.e., the Second Law.* The term 'external' is related to the surroundings of any subsystem. Equilibration of an isolated sample consisting of several subsystems is thus included. The inclusion of x = entropy and f = temperature in $\mathrm{d}U = f\mathrm{d}x + \ldots$ implies that a thermodynamic TS cycle, in one thermodynamic dimension, which formally involves 'no work' also produces entropy for $C_p'' > 0$ (3.81). [In principle, the term 'no work' is misleading, because every thermodynamic cycle of a freely fluctuating subsystem is necessarily related to more than one thermodynamic dimension. In two dimensions, the cycle has the same area in TS and pV diagrams since the Jacobian has value $|\partial(p,V)/\partial(T,S)| = 1$ (First Law).] Moreover, the FDT holds also for temperature fluctuations:

$$\Delta T^2(\omega) = k_{\mathrm{B}}T\,\frac{K_T''(\omega)}{\pi\omega}\,, \quad K_T^* = \frac{T}{C_p^*}\,, \tag{3.183}$$

with C_p^* in one thermodynamic dimension for an isobaric calorimetric experiment on a macroscopic sample.

Now for the subtle point. On the one hand, we have a succession of QM self-experiments in any subsystem. On the other hand, the spontaneous internal fluctuations are, of course, reversible, in spite of the time order for any self-experiment. This can be seen explicitly in the framework of our formulas. Consider an internal 'Boltzmann spike', i.e., a spontaneous, large change of an 'internal disturbance energy' $\overline{E'}(t)$, for observable x_B fluctuations with no external disturbance. We have a 'complementary' situation with only one B observable x_B, and one variable with no B, e.g., f. From thermodynamics,

$$\dot{\overline{E'}} \equiv \frac{\mathrm{d}\bar{E}'}{\mathrm{d}t} = -\mathrm{E}\left[\Delta x_\mathrm{B}\frac{\mathrm{d}f}{\mathrm{d}t}\right] . \tag{3.184}$$

Using the FDT, we find

$$\dot{\overline{E'}} = -\frac{\mathrm{d}}{\mathrm{d}t_f}\mathrm{E}\left[\Delta x_\mathrm{B}\Delta f\right] \overset{\mathrm{FDT}}{=} -\frac{\mathrm{d}}{\mathrm{d}t_f}\overline{E}\,\epsilon(t_x - t_f) = \overline{E}\delta(t_x - t_f) , \tag{3.185}$$

and by integration,

$$\int \mathrm{d}t_x\, \dot{\overline{E'}} = \overline{E} = \text{const.} , \tag{3.186}$$

i.e., there is no irreversibility. The commutation of E and $\mathrm{d}/\mathrm{d}t_f$ is permitted because the time used here corresponds to the apparatus clock time (A) aspect of the subsystem. We may thus make the following claim: *although any QM self-experiment has a time order, the succession of self-experiments is thermodynamically reversible for spontaneous equilibrium fluctuations.*

Comparing the two above statements, and referring to (3.181) and to the above example, we see of course that an external noncomplementary disturbance is needed for irreversibility. Such an external noncomplementary disturbance, Δf_B or, in other cases, Δx_B, should not be confused with the spontaneous QM self-experiments.

The response is linear as long as the disturbance for the smallest representative subsystem does not exceed the f and x amplitudes of its spontaneous fluctuations. However, an external disturbance on a subsystem induces irreversibility, even if it is arbitrarily small compared to the subsystem fluctuation. *Although internal (f) and external (f_B) forces – the same holds for x and x_B – are of the same physical nature in the linear region, it is the noncomplementarity connected with Δf_B and Δx_B which is responsible for irreversibility* (see Sect. 3.4.4).

From the Bochner–Khintchin theorem, we see further that: *although the QM Hamiltonian of a subsystem is symmetric with respect to time reversal, for real thermodynamic subsystems (i.e., with a succession of QM self-experiments) the spontaneous recurrence of a state with smaller internal entropy is not possible.*

Since for mechanical systems without QM self-experiments such a recurrence must occur, we may ask how small the information about the initial state in a chaotic system really is. In a stable mechanical chaotic system with positive Ljapunov exponents, the actual information about the initial state is exponentially small. From the above arguments, we see that it is this small piece of information that is demolished by state collapses at the self-experiments, whereas the correlations of real thermodynamic subsystems are not diminished by the self-experiments. In other words, the spurious information about the initial state in a chaotic subsystem is too small for a QM experiment.

As mentioned above, the succession of self-experiments in real subsystems introduced a third aspect, the vital aspect (C), into the time concept, in addition to the classical time coordinate in $\psi(t)$ (A) and the correlation aspect (B). Vitality of time means that the hypothetical Hamiltonian evolution, invariant against time reversal, is continually ruptured by the succession of QM self-experiments. It is this vitality of time that allows a non-trivial local discussion of the time arrow in the length and time scale of representative subsystems. The self-experiments in different subsystems can define a common time arrow since the FDT holds for any, i.e., also for large subsystems containing the others. The frequency of self-experiments does not have a measurable effect since an arbitrary number of transition states can be inserted in a convolution or a Feynman path integral. From a mathematical point of view, this is an example of the Smirnov–Kolmogorov stability of stochastics.

Neither the expanding universe nor the violation of T invariance for elementary particles is necessary for understanding the Second Law and a common time arrow for subsystems.

The foundation of the Second Law by the FDT is valid as long as we can define the local equilibrium of the smallest representative subsystems. This includes a very large class of non-equilibrium situations, containing at least the complete surroundings of the equilibrium state.

The reader may ask why such a lengthy description of the FDT is appropriate in a book on the glass transition. I would answer that this treatment of FDT justifies the existence of temperature fluctuations for the calorimetric determination of the characteristic length, (2.76)–(2.77). The temperature fluctuation is not an ad hoc assumption but is based on thermodynamics, in particular, on thermodynamic systems defined 'before' the mechanics of particles and fields is inserted. Furthermore, the discussion of the requirements for a representative subsystem to act as an apparatus (Sect. 3.4.7) shows that small characteristic lengths down to the cage door (Figs. 2.33 and 2.34a) get a definite meaning. This discussion of the latter aspect will be continued in Sect. 3.5 (Table 3.2).

3.5 Discussion of Functional Independence

Statistical independence of freely fluctuating subsystems was discussed in relation to spatially neighboring subsystems (Sect. 3.1). The functionality

Table 3.2. Attempt number and event density for several typical processes in the Arrhenius diagram of Fig. 2.32

Process	Time [s]	Length [nm]	Velocity [m/s]	Diffusivity [m^2/s]	Event density [$s^{-1}nm^{-3}$]	Attempt number[a] [1]
c	10^{-12}	0.1	10^2	10^{-8}	10^{15}	–
β	10^{-6}	0.5	5×10^{-4}	2.5×10^{-13}	8×10^6	1.3×10^8
α	10^{-1}	3	3×10^{-8}	9×10^{-17}	3×10^{-1}	2.2×10^7
ϕ	10^{+6}	100	10^{-13}	10^{-20}	10^{-12}	3.7×10^{11}

[a] Of the preceding process.

of subsystems was defined as their relation to Fourier components in the dispersion zone under consideration (Sect. 2.2.5). We now ask about temporal and spatiotemporal relations between different dispersion zones in a given state.

We may ask whether or not the events of the temporally preceding process can define the probability space for the fluctuating pattern of dynamic heterogeneity for the actual process. In other words, whether the pixel density in $(\boldsymbol{r}, t)$ is high enough to define the mobility field $\log \omega(\boldsymbol{r}, t)$ used to represent the pattern. The spatial scale for the pattern of the α process, for example, was defined by the characteristic length ξ_α, the average diameter of cooperatively rearranging regions (CRRs). The question about the probability space seems crucial for understanding the dynamic glass transition because the calorimetric determination of $V_\alpha = \xi_\alpha^3$ near the crossover region yields characteristic volumes much smaller than the first molecular coordination shell: V_α decreases down to the volume of one particle. This was interpreted by the restriction of cooperativity volume to the actual α or a process in the cage door (Sect. 2.5.3).

Consider an isothermal section of the general equilibrium Arrhenius diagram Fig. 2.32. Below the crossover temperature, we find for decreasing frequencies: vibrations and collisions at IR and Raman frequencies, cage-rattling (and boson peak) at terahertz, the Johari–Goldstein β process, e.g., at megahertz, the cooperative α process, e.g., at 10 Hz, and the ultraslow Fischer ϕ modes, e.g., at microhertz. For a section above the crossover temperature, we find the Williams–Götze a process instead of the (α, β) pair, and ϕ some frequency decades higher than before.

We shall now discuss three methodological questions.

(i) Definiteness of Pattern for Dynamic Heterogeneity. The thermal velocities of reasonably light molecules or monomeric units are of order $v_T = 100$ m/s $= 0.1$ nm/1 ps. For the relaxations mentioned above, in the special example, the velocities and diffusivities can be estimated from the ratio of characteristic length divided by time, and length squared divided by time, respectively (Table 3.2).

The actual spatiotemporal event density (reciprocal volume times reciprocal time), and the number of attempts that are provided by the preceding process to represent the pattern of the actual process, are also listed in Table 3.2. We assume that the events of the preceding process define the attempts (= pixels) of the actual process. The attempt number is defined by the volume ratio times the time ratio.

The velocities, diffusivities, and event densities are well ordered. A large temporal and spatial *density of attempts* is guaranteed by the preceding process to make a quasi-continuous probability theory of events for the spectral densities of the actual process, i.e., for β, α, ϕ (and a) processes (Fig. 3.1c). For all three processes (β, α, ϕ, and also for a) the pixel density is large enough to define a quasi-continuous fluctuating spatiotemporal pattern of dynamic heterogeneity. This pattern is factual in space and time if the actual process is itself associated with a distribution of many partial steps, e.g., near the Glarum–Levy defects.

These results are also a consequence of general scaling applied to a comparison of the different dispersion zones. Longer lengths correspond to longer times (first columns of Table 3.2). The results seem important because, as mentioned above, abstract constructions via the mobility field are used for the definition of a characteristic length for the α process, for example.

(ii) Repeated Randomness Assumption. In certain problems of statistical physics, the *repeated randomness assumption* is made for a preceding, faster process [411]. It is considered as miraculous that the enormous number of 'microscopic' variables of the faster process can be eliminated for description of the slower and larger actual process. The randomness assumption is conventionally considered as inescapable: the microscopic variables vary so rapidly that they are able to reach their equilibrium distribution almost instantaneously. Investigating the actual process, the 'macroscopic' variables are not fixed but vary slowly. One seems to be forced, and indeed to be able, to repeatedly readjust the assumed randomness of the faster process.

This repeated randomness assumption, however, should not be associated with statistical independence in the situation of Table 3.2. The cooperative α process and the local β process cannot be considered a priori as statistically independent. Certain physical relations between the α and β processes are expected because both use the α process cages (Fig. 2.33). Although physical relations are not excluded for statistical independence (as several times mentioned above), independence would restrict the situation more than necessary for data evaluation (Example 2 in Sect. 3.1). In order to separate the α and β processes, a precondition for later finding the interrelations between them, we need only identify each of them by an (adjustable) parameter set. The basis for this method was the ω identity of the FDT.

(iii) Terminology. We should distinguish *functional independence* from statistical independence as generally introduced in Sect. 3.1. As mentioned above, the term 'functional' is used here to mean "related to the Fourier components of the actual dispersion zone" (Sect. 2.2.5). Functional independence is used when self-governing or self-determining features of a dispersion zone are stressed. For example, consider hole burning for the α process (Sect. 2.2.5). Heating shifts the frequency spot together with the α process modes along the dispersion zone according to the WLF equation, leaving a hole at the original spot in the Arrhenius diagram. The relevant time scale of the α process is always $\tau_\alpha(T)$. It cannot be accelerated by faster processes (e.g., by phonon thermal conductivity) and cannot be decelerated by slower processes. Similarly, the temperature fluctuation functional to the α process [e.g., δT in (2.76), (2.77) and (2.79)] cannot be quenched by the large phonon thermal conductivity. The α process dispersion zone may be influenced by a parameter change for the other processes, but of course not by the existence of the other processes. It may be possible to describe a certain relationship between the dispersion zones by their high- and low-frequency wings, respectively (Sect. 2.4.5).

This section is concluded with two examples which illustrate the value and limits of the functional independence concept.

Example 1: Physical Relations Between the Cooperative α Process and Fischer Modes ϕ. We start from the common property that a CRR has a Levy distribution for extensive spectral densities. This follows from the representativeness theorem (Sect. 3.1). For α functional CRRs, an internal treatment (Fig. 2.17a, Sect. 2.2.7) yields a relation to the size and distribution of free volume causing a dynamic heterogeneity pattern (Glarum–Levy defects) on the ξ_α length scale. The clue is the Levy property that for $\alpha < 1$, one of the many partial systems inside a CRR is preponderant (Sect. 4.3). For the ϕ pattern, in an external treatment (see also Fig. 2.17a), the same Levy property must now be applied to the set of many CRRs (not to partial volumes of one as for α). One CRR of a larger set must be preponderant. This generates a certain contrast and a new characterizing length, possibly as the average distance between the preponderant CRRs.

The interrelation between the two zones is defined by continuity under a virtual length scale variation: the same Levy exponent and the same Levy distribution must be used for the frontier between the two kinds of mode, i.e., for one CRR. Note that a different situation for statistical independence is used for each Levy distribution: within a CRR for α (Sect. 4.3, leading to a stretched potential), and between CRRs for ϕ (Sect. 4.4, leading to a Debye relaxation). However, the repeated randomness and mutual statistical independence assumptions are not explicitly used to find the relation between the α and ϕ processes.

Example 2: Dynamic Heterogeneity for the Williams–Götze Process (a Process)? Calorimetric experiments carried out to date indicate CRRs of order the particle size above the scenario I crossover region [244,629]. The attempt density estimate from the c process allows (Table 3.2) a quasi-continuous description of such fine-grained dynamic heterogeneities. On the other hand, the concept of homogeneity in a liquid may neglect molecular structure. Dynamic homogeneity would also correspond to the equivalence of all particles in the idealized mode-coupling theory. It may be a matter of opinion to speak in this borderline case of dynamic heterogeneity. The question may change for crossover scenario II if the cage really turns out to be larger than the first coordination shell, as suggested for scenario I (Fig. 2.33).

3.6 Levy Distribution

This section is concerned with the mathematical details of stable or limit distributions with Levy exponents $\alpha < 2$, here called *Levy distributions*. We continue the relevant discussions of nonexponentiality (Sect. 2.2.4), dynamic heterogeneity (Sect. 2.2.5), freely fluctuating subsystems, especially the representativeness theorem (Sect. 3.1), and the fluctuation–dissipation theorem (FDT, Sect. 3.4). A model for the slowing down mechanism of the α process based on redistribution of free volume inside a cooperatively rearranging region (CRR) – resulting in Glarum–Levy defects (Sects. 2.2.4–2.2.5) – will be described in Sect. 4.3. The application to Fischer modes ϕ will be described in Sect. 4.4.

The description is restricted to the spectral density and correlation function of the Levy distribution for stationary fluctuations. Problems of temporal evolution after parameter changes are not considered.

Kohlrausch Function for Retardation. Let us repeat the set of formulas from linear response and FDT adapted to a Kohlrausch function with exponent $\beta_{\mathrm{KWW}} = \alpha$, for a compliance (2.17) with finite equilibrium value $J_e = \lim J(t) > 0$ when $t \to \infty$:

$$J''(\omega) = \pi\omega\Delta x^2(\omega)/k_B T\,, \tag{3.187}$$

$$J_e - J(t) = \Delta x^2(t)/k_B T\,, \tag{3.188}$$

$$J_e - J(t) = \Delta J \exp(-t^\alpha)\,, \tag{3.189}$$

$$G^* = 1/J^*\,, \tag{3.190}$$

$$G''(\omega) = \pi\omega\Delta f^2(\omega)/k_B T\,, \tag{3.191}$$

$$J''(\omega,\alpha) = \omega \int_0^\infty \mathrm{d}t \cos(\omega t)\exp(-t^\alpha)\,, \quad \text{for} \quad \Delta J = 1\,, \tag{3.192}$$

$$x = \omega t \, , \quad y = \alpha \ln \omega \, , \tag{3.193}$$

$$J''(y, \alpha) = \alpha \mathrm{e}^{-y} \int_0^\infty \mathrm{d}x \sin(x) \exp(-x^\alpha \mathrm{e}^{-y}) x^{\alpha - 1} \, , \tag{3.194}$$

$$J_r''(y, \alpha) \equiv 2J''(y, \alpha)/\alpha \, , \tag{3.195}$$

$$J_r(y, 1) = \frac{2\mathrm{e}^{-y}}{1 - \mathrm{e}^{-2y}} \quad \text{for } \alpha = 1 \, , \tag{3.196}$$

$$J_r(y, \alpha \to 0) = \pi \exp(-y - \mathrm{e}^{-y}) \quad \text{for } \alpha \to 0 \, . \tag{3.197}$$

We start from the assumption (3.19) that the spectral density for extensive fluctuations, $\Delta x^2(\omega)$, is a Levy distribution. More precisely, $\Delta x^2(\omega)$ is a probability density function with measure $\mathrm{d}\omega$. The FDT (3.187) shows that the loss compliance divided by frequency $J''(\omega)/\omega$ is also a Levy distribution. Since, for a classical glass transition, $\Delta x^2(\omega)$ is a symmetric function, $\Delta x^2(\omega) = \Delta x^2(-\omega)$, the characteristic function of the $\Delta x^2(\omega)$ distribution is a Kohlrausch function, with Levy exponent α. As the correlation function $\Delta x^2(t)$ corresponds to the characteristic function [Fourier transform (2.72) or (3.127)–(3.130)], the compliance in the time domain (3.188) is a Kohlrausch function [(3.189), $\beta_{\mathrm{KWW}} = \alpha$]. The conjugate modulus G^* (3.190) cannot therefore be an exact Kohlrausch function for $\alpha < 1$, and the spectral density of conjugate intensive variables, $\Delta f^2(\omega)$, cannot be an exact Levy distribution (3.191).

Let us now return to the compliance for the Levy distribution, normalized to $\Delta J = 1$ (3.192). Although effective computer programs for such functions are available [633], it seems useful for illustration to have graphs and explicit formulas at hand as well (below). Introduce reduced variables x for a reduced frequency and y for a Levy-reduced mobility (3.193). We see from a partial integration (3.194), advantageous for Fourier transformation, that the dynamic compliance J^* in the frequency domain depends not only on $y = \alpha \ln \omega$, but also on the Levy exponent α, although the correlation function in the time domain depends only on one variable, e.g., $\alpha \ln t$. To include the $\alpha \to 0$ limit, the loss compliance can be reduced by α (3.195). Then, besides the Lorentz line for the Cauchy distribution (3.196) at $\alpha = 1$, we also get a simple formula for the $\alpha \to 0$ limit (3.197) [630], related to the double exponential distribution.

The basic idea that the widely used Kohlrausch function for fitting retardation is related to a Levy distribution has been reviewed by Shlesinger [631]. Graphs related to the Levy distribution or the corresponding compliances can be found in [632] and [113], respectively. Modern Fast Fourier Transformations (FFTs) are described in [633,634]. The finiteness of all moments on the $\ln \omega$ axis was first described in [115,635]. For our own mathematical description, we mainly follow [279], but sometimes [636]. The relationship between the Williams–Götze process and Levy distributions is described in [33,637].

Preliminaries for the Levy Distribution. The relationship between spectral densities and probability distribution densities is suggested by the non-negativity of the former, $\Delta x^2(\omega) \geq 0$ (Fig. 3.1c). The use of a Levy distribution is more than a mere possibility among many others. A Levy distribution is distinguished by three particularities: generality, strange properties, and specific difficulties.

(i) Generality. The Levy distribution is a limit distribution, i.e., a distribution that is composed of a sum of a large number of equivalent statistically independent components. We all know the Gaussian distribution with this property. The latter has a finite dispersion, whereas the Levy distribution does not. The Gaussian distribution is the extreme case ($\alpha = 2$) of the general case of Levy distributions with exponent $2 > \alpha > 0$. The generality results from the observation that there are no other limit distributions. The mathematical generality corresponds to their general occurrence in nonconventional liquid dynamics for $\alpha \leq 1$ (representativeness theorem Sect. 3.1).

(ii) Properties. For $\alpha < 1$, the relevant case in glass transition dynamics (Sect. 3.1), we have the property that the maximal component is preponderant over any of the other components. This seems strange since the Gaussian distribution has the distinct property that no component is preponderant. In the limit, all components of a Gaussian distribution contribute infinitesimally to the whole. This is not so for a Levy distribution in the case $\alpha < 1$. As the link between mathematics and liquid dynamics is made via the spectral density, we expect a preponderance of one component in an otherwise 'democratic' ensemble of equivalent components to lead to a definite dynamic heterogeneity for $\alpha < 1$. Coupling to spatial aspects such as redistribution of free volume inside a CRR inevitably leads to the picture given by Glarum–Levy defects. The defect with most free volume corresponds to the preponderant component.

(iii) Difficulties. We mention three difficulties which cannot be handled as simply as in many other probability distributions. Firstly, although the characteristic function in the time domain is the Kohlrausch function, it does not seem possible to find a closed analytical representation in the frequency domain, i.e., a simple formula for the Levy distribution $\Delta x^2(\omega)$. Exceptions are the Gaussian distribution ($\alpha = 2$), the Cauchy distribution [$\alpha = 1$, Sect. 2.2.4, (3.196)], and the $\alpha \to 0$ limit related to the double exponential distribution (3.197). A complicated but explicit formula can also be found for $\alpha = 1/2$ [279, 632]. The mathematical

analysis for the other α values is facilitated by the use of regularly varying functions which control analytical uncertainties.
Secondly, the natural connection between special cases of the Levy distribution, $\alpha = (2, 1, 0)$, and the general α case is veiled or hidden by the widespread need to split mathematical propositions, proofs, and approximations into cases for $\{\alpha = 2,\ 2 > \alpha > 1,\ \alpha = 1,\ 1 > \alpha > 0\}$. An example is the initial slope of the Kohlrausch correlation function,

$$-\frac{\mathrm{d}}{\mathrm{d}t}\mathrm{e}^{-t^\alpha}\bigg|_{t=0} = \mathrm{e}^{-t^\alpha}t^{\alpha-1}\alpha\bigg|_{t=0} = \begin{cases} 0 \text{ for } \alpha > 1\,, \\ 1 \text{ for } \alpha = 1\,, \\ \infty \text{ for } \alpha < 1\,. \end{cases} \quad (3.198)$$

The infinite slope for $\alpha < 1$ reflects the high-frequency tail of the Levy spectral density caused by the preponderant Levy component.
Thirdly, as a function of its natural ω variable, the Levy distribution $\Delta x^2(\omega)$ has no dispersion ($\overline{\omega^2} = \infty$) for $\alpha < 2$ and it even has no expectation value ($\overline{\omega} = \infty$) for $\alpha \leq 1$. This is unusual for anyone familiar with other probability distributions. However, both expectation and dispersion exist for the $\log\omega$ measure ($\overline{\log\omega} < \infty$, $\delta\log\omega < \infty$, Sect. 2.2.4, Fig. 2.14c). In the linear-time domain, all moments of the Kohlrausch correlation function $\Delta x^2(t)$ [characteristic function of the Levy distribution $\Delta x^2(\omega)$] do in fact exist (2.18).

A situation appropriate to the application of a Levy distribution is characterized by the following properties (*Levy situation*). We have many subensembles $(1, 2, 3, \ldots, n)$ that are mutually equivalent (democracy) and statistically independent. We consider additive stochastic variables $(\mathbf{X}_1, \mathbf{X}_2, \mathbf{X}_3, \ldots, \mathbf{X}_n)$ where the sum

$$\mathbf{S}_n = \mathbf{X}_1 + \mathbf{X}_2 + \mathbf{X}_3 + \ldots + \mathbf{X}_n \quad (3.199)$$

is of central interest (composition B of Table 3.1). Moreover, the large numbers limit, formally $n \to \infty$, is considered as representative for the problem, i.e., we are in the domain of attraction of a limit distribution. Moreover, the situation must be robust against the combination of subensembles or, inversely, against the partition of the whole system into subensembles, i.e., we need some arbitrariness in additive combinations of $\mathbf{X}_1, \mathbf{X}_2, \ldots, \mathbf{X}_n$. Then we get exactly two possibilities: the existence of a dispersion in the $\mathrm{d}x = \mathrm{d}\omega$ measure, i.e., a variance, leads to a Gaussian distribution ($\alpha = 2$); otherwise, the non-existence of a dispersion leads to a Levy distribution ($2 > \alpha > 0$). We should remember that the relaxation character of dynamics restricts the Levy exponent to $\alpha \leq 1$ (Sect. 3.1).

The difficulty involved in such an analysis is rewarded by concrete mathematical statements. We get a definite distribution $\Delta x^2(\omega)$ that depends on a single parameter α for the symmetric case, $\Delta x^2(\omega) = \Delta x^2(-\omega)$. This corresponds to similarity as defined in Sect. 2.1.1. The size dependence is scaled by a simple power law $n^{1/\alpha}$. The characteristic function is a stretched exponential or Kohlrausch function $\exp(-t^\alpha)$. In spite of the democracy (Sect. 2.2.4) manifested by the sum (3.199), we automatically get a dynamic heterogeneity for $\alpha < 1$, because the maximal component is preponderant. There are many other useful and interesting properties.

Levy Limit Stability. We partly follow [279, 636]. We must first define what is meant by 'equivalent subensembles' of a Levy situation. We use the symbol $\stackrel{d}{=}$ for this equivalence. Hence,

$$\mathbf{U} \stackrel{d}{=} \mathbf{V} \tag{3.200}$$

means that the random variables $\mathbf{U}$ and $\mathbf{V}$ have the same distribution R. Likewise, $\mathbf{U} \stackrel{d}{=} a\mathbf{V} + b$ means that the distributions of $\mathbf{U}$ and $\mathbf{V}$ differ only by location and scale parameters. The latter are defined by the statement that two distributions p are said to be of the same type if

$$p_2(x) = p_1(ax + b) \quad \text{with} \quad a > 0\,. \tag{3.201}$$

We refer to a as the scale factor, and b as the centering or location constant. In our context, $p(x) = \Delta x^2(\omega)$, the spectral density. Since the classical spectral density is always symmetrical, $\Delta x^2(\omega) = \Delta x^2(-\omega)$, centering does not play any role here. "Our subensembles $(1, 2, \dots)$ are equivalent" means therefore that $\mathbf{X}_1, \mathbf{X}_2, \dots$ have a common distribution $\Delta x^2(\omega)$ without the need to think about centering. We thus denote by a general $\mathbf{X}$ and by components $\mathbf{X}_1, \mathbf{X}_2, \dots$ mutually independent random variables with a common distribution R, and by $\mathbf{S}_n$ the sum (3.199), $\mathbf{S}_n = \mathbf{X}_1 + \mathbf{X}_2 + \ldots + \mathbf{X}_n$.

We now list some more or less interchangeable definitions for Levy stability. Note that we used the term Levy distribution for a stable distribution with exponent $\alpha < 2$.

(i) The distribution R is stable if for each n there exist constants $c_n > 0$, γ_n such that

$$\mathbf{S}_n \stackrel{d}{=} c_n\mathbf{X} + \gamma_n\,. \tag{3.202}$$

It does not suffice that (3.202) holds for $n = 2$, but it suffices that (3.202) holds for $n = 2$ and 3. The constants c_n are called norming constants. For Levy exponent $\alpha \neq 1$ (see below), the centering constants b of (3.201) can be chosen so that $\gamma_n = 0$ in (3.202).

(ii) The robustness requirement of a Levy situation is expressed by the word 'arbitrary' in the next definition. The distribution R is stable if and only if, for arbitrary constants c_1, c_2, there exist constants c and γ such that

$$c_1\mathbf{X}_1 + c_2\mathbf{X}_2 \stackrel{d}{=} c\mathbf{X} + \gamma\,. \tag{3.203}$$

(iii) The next definition refers to convolutions (3.18). Higher-dimensional convolutions, with n components, are denoted by $p^{*n} = p * p * \ldots * p$ with n 'factors'. A probability density is stable if, for every $n \geq 2$, there exist arbitrary constants $a_n > 0$ and b_n such that

$$p(x) = p^{*n}(a_n x + b_n) \,. \tag{3.204}$$

(iv) The next definition refers to characteristic functions $f(t)$ (3.22) for probability densities $p(x)$,

$$f(t) = \int_{-\infty}^{\infty} p(x) \exp(itx) \, \mathrm{d}x \,. \tag{3.205}$$

A characteristic function $f(t)$ is stable if and only if there exist arbitrary constants c_n and $a_n > 0$ such that, for every $n \geq 2$,

$$f(t) = \exp(it\, c_n)\, f^n(t/a_n) \,, \tag{3.206}$$

where the a_n are the same constants as in (3.204), $c_n = -b_n/a_n$, and f^n means the nth power of f, corresponding to p^{*n} for p (3.23).

Properties of Levy Distributions. The *Levy exponent* α is introduced by scaling properties. Referring to (3.202), we have:

Property 1a. The norming constants have the form

$$c_n = n^{1/\alpha} \quad \text{with} \quad 0 < \alpha \leq 2 \,. \tag{3.207}$$

Referring to (3.204) and (3.206), we have:

Property 1b. The scale factors have the form $a_n = n^{1/\alpha}$.

It is a simple consequence of this form that $a_n \to \infty$ and $a_n/a_{n+1} \to 1$ for $n \to \infty$.

Note that the sum $\mathbf{S}_{m+n}$ is the sum of the independent random variables $\mathbf{S}_m$ and $\mathbf{S}_{m+n} - \mathbf{S}_m$ distributed, respectively, as $c_m\mathbf{X}$ and $c_n X$. For symmetric Levy distributions ($\gamma_n = 0$), this yields

$$c_{m+n}\mathbf{X} \stackrel{d}{=} c_m\mathbf{X}_1 + c_n\mathbf{X}_2 \,, \tag{3.208}$$

which means, after norming, that

$$s^{1/\alpha}\mathbf{X}_1 + t^{1/\alpha}\mathbf{X}_2 \stackrel{d}{=} (s+t)^{1/\alpha}\mathbf{X} \,, \tag{3.209}$$

for arbitrary $s > 0, t > 0$. This implies in general that all linear combinations $\tilde{a}_1\mathbf{X}_1 + \tilde{a}_2\mathbf{X}_2$ belong to the same type. Equation (3.209) is the generalization of the Gaussian distribution property ($\alpha = 2$) that the dispersions are additive: $\delta x_1^2 + \delta x_2^2 + \ldots + \delta x_n^2 = \delta s_n^2$, where δs_n^2 is the total dispersion of the sum.

The concept of *preponderant component* is important for understanding glass-transition dynamic heterogeneity.

Property 2. The maximal term $\mathbf{M}_n = \max[\mathbf{X}_1, \mathbf{X}_2, \ldots, \mathbf{X}_n]$ is likely to grow exceedingly large and to receive a preponderant influence from the sum $\mathbf{S}_n$.

Consider a Levy distribution with (3.209) and Levy exponent $\alpha < 1$. The average $(\mathbf{X}_1 + \ldots + \mathbf{X}_2)/n$ (note that there is no expectation value for $\alpha < 1$) has the same distribution as $\mathbf{X}_1 n^{-1+1/\alpha}$, and the last factor tends to ∞. Roughly speaking we can say that the average of n variables is likely to be considerably larger than any given component $\mathbf{X}_k$. This is possible only if Property 2 holds. In the case of positive variables such as ω in $\Delta x^2(\omega)$, the expectation of the ratio $\mathbf{S}_n/\mathbf{M}_n$ tends to $1/(1-\alpha)$. Examples: for $\alpha = 1/2$ we have $1/(1-\alpha) = 2$, and the maximal term is expected to have the 'same effect' as the rest of the sum; for $\alpha \to 1$ we have $1/(1-\alpha) \to \infty$, and the preponderance effect gets lost.

Let us now consider the generalization of the central limit theorem. Let $\mathbf{X}_1, \ldots, \mathbf{X}_n$ be mutually independent variables with a common distribution having zero expectation and unit variance or dispersion. The central limit theorem asserts that the distribution $n^{-1/2}\mathbf{S}_n$ tends to the Gaussian normal distribution. This theorem proves that the normal distribution is the only stable distribution with variance. For distributions without variance, similar theorems may be formulated, but the norming constants must be chosen differently. The interesting point is that all Levy distributions and no others occur as such limits:

Property 3. A distribution possesses a domain of attraction if and only if it is Gaussian or Levy stable.

As a physical example concerning the dynamic glass transition, consider the 'stabilization' of the 'frequency breakthrough' at a concentration of 'too much' free volume in Sects. 2.1.3 and 2.2.5-2.2.6 (Glarum–Levy defect of the dynamic glass transition). We call the breakthrough situation 'unstable' if increasing concentration of free volume in a partial system leads to increasing local mobility there. We can imagine that the 'additive' participation of equivalent neighboring partial systems makes the situation 'stable'. Then we get a Levy situation. If the dynamics is 'overdamped', we have $\alpha \leq 1$. This puts the situation inevitably in the domain of attraction of a Levy distribution with a preponderant component for $\alpha < 1$, i.e., a dynamic heterogeneity with a 'defect' consisting of a partial system with preponderant mobility and free volume.

The application of 'regularly varying functions' allows us to determine the behavior of $p(x)$ for large x. A regularly varying function L in the t interval from 0 to ∞ is defined by the property

$$\frac{L(xt)}{L(t)} \to 1 \quad \text{for every } x \text{ and for } t \to \infty\,. \tag{3.210}$$

Property 4. For large x, the probability density of a Levy distribution behaves asymptotically as $p(x) \sim 1/x^{1+\alpha}$ for $\alpha < 2$.

For $\alpha < 1$, we can even find a converging series expansion [279]. Property 4 means that the spectral density $\Delta x^2(\omega)$ and the loss compliance $J''(\omega)$ behave as

$$\Delta x^2(\omega) \sim 1/\omega^{1+\alpha}\,, \quad J''(\omega) \sim 1/\omega^{\alpha} \tag{3.211}$$

for large ω (Fig. 2.2e with $\alpha = \beta_{\mathrm{KWW}}$). This behavior leads to:

Property 5. A distribution belonging to a domain of attraction with index α (= Levy exponent) possesses absolute moments $m_\beta = \int |x|^\beta\, p(x)dx$ of all orders $\beta < \alpha$. If $\alpha < 2$, no moments of order $\beta > \alpha$ exist.

This means that, for Levy distributed symmetric spectral densities $\Delta x^2(\omega)$, $\Delta x^2(-\omega) = \Delta x^2(+\omega)$, with exponent α in the dω measure,

$$\overline{\omega^\beta} \quad \begin{cases} \text{exists for } \beta < \alpha\,, \\ \text{does not exist for } \beta > \alpha\,. \end{cases} \tag{3.212}$$

For $\alpha < 1$, we have no expectation ($\beta = 1$, $\bar{\omega} = \infty$) and no variance, i.e., no dispersion ($\beta = 2$, $\overline{\omega^2} = \infty$).

The uniqueness of the Kohlrausch function for compliances in the time domain follows from the symmetry of $\Delta x^2(\omega)$ and from:

Property 6. All symmetric Levy distributions have characteristic functions of the form $f(t) \propto \exp(-a\,|t|^\alpha)$ with the exponent α, $0 < \alpha \leq 2$, and a constant a, $a > 0$.

As an example, a simple understanding of the scaling property, Property 1b, $a_n = n^{1/\alpha}$, follows for symmetric distributions with such a characteristic function from

$$f^n\left(\frac{t}{n^{1/\alpha}}\right) \equiv \left[f\left(\frac{t}{n^{1/\alpha}}\right)\right]^n = \mathrm{e}^{-t^\alpha} \equiv f(t)\,, \quad t > 0\,. \tag{3.213}$$

Note that the general similarity theorem for Fourier transforms of a function $g(x)$, $f(t) = f\{g(x)\}(t)$, reads

$$f\{g(\tilde{a}x)\}(t) = (1/\tilde{a})f(t/\tilde{a})\,. \tag{3.214}$$

Note further, that the class of 'infinitely divisible' characteristic functions is larger than the class of characteristic functions for Levy distributions. The former contains, for example, the Poisson distribution, which is not a Levy distribution. A characteristic function is infinitely divisible if and only if, for each n, there exists a characteristic function f_n such that $f_n^n = f$. Examples:

- the symmetric Levy distribution with (3.213) is obviously infinitely divisible;
- the Levy situation of Fig. 4.2b for large m corresponds to infinite divisibility [(3.204) and (3.206)].

As mentioned above, although we have no moments in the frequency domain, $\overline{\omega^\beta}$ for $\beta > \alpha$, all moments of the Kohlrausch function exist in the time domain with measure dt ($\overline{\tau^\beta} < \infty$, $0 < \beta < \infty$). In addition, an important point for application of the fluctuation–dissipation theorem, all moments of Levy distributions exist if the d$\log\omega$ measure is used, i.e., $\overline{(\log\omega)^\beta} < \infty$ for $0 < \beta < \infty$.

Diagrams and Moments for Kohlrausch Susceptibilities. The aim in this section is illustration. For pragmatic aspects (e.g., computer programs), we refer again to [633]. The $G(t)J(t)$ problem is also discussed in [114].

A *susceptibility* is either a compliance

$$J(\log t)\ , \quad J^*(\log\omega) = J'(\log\omega) - iJ''(\log\omega)$$

or a modulus

$$G(\log t)\ , \quad G^*(\log\omega) = G'(\log\omega) + iG''(\log\omega)\ .$$

These are connected by $J^*G^* = 1$. A *Kohlrausch susceptibility* is defined by the compliance $J(\log t)$ being a Kohlrausch function, i.e., a stretched exponential in the time domain (2.17), viz.,

$$J_e - J(\xi) = \Delta J \exp(-e^\xi)\ , \quad \xi = \alpha \ln t\ , \quad \alpha = \beta_{\text{KWW}}\ , \quad 0 < \alpha \le 1\ , \tag{3.215}$$

where ln is the natural logarithm and $\exp x = e^x$. The Kohlrausch exponent β_{KWW} is here identified with the Levy exponent α. $J''(\omega,\alpha)$ is equivalent to $\omega p(\omega,\alpha)$ with $p(\omega,\alpha)$ the symmetric Levy distribution (density). The preference for logarithmic d$\ln\omega$ measure is motivated by the FDT (2.75). Although the Levy distribution $p(\omega,\alpha)$ has no moments of order $\beta > \alpha$ in the dω measure (3.212), all moments of $J''(\log\omega)$ and $G''(\log\omega)$ exist in the d$\log\omega$ measure, similarly to the densities.

As mentioned above, the calculation is based on (3.194) and (3.195), both from (3.192) [115], see also [630,634,635]. Although the Kohlrausch function in the time domain is scaled by $\xi = \alpha\ln t$ (3.215), the reduced dynamic compliances depend on two variables α, $y = \alpha\ln\omega$, and the reduced moduli on three,

$$\alpha\ , \quad y = \alpha\ln\omega\ , \quad \gamma = \Delta J/J_g\ . \tag{3.216}$$

The dynamic susceptibilities and dynamic moduli in the time domain are not strictly scaled by y and by ξ, respectively. However, Fig. 3.7 shows that the important range in the reduced frequency $y = \alpha\ln\omega$ and reduced time $\xi = \alpha\ln t$ is between $-4 \lesssim y, \xi \le +4$, independent of α after the reduction.

Analytical solutions were quoted for $\alpha = 1$ (3.196) and $\alpha \to 0$ (3.197). For $\alpha = 1/2$ we have, for $J''_T(y,\alpha) \equiv 2J''(y,\alpha)/\alpha$ (3.195),

$$J_r''(y,\alpha=1/2) = -\mathrm{Im}\left[4\exp\left[-y-\frac{\mathrm{i}}{4}\mathrm{e}^{-2y}\right]\sqrt{\pi}\frac{(1-\mathrm{i})}{2\sqrt{2}}\mathrm{erfc}\left[\frac{1-\mathrm{i}}{2\sqrt{2}}\mathrm{e}^{-y}\right]\right], \tag{3.217}$$

where erfc is the complementary error function. The asymptotic slopes are (Fig. 2.2e)

$$\frac{\mathrm{d}\ln J''}{\mathrm{d}\ln\omega} = 1 \quad \text{for small } \omega\,, \tag{3.218}$$

$$\frac{\mathrm{d}\ln J''}{\mathrm{d}\ln\omega} = -\alpha \quad \text{for large } \omega\,. \tag{3.219}$$

The asymptotes themselves are

$$J''(y,\alpha) \sim \exp(y/\alpha)\Gamma(1+1/\alpha) = \omega\Gamma(1+1/\alpha)\,, \quad \text{for } y \ll -1\,, \tag{3.220}$$

$$J''(y,\alpha) \sim \frac{\alpha\mathrm{e}^{-y}\pi}{2\Gamma(1-\alpha)\sin[(1-\alpha)\pi/2]}\,, \quad \text{for } y \gg +1\,. \tag{3.221}$$

The loss compliance has the usual maximum (Fig. 3.7a, b) with a strange loop in the reduced maximum frequency $y_{\max} = \alpha\ln\omega_{\max}$ (Fig. 3.7c, d). This maximum frequency can be approximated (Fig. 3.7d) by a parabola,

$$y_{\max}(\alpha) = 0.60607\alpha(\alpha-1)\,, \quad \ln\omega_{\max} = 0.60607(\alpha-1)\,. \tag{3.222}$$

The maximum value of the reduced loss compliance $J''_{r,\max} \equiv 2J''_{\max}/\alpha$ is a slowly varying function of the exponent α (Fig. 3.7e). The reduced dynamic storage compliance $J'(y,\alpha) - J_{\mathrm{g}}$, for $\Delta J = 1$, is displayed in Fig. 3.7f. The upper part of this figure shows the usual step but also without strict scaling

▶

Fig. 3.7. Kohlrausch susceptibilities. (**a**) Kohlrausch loss susceptibility $J''(y,\alpha)$, according to (3.192), as a function of $y = \alpha\ln\omega$, parameter $\alpha = 0.1, 0.3, 0.5, 0.7$ 0.9, 1.0, log–log plot. (**b**) Reduced Kohlrausch loss susceptibility $\tilde{J}''(y,\alpha)$, (3.192) additionally scaled to $\tilde{J}''_{\max} = 1$, as a function of $y = \alpha\ln\omega$, parameter $\alpha =$ 0, 0.1, 0.3, 0.5, 0.7, 0.9, 1.0. (**c**) The same, but zoomed near the maxima. (**d**) Shift in the reduced maximum frequency, $y_{\max} = \alpha\ln\omega_{\max}$, as a function of parameter α. The *points* are from the Fourier transform and the *curve* is from (3.222). (**e**) Shift in the reduced (3.193) maximum value $J''_{r,\max} = 2J''_{\max}/\alpha$ as a function of α. (**f**) Step in the storage dynamic compliance, $J'(y,\alpha) - J_{\mathrm{g}}$, as a function of reduced frequency $y = \alpha\ln\omega$. Step height $\Delta J = 1$. The α parameters are the same as in (**b**) and (**c**). *Upper part*: semilog representation. *Lower part*: log–log representation

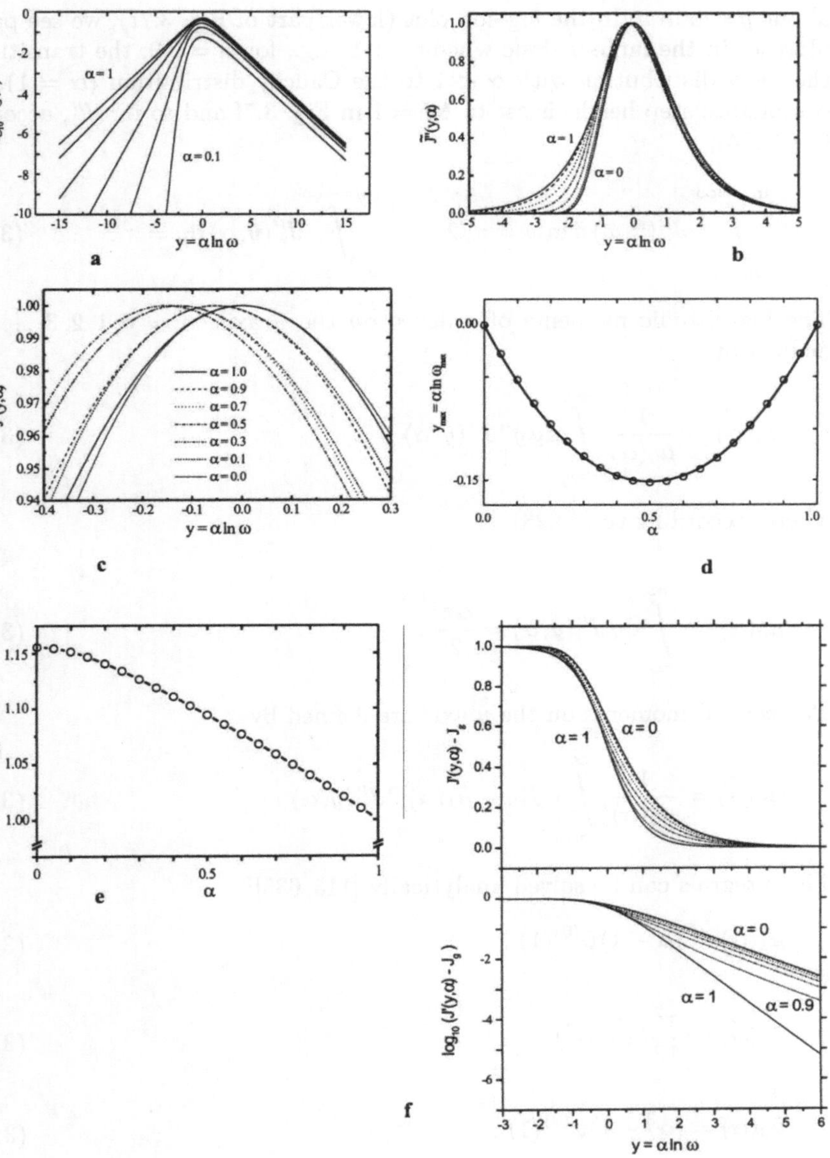

α = 1
α = 0.1
$\log_{10} J''(y,\alpha)$
$y = \alpha \ln \omega$
a
α = 1
α = 0
$\tilde{J}''(y,\alpha)$
$y = \alpha \ln \omega$
b
α = 1.0
α = 0.9
α = 0.7
α = 0.5
α = 0.3
α = 0.1
α = 0.0
$\tilde{J}''(y,\alpha)$
$y = \alpha \ln \omega$
c
$y_{max} = \alpha \ln \omega_{max}$
α
d
$2 J''_{max} / \alpha$
α
e
α = 1
α = 0
$J'(y,\alpha) - J_g$
α = 0
α = 1
α = 0.9
$\log_{10} (J'(y,\alpha) - J_g)$
$y = \alpha \ln \omega$
f

about $y = \alpha \ln \omega$. In the log–log plot (lower part of Fig. 3.7f), we see particularities in the large-y slope when $\alpha \to 1$, e.g., for $\alpha = 0.9$, the transition of the Levy distribution with $\alpha < 1$ to the Cauchy distribution ($\alpha = 1$). The compliance step height is set to $\Delta J = 1$ in Fig. 3.7f and so for J'', according to (3.194),

$$\int\limits_{\ln\omega=-\infty}^{\ln\omega=\infty} J''(\ln\omega)\,\mathrm{d}\ln\omega = \pi/2\ , \qquad \int\limits_{y=-\infty}^{y=+\infty} J''_r(y,\alpha)\,\mathrm{d}y = \pi\ . \tag{3.223}$$

The logarithmic moments of order n on the y axis, $n = 0, 1, 2, 3, \ldots$, are defined by

$$\mu_n(\alpha) = \frac{1}{\mu_0(\alpha)} \int\limits_{-\infty}^{\infty} \mathrm{d}y\, y^n J''(y,\alpha)\ , \tag{3.224}$$

where, according to (3.223),

$$\mu_0(\alpha) = \int\limits_{-\infty}^{\infty} \mathrm{d}y\, J''(y,\alpha) = \frac{\alpha\pi}{2}\ . \tag{3.225}$$

The central moments on the y axis are defined by

$$\tilde{\mu}_n(\alpha) = \frac{1}{\mu_0(\alpha)} \int\limits_{-\infty}^{\infty} \mathrm{d}y\, [y - \mu_1(\alpha)]^n J''(y,\alpha)\ . \tag{3.226}$$

The integrals can be solved analytically [115, 635]:

$$\mu_1(\alpha) = (\alpha - 1)\psi^{(0)}(1)\ , \tag{3.227}$$

$$\tilde{\mu}_2(\alpha) = \frac{\pi^2}{12}(2 + \alpha^2)\ , \tag{3.228}$$

$$\tilde{\mu}_3(\alpha) = (\alpha^3 - 1)\psi^{(2)}(1)\ , \tag{3.229}$$

$$\tilde{\mu}_4(\alpha) = \frac{\pi^4}{240}(36 + 20\alpha^2 + 19\alpha^4)\ , \tag{3.230}$$

$$\tilde{\mu}_5(\alpha) = (\alpha - 1)\frac{5\pi^2}{6}(2 + 2\alpha + 3\alpha^2 + \alpha^3 + \alpha^4)\psi^{(2)}(1) + (\alpha^5 - 1)\psi^{(4)}(1)\ , \tag{3.231}$$

with

$$\psi^{(0)}(1) \approx -0.5772157 = -C\ , \tag{3.232}$$

$$\psi^{(2)}(1) \approx -2.4041138\ , \tag{3.233}$$

$$\psi^{(4)}(1) \approx -24.8862661\ , \tag{3.234}$$

the (nth) derivatives of the digamma function $\psi(x) = \Gamma'(x)/\Gamma(x)$ at $x = 1$.

As the logarithm of frequency is reduced by α, $y = \alpha \ln \omega$, the variation of the central moments with α is moderate (Figs. 3.8a and b). $\tilde{\mu}_2$ increases from about 1.6 to 2.5, and $\tilde{\mu}_3$ deceases from about 2.4 to zero, as α varies between $\alpha = 0$ and $\alpha = 1$. To get absolute central moments ($\mathrm{d}\ln\omega$ measure instead of $\mathrm{d}y = \alpha \mathrm{d}\ln\omega$), the $\tilde{\mu}_n$ values must be multiplied by $(1/\alpha)^n$, e.g., by 4 for $\alpha = 0.5$ and $n = 2$.

The old problem of approximate (Fig. 2.2e) mutual correspondence between a Havriliak–Negami (HN) function, (2.14)–(2.16), and a Kohlrausch function [638] can be rationalized by the application of moments. The problem is that, for given temperature and step height ΔJ, HN has 3 parameters $\{\omega_{\max}, b, g\}$, where b and g are the wing (or shape) parameters, whereas Kohlrausch has only two $\{\omega_{\max}, \alpha\}$. One can try to find corresponding HN parameter sets from different requirements, e.g., the same set $\{\omega_{\max}, \tilde{\mu}_n, \tilde{\mu}_{n'}\}$, $n \neq n'$, $n \neq 0$, $n' \neq 0$, or the same set of three moments as that of a given Kohlrausch susceptibility. Of course, the results depend on the choice, and the choice may depend on the aim.

As mentioned above, the modulus curves depend on an additional parameter. We select reduction with the glass compliance $J_\mathrm{g} = 1/G_\mathrm{g}$,

$$\gamma = \frac{\Delta J}{J_\mathrm{g}}\ , \quad J_\mathrm{e} = J_\mathrm{g} + \Delta J\ , \tag{3.235}$$

and normalize the modulus step finally to $\Delta G = 1$ for G' and the peak height to $G''_{\max} = 1$ for G''. Real and imaginary parts of such normalized dynamic moduli $G^* = G' + \mathrm{i}G''$ as functions of the reduced frequency $y = \alpha \ln \omega$ depend weakly on γ between $\gamma = 0.1$ and $\gamma = 0.5$ (Fig. 3.8c). In the time domain, the representation is chosen as the product (not convolution) form, $G(t)J(t)$ (Fig. 3.8d). At least in the $\gamma = 0.1$–0.5 and $\alpha = 0.3$–1.0 range, the minimum is approximately scaled by [115, 634]

$$[G(t)J(t) - 1]_{\min} \approx -(0.10\text{–}0.15)\alpha^2\gamma^2\ , \tag{3.236}$$

where the weakly varying prefactor is 0.10 for $\gamma = 0.5$ and 0.15 for $\gamma = 0.1$. Figure 3.8d and (3.236) are useful for an approximation to $G(t)$ when $J(t)$ corresponds to a Kohlrausch function [e.g., $J = 1 + \gamma(1 - \exp(-t^\alpha))$],

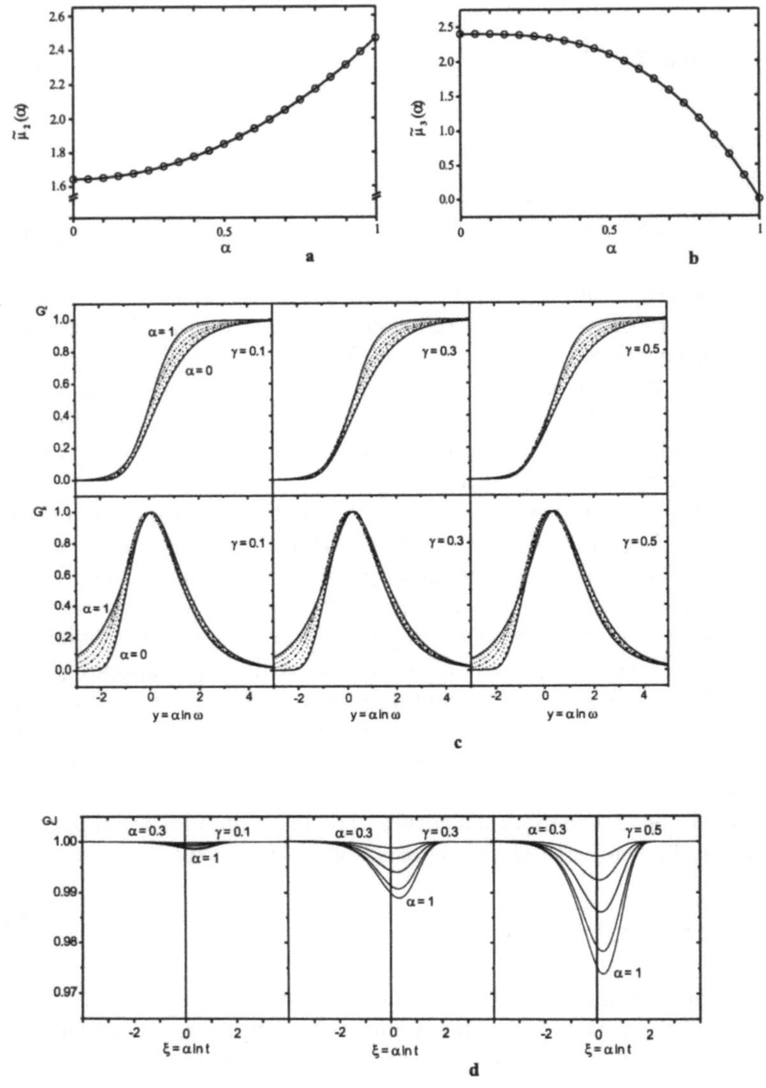

Fig. 3.8. Kohlrausch susceptibilities (continued). (**a**) Second central moment of $J''(y)$, according to (3.226), about $y = \alpha \ln \omega$ as a function of α. The *points* are Fourier transform data, and the *curve* is from (3.228). (**b**) The same for the third central moment (3.229). (**c**) Normalized real (G' for $\Delta G = 1$) and imaginary part (G'' for $G''_{\text{max}} = 1$) of modulus as a function of reduced frequency $y = \alpha \ln \omega$, calculated from $G^* = 1/J^*$ with J^* the Kohlrausch compliance ($\gamma = \Delta J/J_g$, index g = glass for $\tau \to 0$ or $\omega \to \infty$). Parameters $\alpha = 0, 0.1, 0.3, 0.5, 0.7, 0.9, 1.0$. (**d**) Product $G(t)J(t)$ as a function of reduced time $\xi = \alpha \ln t$ for $\gamma = 0.1$, $\gamma = 0.3$, $\gamma = 0.5$. Parameters $\alpha = 0.3, 0.5, 0.7, 0.9, 1.0$

$$G(t) = \frac{G(t) \cdot J(t)}{J(t)} . \tag{3.237}$$

The result may profit from a $G(t) \cdot J(t) \approx 1$ approximation. For example, when $\alpha = 0.7$ and $\gamma = 0.2$, the minimal value of (3.236) is about 0.003, i.e., the $G(t) \approx 1/J(t)$ approximation is better than 1%.

4. Slowing Down Mechanisms

The basic phenomenon of the dynamic glass transition is the more or less continuous slowing down of the relevant molecular motion as expressed for example by increasing viscosity for falling temperatures (Sect. 2.1.1) or by the declining traces of different processes in the Arrhenius diagram (e.g., a, β, α and ϕ in Fig. 2.32). These processes were described in Chap. 2, building up to a preliminary physical picture in Sect. 2.5.3. A series of phenomenological concepts was discussed in Chap. 3. Nonconventional aspects were included since I think that the cooperative α process and the ultraslow ϕ Fischer modes cannot be explained by conventional theories of liquid dynamics. The aim of the present chapter is to illustrate mathematical structures that could reflect the slowing down.

(i) Idealized mode-coupling theory (MCT) near its critical temperature of arrest for the Williams–Götze process (high-temperature a process in the warm liquid, Sect. 4.1).
(ii) Activated transition-state picture for the Johari–Goldstein β process (Sect. 4.2).
(iii) Glarum–Levy model for the cooperative α process at low temperatures, in the cold liquid (Sect. 4.3).
(iv) An external Levy model diffusion regime, to estimate temporal and spatial scales of Fischer modes ϕ (Sect. 4.4).

Note the difference between the methods. This characterizes the present theoretical situation, namely, that the conventional theory of liquid dynamics does not seem to provide a common basis for explaining all four processes. A way out was described speculatively in Sect. 3.2.3, by the desired construction of a microscopic von Laue distribution which can integrate aspects of Levy distributions for mobility and free volume. This distribution seems to constitute the common denominator for features lying beyond the statistics of the Gaussian distribution.

The mathematical structures are robust against variations in the details. This is necessary for the dynamic glass transition because of its dynamic generality compared with the multifarious molecular structures on the same length scales. The advantage of robustness for our representation is that the simplest mathematical structure is representative for the process considered.

The 'disadvantage' of robustness, on the other hand, is that the operative possibilities for modifications are restricted if we intend to make variations that concern the basic mathematical structures. The latter are of the type of Thom's nonlinear catastrophe underlying the critical point of MCT for the a process, and also the fixed dimensionality of the molecular energy mountain between the bowls forming the energy barrier for the β process, and the property that all limit distributions (limit for increasing cooperativity) are in the domain of attraction of Levy distributions for the α process and ϕ process dynamics.

The last section (Sect. 4.5) is devoted to modifications in the dynamics (acceleration and spreading Fig. 2.4b) caused by partial freezing near and below the glass temperature T_g, i.e., at the thermal glass transition. Firstly, a solution for the preparation time problem (τ_{prep} of Sect. 2.1.4) is suggested by an abstract treatment of a relation between the frequency of a given mode [e.g., at the maximum of the imaginary part of the equilibrium dynamic heat capacity, $C''_{p,\max}(T, \log\omega)$] and the cooling rate $\dot{T} = \mathrm{d}T/\mathrm{d}t$ giving the calorimetric glass temperature at this temperature, $T_g = T$ (Sect. 4.5.1). Secondly, the nonlinear effect of the frozen parts on retardation (entropy or volume recovery) is described by means of the Narayanaswamy–Moynihan model (Sect. 4.5.2), followed by the example of the Kovacs expansion gap (Sects. 4.5.3). The section closes with a short discussion of physical aging (Sect. 4.5.4).

4.1 Mode Coupling

Physical aspects of the mode-coupling approach to the Williams–Götze process were described and discussed in Sect. 2.3.2. We present here the mathematical core of the slowing down process in a simplified variant of idealized mode-coupling theory MCT [639]. The basic ideas come from Götze [373,640]. Further developments are described in later reviews [33, 47, 376, 641, 642].

The combination of the Langevin equation (2.100) with the MCT substitutions for its random force and memory, (2.101) and (2.102), was called the mode-coupling equation in Sect. 2.3.2. We start with its Laplace transform

$$\phi_q(z) = \frac{-1}{z + \dfrac{-\Omega_q^2}{z + \mathrm{i}\nu_q + \Omega_q^2\, m_q(z)}} , \tag{4.1}$$

where ϕ_q is the complex Fourier component of the spatial density correlator (2.99) at wave vector $Q = q$, z is the complex frequency resulting from a temporal (t) Laplace transformation

$$\mathfrak{L}\,[\phi(t)](z) = \mathrm{i}\int\limits_0^\infty \mathrm{d}t\, \mathrm{e}^{\mathrm{i}zt}\phi(t) , \tag{4.2}$$

Ω_q is the frequency scale of the molecular (or colloidal) transient, ν_q a Newtonian friction, and $m_q(z)$ the memory kernel. The last two objects are contained in the memory function $\gamma(t)$ of (2.100). The truncated continued-fraction form of (4.1) is typical for the MCT closure of the Langevin equation treatment in liquid dynamics. For closure, we put here [compare with (2.103)] a simple multiplicative mode coupling, i.e., control of memory by the modes themselves,

$$m_q(t) = \sum_{kp} V_{qkp}\phi_k(t)\phi_p(t) , \tag{4.3}$$

where the vertices V_{qkp} express static correlations in the memory. For simplicity we consider the wave vectors $\{qkp\}$ to be sets of discrete numbers. This introduces some algebra into the treatment. So V_{qkp} is a (k,p) matrix for given q. Using detailed methods of MCT for simple-liquid dynamics [276,359], the vertices can be calculated as functions of state for given intermolecular or intercolloidal potentials. The non-negativity obtained from such calculations,

$$V_{qkp} \geq 0 , \tag{4.4}$$

ensures the dynamic stability of the approach $[m_q(t) \geq 0]$. We introduce the nonergodicity parameter f_q and the corresponding difference part of the correlator, $g_q(t)$, by

$$\phi_q(t) = f_q + (1-f_q)^2 g_q(t) . \tag{4.5}$$

The quadratic factor $(1-f_q)^2$ is chosen for a simpler reduction of the nonlinear equations. Then, from (4.1), for small g_q (i.e., near the arrested state), we get to second order in g_q

$$\begin{aligned} \frac{f_q}{1-f_q} - \sum_{kp} V_{qkp}\, f_k f_p + \sum_k [C_{qk} - \delta_{qk}] g_k \\ = -(1-f_q)\,[z g_q(z)]^2 - \sum_{kp} C_{qkp} z \mathfrak{L}\,[g_k(t) g_p(t)]\,(z) \\ + \frac{z(\mathrm{i}\nu_q + z)}{\Omega_q^2} + O(g^3) , \end{aligned} \tag{4.6}$$

where the matrix C_{qk} is defined by

$$C_{qk} = \sum_p V_{qkp}(1-f_k)^2 f_p , \tag{4.7}$$

and C_{qkp} are, of course, renormalized vertices,

$$C_{qkp} = V_{qkp}(1-f_k)^2(1-f_q)^2 . \tag{4.8}$$

The dynamic critical point (in the crossover region) is defined by critical V^{c}_{qkp} and f^{c}_{q} values such that

$$\frac{f^{c}_{q}}{1-f^{c}_{q}} = \sum_{kp} V^{c}_{qkp} f^{c}_{k} f^{c}_{p} \tag{4.9}$$

and

$$\det[C^{c}_{qk} - \delta_{qk}] = 0 , \tag{4.10}$$

from the first line of (4.6). Equation (4.10) means that the critical $C_{qk} = C^{c}_{qk}$ matrix has eigenvalue 1.

The solvability condition of the MC equation (4.6) needs some algebraic preliminaries. The non-negativity of the C_{qk} matrix,

$$C_{qk} \geq 0 \quad \text{(all elements)} \ , \tag{4.11}$$

follows from (4.4) and (4.7). The algebraic consequences of (4.11) are summarized by the Perron–Frobenius theorem [643]: we can construct a space R^h with dimension h equal to the dimension of the discrete wave vector sets, where C_{qk} is considered as an operator with non-degenerate positive eigenvalues E and eigenvectors e_q :

$$e_k > 0 \, , \quad \sum_{k} C_{qk} e_k = E e_q \ . \tag{4.12}$$

In the dual space $\hat{R}^h$,

$$\hat{e}_q > 0 \, , \quad \sum_{q} \hat{e}_q C_{qk} = E \hat{e}_k \ . \tag{4.13}$$

The metric in R^h is chosen as

$$\sum_{q} \hat{e}_q e_q = \sum_{q} (1 - f^{c}_{q}) e_q e_q = 1 \ . \tag{4.14}$$

The linear term in the first line of (4.6) vanishes for

$$g_q(t) = e_q g(t) \ , \tag{4.15}$$

and this equation introduces the 'scalar' function $g(t)$ as the subject of our continued investigations.

After these algebraic preliminaries, the solvability condition of (4.6) for getting nontrivial $g(t)$ functions is that the right-hand side of (4.6) (second plus third line) should be orthogonal to the dual eigenvector $\hat{e}_q$. This implies, for the critical state,

$$-[zg(z)]^2 - \lambda z \mathfrak{L}\left[g^2(t)\right](z) + O(g^3, z\nu/\Omega^2) = 0 \, , \tag{4.16}$$

with the coupling constant

$$\lambda \stackrel{\text{def}}{=} \sum_{qkp} \hat{e}_q C^{c}_{qkp} e_k e_p \ . \tag{4.17}$$

The long-time solution of the nonlinear equation (4.16) is

$$g(t) = (t/t_0)^{-a} \ , \tag{4.18}$$

where the exponent a is determined by the coupling constant λ,

$$\lambda = \frac{\Gamma^2(1-a)}{\Gamma(1-2a)} \ . \tag{4.19}$$

The generation of gamma functions Γ from power laws like (4.18) was explained in the context of (2.112).

The reduction theorem of idealized MCT concerns the behavior away from (but near) the critical point of dynamic arrest,

$$\Delta_{qkp} \neq 0 \quad \text{but} \quad f_q = f_q^{c} \ , \tag{4.20}$$

where the vertex differences Δ_{qkp} determine the 'distance' to the critical point,

$$\Delta_{qkp} = V_{qkp} - V^{c}_{qkp} \ . \tag{4.21}$$

Using (4.5) and (4.20),

$$\phi_q(t) = f_q^{c} + (1 - f_q^{c})^2 e_q g(t) \ ,$$

the solvability condition away from the critical state is

$$\sigma - [zg(z)]^2 - \lambda z \mathfrak{L}\left[g^2(t)\right](z) = O(g^3, z\nu/\Omega^2) \ . \tag{4.22}$$

Comparing with (4.16), we see that the zero at the critical point is substituted by a distance parameter σ away from the critical point,

$$\sigma = \sum_{qkp} \hat{e}_q \Delta_{qkp} (1 - f_k^{c})^2 (1 - f_q^{c})^2 \ , \tag{4.23}$$

with $\sigma \to 0$ for approach to the critical point.

We obtain two solutions for the quadratic part of (4.22),

$$g(t) = \sqrt{|\sigma|}\, g_{\pm}(t) \ , \tag{4.24}$$

where

$$O(g^3, z\nu/\Omega^2) \to 0 \quad \text{for} \quad \sigma \to 0 \quad \text{and} \quad z \to 0 \ . \tag{4.25}$$

The asymptotic solution for large times t is therefore (also for $\sigma \neq 0$)

$$g(t) \sim (t/t_1)^{-a} , \tag{4.26}$$

with

$$t_1 = t_0 \tag{4.27}$$

from matching at the critical result (4.18).

The $g_+(t)$ function is for $\sigma > 0$ and corresponds to the cage (c) process wing, whilst $g_-(t)$ is for $\sigma < 0$ and corresponds to the high-temperature (a) process wing, both near the susceptibility minimum of Fig. 2.21e (Sect. 2.3.2). Equation (4.22), the *reduction theorem* equation, is formally solved by scaling laws with index λ,

$$g(t) = \sqrt{|\sigma|} \cdot g_\pm^\lambda(t/t_\sigma) . \tag{4.28}$$

A square-root distance factor for the static correlations, depending on the state, is separated from the scaled time functions. The reduction times t_σ express different scaling laws for $\sigma > 0$ and $\sigma < 0$. Solutions to (4.22) are scaled by

$$t_\sigma = t_0 \left|\sigma\right|^{-1/2a} , \tag{4.29}$$

and $g_\pm^\lambda(z)$ is a solution of the correspondingly reduced (4.22),

$$\mp 1 - [z g_\pm^\lambda(z)]^2 - \lambda z \mathfrak{L}\left[(g_\pm^\lambda(t))^2\right](z) = 0. \tag{4.30}$$

For $\sigma > 0$, the asymptote of the first solution is

$$g_+^\lambda(t \gg t_\sigma) \sim A t^{-a} , \tag{4.31}$$

with exponent $-a$, whereas for $\sigma < 0$, (4.22) cannot be solved for $z \to 0$ (i.e., $t \to \infty$) with a finite function, $\left|g_-^\lambda(t \to \infty)\right| < \infty$. For $\sigma < 0$, we must use the second solution (Fig. 2.21c and d) of (4.22),

$$g_-^\lambda(t \to \infty) \sim -B t^b , \tag{4.32}$$

with the exponent b from

$$\lambda = \frac{\Gamma^2(1+b)}{\Gamma(1+2b)} , \tag{4.33}$$

to be compared with (4.19). For further development of the analysis, the reader is recommended to consult the above literature.

The explanation for slowing down suggested by idealized MCT is as follows. Consider falling temperatures. To explain the dramatic slowing down of the Williams–Götze a process it is sufficient to assume that the vertices

V_{qkp} are continuously increasing functions. This property is transferred to the reduced vertices C_{qkp} and to the coupling constant λ. Transferred to the cage picture, this means that escape through the cage door becomes increasingly difficult. The 'quadratic' reduction theorem shows that a fold-type critical state will be reached at $T = T_c$, where the particle in the cage is arrested. This critical temperature T_c is expected in the crossover region. Approach to the arrest is accompanied by a dramatic increase in the t_σ reduction time scale (4.29), because the distance parameter σ decreases and has a simple zero at T_c, $|\sigma| \sim |T - T_c|$. The singularity is governed by the square root factor $\sqrt{|\sigma|}$ for $g(t)$ in (4.28) – a reflection of the underlying fold catastrophe – and, let us repeat this important point, by the singularity of the t_σ reduction times in (4.29), $t_\sigma \to \infty$ for $T \to T_c$ (for both a and c process wings [644]). The calculations from the reduction theorem (4.22) to the scaling equation (4.29) are summarized by (2.108). The specific scaling comes formally from the insertion of power laws in equations of the (4.22) type. For the glass-forming salt melt CKN, for instance, $\lambda = 0.81$ and $a = 0.27$, so that the slowing down of the a process is of order four time decades from about $T_c + 100$ K down to the crossover region with $T_c \approx 100°$C.

The von Schweidler law (4.32) is completed to a Kohlrausch law for large wave vectors q. This was first observed by Fuchs [282]. The connection with a Levy distribution is mediated by the many (q, k, p) values that contribute via the vertices V_{qkp} to the coupling constant λ. This means that, for the short distances in the cage door corresponding to the large q values, escape of the particle is connected with the effect of many independent 'collisions' from the equivalent particles cooperatively forming the cage door. Since $b = \alpha$, $b < 1$ implies a Levy exponent $\alpha < 1$. This in turn implies a preponderant Levy component (Sect. 3.6). A preponderant Levy component from $\alpha < 1$ was also used for the Glarum–Levy model of the α process, below the crossover. The Fuchs observation is therefore important for discussion of continuity and discontinuity of (a, β, α) processes in the crossover region.

On the one hand, a preponderant collision means that the escape event may be, to a certain extent of order $[1 - \alpha]$, concentrated in space and time. This is similar to an activated process to overcome a fixed energy barrier, as typical for the β process. The latter event (Sect. 4.2) is completely definite in space and time: via a transition state, the movement is from one energy valley to the other. We therefore expect a certain continuity between the a process above and the β process below the crossover, also for small cages (scenario I).

On the other hand, having a distribution means that cage escape cannot be completely described by a concentrated 'hopping' over a localized barrier (as thought before MCT was introduced for the a process). Instead, to a certain extent of order $[\alpha]$, cage escape is a distributed process with many components. At lower temperature, for the α process below the crossover, more and more particles are involved in the Levy process forming some kind of cooperativity shell around the cage (Sect. 4.3). For small cages, the onset of a cooperativity shell is reflected by a discontinuity between the end of the a trace and the start of the α trace. This 'Levy overlap' between a and α process principles ensures that the α process onset (Sect. 2.5.3) is not too far below the $a\beta$ trace in the crossover region for small a process cages (scenario I). Larger a process cages, made from more particles, as expected for scenario II, diminish the discontinuity between a and α and enlarge the difference between the a and β processes, e.g., between their (apparent) activation energies.

Let us finally speculate in detail about the issue of extended MCT from this point of view. 'Extended' refers to an attempt to continue MCT methods below the critical point in the crossover region (Sect. 2.3.2). Heat capacity spectroscopy on the few substances so far measured above the crossover for scenario I [209] indicates large functional temperature fluctuations up to $\delta T = 20$ K leading to small and approximately constant cooperativities of order $N_a \approx 1$ particles. This corresponds to the a process diffusion through the cage door volume. It is further indicated that in a certain temperature interval of the crossover region, all three processes (a, β, α) can be observed simultaneously (Fig. 2.34b) [209, 713]. If such large spatiotemporal temperature fluctuation really exists, we can imagine sometimes being at a temperature 20 K higher and having a well defined a process, and sometimes being at a temperature 20 K lower and having well defined, separated β and α processes. The large temperature fluctuation would thus explain the occurrence of three traces in Fig. 2.34b.

On the other hand, MCT is implicitly based on the Gibbs treatment with sharp subsystem temperature from the Zeroth Law: $\lambda = \lambda(T)$ and $\sigma = \sigma(T)$ were considered to be exact functions of temperature. It seems impossible to model simultaneously all three processes (a, β, α) with such different properties in a small mobility interval (Sect. 2.5.3) using probability distributions having a sharp temperature. I think, if at all, that extended MCT should be based on distributions with sufficiently large temperature fluctuation. It is not clear if and how this can be done. Far above the crossover region, for the a process, such a distribution is not necessary because the cage-door situation is not qualitatively changed with temperature there ($N_a \approx$ const.).

Let us repeat this speculative but important issue in other words. I refer now to a similar discussion in Sect. 2.5.3, in the context of Fig. 2.34b. The desired statistical distribution with temperature fluctuation that can be substituted for the Gibbs distribution (having no temperature fluctuation) has not yet been invented. The problem was addressed by using a microscopic von Laue distribution (Sect. 3.2.3). What could be the difference between a full fluctuation including temperature fluctuation and a Gibbs-type energy fluctuation with no temperature fluctuation? Compare with the discussion in the last two paragraphs of Sect. 3.2.2. Consider the crossover region whose temperature range may be smaller than or equal to the functional temperature fluctuation there. Recall the meaning of functionality being the relationship to Fourier components of a given dispersion zone or a given process under consideration. I think that the Gibbs energy fluctuation (3.29) mainly reflects smearing-out of trends, and smoothing of topographic details of traces in the crossover region of the Arrhenius diagram, whereas the full fluctuation (3.36), including the true temperature fluctuation, reflects the changed dynamic state, sharpening the Arrhenius-diagram topography in the crossover region.

Summarizing this speculation, temperature fluctuation means that the high-temperature state with the a process and the low-temperature state with the separated α and β processes are partly covered by the fluctuation. Temperature fluctuation should thus reflect the different state qualities at $T + \delta T$ and $T - \delta T$. This means that the Gibbs energy fluctuations could live with a simple $a \to \beta$ transition, whereas a distribution with full fluctuation, including δT, could reflect all three processes simultaneously. A picture of such a dynamic heterogeneity in the crossover region was represented in Fig. 2.34b.

4.2 Activated Rate Processes

The basic concept for explaining the Johari–Goldstein β process is the classical physicochemical rate process for overcoming an energy barrier. This corresponds to the black-and-white depiction ignoring the assistance of the cooperativity shell (Sects. 2.2.1, 2.5.3). On molecular length scales, a fixed local mechanism is described by an energy mountain with constant dimensionality d (Fig. 2.9c) and with two sufficiently deep bowls serving as initial and final states for overcoming the barrier. The x coordinates are mainly angles responsible for local mechanisms.

The robustness of the approach allows us to consider a simple symmetric barrier in one dimension rather than a local mountain (Fig. 4.1a). The two boundary potentials are formally chosen to be high, to focus the model on one local process. The only relevant energy parameter in the simple *barrier model* is the activation energy ε_A.

Consider an ensemble of such situations. The particles have a distribution of energies that is characterized, in equilibrium, by the Boltzmann factor

$$w(x) \propto \exp\left[-\frac{\varphi(x)}{k_B T}\right] . \tag{4.34}$$

The phase space factor is neglected for simplicity. Kinetically, a particle at low energy has a high oscillation frequency (ν_0) given by the curvature of $\varphi(x)$ in the valleys (Fig. 4.1b). Having an energy distribution, however, means that we must also consider higher energies. We have a certain frequency ν for overcoming the barrier. The success-to-attempt ratio is a normalized frequency that can be related to the Boltzmann probability factor by

$$\frac{\nu}{\nu_0} = \frac{\text{frequency per time}}{(\text{frequency per time})_0} = \frac{w(\nu)}{w(\nu_0)} = \frac{w}{w_0} = \exp\left[-\frac{\varepsilon_A}{k_B T}\right] . \tag{4.35}$$

A mechanism obeying

$$\ln \nu = \ln \nu_0 - \frac{\varepsilon_A}{k_B T} \tag{4.36}$$

is called an *Arrhenius process*. This is a straight line in a $\ln \nu$–T^{-1} plot (Arrhenius diagram) with the intercept $\ln \nu_0$ for $T^{-1} = 0$ ($T = \infty$) determining the *prefactor* ν_0 of the exponential and the slope $-\varepsilon_A/k_B$ determining the *activation energy* ε_A. It is the Boltzmann factor that introduces the logarithm of frequency (the mobility) here.

The experimental response of this process is an exponential decay (Debye relaxation, Sects. 2.1.2, 2.2.4, and 3.3). The thermal fluctuation can be imagined as a fluctuation in the valley depths. According to the FDT, the response can be calculated from the system answer to a step perturbation (Fig. 4.1c), switching on a valley difference $\Delta\varepsilon$ at time $t = 0$. The new equilibrium can be calculated from the new Boltzmann factors, which are now different for the two valleys. We characterize the energy distribution by horizontal lines for considerable occupation numbers (N_i, $i = 1, 2$). In the old equilibrium ($\Delta\varepsilon = 0$, $t \leq 0$), we had

$$N_1 = N_2 , \quad \Delta N \equiv N_2 - N_1 = 0 . \tag{4.37}$$

The new equilibrium ($\Delta\varepsilon > 0$, $t \to \infty$) is

$$\frac{N_1}{N_2} = \frac{w_1}{w_2} = \frac{\exp[-(\varepsilon + \Delta\varepsilon/2)/k_B T]}{\exp(-[\varepsilon - \Delta\varepsilon/2)/k_B T]} \approx 1 - \frac{\Delta\varepsilon}{k_B T} , \tag{4.38}$$

where the approximation is for

$$\frac{\Delta\varepsilon}{k_B T} \ll 1 . \tag{4.39}$$

The transition behavior is calculated from a balance: change = access − loss. A kinetic equation for this structure that uses transition probabilities is called a master equation. Then

$$\frac{d(N_2 - N_1)}{dt} = w_{12} N_1 - w_{21} N_2 , \tag{4.40}$$

where access to 2 is given by the $1 \to 2$ transition probability w_{12} times the total occupation number of 1, N_1, and so on. The transition probabilities are calculated from success-to-attempt ratios again,

$$w_{12} = \nu \exp\left[-\left(-\frac{\Delta\varepsilon}{k_B T}\right)\right] > w_{21} = \nu \exp\left[-\frac{\Delta\varepsilon}{k_B T}\right] , \tag{4.41}$$

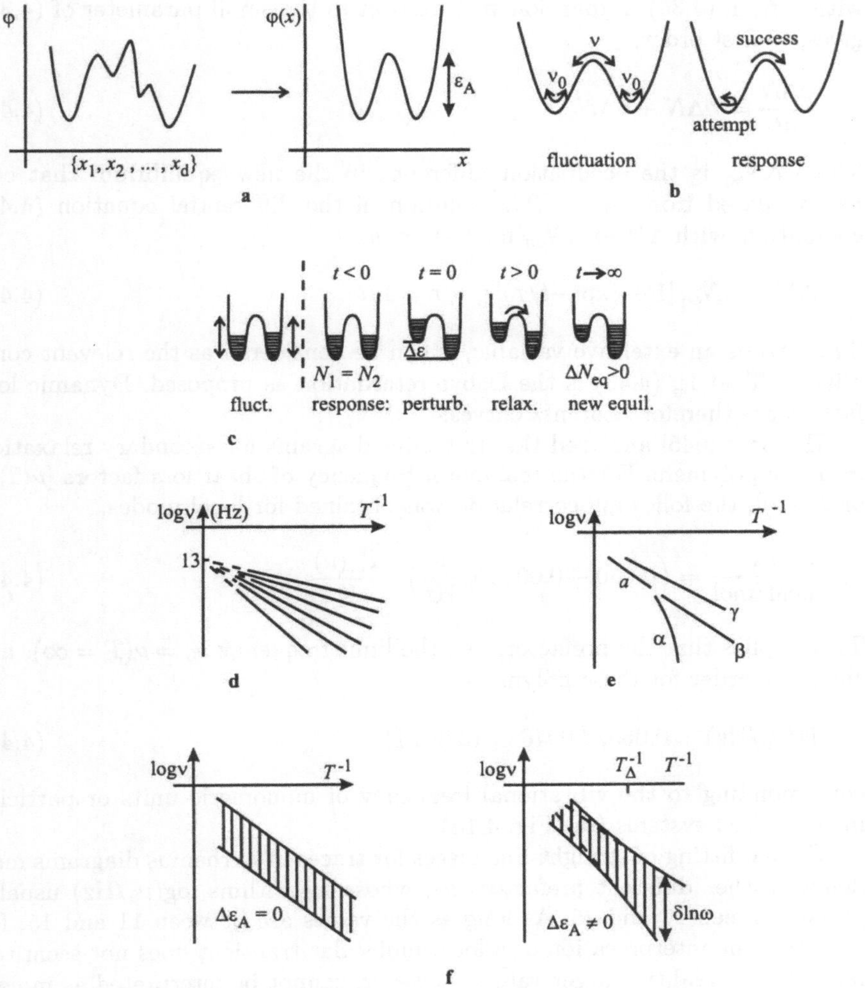

Fig. 4.1. Barrier model. **(a)** Simplification of an energy mountain $\varphi(x_1, x_2, \dots, x_d)$ in d dimensions (schematic) to a one-dimensional symmetrical energy barrier $\varphi(x)$ with activation energy ε_A. **(b)** Fluctuation and response pictures are equivalent according to the fluctuation–dissipation theorem FDT. **(c)** Calculation of response. **(d)** Heijboer's Arrhenius-plot correlation for local modes of several polymers. **(e)** γ relaxations are independent local barrier processes which do not meet the Williams–Götze *a* process at high temperatures. **(f)** Dispersion of local mechanisms with (*right*) and without (*left*) distribution of activation energy $\Delta\varepsilon_A$

with ν from (4.36). Expansion with respect to the small parameter of (4.39) gives, to first order,

$$\frac{\mathrm{d}\Delta N}{\mathrm{d}t} = \nu\Delta N + \nu\Delta N_{\mathrm{eq}} \,, \tag{4.42}$$

where ΔN_{eq} is the occupation difference in the new equilibrium that can be calculated from (4.38). The solution of the differential equation (4.42) compatible with $\Delta N \to \Delta N_{\mathrm{eq}}$ for $t \to \infty$ is

$$\Delta N = \Delta N_{\mathrm{eq}}\,[1 - \exp(-t/\tau)] \;, \quad \tau = 1/\nu \,. \tag{4.43}$$

Since ΔN is an extensive variable, it can be considered as the relevant compliance. That is, (4.43) is the Debye retardation as proposed. Dynamic loss factors are therefore Lorentz curves.

Heijboer [645] analyzed the Arrhenius diagrams for secondary relaxation in many polymers. For the maximum frequency of shear loss factors [$\nu(T_{\mathrm{m}})$ or $T_{\mathrm{m}}(\nu)$], the following correlation was obtained for local modes,

$$\frac{\varepsilon_{\mathrm{A}}}{\mathrm{kcal/mol}} = \left(0.060 - 0.0046 \lg \frac{\nu}{\mathrm{Hz}}\right) \cdot \frac{T_{\mathrm{m}}(\nu)}{\mathrm{K}} \,. \tag{4.44}$$

This implies that the prefactor, i.e., the limit frequency $\nu_0 = \nu(T = \infty)$, has the same order for these polymers

$$\lg(\nu_0/\mathrm{Hz}) = 0.060/0.0046 = 13.0(\pm 1) \,, \tag{4.45}$$

corresponding to the vibrational frequency of monomeric units or particles in condensed systems (see Fig. 4.1d).

Formal fitting of straight-line pieces for traces in Arrhenius diagrams may deliver rather different prefactors ν_0, whose logarithms $\log(\nu_0/\mathrm{Hz})$ usually range between 11 and 25. As long as the values are between 11 and 15, for example, the interpretation of a local molecular transient does not seem too far from the reality. Larger values, however, cannot be interpreted as molecular frequencies. The traces do not permit a linear extrapolation to $T^{-1} \to 0$ ($T \to \infty$). Radical changes in the molecular process have to be expected, especially if no phase transition occurs at higher temperatures and thermal degradation is ignored. Most plausibly, we expect the process trace to curve in the Arrhenius diagram at higher temperatures, above the experimental window, in such a way that a reasonable molecular prefactor can really be identified. Two possibilities are suggested.

- A hindered glass transition in confined geometry (Fig. 2.25c) may be caused by nanometer structural or morphological peculiarities in the sample itself. This would mean that, at high (unobserved) temperatures, the characteristic length of the cooperative process becomes smaller than the confinement scale. We then have a 'free' dynamic glass transition with the desired, curved trace. In the experimental window, however, we observe a fixed mechanism with many degrees of freedom corresponding to the cooperativity at T_D of Figs. 2.25a–c for the hindered glass transition. An example of such behavior is provided by the γ relaxations in polyethylene [7].

- The straight line is an α process piece with low Vogel temperature T_0 (Figs. 2.1c and e).

The barrier models were qualified in the 1930s by Eyring and Kramers [646,647]. They considered length and angle ranges (phase space) in addition to energy, e.g., a simple energy mountain over two coordinates x_1 and x_2 (Fig. 2.9c). The barriers may turn into saddles that can be connected with a statistical weight (transition state). The activation variable in the exponent gets a thermodynamic meaning that is expressed by an additional activation entropy S_A, for example. The universal Eyring formula for the attempt frequency, $\nu_0 = 2k_BT/\hbar$ (from $\hbar\omega \approx k_BT$), was modified by Kramers to include properties of the potential function $\varphi(x_1, x_2)$. He analyzed the activation on the basis of his generalized Langevin equation.

In polymers, for example, activation models with a few x coordinates are applied to secondary relaxations. We may distinguish motion of side groups and local modes of the chain itself. Both are Arrhenius processes, i.e., straight lines in an Arrhenius diagram (Figs. 4.1e and 2.32b). In amorphous polymers, those Arrhenius processes which find their own limit frequencies above and independently from the Williams–Götze a process and its molecular transient are called γ *processes*. Those which tend to merge with the α process in the crossover region are called β processes. Usually, the side groups also participate in the β process [645].

Experience shows that the dispersion of Arrhenius processes is usually much wider than the 1.14 logarithmic decades of the Lorentz line for a Debye relaxation. This is linked with a distribution of activation energies of spatially separate local processes, parameterized for example by an interval $\Delta\varepsilon_A$ or a Gaussian distribution caused by the disorder of real ($T < T_g$) or thermokinetic ($T > T_g$) structure. Below a certain temperature (T_Δ, from $\varepsilon_A/k_BT_\Delta \approx 1.14$), the width of the distribution is then determined by $\Delta\varepsilon_A$ (Fig. 4.1f),

$$\delta \ln\omega \approx \frac{\Delta\varepsilon_A}{k_BT} \propto \frac{1}{T}\,, \quad T < T_\Delta\,. \tag{4.46}$$

The proportionality to $1/T$ is only valid if $\Delta\varepsilon_A$ is independent of T. Usually, this condition is not strictly fulfilled because $\Delta\varepsilon_A$ is heavily influenced by the density. The strength of this influence comes from the steepness of repulsive intermolecular potentials [648]. Roughly speaking, a 5% increase in interparticle distances is sufficient to transform the broad dispersion into a Lorentz line. Moreover, the real mechanism is influenced by the molecular environment.

When discussing the physical pictures (Sect. 2.5.3), it was mentioned that more or less all molecules of a cooperatively rearranging region CRR for the α process participate in the β process [592], a few with large-angle jumps, the rest with a distribution of small-angle jumps. The barrier model is, as mentioned above, a black-and-white depiction of this situation, counting only the large-angle jumps.

The slowing down of the activation process is thus explained by the exponential decrease of the success-to-attempt rate for decreasing temperature (4.35)–(4.36). Arbitrarily low frequencies and, correspondingly, arbitrarily long relaxation times, can be reached for high barriers or low temperatures (Fig. 4.1b). The difference to cage escape for the high-temperature a process was indicated at the end of Sect. 4.1: the Levy distribution for the a process does not allow interpretation of cage escape by a simple barrier with a fixed mechanism of given dimensionality. The difference to the α process slowing-down is increasing dimensionality of the landscape mountain responsible for the latter, caused by increasing cooperativity for lower temperatures.

4.3 Fluctuating Free Volume. Glarum–Levy Defects

This section gives a phenomenological approach to the cooperative α process of the dynamic glass transition in cold liquids [7, 226, 296, 298, 576, 649]. The basic picture is the ω part of the Fig. 2.15a partition for dynamic heterogeneity, first published in 1978 (Fig. 4.2a) [138].

To explain Glarum–Levy defects (Sects. 2.2.4–2.2.5) and fluctuating free volume, we refer to the introduction of the free volume concept in Sect. 2.1.3 and to the attempt to show how the mathematical term 'stable' for Levy distributions (Sect. 3.6) can be connected with the physical Glarum model for the α process. The situation of Fig. 4.2a was said to be dynamically 'unstable' against local breakthroughs to high mobilities ($\log \omega_i$) if a lot of free volume is concentrated in one partial system. A democratic arrangement of partial systems in a cooperatively rearranging region CRR tends to such concentrations because the mobility of one partial system is preponderant in overdamped (relaxation) dynamics, i.e., for Levy exponents $\alpha < 1$. The situation is 'stabilized' by limit distributions (Levy distributions) resulting in a control of the preponderant partial system with extraordinarily high mobility by the other partial systems (Glarum–Levy defect inside any CRR). Scaling of these Levy distributions (Sect. 3.6) requires a free volume parameter that is invariant against the arbitrariness of the partition of a CRR into partial systems. This invariant free volume will be called 'fluctuating free volume' and will be calculated from the model of Fig. 4.2a by using a model for cooperativity ('minimal coupling'). An important result is a formula (4.65) for the physical meaning of the Levy exponent α. It describes the log–log sensitivity to the way increase in free volume permits smaller cooperativities.

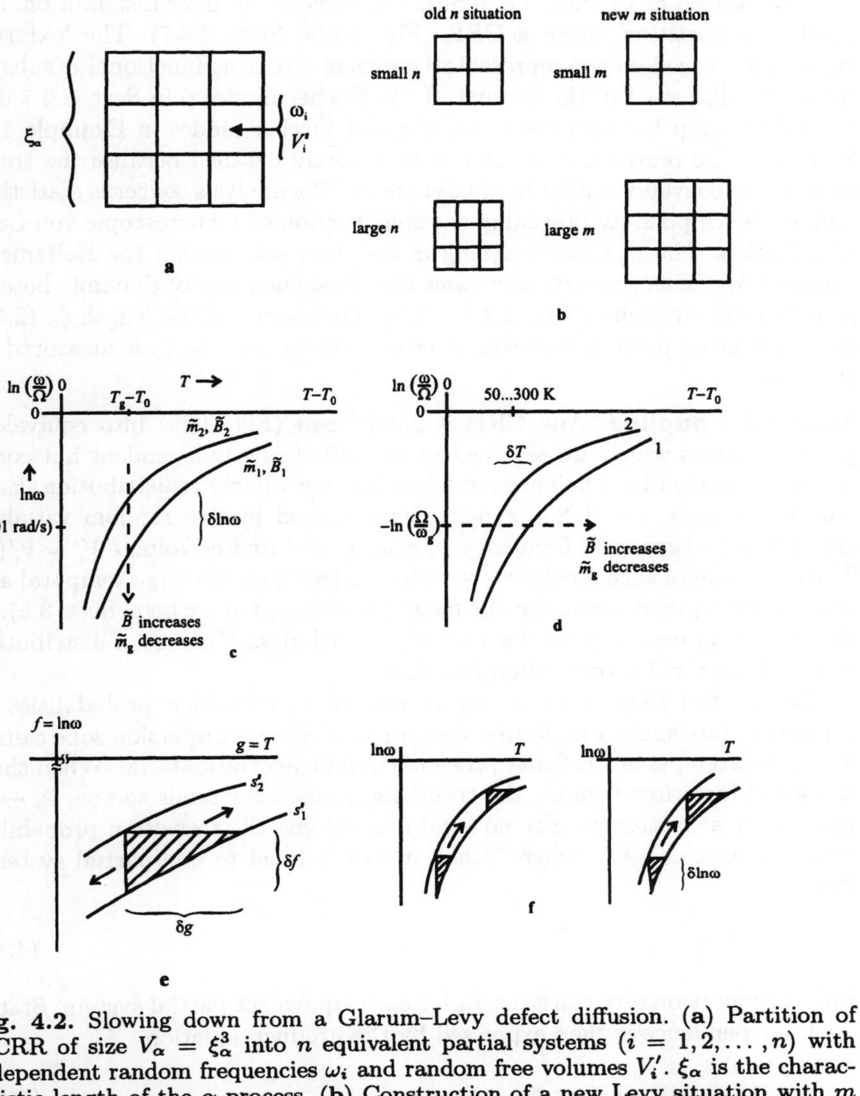

Fig. 4.2. Slowing down from a Glarum–Levy defect diffusion. **(a)** Partition of a CRR of size $V_\alpha = \xi_\alpha^3$ into n equivalent partial systems ($i = 1, 2, \ldots, n$) with independent random frequencies ω_i and random free volumes V_i'. ξ_α is the characteristic length of the α process. **(b)** Construction of a new Levy situation with m equal partial systems for discussion of different CRR sizes. **(c)** Isothermal section at $T = T_g$ of two WLF curves with common asymptotes separated by one dispersion δT. $\tilde{B}_1 > \tilde{B}_2$, $\tilde{B}_1 = \tilde{B}_2 + \delta\tilde{B}$, $\tilde{m}_1 < \tilde{m}_2$, $\tilde{m}_1 = \tilde{m}_2 - \delta\tilde{m}$. **(d)** Isochronous section of the two WLF curves in **(c)**. **(e)** Finding the set of curves $f(g, s')$ for which the triangles between locally neighboring curves remain of constant area $\delta f \cdot \delta g/2$ when shifted along the curves. **(f)** Comparison of the difference between neighboring curves with invariant triangle area (*left*) and increasing triangle area (*right*, for approximately constant $\delta \ln \omega$)

This section deals with the 'internal treatment' by Levy distribution, applied to a partition inside a CRR (Fig. 2.17a, Sect. 2.2.7). The 'external treatment', based on the representativeness of CRRs as functional α subsystems, was discussed in the context of the Fischer modes ϕ in Sect. 2.2.7 and the relationship between the α process and Fischer modes in Example 1 of Sect. 3.5. The representation here is only partly detailed because the treatment is nonconventional for liquid dynamics. The analysis concerns statistical independence possibly preceding the construction of a microscopic von Laue distribution. The minimal coupling model does not contain the Boltzmann constant k_B. This property expresses the disengagement of dynamic heterogeneity from structure (Sect. 2.2.5–2.2.6). The characteristic length ξ_α (2.76) of the resulting pattern, however, does contain k_B because ξ_α is measured by the FDT.

Minimal Coupling. Any CRR is partitioned (Fig. 4.2a) into equivalent partial systems which are assumed to be statistically independent but cooperatively coupled by a balance condition for free volume redistribution. Each partial system i, $i = 1, 2, \ldots, n$, is characterized by two random variables depending on time t: its frequency $\omega_i = \omega_i(t)$ and its free volume $V_i' = V_i'(t)$. The definition of such stochastic variables is based on the large temporal and spatial density of attempts for the faster processes (β and c here, Sect. 3.5). In the stationary case, any stochastic variable includes, of course, a distribution of amplitudes and a correlation function.

The partial frequencies ω_i are connected to transition probabilities $\tilde{\omega}_i$ expressing rare success as Fourier components in the α dispersion zone caused by many attempts of the faster processes mobilizing the material. When there is a lot of local free volume, the transition probability tends to one, $\tilde{\omega}_i \to 1$, success for any attempt. For no local free volume the transition probability tends to zero, $\tilde{\omega}_i \to 0$, where 'local' means 'related to one partial system'. Put

$$\tilde{\omega}_i = A_i \omega_i \, , \tag{4.47}$$

with a proportionality constant A_i for each equivalent partial system. Statistical independence is then expressed by the product equation

$$\omega_\alpha = \text{const.} \times \omega_1 \times \omega_2 \times \ldots \times \omega_n \, . \tag{4.48}$$

This equation corresponds to the 'and' probability $P(AB) = P(A) \cdot P(B)$ (3.15). The index i is not for a concrete ω_i value but for the stochastic ω_i variable of the ith partial system, corresponding to the attachment of individual frequency variables to each partial system *before* the frequencies are identified with a common frequency variable ω by the ω identity of the fluctuation–dissipation theorem FDT [cf. (3.15), and (3.17) for the external treatment]. *After* the variables are identified ('measured', Sect. 3.4), an n-fold convolution will be obtained and will be used for the internal Levy treatment, analogously to the transformation from (3.17) to (3.18) for the external treatment.

The *minimal coupling hypothesis* assumes that the control of local frequency ω_i by free volume is local, via the individual free volume V_i' of the partial system considered,

$$\omega_i = \omega_i(V_i') , \quad \text{same } i , \quad i = 1, 2, \dots , n . \tag{4.49}$$

Competition between the partial volumes V_i is regulated in a balance region of free volume redistribution. This region is assumed to be the smallest representative functional subsystem, the cooperatively rearranging region CRR:

$$V_\alpha' = V_1' + V_2' + \dots + V_n' . \tag{4.50}$$

The treatment of all partial systems is thus completely equivalent, i.e., democratic (Sect. 2.2.4). Equations (4.49) and (4.50) express the cooperativity inside a CRR: changes in the individual mobility $\log \omega_i$ in one partial system will be randomly balanced by changes in other partial systems. This coupling is called 'minimal' by analogy with gauge field theory where coupling is delegated to the underlying space–time connections, here thermodynamically represented by the free volume variables $V_i'(t)$ of the partial systems and their balance.

Minimal coupling seems the appropriate generalization for the disengagement of glass transition dynamics from the multifarious structures of glass formers. A further generalization would be the application of a correspondingly 'free entropy' S_i'. The corresponding balance equation,

$$S_\alpha' = S_1' + S_2' + \dots + S_n' , \tag{4.51}$$

merely expresses the microscopic reversibility of fluctuations. We would get a theory that is completely based on heat alone, the $T\mathrm{d}S$ term of the fundamental form of the First Law. Heat is, so to say, the total triumph of amorphicity over structure. The German philosopher G.W.F. Hegel wrote: "Wärme ist das Sich-Wiederherstellen der Materie in ihrer Formlosigkeit, ihre Flüssigkeit, der Triumph ihrer abstrakten Homogenität über die spezifischen Bestimmtheiten." [Heat is the restoration of matter into its shapelessness, its liquid, the triumph of its abstract homogeneity over specific determination.]

Having two different activities such as free volume and 'free entropy' we may ask whether or not they lead to the same CRRs. If the primary construction is the mobility pattern of dynamic heterogeneity related to thermodynamic fluctuations in density and entropy, respectively, the characteristic volumes V_α calculated from compressibility (2.78) or from calorimetry (2.76) must be exactly the same. If not, the interference pattern, unavoidable between two different dynamic heterogeneities, would contradict their nonzero coupling from $\overline{\Delta V \Delta S} = k_\mathrm{B} T (\partial V / \partial T)_p$, (3.38). In the following, we consider mainly the free volume variant.

The functional form of $\omega_i(V_i)$ is completely defined by the three equations (4.48), (4.49), and (4.50). They form a set of functional equations with the unique solution

$$\tilde{\omega}_i = (\tilde{\omega}_{M_n})^{1/n} \exp \frac{V_i'}{v_\mathrm{f}} , \tag{4.52}$$

$$\tilde{\omega}_\alpha = \tilde{\omega}_{M_n} \exp \frac{V_\alpha'}{v_\mathrm{f}} = \tilde{\omega}_{M_n} \exp \frac{V_1' + V_2' + \ldots + V_n'}{v_\mathrm{f}} , \tag{4.53}$$

where $\tilde{\omega}_{M_n}$ is a constant depending on the partition (M_n), and v_f a scaling parameter of the CRR not depending on the partition, i.e., invariant against the partition. The latter is called *fluctuating free volume*. It regulates the change of local mobility with the change of locally available free volume, inside the CRR:

$$\Delta \ln \omega = \frac{\Delta V'}{v_\mathrm{f}} . \tag{4.54}$$

A corresponding quantity s_f for the entropy variant (4.51) could be called *fluctuating free entropy*. The occurrence of the logarithm of ω in (4.54) fits in well with the $\log \omega$ measure from the FDT (2.75).

Equation (4.52) holds for any partial system in any partition M_n, with $n = 1, 2, 3, \ldots$, and (4.53) explicitly includes the trivial partition $n = 1$. The fluctuating free volume v_f is therefore a general norming or scaling parameter for arbitrary partitions of a CRR. This will be important for application of the Levy distribution.

The relation of fluctuating free volume to conventional free volume $[v_\mathrm{f}^\mathrm{conf} \sim (T - T_0)]$ is discussed in [296]. The two are different, that is, $v_\mathrm{f} \neq v_\mathrm{f}^\mathrm{conf}$, despite the fact that an equation of type (4.54) is formally used in both treatments.

Levy Treatment of Dynamic Heterogeneity. We show that minimal coupling is a Levy situation. The dynamic heterogeneity follows from Property 2 (Sect. 3.6) of the Levy distribution. The preponderant Levy component for $\alpha < 1$ is identified with the partial system with extraordinarily high mobility and large concentration of free volume (Glarum–Levy defect).

Firstly, we construct a map of minimal coupling in a Levy situation. As mentioned above, the product equation of minimal coupling (4.48), before the identification of the partial frequency variables $\{\omega_i\}$ by the ω identity of FDT, is equivalent to the convolution equation (3.204) of the Levy distribution,

$$p(x) = p^{*n}(a_n x + b_n) , \tag{4.55}$$

after identifying the partial frequencies with a common frequency variable ω, $\{\omega_i\} \to \omega$. Depending on the kind of compliance (activity) measured by the FDT, we can (Fig. 3.1c) identify $p(x)$ with the spectral density of the corresponding extensive variable, V' or S', e.g.,

$$x \propto \omega \, , \tag{4.56}$$

$$p(x)\,\mathrm{d}x \propto \Delta V'^2(\omega)\,\mathrm{d}\omega \, . \tag{4.57}$$

Equation (4.57) can be grasped via the measuring process (experiment) of volume fluctuation by the FDT. The stochastic variable $\omega_i(t)$ was defined in terms of a success-to-attempt ratio. The success produces (via the Wiener–Khintchin theorem) a contribution to the partial spectral density $V_i'^2(\omega_i)$. This contribution is called an event in the corresponding histogram. The ω_i event realizes its contribution to the susceptibility at the common frequency $\omega = \omega_i$ obtained from the ω identity, $\{\omega_i\} \mapsto \omega$. Since ω is a variable, higher ω_i values mean contributions at higher ω, etc. Figure 3.1c is thus the reduction of the partial histograms to the common histogram on the common ω axis.

Furthermore, the solution (4.53) of the functional equation for minimal coupling is equivalent to the Levy sum equation (3.202),

$$\mathbf{S}_n \stackrel{d}{=} c_n \mathbf{X} + \gamma_n \, , \tag{4.58}$$

where $\mathbf{S}_n = \mathbf{X}_1 + \mathbf{X}_2 + \ldots + \mathbf{X}_n$. After transforming (4.53) into logarithmic form, we can identify (map)

$$\mathbf{X}_i = V_i'(t) \, , \tag{4.59}$$

$$c_n = a_n = 1/v_\mathrm{f} \, , \tag{4.60}$$

$$b_n = \ln(\omega_{M_n})^{1/n} \, . \tag{4.61}$$

Since in the classical (non-quantum mechanical) case the spectral density is symmetrical (3.144), the centering constants are not important for the following. The missing n index at v_f from (4.60) will be discussed in the context of (4.62) and (4.63) below. The robustness necessary for a Levy situation [see the comments concerning (3.199) and (3.203)] is guaranteed by the arbitrariness of the partition in Fig. 4.2a. The internal Levy situation inside any CRR is consistent with the external Levy situation between different CRRs caused by their representativeness.

This concludes the map of the minimal coupling model into a Levy situation. The main result is (4.60) formulated as a proposition for the map:

Proposition. *The invariant fluctuating free volume parameter of minimal coupling can be identified with the reciprocal norming constant or scaling factor of the Levy distribution.*

Secondly, we estimate the density contrast of dynamic heterogeneity of Glarum–Levy defects. The quantification of dynamic heterogeneity for $\alpha < 1$ is based on the above proposition. From the scaling properties, Properties 1a and 1b (3.207) of the Levy distribution, we obtain formally, with the identification (4.60),

$$v_\mathrm{f} \overset{?}{\sim} n^{-1/\alpha} . \tag{4.62}$$

This relation seems to contradict the independence of v_f from the partition M_n of a CRR. The appropriate Levy situation is constructed as follows. Above we started from a CRR of given size, and the partition gives partial systems of decreasing size for increasing n. Now we start from a partial system of given size and construct CRRs of increasing sizes for increasing numbers m of them (Fig. 4.2b). Increasing m, for example, could correspond to decreasing temperature T, since the CRR volume V_α increases there. The arbitrariness of the n situation is transferred to the m situation by the arbitrariness in choosing the size of the partial system. Then the proportionality in (4.62) reads

$$v_\mathrm{f} \propto m^{-1/\alpha} . \tag{4.63}$$

Since m is now proportional to the characteristic size V_α of the subsystem, or to the cooperativity N_α expressing the number of particles in V_α, we obtain

$$v_\mathrm{f} \propto N_\alpha^{-1/\alpha} . \tag{4.64}$$

The fluctuating free volume decreases with the size of cooperativity, dramatically for large cooperativities N_α and small Levy exponents α. Equation (4.64) permits a physical interpretation of the Levy (= Kohlrausch) exponent α:

$$\boxed{\alpha = -\frac{\mathrm{d}\log N_\alpha}{\mathrm{d}\log v_\mathrm{f}}} . \tag{4.65}$$

Equation (4.65) seems to be one of the most important formulas in the unconventional parts of our book. The minus sign in (4.65) expresses the fact that enlargement of free volume diminishes the need for large cooperativities. This Levy exponent for the glass transition describes the log–log sensitivity of cooperativity to free volume. Defining the reciprocal fluctuating free volume as *restrictiveness*

$$r_v = 1/v_\mathrm{f} , \tag{4.66}$$

we get

$$\alpha = +\frac{\mathrm{d}\log N_\alpha}{\mathrm{d}\log r_v} . \tag{4.67}$$

This Levy exponent then describes the logarithmic sensitivity of cooperativity to the restrictiveness of the situation, or conversely, the positive effect of restrictiveness on the size of the cooperativity unit.

An analogous formula is obtained from a treatment of 'free entropy' s_{f} instead of free volume,

$$\alpha^{(s)} = -\frac{\mathrm{d}\log N_\alpha}{\mathrm{d}\log s_{\mathrm{f}}} . \tag{4.68}$$

The sensitivity of N_α to s_{f} may be different from its sensitivity to v_{f}, and so on: the Levy exponent may depend on the activity, $\alpha^{(s)} \neq \alpha^{(v)} \neq \ldots$. An example is the different Kohlrausch exponent for coherent and incoherent dynamic neutron scattering (Fig. 2.19d).

The size of the Levy exponent α for the glass transition is bounded above and below. The α values are always smaller than or equal to unity, $\alpha \leq 1$, because otherwise, as mentioned above (Sect. 3.1), the relaxation character of the α process would be destroyed by resonance components excluded in the gigahertz region and below. Overdamped dynamics seems necessary for the occurrence of a preponderant component: the corresponding Glarum defect is related to the diffusion character of its motion. Underdamped dynamics would be 'too fast' and would crush the defect. The upper limit is thus given by the Cauchy distribution corresponding to Debye relaxation. A lower limit is indicated by the extreme smallness of fluctuating free volume v_{f} for small α at large N_α. In our model, v_{f} monitors the α process, and this means that v_{f} cannot be too small because other sources of mobility, e.g., wing processes possibly neglected in the model, can take control if v_{f} becomes too small. This effect was called exhaustion in Sect. 2.5.1. Typical α values for the α process are between $\alpha = 0.4$–1.0, depending on temperature and activity. The α value $\alpha = 0.4$ seems to be a lower limit because v_{f} values smaller than $N_\alpha^{-1/\alpha} \approx 10^{-5}$, for $N_\alpha = 100$ and $\alpha = 0.4$, seem too small to control such a large cooperativity. Smaller fitted α values, $\alpha < 0.4$, may indicate either a small cooperativity N_α or formally, for an unmotivated fitting, a spatial heterogeneity of glass temperatures, as for proteins [717] in the native state or for other nanophase separations.

In a dynamic approach to the glass transition, exhaustion can also smooth the extrapolated temperature behavior of configurational entropy near the Vogel–Kauzmann temperature.

The contrast of density heterogeneities 'caused' by the dynamic heterogeneity of the glass transition can be estimated from (4.64). In our approach, the dynamic heterogeneity for $\alpha < 1$ is generated by the inevitable existence of a preponderant Levy component. The identifications (4.56) and (4.57) imply that this component corresponds to a partial system with extraordinarily high frequency and concentration of free volume: inside each CRR, there exists an island of mobility and free volume, called a Glarum–Levy defect (Sect. 2.2.4). The order of magnitude for the *density contrast* C^* generated by the dynamic heterogeneity is estimated from (4.64) putting $C^* = 1$ for $N_\alpha = 1$, as for an a process cage door. Then

$$C^* = 1/N_\alpha^{1/\alpha} , \tag{4.69}$$

with α the Levy exponent for free volume. This contrast is large in the crossover region and very small at the glass temperature, e.g., $C_g^* \approx 5 \times 10^{-4}$ for $N_\alpha^g = 100$ and $\alpha_g = 0.6$. Such a contrast seems too small for present scattering methods. The contrast increases with increasing temperature because N_α decreases and, often, α increases, so that there is a certain chance of detecting the 'density pattern of the glass transition' by scattering halfway between T_g and the crossover temperature. The experimental problem for finding a corresponding structure is to exclude thermodynamic nanophase separation. In the crossover region, the determination of the α process pattern is difficult because the characteristic length is small (0.5 nm) and an additional length of dynamic heterogeneity is induced by the occurrence of the three 'dynamic phases' in Fig. 2.34a and b. If the cooperativity at the glass temperature T_g is not too small, e.g., if $N_\alpha(T_g) > 15$, the contrast between the frozen cooperativity shell and the island of mobility (the latter was the Glarum–Levy defect at $T > T_g$) can be amplified at falling temperatures $T < T_g$ by the vault effect (Sect. 2.2.6).

To sum up, the large dynamic heterogeneity of order $\delta \ln \omega \approx 1/\alpha$ is accompanied by an extremely small density contrast of order $C^* = N_\alpha^{-1/\alpha}$. This small contrast facilitates the input of a Levy part into the finite differences Δ of the von Laue fluctuation formula (3.34) for construction of a microscopic von Laue distribution with contributions beyond the Gaussian distribution (Sect. 3.2.3). The structure of the Levy situation does not allow a Taylor series expansion about the small v_f parameter (4.53). A series expansion about the small contrast parameter C^*, however, seems possible.

Conservative Levy Society. We may think of human history as a succession and parallelism of productive dynamics, inequality, catastrophes, and wars, for instance. I do not understand how society functions and whether or not it can be sensibly controlled. We may ask why a democracy of equivalent people is so heterogeneous, and how the heterogeneity can be controlled. I therefore seek an example in which the explanation for a simple heterogeneity is highly nontrivial. Such an example is the Levy situation with Levy exponents $\alpha \leq 1$. The term 'conservative' is important for competition and will be used for the fact that important things must also be balanced in society, like free volume for the glass transition. I will discuss one dimension (e.g., free volume), whilst the complications due to different activities, obvious in a pluralistic society, will not be discussed.

It may therefore be interesting to have a dictionary relating α process concepts of the dynamic glass transition to features of a democratic society with large dynamic heterogeneity, e.g., a map (homomorphism) of society onto the Levy situation (*Levy society*). The statistical independence of subsystems and of partial systems can correspond to the concept of freedom in society. The free volume thus corresponds to a balance quantity that can be (re)distributed. The fluctuating free volume corresponds to a quantitative and invariant measure of freedom. The small contrast for the free volume pattern in (4.69) corresponds to the small actual freedom in the context of large units (N_α), in spite of the large dynamic heterogeneity. The large 'start velocity' of a Levy situation $\sim (1/t)^{1-\alpha}$ for $\alpha < 1$ at small t (3.198) may be used to characterize dynamics in society. The preponderant Levy component, represented for example by the 'unlimitedly' rich high-frequency tail of the Levy distribution, corresponds to the high productivity of society. Hence, socialism corresponds to a Levy exponent $\alpha > 1$, and a homogeneous, undynamic, and unproductive society, whilst American capitalism corresponds to small $\alpha < 1$, a heterogeneous, dynamic, and productive society, and Rhenish (German) capitalism to moderate α, but $\alpha < 1$. The $\alpha < 1$ relation seems desirable to get large enough driving forces, but α should not be too small to limit dynamic heterogeneity. The cooperativity size could correspond to the size of society's units (e.g., production or management units for multinationals, or state size and power for federalism as another activity). The breakthrough instability at high mobility for the glass-transition Levy situation, i.e., the trend to higher mobility $\log \omega$ (Sect. 2.1.3), could correspond to the psychological nature of human beings insofar as this is relevant to society, e.g., in politics. (The German historian Leopold von Ranke wrote: "Allein es liegt nun einmal nicht in der Natur des Menschen, sich mit einem beschränkten Gewinn zu genügen, und die siegreiche Menge wird niemals verstehen, innezuhalten" [650].) [It is just not in the nature of human beings to be satisfied by a limited gain, and the victorious crowd will never know when to stop.] In democratic situations, this instability is temporarily stabilized by the Levy distribution. The term 'temporarily' is related to the evolution of society, i.e., to a slow variation of parameters.

Apart from the thin cover of civilization, one of the most important, perhaps the only parameter that is really available for controlling an evolutionary Levy society is the Levy exponent α (4.65) and (4.67). In physical terms, the exponent α rationalizes the following relation: enlargement of free volume diminishes the need for large cooperativities. In general terms, the exponent α expresses the percentage change in unit size caused by a percentage change in restrictiveness that forces society to enlarge the unit size to achieve large enough efficiency in face of competition. The point is to get a qualitative grasp of the relationship between large dynamic heterogeneity $(1/\alpha)$, unit size (N_α), and residual invariant freedom (v_f, C^*) in such a framework [(4.64) and (4.69)]. Laws and decrees should be regularly invented and pushed through to control the size of cooperativity (units) by fine-tuning appropriate restrictions for residual freedom, without scarifying the optimal $\alpha < 1$ condition, as long as society has not evolved into a stationary, perhaps periodic state, but one that is worth living in (utopia). The problem of several dimensions mentioned above (in physical terms, of different activities such as free volume, 'free entropy' and other 'free entities', including their interdependencies) also seems important, but is only mentioned in passing here.

Ideal WLF Equation. The following is a listing of WLF equation properties based on the hyperbolic form of this equation, (2.24)–(2.27), and described in natural logarithms to suppress the factor $\ln 10 \approx 2.3$.

$$\tilde{B} = 2.3B\,, \quad \tilde{m}_{\mathrm{g}} = 2.3m(T_{\mathrm{g}})\,, \tag{4.70}$$

$$(T - T_0)\ln(\Omega/\omega) = (T_{\mathrm{g}} - T_0)\ln(\Omega/\omega_{\mathrm{g}}) = \text{ const. } = \tilde{B} = \tilde{B}_{\mathrm{g}}\,, \tag{4.71}$$

$$\{\Omega, T_0, \tilde{B}\}\,, \tag{4.72}$$

$$T_{\mathrm{g}} \longleftrightarrow T\,, \quad \ln\omega_{\mathrm{g}} \longleftrightarrow \ln\omega\,, \tag{4.73}$$

$$\frac{\mathrm{d}T}{\mathrm{d}\ln\omega} = \frac{T - T_0}{\ln(\Omega/\omega)}\,, \tag{4.74}$$

$$\tilde{m}_{\mathrm{g}} = \left.\frac{\mathrm{d}\ln\omega}{\mathrm{d}\ln T}\right|_{T_{\mathrm{g}}} = T_{\mathrm{g}}\left.\frac{\mathrm{d}\ln\omega}{\mathrm{d}T}\right|_{T_{\mathrm{g}}}\,, \tag{4.75}$$

$$\tilde{m}_{\mathrm{g}} = \ln(\Omega/\omega_{\mathrm{g}})\frac{T_{\mathrm{g}}}{T_{\mathrm{g}} - T_0}\,, \tag{4.76}$$

$$\tilde{m}_{\mathrm{g}} = \frac{T_{\mathrm{g}}}{(T_{\mathrm{g}} - T_0)^2}\tilde{B} = \ln(\Omega/\omega_{\mathrm{g}}) + \frac{T_0}{\tilde{B}}\ln^2(\Omega/\omega_{\mathrm{g}})\,, \tag{4.77}$$

$$\frac{\mathrm{d}\tilde{m}_{\mathrm{g}}}{\mathrm{d}\tilde{B}} = -\frac{T_0}{\tilde{B}^2}\ln(\Omega/\omega_{\mathrm{g}})\,, \tag{4.78}$$

$$T_{\mathrm{g}} - T_0 = \frac{\tilde{B}}{\ln(\Omega/\omega_{\mathrm{g}})}\,, \quad \frac{T_{\mathrm{g}} - T_0}{T_{\mathrm{g}}} = \frac{1}{m_{\mathrm{g}}}\ln(\Omega/\omega_{\mathrm{g}})\,, \tag{4.79}$$

$$\frac{\delta\ln\omega}{\delta T} = \frac{\mathrm{d}\ln\omega}{\mathrm{d}T} = \frac{\ln(\Omega/\omega)}{T - T_0}\,, \tag{4.80}$$

$$\left.\frac{\delta \ln \omega}{\delta T}\right|_{T_g} = \frac{\tilde{m}_g}{T_g}, \tag{4.81}$$

$$\delta T \cdot \delta \ln \omega = \text{const.}' = \tilde{C}_2, \tag{4.82}$$

$$\delta \ln \omega \equiv \delta^{2-1} \ln \omega = \frac{\tilde{a}}{T_g - T_0} = \frac{\tilde{a} \ln(\Omega/\omega_g)}{\tilde{B}}, \tag{4.83}$$

$$\delta T \equiv \delta^{1-2} T = \tilde{b}(T_g - T_0) = \frac{\tilde{b}\tilde{B}}{\ln(\Omega/\omega_g)}, \tag{4.84}$$

$$\frac{\tilde{a}}{\tilde{b}} = \left.\frac{\delta \ln \omega}{\delta T}\right|_{T_g} (T_g - T_0)^2 = m_g \frac{(T_g - T_0)^2}{T_g}, \tag{4.85}$$

$$\delta \tilde{B} \equiv \delta^{1-2} \tilde{B} = (T_g - T_0)\delta \ln \omega = \ln(\Omega/\omega_g)\delta T, \tag{4.86}$$

$$\delta \tilde{m} \equiv \delta^{1-2} \tilde{m}_g = \frac{T_g}{T_g - T_0} \delta \ln \omega = \ln(\Omega/\omega_g) \frac{T_g}{(T_g - T_0)^2} \delta T, \tag{4.87}$$

$$\tilde{C}_2 = \tilde{a}\tilde{b} = \frac{(\delta \tilde{B})^2}{\tilde{B}} = \frac{(T_g - T_0)^2}{T_g} \frac{(\delta \tilde{m})^2}{\delta \tilde{m}}, \tag{4.88}$$

$$F_g = \frac{T_g}{T_g - T_0} = \frac{\tilde{m}_g}{\ln(\Omega/\omega_g)}. \tag{4.89}$$

The hyperbolic form of (4.71) becomes evident for $\tilde{f}\tilde{g} = \tilde{f}_g \tilde{g}_g = \text{const.}$, with the factors $\tilde{f} = -\ln(\omega/\Omega) = +\ln(\Omega/\omega)$ and $\tilde{g} = T - T_0$. The constant is called a *global invariant*, $\tilde{B} = \tilde{B}_g$. The WLF equation has three parameters (4.72): the frequency asymptote Ω, the temperature asymptote (Vogel temperature T_0), and the global invariant $\tilde{B}$. The asymptotes are chosen as coordinate axes in Figs. 4.2c and d. The hyperbola is symmetrical between the current point $(\ln \omega, T)$ and the reference point $(\ln \omega_g, T_g)$ (4.73). All the following equations can thus be read as dependence on $\ln \omega$ and T rather than $\ln \omega_g$ and T_g.

The derivative along the WLF curve is equal to the quotient of the hyperbolic factors, $\mathrm{d}\tilde{f}/\mathrm{d}\tilde{g} = \tilde{f}/\tilde{g}$, a typical hyperbolic property (4.74). The steepness index of fragility, briefly the *fragility* $\tilde{m}_g$, not forgetting the ln 10 factor of (4.70), is the logarithmic derivative (4.75), which can also be expressed by the factor quotient (4.76). The Arrhenius process ($T_0 = 0$) has the smallest fragility, $\tilde{m}_g$(Arrhenius) $= \ln(\Omega/\omega_g)$; this $m_g = \tilde{m}_g/2.3 = \log(\Omega/\omega) = \log(2\pi\nu_0/\omega_g)$ usually lies between 15 and 20. One factor in (4.75) can be substituted by the global invariant $\tilde{B}$ (4.77). Increasing global invariant $\tilde{B}$ means decreasing fragility $\tilde{m}_g$ (4.78). The difference between glass temperature T_g and Vogel temperature T_0, for a given frequency asymptote Ω, can be expressed by $\tilde{B}$ or $\tilde{m}_g$ (4.79). The difference quotient formed by the dispersions $\delta \ln \omega$ and δT (2.138) can be approximated by the derivative, by the factor quotient (4.80), and by the fragility $\tilde{m}_g$ (4.81).

Besides the global invariant $\tilde{B}$, $\tilde{f} \cdot \tilde{g} = \tilde{B} =$ const., the set of hyperbolas with common asymptotes also has a *local invariant* $\tilde{C}_2 = \delta T \cdot \delta \ln \omega$, corresponding to $\delta \tilde{f} \cdot \delta \tilde{g} =$ const. (4.82). The dispersions will be defined as positive [see (4.83), (4.84) with the proportionality constants $\tilde{a}$, $\tilde{b}$, and Figs. 4.2c and 4.2d]. The quotient of the proportionality constants may be expressed in terms of the dispersion ratio or the fragility (4.85). A neighborhood of hyperbolas can be described by the dispersions $\delta \ln \omega$ and δT, but also by corresponding differences in the global invariant $\delta \tilde{B}$ and fragility $\delta \tilde{m}$ [(4.86) and (4.87)]. The relation between the local invariant $\tilde{C}_2$ and the global invariant is quadratic in $\tilde{B}$ (4.88). We may also define a global measure of fragility, F_g, suited to the WLF equation (4.89).

WLF and Related Equations from Minimal Coupling. The aim in this section is to show that the local invariance (4.82) is not only necessary but also (with the exception of a set of straight lines in the $\ln \omega$–T plot) sufficient for the set of hyperbolas as representatives for the WLF equation.

The motif for the local invariant is cooperativity expressed by the independence equation (4.48) of minimal coupling. Note that there are no Levy problems with logarithmic dispersions (Fig. 2.14c and Sects. 2.2.4 and 3.6). From the product equation (4.48), leading to the above Levy treatment of Glarum defects, we obtain in the *logarithmic measure* the following additivity relation for the partial mobility dispersions,

$$\delta \ln \omega_\alpha = \delta \ln \omega_1 + \delta \ln \omega_2 + \ldots + \delta \ln \omega_n \,. \tag{4.90}$$

In the extensive treatment (right-hand side of Fig. 4.2b), we get $\delta \ln \omega \equiv \delta \ln \omega_\alpha \propto N_\alpha^{1/2}$. On the other hand, if the temperature fluctuation δT can be grasped by the conventional additivity concept, then $\delta T \propto N_\alpha^{-1/2}$, because temperature is an intensive thermodynamic variable. Thus,

$$\delta T \cdot \delta \ln \omega \propto N_\alpha^{-1/2} \cdot N_\alpha^{+1/2} \sim \text{const.} \tag{4.91}$$

Assume for the time being that the proportionalities in (4.91) do not depend on temperature (‘*ideal WLF*’ situation). Then

$$\delta T \cdot \delta \ln \omega = \text{const.} \tag{4.92}$$

Our task is now one of differential geometry (Fig. 4.2e), namely, to find all sets of curves $f(g, s')$ with the local (i.e., more than differential) condition $\delta f \cdot \delta g = \text{const.}$ under a shift along the curves, where f is the mobility $\ln \omega$, g is the temperature T, and s' is the curve parameter (e.g., $\tilde{B}$ or $\tilde{m}$, if hyperbolas are indeed obtained).

Consider two neighboring curves $(1, 2)$ such that a Taylor series expansion to second order in $(s'_1 - s'_2)^2$ is sufficient. Then

$$\delta f \cdot \delta g = K(s'_1 - s'_2)^2 , \tag{4.93}$$

with the local invariant K by shifting along the curves. Using the relation $(\partial f/\partial g)(\partial g/\partial s)(\partial s/\partial f) = -1$ and scaling the s' parameter to s by

$$\mathrm{d}s = \sqrt{K}\,\mathrm{d}s' , \tag{4.94}$$

we get a nonlinear partial differential equation for $f(g, s)$,

$$\left(\frac{\partial f}{\partial s}\right)^2 = \frac{\partial f}{\partial g} . \tag{4.95}$$

Excluding any set of parallel straight lines on an fg plot, which are a trivial solution of (4.92), and using the locality of condition (4.92), $\delta f \cdot \delta g = \text{const.}$, the method of characteristics [651] gives [298] the unique solution

$$(f - f_0)(g - g_0) = C(s) , \tag{4.96}$$

with (f_0, g_0) constants and $C(s)$ a function of s. The straight lines are excluded because of their finite values for mobility $f = \ln \omega$ at $T = 0$ and their infinite values for mobility at $T \to \infty$. Putting $f_0 = \ln \Omega$, $g_0 = T_0$, and $C(s) = \tilde{B}$ in (4.96), we do indeed get the set of ideal WLF hyperbolas (4.71). This proves the above claim that the existence of the local invariant $(\delta T \cdot \delta \ln \omega)$ is sufficient for WLF.

Let us now discuss this result. Arrhenius processes, straight lines in the Arrhenius diagram (mobility vs. reciprocal temperatures, $\ln \omega$ vs. $1/T$), are included in the set of hyperbolas for $T_0 = 0$. Negative Vogel temperatures $T_0 < 0$ are excluded by the landscape interpretation of T_0, Fig. 2.24f, $T_0 \sim \varepsilon$, T_0 being a measure for the roughness of the landscape.

The main result from our approach to the ideal WLF equation, however, is a formula for the ideal WLF temperature dependence of cooperativity, $N_\alpha(T)$, provided that the local invariant (4.92) does not depend on temperature. From

$$\delta \ln \omega \sim \sqrt{N_\alpha} , \quad \delta T \sim 1/\sqrt{N_\alpha} , \tag{4.97}$$

and with the hyperbolic properties (4.83)–(4.84), we get from (4.92)

$$N_\alpha(T) \propto \frac{1}{(T - T_0)^2} , \quad \xi_\alpha = N_\alpha^{1/3} \propto \frac{1}{(T - T_0)^{2/3}} , \tag{4.98}$$

for the ideal WLF case. The cooperativity steeply increases with decreasing temperatures. The $T_0 \to 0$ limit for very strong glasses does not lead to an Arrhenius model with zero or constant cooperativity; in fact, $N_\alpha(T)$ remains proportional to $1/T^2$. The temperature dependence of the fluctuating free volume v_f is calculated from (4.64) with $\delta \ln \omega \approx 1/\alpha$, (4.84), and (4.98), giving

$$\frac{v_\mathrm{f}(x)}{v_\mathrm{f}(x=1)} = x^{2/(\alpha_\mathrm{on} x)} , \tag{4.99}$$

$$x = \frac{T - T_0}{T_\mathrm{on} - T_0} , \quad 0 < x \leq 1 , \tag{4.100}$$

where the constant α_on is the Levy (Kohlrausch) exponent for the α process at the α onset ($x = 1$) and x a reduced temperature varying linearly between 0 and 1 in the temperature range between the Vogel temperature T_0 and the onset temperature T_on in the crossover region.

The assumption that the local invariant does not depend on the state is surely too strong to be realistic. A more moderate assumption would be that the local invariant is approximately constant inside the temperature range for the various processes. This assumption corresponds to different WLF (or Arrhenius, or even straight line) type pieces for the different temperature ranges (Fig. 2.1c).

Heat capacity spectroscopy experiments show [216] that the local invariant $\delta T \cdot \delta \ln \omega$ is not constant even inside the α process range. Although $\delta \ln \omega$ (or the Levy exponent $\alpha \approx 1/\delta \ln \omega$) does not vary very much in the frequency range between millihertz and kilohertz for the three polymers so far investigated ($\delta \ln \omega \approx$ 3–2), the temperature fluctuation increases systematically (δT = 3 K to 12 K). This means that the local 'invariant' increases from about 5 to about 25, $\delta T \cdot \delta \ln \omega \approx A + BT$, $B > 0$. This may perhaps be interpreted as a far-reaching effect of the crossover region in the direction of lower temperatures, as will be described in the next section.

Geometrically, the increase in $\delta \ln \cdot \delta \ln \omega$ implies that the triangle area between neighboring curves increases (Fig. 4.2f). From this picture, we see that the asymptotic frequency $\ln \Omega$ for the α process may reach high values if the curves are still adjusted by a hyperbolic (WLF) piece. One third of the substances in Table 2.10 have values $\log(\Omega/\mathrm{rad\,s^{-1}}) > 15$. The value 15 seems to be the upper limit for any direct physical interpretation of $\log \Omega$ in terms of frequencies of the molecular transient. Any WLF equation seems, therefore, to be some kind of approximation. The meaning of their parameters, in particular, their asymptotes, should not be overestimated.

Dispersion of Dynamic Heterogeneity for the α Process. The distribution of CRR sizes about the average becomes large in the crossover region. We try to estimate this distribution by phenomenological arguments.

The dispersion of cooperativity, $\delta N_\alpha(T)$, is calculated from a 'cooperativity–temperature equivalence' along the α dispersion zone,

$$\frac{\delta N_\alpha(T)}{\delta T(T)} = \left|\frac{\mathrm{d}N_\alpha}{\mathrm{d}T}\right| . \tag{4.101}$$

This equation is analogous to (2.138) and (4.80) for the mobility $\ln\omega$ expressing the time–temperature equivalence. The cooperativity size is considered as a spatial representative for dynamic heterogeneity (Fig. 2.34a).

Consider first the ideal WLF behavior. We obtained

$$N_\alpha \propto \frac{1}{(T-T_0)^2} , \quad \delta T \propto T - T_0 . \tag{4.102}$$

From the first formula of (4.102), we get $\mathrm{d}N_\alpha/\mathrm{d}T \propto -1/(T-T_0)^3$, so that

$$\delta N_\alpha(T) \propto \frac{1}{(T-T_0)^2} \quad \text{and} \quad \frac{\delta N_\alpha(T)}{N_\alpha(T)} \approx \text{const.} \tag{4.103}$$

The relative dispersion of the cooperativity (i.e., the relative distribution of CRR sizes) is constant for the ideal WLF behavior of the dispersion zone. Numerical values will be estimated below.

Secondly, we consider the region near the α onset in the crossover region. In general, small $N_\alpha \to 1$ values near the onset will generate large relative fluctuations $\delta N_\alpha/N_\alpha$. Equation (4.103) is therefore expected to be violated at temperatures near the crossover region.

The observation of a square-root behavior below an onset temperature T_{on} in the crossover region, $N_\alpha^{1/2} \propto (T_{\text{on}} - T)$ (for the few substances where the crossover region has so far proved accessible to heat capacity spectroscopy), was explained by a dynamic Landau expansion [576]. Consider the free energy increase f_α across the α dispersion zone, i.e., between the glass zone and the liquid (or flow) zone of the equilibrium liquid state (not between the liquid and crystal state, Fig. 2.11c). Take the cooperativity N_α as the control parameter and $\delta S^2/N_\alpha = k_B \Delta C_p$ as the order parameter of the Landau expansion,

$$f_\alpha = f_\alpha(N_\alpha , \delta S^2/N_\alpha) . \tag{4.104}$$

Equation (4.104) describes a situation with variable cooperativity (CRR size) where the fluctuation dominates due to the *smallness of CRRs*. A Landau expansion of f_α around the α process onset gives

$$N_\alpha^{1/2} \propto (T_{\text{on}} - T) \quad \text{and} \quad \Delta C_p \propto (T_{\text{on}} - T) \tag{4.105}$$

below the onset temperature $T < T_{\text{on}}$. So far this has been observed experimentally for the splitting scenario I of Sect. 2.2.1. For scenario II, only the first equation of (4.105) is sometimes indicated by experiment [217]. The following will therefore be restricted to scenario I. From the first equation of

(4.105), we obtain $\mathrm{d}N_\alpha/\mathrm{d}T \propto T_{\text{on}} - T$. From the second equation of (4.105), and (2.77) for determination of average CRR size, $\bar{N}_\alpha = N_\alpha$,

$$N_\alpha = \frac{RT^2\Delta(1/c_V)}{M_0(\delta T)^2} \approx \frac{RT^2\Delta c_p}{\bar{c}_p^2 M_0(\delta T)^2}, \tag{4.106}$$

we get a square-root divergence for the temperature fluctuation in scenario I, viz.,

$$\delta T \propto (T_{\text{on}} - T)^{-1/2}. \tag{4.107}$$

Hence, from the equivalence formula (4.101),

$$\delta N_\alpha(T) \propto (T_{\text{on}} - T)^{1/2}, \tag{4.108}$$

$$\frac{\delta N_\alpha(T)}{N_\alpha(T)} \propto (T_{\text{on}} - T)^{-3/2}. \tag{4.109}$$

The relative dispersion of the N_α distribution tends to a singularity near the onset, as depicted in Fig. 2.34a. Since for a y distribution with $y > 0$, the relative dispersion is usually bounded by a limit of order 1, $\delta y/\bar{y} \lesssim 1$, we treat the singularity problem below with a cutoff method.

Equation (4.108) corresponds to a definite deviation from the WLF equation. According to (4.90) for 'the new m situation' of Fig. 4.2b, $\delta \ln \omega_\alpha$ becomes small for small N_α (both of order 1). Since $\delta \ln \omega_\alpha$ cannot be smaller than $\delta \ln \omega_\alpha \approx 1$ ($\alpha = 1$ corresponds to a Debye relaxation), we put $\delta \ln \omega \approx 1/\alpha \approx 1$ and obtain from (4.107) in the crossover region ($T < T_{\text{on}}$),

$$\frac{\mathrm{d}\ln\omega}{\mathrm{d}T} \approx \frac{\delta \ln \omega}{\delta T} \propto (T_{\text{on}} - T)^{1/2}. \tag{4.110}$$

The slope of the α dispersion zone near the onset is thus expected to become smaller than from a large-scale α process-fitting by the WLF equation.

To get a closed analytical approximation for the temperature dependence of $N_\alpha(T)$ and its dispersion $\delta N_\alpha(T)$ for the α process, we try an interpolation between the ideal WLF behavior at low temperatures [small x in (4.100)] and the onset behavior near $x \approx 1$ but $x \leq 1$. Since $N_\alpha^{1/2} \propto 1/x$ for ideal WLF and $N_\alpha^{1/2} \propto (1-x)$ near onset, we interpolate the square root of cooperativity by

$$N_\alpha^{1/2}(x) = \frac{A(1-x)}{x}, \tag{4.111}$$

with A an individual parameter. Hence,

$$\frac{\mathrm{d}N_\alpha(x)}{\mathrm{d}x} = -\frac{2A(1-x)}{x^3}. \tag{4.112}$$

Interpolating similarly $\delta N_\alpha(x)$ and δx, we obtain for the dispersions of N_α and reduced temperature x,

$$\delta N_\alpha(x) = \frac{A'(1-x)^{1/2}}{x^2} \,, \quad \delta x = \frac{A''x}{(1-x)^{1/2}} \,, \tag{4.113}$$

with two new parameters, A' and A''. Comparing

$$\frac{\delta N_\alpha(x)}{\delta x} = \frac{A'}{A''} \frac{1-x}{x^3} \tag{4.114}$$

with $|\mathrm{d}N_\alpha(x)/\mathrm{d}x|$ from (4.112), we see that

$$2A = \frac{A'}{A''} \,. \tag{4.115}$$

The interpolated phenomenological $\delta N_\alpha/N_\alpha$ behavior is therefore determined by two parameters.

Let us conclude this interpolation with a numerical estimate. Heat capacity spectroscopy [215, 216] gives A values in the range between $A = 2$ and $A = 15$. Take $A = 10$. The crossover singularity will be treated by a cutoff method, as mentioned above. A minimal N_α value $N_\alpha^{\min}$ is assumed for real α process cooperativity. We must, of course, avoid $N_\alpha = 0$. Assume a breakdown of our approach near the onset singularity at $x = 1 - \varepsilon$, $0 < \varepsilon \ll 1$ with ε a small cutoff parameter for the singularity,

$$\frac{\delta N_\alpha}{N_\alpha} \propto \frac{1}{(1-x)^{3/2}} = \frac{1}{\varepsilon^{3/2}} \,. \tag{4.116}$$

According to our physical pictures in Sect. 2.5.3, the minimal characteristic volume for the α process, from factual entropy fluctuation, has the size of the α or a cage door (Sect. 2.5.3). This means that $N_\alpha^{\min}$ is of order one. Hence we put for the estimate

$$\delta N_\alpha^{\min} = N_\alpha^{\min} = 1 \quad \text{at} \quad x = 1 - \varepsilon \,. \tag{4.117}$$

Then

$$N_\alpha^{1/2}(\varepsilon) = A\varepsilon \,, \quad \delta N_\alpha^{1/2}(\varepsilon) = A'\varepsilon^{1/2} \,, \quad \delta x(\varepsilon) = A''\varepsilon^{-1/2} \,. \tag{4.118}$$

After some arithmetic, for $\varepsilon = 0.1$ (i.e., both the temperature difference $T_{\text{on}} - T$ and the temperature fluctuation δT of order 10 K) in the $x \lesssim 0.9$ range, but not too far below the onset, we get

$$\delta N_\alpha \approx \frac{3(1-x)^{1/2}}{x^2} \,, \quad \frac{\delta N_\alpha}{N_\alpha} \approx \frac{0.03}{(1-x)^{3/2}} \,, \quad \delta x \approx \frac{0.15x}{(1-x)^{1/2}} \,. \tag{4.119}$$

From $\delta N_\alpha/N_\alpha \approx 1$, according to the cutoff (4.117) of the singularity (4.116), we obtain $\delta N_\alpha/N_\alpha \approx 0.1 \approx$ const. (4.103) for $x \approx 1/2$ from the second equation of (4.119), i.e., near the glass temperature T_g for a typical glass former with the onset region in the mega-to-giga hertz frequency range. The distribution near the onset with $\bar{N}_\alpha = 1$, $\delta N_\alpha / N_\alpha \approx 1$ seems to be centered near $N_\alpha = 1$, whereas the distribution near T_g is centered on $N_\alpha = \bar{N}_\alpha$ of order 100 with a relatively small dispersion δN_a of order 10. This proves the distribution sketch for $\varphi(x)$ in Fig. 2.34a.

The distribution of CRR sizes is thus relatively wide near the crossover region. This complicates the determination of small cooperativities there by independent methods. At the glass temperature, however, the distribution is concentrated near the large average CRR size there with a dispersion of order ten percent.

Slowing Down Mechanism. The slowing down of the ideal WLF situation corresponds to the qualitative argumentation of Adam and Gibbs [122]: the α process trace in the Arrhenius diagram is curved because the cooperativity increases with falling temperature (4.102). A lot of details of such a non-Arrhenius type behavior can be illustrated by the fluctuation approach to the α process described above.

The conventional free volume ($v_f^{\mathrm{conf}} \sim T - T_0$ [118]) of classical arguments is now substituted by the fluctuating free volume. The small density contrast connected with the latter (4.69) corresponds to the disengagement of general dynamics from the individual, multifarious structures of glass formers. How these individualities still affect the fragility will be discussed below [see (5.2)].

The general WLF parameter finally responsible for the non-Arrhenius curvature is the Vogel temperature T_0. As mentioned in Sect. 2.1.1, T_0 should be understood as a parameter describing the physical situation in the range of validity of the WLF equation, rather than a parameter from an extrapolated asymptotic limit-temperature phenomenon which is never accessible by experiment on any given time scale. At low temperatures, the fluctuating free volume v_f becomes too small to control the α process, even far above the Vogel temperature T_0 [exhaustion as a consequence of (4.99)]. We discussed $k_B T_0$ as the roughness of the energy landscape (Fig. 2.24f). The WLF curvature means that the system point in the representative CRR landscape needs increasingly more time to realize the Glarum–Levy defect in a CRR of increasing size, more time than would be needed for a complicated barrier process in a CRR of constant size.

The broad distribution of small CRRs in the crossover region with a large temperature fluctuation δT seems a challenge for statistical physics. It will probably require the construction of a microscopic von Laue distribution.

4.4 Fischer Modes

The physical description of the ultraslow Fischer modes ϕ was based on the relationship between external and internal treatments by Levy distributions

(Fig. 2.17a, Sect. 2.2.7). The Glarum–Levy defect diffusion model for the cooperative α process follows from the internal treatment. The cooperativity is restricted to partial systems inside a cooperatively rearranging region CRR by the model of minimal coupling (Sect. 4.3). The speckle model for the Fischer ϕ modes follows from the external treatment. Cooperativity between different CRRs is not necessary. The relationship between the ϕ and α processes (Example 1 of Sect. 3.5) was mediated by the representativeness theorem (Sect. 3.1). This relationship follows from a continuity argument which in turn results from the equality of the Levy exponents α of the two treatments for each given activity. Slowing down of the Fischer modes will thus be related to slowing down of the α process with falling temperatures. The ϕ and α traces in the Arrhenius diagram are not parallel, however, and the mobility distance depends on the Levy exponent $\alpha(T)$ which may vary with temperature. Some details of the following reasoning are related to difficult problems in our nonconventional dynamic approach to liquid dynamics and it has not yet been possible to clarify them. I think, however, that the argument as a whole is consistent.

We consider four groups of arguments.

First Group: Existence and Significance of ϕ Speckles. The reader is referred to Sect. 2.2.7. Look again at the right-hand side of Fig. 2.17a (Fig. 4.3a). We ask whether the representativeness theorem implies a preponderant CRR on the time scale τ_ϕ of the ϕ process. The theorem states that the spectral density of all subsystems larger than or equal to an average CRR (as measured via a dynamic compliance J^* by the FDT) is the same and has the shape of a Levy distribution with a $J''(\log\omega)$ loss peak in the α dispersion zone (time scale τ_α). Applying this theorem to the much slower τ_ϕ scale would mean being at large negative $y = \alpha\ln(\omega\tau_\alpha)$ values, e.g., in the far left wing in a y range near $y_\phi = -5$ (Figs. 3.7a, b).

To construct a Levy situation significantly applicable to this τ_ϕ or y_ϕ ϕ process range, we need a unique and significant relation between mobility $\log\omega$ and free volume V' for this y_ϕ range. Such a relation will be called a mobility function $\log\omega(V')$. From the Levy distribution compliance picture Fig. 3.7a, we see that the loss part $J''(\log\omega,\alpha)$ wing depends monotonically on ω and α. Since for the α process y_α range (peak of J'' in Fig. 3.7b) the mobility function $\log\omega(V')$ was monotonically increasing (4.54) and since the α-exponent spreading of the J'' peak wing at y_ϕ is still monotonic and significant, we conclude that the representativeness theorem of the α process range also mediates a unique and monotonic mobility function $\log\omega(V')$ for the very small free volume V' variations in the ϕ process range. This seems a reliable condition for a ϕ functionality based on the very small part of free volume V' variation relevant for the ϕ process mobility y_ϕ. If the ϕ process range can be restricted to a certain t_ϕ or y_ϕ interval – this will be shown in the fourth group of arguments below – then the Levy situation of the

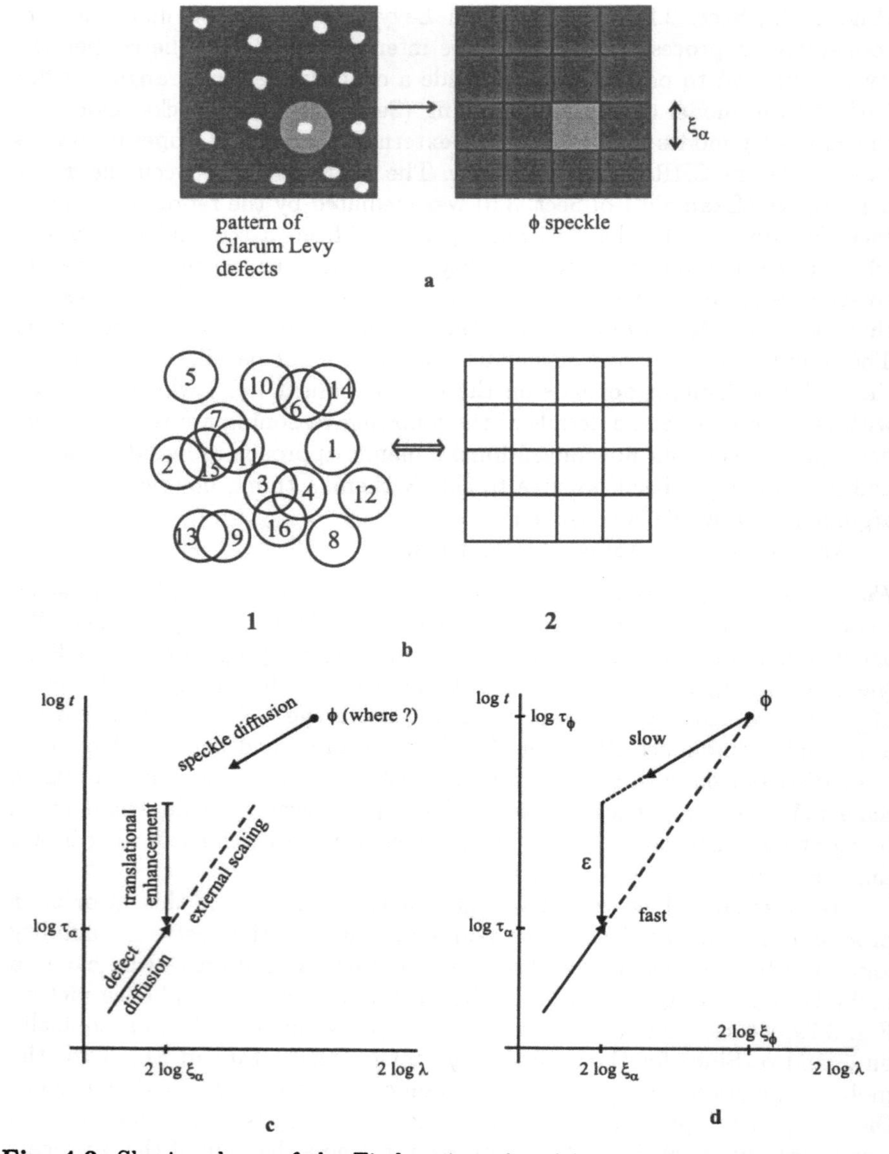

Fig. 4.3. Slowing down of the Fischer ϕ modes. (**a**) ϕ process speckle as a consequence of the representativeness theorem. The ϕ speckle corresponds to the preponderant component of the external Levy treatment. (**b**) ϕ process speckles (1) and additive partition in CRRs (2) according to the Riesz representation theorem (Sect. 3.1, Fig. 3.1b). (**c**) Elements needed for determination of time t and length λ scales of Fischer modes ϕ. (**d**) Triangle construction in a $\log t$–$\log \lambda$ plot (diffusion plot) for finding the length ξ_ϕ and time τ_ϕ scales of Fischer modes ϕ. ε is the translational enhancement for the α process

representativeness theorem can be consistently and significantly applied to the ϕ process.

We consider therefore a set of CRRs and apply the arguments of the representativeness theorem to the expected ϕ process range. We get a significant influence of the preponderant component. Its properties are estimated from the y scaling of Fig. 3.7b and the existence of a significant mobility function $\log \omega(V')$: the preponderance, originally attributed to the maximum and the near high-frequency wing of the α process $J''(y, \alpha)$ peak, is also attributed to the far low-frequency wing. For the preponderant component, we expect a large concentration of the very small free volume functional to the ϕ process range. This defines the speckle as a CRR being the Levy component with preponderant concentration of free volume.

Second Group: Two Kinds of Diffusion. (i) *Speckle diffusion.* The random light up fluctuation of speckles (Fig. 4.3b, from Fig. 2.18c) at a random set of space–time points,

$$\{(\boldsymbol{r}_1, t_1), (\boldsymbol{r}_2, t_2), \ldots\}, \tag{4.120}$$

can be connected with diffusion via the FDT (Sect. 2.2.7). We need the Riesz representation theorem for a map to nonoverlapping space-filling subsystems. This is because, in order to make a 'thermodynamic definition', i.e., a measuring process with FDT subsystem apparatus (Sect. 3.4.7), any diffusion in a liquid needs such a σ additivity of probability theory for fixing extensive variables. An example is dynamic light scattering in a sample, registered by photon correlation spectroscopy. Then we argue as follows. These subsystems themselves (e.g., CRRs) need not diffuse mutually (need not change their neighboring CRRs) but can be considered as fixed in space, at least approximately in the ϕ process range. It is only its property (preponderance: yes or no) that changes. The situation on large scales is similar to spin diffusion. Since such diffusion can be described by hydrodynamic methods, we get the normal dispersion law in time t and spatial scale λ for speckle diffusion,

$$\text{speckle diffusion:} \quad t \sim \lambda^2 . \tag{4.121}$$

(ii) Glarum–Levy *defect diffusion.* On the comparably short spatial scale of the α process inside each CRR, in the internal treatment, defects effect a real Glarum diffusion, changing their molecular environment. Note that defect diffusion may differ from particle diffusion. To find an approximate dispersion law for defect diffusion, we start from the normal diffusion according to the treatment in Sect. 3.1. Calculating the reciprocal dynamic structure factor as a function of the scattering vector Q and complex frequency z from the differential equation (3.10), we get

$$S^{-1}(Q, z) = \frac{z + \mathrm{i}DQ^2}{\mathrm{i}k_\mathrm{B}T\chi}, \tag{4.122}$$

where χ is the boundary susceptibility defined by the FDT (3.11) and D the diffusion coefficient. This gives the normal dispersion law $\omega/Q^2 = D$. From the standpoint of the Levy distribution, (4.122) corresponds to a Cauchy distribution for the spectral density of a dynamic variable (Levy exponent $\alpha = 1$). In a certain, rough approximation (Fig. 3.7b is scaled by $y = \alpha \ln \omega = \ln \omega^\alpha$), the other Levy distributions ($\alpha < 1$) correspond to a Cauchy distribution scaled by

$$\omega \to \omega^\alpha . \tag{4.123}$$

To this approximation, with $t \approx 1/\omega$ in the Levy dispersion law, this results in

$$\text{approximate defect diffusion:} \quad t \sim \lambda^{2/\alpha} . \tag{4.124}$$

Such a dispersion law for *sublinear diffusion* has actually been observed by dynamic neutron scattering [281], although for the high-temperature *a* process (Sect. 2.3.1).

Consider now a large pattern with several diffusing Glarum–Levy defects (Figs. 2.17a and 4.3a). Use the continuity argument between internal and external treatments as a connection between the α and ϕ processes (Sect. 3.5). Then, as long as no characteristic length occurs for $\lambda > \xi_\alpha$, the dispersion law for defect diffusion (4.124) with the same Levy exponent α may also be used to scale the external treatment.

Third Group: Construction of a ϕ Situation in the Diffusion Plot. We have three experimentally confirmed elements for finding the ϕ process in a space–time plot (logarithm of typical length scale λ, logarithm of typical time scale t: the $\log \tau$–$\log \lambda$ plot is a diffusion plot). They are represented as arrows in Fig. 4.3c:

(i) the speckle diffusion with slope $\mathrm{d}\log t/\mathrm{d}\log\lambda = 2$, i.e., the diffusion modes D of Fig. 6.1,
(ii) the defect diffusion with a larger slope for $\alpha < 1$, $\mathrm{d}\log t/\mathrm{d}\log\lambda = 2/\alpha$,
(iii) the translation enhancement of the α process, giving a vertical distance ($\Delta \log t = \log \varepsilon$) between defect diffusion at the lower end, at $\lambda = \xi_\alpha$ (translation), and speckle diffusion at the upper end, at $\lambda = \xi_\alpha$ (no translation).

A qualitative measure for ε is the mobility dispersion of the α process. Taking the half-width of a Kohlrausch loss susceptibility and referring to Table 2.3, we get $\varepsilon = 10^{\Delta \log_{10} \omega} \approx 10^{c/\alpha}$, $\log_{10} \varepsilon \approx c/\alpha$ with $c \approx 1.1$ weakly dependent on α.

To find the time and length scale for the Fischer modes, t_ϕ and λ_ϕ, we construct a triangle from the elements using the external ($\lambda > \xi_\alpha$) scaling for the defect diffusion (4.124). The ϕ angle of the triangle corresponds to the Fischer modes (Fig. 4.3d).

From this construction we calculate the ratio of length and time scales for the Fischer process related to the α process:

$$\xi_\phi/\xi_\alpha \approx 10^{c/2(1-\alpha)} , \quad (4.125)$$

$$\tau_\phi/\tau_\alpha \approx 10^{c/\alpha(1-\alpha)} . \quad (4.126)$$

A typical example is $\alpha = 0.7$ which gives, for $c = 1.1$, the values $\xi_\phi/\xi_\alpha \approx 10^{1.8} \approx 70$, i.e., $\xi_\phi \approx 105$ nm for $\xi_\alpha \approx 1.5$ nm, and $\tau_\phi/\tau_\alpha \approx 10^{5.24} \approx 1.7 \times 10^5$, in qualitative agreement with experiment [306].

Fourth Group: Observables of the Diffusion Plot Fig. 4.3d. In the ϕ process range, as long as the structure factor is much larger than in the plateau range (Figs. 2.17a, b, Sect. 2.2.7), we observe the speckle gas, e.g., by photon correlation spectroscopy PCS. Speckle diffusion has the dispersion law (4.121). In the plateau range, when the structure factor has the small value of (2.82), we observe the defect gas reduced by the α cooperativity, e.g., by means of Brillouin light scattering. The defect diffusion should have the dispersion law (4.124). In the ϕ process range, the defect diffusion (making the relation between the α and ϕ processes) is hidden below the large structure factor of the speckle gas.

The speckles themselves, with individual frequency or time variables, can be directly observed as laser speckles [344]. This kind of light scattering in the 100 nm scale goes beyond the FDT with the ω identity, which identifies the ω variables of individual speckles, as for diffusion.

To get a ϕ situation with an experimental length scale ξ_ϕ, the relation between speckle diffusion and virtual defect diffusion must change beyond the ϕ angle of our Fig. 4.3d triangle. At the ϕ angle, the translational enhancement is exhausted and the virtual defect diffusion (between the CRRs) would become slower than the speckle diffusion. Beyond the ϕ angle, the faster speckle diffusion would destroy the origin for speckle formation, defect diffusion across the CRR boundaries. We have no measurable dynamic fluctuation for length and time ranges beyond, $\lambda > \xi_\phi$, $t > \tau_\phi$. Below the ϕ angle ($\lambda < \xi_\phi$, $t < \tau_\phi$), the faster defect diffusion averages the dynamic α heterogeneity and we get 'homogeneous' speckles without visible defects on the speckle time scale (Fig. 4.3a).

The slowing down of the ϕ process is thus connected with the slowing down of the α (and a) process via (4.126). The τ_ϕ/τ_α ratio increases (i.e., the ϕ process becomes slower) with larger Levy exponent α as long as $\alpha > 0.5$. Formally, for $\alpha \to 1$, the ϕ process would become undetectable (too slow, with modes too large). Since α usually decreases with decreasing temperature, the slowing down of the ϕ process with temperature is weaker than that of the α process, as long as $\alpha > 0.5$. The ratio of characteristic lengths increases with increasing α. The typical decrease in $\xi_\phi(T)$ with temperature far above T_g (Fig. 2.17f) seems to result from competition between weak $\alpha(T)$ exponent increase and strong $\xi_\alpha(T)$ length decrease.

4.5 Modifications Due to Partial Freezing-in

This section is concerned with the conceptual and mathematical background for acceleration effects connected with freezing-in. Acceleration means that relaxation below the glass temperature $T \lesssim T_g$ is usually faster than the equilibrium relaxation at the same temperature T would be. Deceleration is rare and mainly restricted to situations where the fictive temperature is lower than the actual temperature, $T_f < T$. This section continues the discussion of the thermal glass transition (Sect. 2.1.4), structural relaxation (Sect. 2.1.5), possible motions in the glass structure (Sect. 2.2.6), and how relaxations between the dispersion zones (Sect. 2.4.5) are reflected in the glass state.

The general and difficult question of a preparation time τ_{prep} will be developed up to a simplified calculation for a cooling process across the thermal glass transition at T_g (Sect. 4.5.1). After that, a standard formulation of the Narayanaswamy procedure (the Narayanaswamy–Moynihan model) for structural relaxation will be presented (Sect. 4.5.2). As an example, we discuss the Kovacs expansion gap again (Sect. 4.5.3), obviously a never-ending story. A short discussion of some concepts for physical aging concludes this section (Sect. 4.5.4).

4.5.1 Preparation Time, Glass Frequency, and Vitrification Rate

We start with a mathematical analysis of an idealized freezing process during simple cooling as described in words in Sects. 2.1.4 (preparation time) and 2.2.6 (2.79).

▶

Fig. 4.4. Calculation of an idealized preparation time for a cooling experiment. (**a**) Comparison between the frequency ω of a real-part $C_p'(T)$ curve from HCS at T_ω and a cooling rate $\dot{T}$ of a $C_p(T)$ curve from DSC at T_g for $T_g = T_\omega$. (**b**) Schematic freezing-in (*hatched*) of a spectrum during cooling. (**c**) Comparison of a heating or cooling drift rate with the temperature fluctuation δT of a CRR (Fourier components in the dispersion zone of the dynamic glass transition). Situations (i), (ii), and (iii) are explained in the text. This picture evidences the fact that δT is the reduction parameter for $\dot{T}$. (**d**) Poles and zeros of the Laplace transform of a modulus $\bar{G}(s)$. (i) For modes with finite α_k amplitudes, $\tau_k \neq \tau_k'$. (ii) For cooperativity modes with infinitesimally small amplitudes, $\alpha_k \to 0$, $\tau_k \to \tau_k'$. (**e**) Cancelling of frozen-in parts (*hatched*) of cooperativity modes for the two versions A, B in the text. *Left*: correlation function for entropy fluctuations $\Delta S^2(t)$. *Right*: corresponding imaginary part of the entropy compliance $C_p''/T \propto C_p''(\log\omega)$. (**f**) Comparison of freezing-in curves with equilibrium curves, considering the effect of partial freezing-in of the slower modes. Variants I and II are discussed in the text

The step in the real part $C_p'(T,\omega)$ from heat capacity spectroscopy (HCS) is characterized by a dynamic glass temperature T_ω, and the step in the freezing-in heat capacity $C_p(T,\dot{T})$ from differential scanning calorimetry (DSC) is characterized by an ordinary glass temperature $T_g(\dot{T})$ (Fig. 4.4a). Liberating the cooling curve, formally, from the 'deformations' due to non-linearities (Fig. 2.5c), we obtain the same curves as a function of T. Shifting

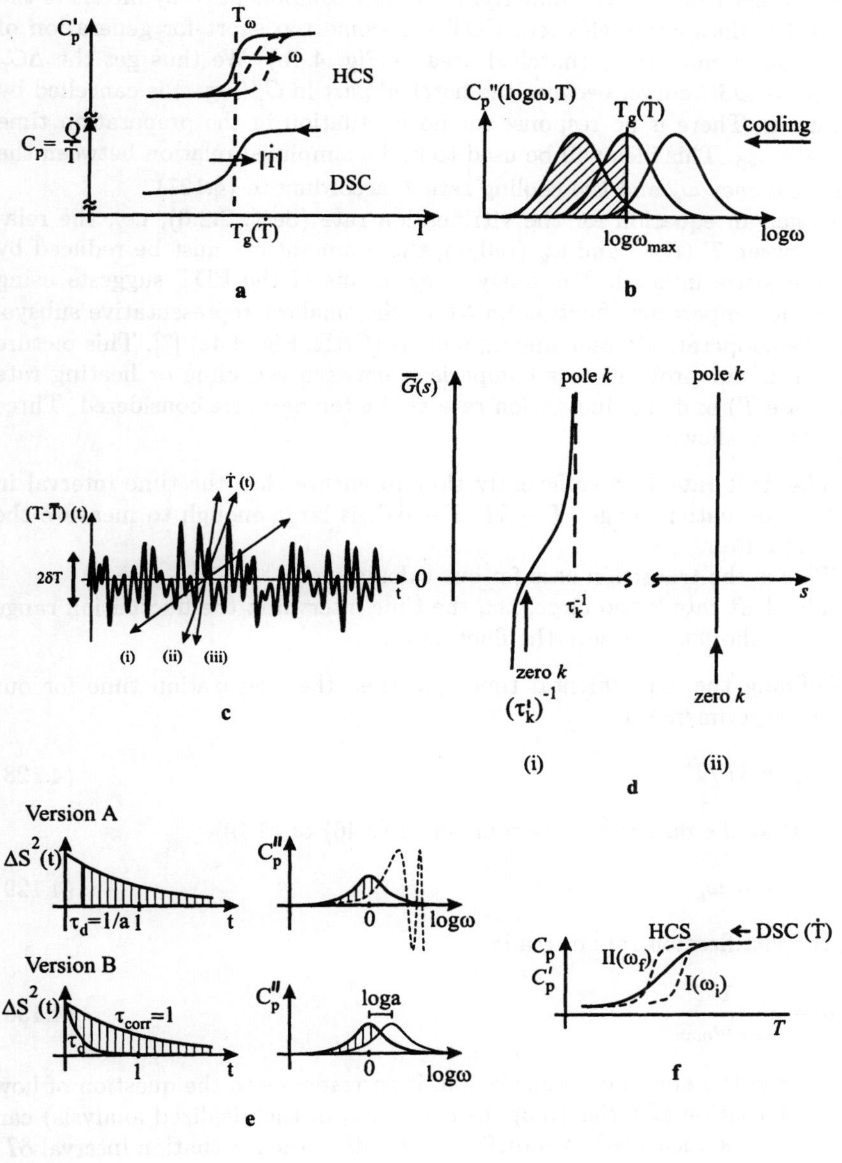

them to congruence at T_g, we obtain the desired relation between frequency ω and cooling rate $\dot{T}$,

$$T_\omega = T_g(\dot{T}) \, . \tag{4.127}$$

Consider now the shift of the imaginary part $C_p''(\log\omega, T)$ with falling temperatures (Fig. 4.4b, cf. also Fig. 2.16c). At a certain frequency $\omega_{\max}$, the measurement times for the underlying Fourier components – by means of the fluctuation–dissipation theorem FDT – become too short for generation of a thermodynamic signal (hatched area in Fig. 4.4b). We thus get the ΔC_p step in the DSC curve, because the hatched part in $C_p''(\log\omega)$ is cancelled by freezing-in. There is no response for no fluctuation in the preparation time interval τ_{prep}. This idea will be used to find a simplified relation between the glass frequency ω_g and the cooling rate $\dot{T}$ according to (4.127).

To get an equation for the vitrification rate (Sect. 2.2.6), i.e., the relation between $\dot{T}$ (K/s) and ω_g (rad/s), the temperature must be reduced by a temperature interval. The analysis by means of the FDT suggests using the mean temperature fluctuation δT of the smallest representative subsystem, the cooperatively rearranging regions (CRR, Fig. 4.4c) [7]. This picture transforms our problem to a comparison between a cooling or heating rate (drift rate $\dot{T}$) and the fluctuation rate at the temperature considered. Three variants are shown.

(i) The drift rate $\dot{T}$ is sufficiently slow to ensure that the time interval in the fluctuation range $(T - \delta T \;\; T + \delta T)$ is large enough to measure the fluctuation.
(ii) This is the transition case for our calculation.
(iii) The drift rate is too large, i.e., the time interval in the fluctuation range is too short to measure the fluctuation.

Defining the experimental time $\tau_{\exp}$ (i.e., the preparation time for our cooling experiment) by

$$\tau_{\exp} = \delta T / \dot{T} \, , \tag{4.128}$$

we see that the numerical a parameter in (2.46) or (2.79),

$$\dot{T} = a \, \delta T \, \omega_g \, , \tag{4.129}$$

i.e., the vitrification rate a, reads

$$a = \frac{1}{\tau_{\exp}\,\omega_{\max}} \, . \tag{4.130}$$

This rate is the aim of our calculation: it corresponds to the question of how fast a fluctuation (Fourier component at $\omega_{\max}$ in the idealized analysis) can be if it is to be measured by a drift $\dot{T} = \mathrm{d}T/\mathrm{d}t$ in the fluctuation interval δT.

The calculation is first confronted with the modulus–compliance displacement problem of Sect. 2.2.4 (Fig. 2.14b). The entropy compliance peak $C_p''(\log\omega)/T$ for given temperature T has a lower maximum frequency $\omega_{\max}$ than the corresponding temperature modulus peak $G_T''(\log\omega)$. The logarithmic frequency shift is of order

$$\Delta\log\omega \approx \log\frac{C_p^{\text{liquid}}}{C_p^{\text{glass}}} \approx \frac{\Delta C_p}{\bar{C}_p} . \tag{4.131}$$

Since a mode like the one needed for calculation of $\omega_{\max}$ in (4.130) (Fig. 4.4b) cannot be simultaneously mobile (compliance) and frozen (modulus), the partition into Debye-spectrum modes with finite amplitudes is not suitable. The other extreme would be to use the ω identity of the FDT alone to define the Fourier components as the modes that are either frozen or mobile. The problem would then be relegated to the spectral densities. This would contradict the expectation of a universal idealized value for a. Moreover, we have to explain how freezing-in of a Fourier component of spectral density will be measured.

The answer should be consistent with the ω identy of the FDT and must escape from the modulus–compliance displacement. Equation (4.131) suggests using Debye modes with infinitesimally small amplitudes because $\Delta\log\omega \to 0$ for $\Delta c_p \to 0$. These modes were called cooperativity modes in Sect. 2.2.4.

On the one hand, their displacement tends to zero. Consider the Laplace transforms of compliance, $\bar{J}(s)$, and modulus, $\bar{G}(s)$, with $\bar{J}(s) = 1/\bar{G}(s)$. Substituting the complex frequency s by the real frequency ω, this reciprocity corresponds to (2.8) for linear response, $J^*(\omega)\cdot G^*(\omega) = 1$. Decomposition into N modes gives

$$\bar{G}(s) = G_0 + \sum_{k=1}^{N} \frac{\alpha_k}{s + 1/\tau_k} . \tag{4.132}$$

The modulus $\bar{G}(s)$ has N simple poles at $s_k = -1/\tau_k$, $k = 1, 2, \ldots, N$, and N simple zeros at τ_k', whereas $\bar{J}(s)$ has its zeros at τ_k and its poles at τ_k' (Fig. 4.4d). For finite amplitudes α_k, we find $\tau_k \neq \tau_k'$, but for infinitesimally small amplitudes ($\alpha_k \to 0$), we see from (4.132) that $\tau_k' \to \tau_k$ because the s region that is influenced by such an $\alpha_k \to 0$ mode is concentrated in the immediate vicinity of the pole (right-hand side of Fig. 4.4d).

On the other hand, each mode has a Debye form (Lorentz line) over $\log\omega$ so that a successive thermodynamic freezing-in can be studied on each single mode. Near the $C_p''(\log\omega)$ maximum, the spectrum can be imagined as a set of equivalent cooperativity modes. The idealized treatment ignores influence from preceding freezing-in of slower modes. In the following, we put $\omega_g = \omega_{\max}$ for this treatment.

In the calculation, we introduce a dimensionless time τ_d for manipulation of the correlation function. The τ_d time is reduced with the correlation time of the exponential decay for the Debye relaxation. For $\tau_\mathrm{d} > t$, the susceptibility is assumed to be observable - the experimental time of (4.128) is sufficiently large to feel the fluctuation – and for $\tau_\mathrm{d} < t$, the susceptibility is assumed not to be observable, because the experimental time is too short. Two versions will be considered (Fig. 4.4e). Version A is the direct cutoff treatment of the underlying (FDT) correlation function,

$$x_\mathrm{d}^2(t) = \begin{cases} x^2(t) & \text{for} \quad t < \tau_\mathrm{d} \,, \\ 0 & \text{for} \quad t > \tau_\mathrm{d} \,. \end{cases} \tag{4.133}$$

Version B is an equivalent shortening of the correlation time ($1 = \tau_\mathrm{corr} \to \tau_\mathrm{d}$) to model the freezing-in cancellation in the correlation function. To calculate the vitrification rate $a = \dot{T}/\omega\delta T$, we must change to the frequency domain.

For both versions, τ_d is determined from the criterion that the half-area of the imaginary part of the corresponding Fourier transform, i.e., the susceptibility, as a function of $\log\omega$ is cancelled by the manipulations of the correlation function, $x^2(t) \to x_\mathrm{d}^2(t)$. The logarithm of ω must be used because $\mathrm{d}\log\omega$ is the time–temperature equivalence partner of $\mathrm{d}T$ in the cooling rate $\dot{T} = \mathrm{d}T/\mathrm{d}t$. As mentioned above, the original, non-manipulated correlation functions of our cooperativity modes are exponential decays with correlation time τ_corr, and the loss susceptibility is a Lorentz line $\sim x/(1 - x^2)$ with $x = \omega\tau_\mathrm{corr}$ and maximum at $\omega\tau_\mathrm{corr} = 1$.

Calculation of version A gives the Fourier transform ($t = \tau/\tau_\mathrm{corr}$, $\omega = \tau_\mathrm{corr} \cdot \omega$)

$$\int_0^{\tau_\mathrm{d}=1/a} \mathrm{e}^{-t} \cos(\omega t)\,\mathrm{d}t = \frac{1}{1+\omega^2}\left[1 + \mathrm{e}^{-\tau_\mathrm{d}}[-\cos(\omega\tau_\mathrm{d}) + \omega \sin(\omega\tau_\mathrm{d})]\right] \,, \tag{4.134}$$

which must be compared with the original Lorentz line (Fig. 4.4e). Version B results in a shift of the Lorentz line as long as the half-area is reached. The results are [652] $a = 5.56$ for A and $a = 5.42$ for B. It seems that the a value obtained is not very sensitive to the freezing model since the two versions could hardly be more different. We therefore find an idealized vitrification rate

$$a = \frac{\dot{T}}{\omega_\mathrm{g}\delta T} = 5.5 \pm 0.1 \,. \tag{4.135}$$

The relatively large a value reflects the transformation of the temporally linear temperature drift rate to the logarithmic partner ($\log\omega$ or $\log t$) in the local time–temperature equivalence.

Although the a value in (4.135) has been qualitatively confirmed by calorimetry [317], mitigating the idealization by using the fact that $T = T_\mathrm{g}$

corresponds approximately to the maximum frequency $\omega_{\max}$ of dynamic $C_p''(\log\omega)$ curves, the influence of the previous freezing-in of slower modes cannot be ignored [653]. The larger freezing-in interval due to the DSC curve deformation makes the practical relation between freezing frequency and cooling rate strongly dependent on the progress of freezing-in (Fig. 4.4f). Variant I is for the initial state with no previous freezing-in, whilst variant II is for the final state with complete freezing-in beforehand. We see that the dynamic glass temperature T_ω (cf. Fig. 4.4a) decreases: $T_{\omega \mathrm{f}} < T_{\omega \mathrm{i}}$. This means that an effective vitrification rate $a_{\mathrm{eff}} = \dot{T}/(\delta T \cdot \omega_{\mathrm{eff}})$, $\omega_{\mathrm{f}} \leq \omega_{\mathrm{eff}} \leq \omega_{\mathrm{i}}$, shifted to the same temperature for comparison, will increase as vitrification proceeds. This corresponds to the general acceleration effect of partial freezing and has been confirmed by Hutchinson in calculations with the Narayanaswamy–Moynihan model [654].

4.5.2 Narayanaswamy Mixing

Structural relaxation below the glass temperature T_{g}, as illustrated in the Arrhenius diagram Fig. 2.4b, has two aspects: acceleration in comparison to the hypothetical equilibrium response at the actual temperature $T < T_{\mathrm{g}}$, and widening of the transformation interval, from glass time to equilibrium, in comparison to the width of the equilibrium dispersion zone. The first aspect was described by a fictive temperature for the recovery, $T_{\mathrm{f}} > T$, whilst the second aspect will be described by iteration loops needed to solve the nonlinear functional equation (2.41),

$$T_{\mathrm{f}}(t) = \mathfrak{F}\,\{T(t'),\, T_{\mathrm{f}}(t')\}\ , \quad t' < t\ , \tag{4.136}$$

with T_{f} occurring on both sides. The $T_{\mathrm{f}}(t')$ function on the right-hand side means that the recovery is also controlled by the history of recovery itself. This effect is missing in the equilibrium linear response (no mixing parameter x, see below). The time t' describes the history up to the actual time t. The physical situation was discussed in Sect. 2.1.5. The relation of T_{f} to the heat capacity C_p is, for the example of entropy recovery,

$$\frac{\mathrm{d}T_{\mathrm{f}}}{\mathrm{d}T}(T) = \frac{C_p(T) - C_p^{\mathrm{glass}}(T)}{C_p^{\mathrm{liquid}}(T) - C_p^{\mathrm{glass}}(T)}\ , \tag{4.137}$$

where the glass index refers to the glass state and the liquid index to the equilibrium liquid zone.

The Narayanaswamy formulas are based on a convolution similar to the linear response and an additive mixing of T and T_{f} in the exponent generated from the inversion of the $\log t$ measure in the time–temperature superposition. The task is, given the history as a $T(t')$ program, to find the actual extensive variable $T_{\mathrm{f}}(t)$.

We start from the material equation for the linear response of an extensive variable $x(t)$ (e.g., x = volume, entropy, shear angle, polarization, or for another activity),

$$x(t) = \int_{-\infty}^{t} \dot{f}(t')\, J(t-t')\,\mathrm{d}t' , \tag{4.138}$$

where $f(t')$ is the time program for an intensive variable [mostly temperature $T(t')$], and J the compliance between x and f. The adaptation to structural relaxation for a partly frozen glass is mediated by (Figs. 2.6a and 2.8b)

$$x_{\mathrm{B}} \to T_{\mathrm{f}} , \quad \dot{f} \to \dot{T} , \quad J \to 1-\phi , \quad (t-t') \to (\zeta - \zeta') , \tag{4.139}$$

where $1-\phi$ is the compliance and $\{\zeta,\ \zeta' = \zeta(t')\}$ the material times.

We introduce a reference time τ_{b} whose equilibrium dependence on temperature $T = T_{\mathrm{b}}$ may be given by a WLF equation (with $\Omega = 1/\tau_0$ the frequency asymptote, where τ_0 is also used to define the time unit, e.g., the second, and with T_0 the temperature asymptote or Vogel temperature),

$$\tau_{\mathrm{b}}(T_{\mathrm{b}}) = \tau_0 \exp \frac{\tilde{\tilde{B}}(T_{\mathrm{g}} - T_0)^2}{T_{\mathrm{b}} - T_0} , \tag{4.140}$$

where

$$\tilde{\tilde{B}} = - \left. \frac{\mathrm{d}\ln\tau}{\mathrm{d}T} \right|_{T_{\mathrm{g}}} . \tag{4.141}$$

The quantity $\tilde{\tilde{B}}$ is the amount of slope along the WLF equation in an Arrhenius plot, not to be confused with the WLF parameters $\tilde{B} = 2.3B$ in (4.70)–(4.71). In fact, $[\tilde{\tilde{B}}] = 1/\mathrm{K}$, whereas $[\tilde{B}] = [B] = \mathrm{K}$. The curvature of the WLF trace is regulated by $T_0 > 0$. Then we get the actual fictive temperature

$$T_{\mathrm{f}}(t) = T(t) - \int_{t_0}^{t} \mathrm{d}T'\phi(\Delta\zeta) , \quad T' = T(t') , \tag{4.142}$$

from a $T(t')$ history starting at t_0 in equilibrium. The compliance $1-\phi$ is taken as a Kohlrausch function [180, 502], i.e.,

$$\phi(\Delta\zeta) = \exp[-(\Delta\zeta)^{\beta_{\mathrm{KWW}}}] , \quad 0 < \beta_{\mathrm{KWW}} \le 1 . \tag{4.143}$$

It is the dependence on only one parameter β_{KWW}, the Levy exponent $\alpha = \beta_{\mathrm{KWW}}$, that makes the formulas so robust over many time decades. The material time is defined by

$$\Delta\zeta = \zeta(t) - \zeta(t') = \int_{t'}^{t} \frac{\mathrm{d}t'}{\tau_s} , \tag{4.144}$$

with Narayanaswamy mixing in the exponential,

$$\tau_s \equiv \tau_0 \exp\left[\tilde{B}(T_g - T_b)^2 \left(\frac{x}{T - T_0} + \frac{1-x}{T_f - T_0}\right)\right] . \tag{4.145}$$

As mentioned above, the mixing parameter x, $0 < x < 1$, combines the influence of actual temperature T and fictive temperature T_f histories, $T = T(t')$, $T_f = T_f(t')$, on the structural relaxation. The exponential mixing reflects the fundamental fact that $\log t$ is the T partner in the time-temperature equivalence. The model with Narayanaswamy mixing and the Kohlrausch function is usually called the *Narayanaswamy–Moynihan model*.

These equations are usually solved by numerical iteration loops. The time history is broken into a sequence of n suitably chosen time intervals k,

$$T_f^{(n)} = T^{(n)} - \sum_{k=1}^{n} \left(\frac{\Delta T}{\Delta t}\right)^{(k)} \phi^{(k)} \Delta t^{(k)} , \quad k = 1, 2, ..., n , \tag{4.146}$$

with

$$\phi^{(k)} = \exp\left[-\left(\frac{1}{\tau_b}\sum_{i=k}^{n} \exp\left\{\tilde{\tilde{B}}(T_g - T_0)^2\left[\frac{1}{T_b - T_0} - \left(\frac{x}{T^{(i)} - T_0} + \frac{1-x}{T_f^{(i-1)} - T_0}\right)\right]\right\}\Delta t^{(i)}\right)^{\beta_{KWW}}\right] , \tag{4.147}$$

where $i = k, k+1, \ldots , n$.

For fixed reference time τ_b (which is a representative for the the experimental time τ_g problem in Sect. 4.5.1), we have four adjustable parameters for an Arrhenius approximation ($T_0 = 0$),

$$\left\{\tau_g , \tilde{\tilde{B}} , x , \beta_{KWW}\right\} . \tag{4.148}$$

The τ_0 parameter is absorbed into the 'glass time' at the glass temperature, $\tau_g = \tau(T_g)$. In an entropy-oriented version of the WLF equation due to Hodge [655],

$$\tau = \tau_0 \exp\left[\frac{Q}{RT(1 - T_0/T_f)}\right] , \tag{4.149}$$

where Q is a constant related to the configurational entropy, the four parameters are

$$\{\tau_g\,,\ Q\,,\ T_0\,,\ \beta_{\mathrm{KWW}}\}\ . \tag{4.150}$$

Equation (4.149) substitutes the slope $\tilde{\tilde{B}}$ and the mixing parameter x by Q and the Vogel temperature T_0.

More flexible variants use T_0 as an additional parameter. The whole equilibrium function $\tau_b(T_b)$ can be globally [656] or successively [657,658] determined, the latter variant by means of an additional loop. The robustness of (4.143) over many time decades allows the determination of the calorimetric equilibrium trace in the Arrhenius plot, $\log \tau_b$ vs. $1/T$, in an informative time interval between about 10^{-4} and 10^3 seconds, from one carefully measured DSC heating curve across the transformation interval.

Nevertheless, the Narayanaswamy formulas must be considered as a merely qualitative reproduction of structural relaxation (Sect. 2.1.5). Different $T(t')$ histories are usually fitted by different parameter sets, and we are confronted with small but systematic deviations. A large class of different $T(t')$ histories can be fitted by one parameter set if, formally, the $t \to \infty$ limit is chosen above the equilibrium. This requires one additional parameter [184, 659]. In practical terms, ΔC_p is properly diminished. As mentioned in Sect. 2.1.5, this need not be connected with the final state, but can also be interpreted as the initial influence of faster relaxations than for α process recovery, e.g., the A1 process, or the survival of free-volume or free-entropy production near the Glarum–Levy defects below T_g, due to the vault effect (Sect. 2.2.6). The time behavior of the faster processes themselves cannot be reproduced with a Kohlrausch relaxation ϕ according to (4.143).

Current discussions of structural relaxation are connected with a possible violation of the fluctuation–dissipation theorem (FDT) in the partially frozen state. One aim is to connect the extensive phenomenological fictive temperature with an intensive temperature (Nyquist factor k_BT) through a more microscopic concept that can, in addition, be partitioned in the frequency or time domain.

Consider structural relaxation after a quench (preparation time τ_{prep}) to a temperature T below the glass temperature. Probe this state by linear response in the time (t) domain or the frequency (ω) domain. Apart from τ_{prep}, we have two times: the waiting time t_w elapsed since the quench, and the probing time $t \approx 1/\omega$. Assume we can find the correlation functions $\Delta x^2(t, t_w)$ or spectral densities $\Delta x^2(\omega, t_w)$ that continue to operate in the partially frozen states, and the corresponding compliances, $J(t, t_w)$ or $J^*(\omega, t_w)$. (The same could be assumed for intensive variables f and their moduli.) To avoid details [660,661], we consider a $t \ll t_w$ regime with fast quasi-stationary fluctuations depending on a parameter t_w. Then there would be no violation of the FDT if (2.70) and (2.71) also remained valid in the partially frozen state changing with t_w. By the differentiated form of the FDT, this would mean that, in the time domain,

$$\dot{J}(t,t_{\mathrm{w}}) \overset{?}{=} \frac{\dot{\Delta x}^2(t,t_{\mathrm{w}})}{k_{\mathrm{B}}T}\ , \tag{4.151}$$

where $\dot{J}(t,t_{\mathrm{w}}) = (\mathrm{d}/\mathrm{d}t)J(t,t_{\mathrm{w}})$ and $\dot{\Delta x}^2(t,t_{\mathrm{w}}) = (\mathrm{d}/\mathrm{d}t)[\Delta x^2(t,t_{\mathrm{w}})]$, or in the frequency domain,

$$J''(\omega,t_{\mathrm{w}}) \overset{?}{=} \frac{\pi\omega\Delta x^2(\omega,t_{\mathrm{w}})}{k_{\mathrm{B}}T}\ , \tag{4.152}$$

for equal temperatures of sample and heat bath, T (FDT) $= T$ (heat bath). Since the partially frozen glass is out of equilibrium, the problem is not trivial. In addition, all variables depend on the parameters of the quench history. After studying the details [660, 661], the case can be discussed for any times $\{t < t_{\mathrm{w}}, t_{\mathrm{w}}\}$.

In the time domain, the problem is tackled by computer simulation in different model glass systems [661–663]. They define a function $\tilde{X}(t,t_{\mathrm{w}})$ by

$$\tilde{X}(t,t_{\mathrm{w}}) \equiv \frac{k_{\mathrm{B}}T\dot{J}(t,t_{\mathrm{w}})}{(\partial/\partial t)\Delta x^2(t,t_{\mathrm{w}})}\ , \tag{4.153}$$

where J is taken, for example, as the incoherent part of the intermediate scattering function [663]. For short times t, they find $\tilde{X} = 1$ (FDT region), whilst for long times, the result is $\tilde{X} < 1$ (a few percent, aging region). $\tilde{X} < 1$ means a smaller response than expected from the remaining fluctuation. They suggest the *effective temperature*

$$T_{\mathrm{eff}}(t,t_{\mathrm{w}}) \equiv \frac{T}{\tilde{X}(t,t_{\mathrm{w}})} \geq T \tag{4.154}$$

as a possible substitute for the fictive temperature T_{f} of the corresponding activity x.

In the frequency domain, the problem is tackled by dielectric noise experiments, e.g., in glycerol near and below T_{g} [664]. The effective temperature is then defined by a formula with the structure

$$T_{\mathrm{eff}}(\omega,t_{\mathrm{w}}) = \frac{\pi\omega\Delta f^2(\omega,t_{\mathrm{w}})}{k_{\mathrm{B}}M''(\omega,t_{\mathrm{w}})}\ , \tag{4.155}$$

where Δf is the voltage noise, for instance, for $\omega \approx 50$ rad/s, and M the corresponding dielectric modulus. The effective temperature is also observed to be a few kelvins higher than the bath temperature T during structural relaxation. For a long wait after structural relaxation, the effective temperature tends to the actual temperature, $T_{\mathrm{eff}} \to T$, as expected.

The trivial interpretation is again that the effective noise temperature is increased in comparison to the heat-bath temperature because the additional free volume of the frozen state contributes more to the fluctuations than to the response. Interpreting the FDT as an experimental equation, this would mean that, for the acceleration after the quench, a certain fraction of the quanta $\hbar\omega$ from fluctuations (Fig. 3.1e) are 'used' to stimulate the structural relaxation and are thus missing from the measured linear response. I expect the effective temperature, similar to the fictive temperature, to depend on the activity used for its definition. The question of what a Zeroth Law thermometer would measure in a theoretical aging situation is discussed in [665].

4.5.3 Example: Kovacs Expansion Gap

Two aspects of the Kovacs expansion gap (Fig. 2.7e, Sect. 2.1.5) are worth mentioning.

(i) The extrapolation of the effective time τ_{eff} for structural relaxation from the $-\delta = 10^{-3}$ expansion recovery range to $-\delta \to 0$ is different from the true $\tau_{\text{eff}}(\delta)$ behavior in the $-\delta = 10^{-4}$ range. This was explained by the long-time tails of the underlying retardation function ϕ (Fig. 2.7f).
(ii) The $-\delta = 10^{-3}$ range extrapolations can lead to different τ_{eff} values depending on the temperature jump size ΔT before the expansion experiments.

More recent analyses of the experimental data [171, 666] confirm the first aspect, but give only weak indications for the second aspect. In this section, we shall investigate whether the Narayanaswamy–Moynihan model (4.146)–(4.147) can produce the gap. For parameters that simulate the behavior of PVAC, the Kovacs–Ritland jump experiments are modelled for a final temperature $T_{\text{final}} = 40°\text{C}$ with starting equilibrium temperatures between 30°C and 50°C (Fig. 4.5a). As expected, the recovery behavior is not symmetrical (nonlinearity, Fig. 4.5b). Some interesting intersections of the contraction recovery curves are indications for the many parameters that influence the structural relaxations at temperatures above the glass temperature T_g.

▶

Fig. 4.5. Example of calculations using the Narayanaswamy–Moynihan model (4.146)–(4.147) with parameters for a PVAC sample: $\tau_B(T_B = 40°\text{C}) = 139$ s, $\tilde{\tilde{B}} = 0.67/\text{K}$, $x = 0.32$, $\beta_{\text{KWW}} = 0.62$, $T_0 = -10°C$. (**a**) $T(t)$ programs to simulate Kovacs jump experiments from equilibrium temperatures 30°C (2.5 K) 50°C to a final temperature $T_{\text{final}} = 40°\text{C} = T_g$. The jump rate was 600 K/min. (**b**) The recovery variable $\delta = (T_f - T_{\text{final}})/T_{\text{final}}$ as a function of waiting time t after the jump. (**c**) Kovacs diagram for the effective time $\tau_{\text{eff}} = -\delta/(\text{d}\delta/\text{d}t)$ as a function of δ, as derived from (**b**). (**d**) $T(t)$ programs modelling the degree of equilibration before the 35°C → 40°C jump for expansion. The cooling rates are $-\dot{T} = 10^{-4}$, 10^{-3}, 0.001, 0.01, 0.1, 0.2, 0.5, 1,2, 5, 10, 20 K/min, the jump rate was again 600 K/min. (**e**) Recovery after the $T(t)$ programs of (**d**). The curve after $\dot{T} = -10^{-4}$ K/min cooling corresponds to the jump curve after $T_{\text{eq}} = 35°\text{C}$ in (**b**). (**f**) Kovacs diagram for the recovery curves of (**d**). (**g**) DSC cooling–heating cycle for $|\dot{T}| = 10$ K/min with our parameter set

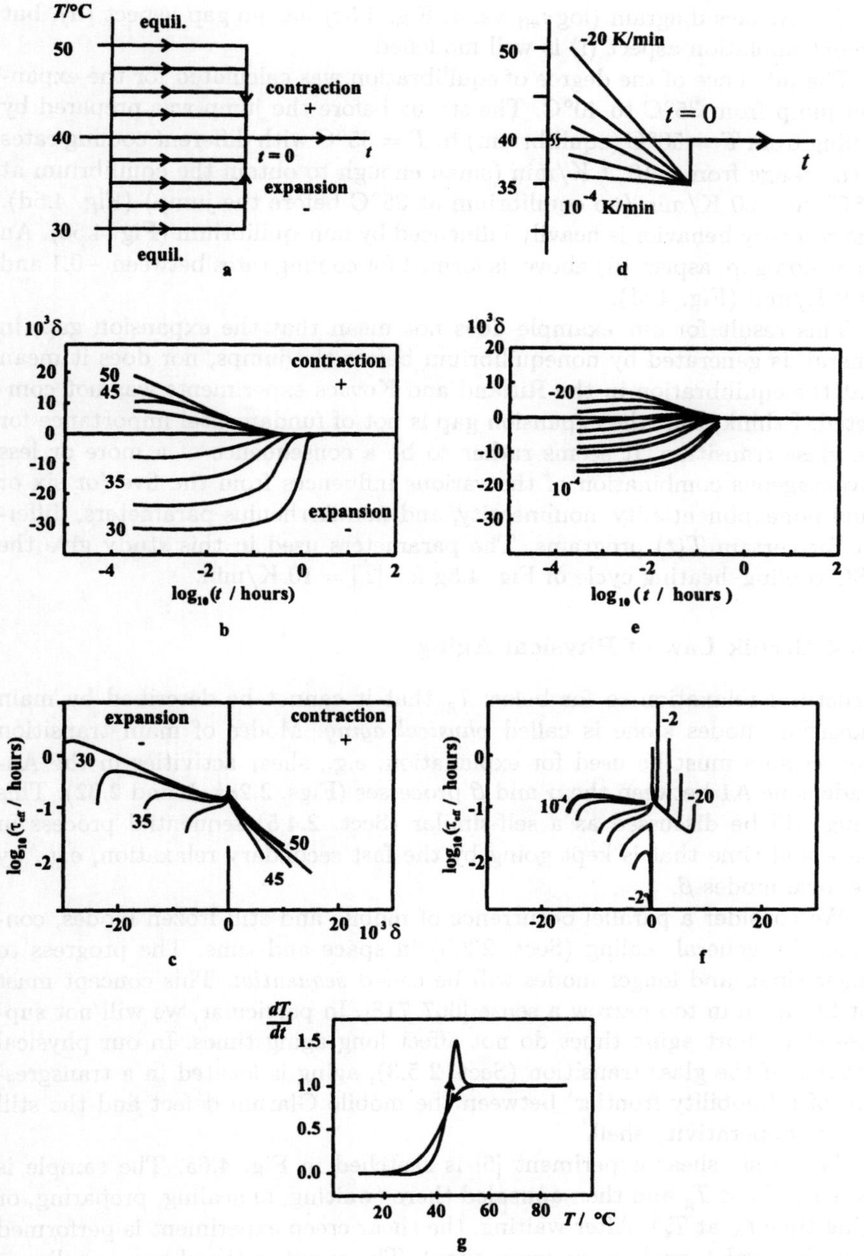
T/°C
equil.
50
40
30
contraction
+
t = 0
t
expansion
-
equil.
a
-20 K/min
50
40
35
t = 0
t
10⁻⁴ K/min
d
10³δ
20
10
0
-10
-20
-30
50
45
35
30
contraction
expansion
-4
-2
0
2
log₁₀(t / hours)
b
-20
10⁻⁴
log₁₀(t / hours)
e
expansion
contraction
log₁₀(τ_eff / hours)
30
35
50
45
-20
0
20
10³δ
c
-2
-20
10⁻⁴
f
dT_f/dt
1.5
1.0
0.5
0.0
20
40
60
80
T / °C
g

The Kovacs diagram ($\log \tau_{\text{eff}}$ vs. δ, Fig. 4.5c) has no gap aspect (ii), but the extrapolation aspect (i) is well modelled.

The influence of the degree of equilibration was calculated for the expansion jump from 35°C to 40°C. The status before the jump was prepared by cooling from $T = 50$°C (equilibrium) to $T = 35$°C with different cooling rates in the range from -10^{-4} K/min (small enough to obtain the equilibrium at 35°C) to -20 K/min (no equilibrium at 35°C before the jump) (Fig. 4.5d). The recovery behavior is heavily influenced by nonequilibrium (Fig. 4.5e). An expansion gap, aspect (ii) above, is formed for cooling rates between -0.1 and -1.0 K/min (Fig. 4.5f).

This result for our example does not mean that the expansion gap, in general, is generated by nonequilibrium before the jumps, nor does it mean that the equilibration in the Ritland and Kovacs experiments was not completed. I think, that the expansion gap is not of fundamental importance for the glass transition. It seems rather to be a consequence of a more or less advantageous combination of the various influences from the five (or six or four) nonexponentiality, nonlinearity, and non-Arrhenius parameters, different for certain $T(t)$ programs. The parameters used in this study give the DSC cooling–heating cycle of Fig. 4.5g for $|\dot{T}| = 10$ K/min.

4.5.4 Struik Law of Physical Aging

Structural relaxation so far below T_g that it cannot be described by main transition modes alone is called *physical aging*. Modes of main transition predecessors must be used for explanation, e.g., shear activities in the Andrade zone A1 between the α and β processes (Figs. 2.28a, b and 2.32). This aging will be discussed as a self-similar (Sect. 2.4.5) sequential process in space and time that is kept going by the fast secondary relaxation, e.g., by the local modes β.

We consider a parallel occurrence of mobile and still frozen modes, connected by general scaling (Sect. 2.2.5) in space and time. The progress to longer times and longer modes will be called *sequential*. This concept must not be taken in too narrow a sense [667, 718]. In particular, we will not suppose that short aging times do not affect long aging times. In our physical pictures of the glass transition (Sect. 2.5.3), aging is located in a transgression of a 'mobility frontier' between the mobile Glarum defect and the still frozen cooperativity shell.

The basic shear experiment [6] is sketched in Fig. 4.6a. The sample is cooled to $T_e < T_g$ and then annealed there (waiting, annealing, preparing, or aging time t_w at T_e). After waiting, the shear creep experiment is performed (time interval t, probing or creep time). The result is the shear compliance $J(t, t_w, T_e)$, usually measured for $t \ll t_w$. Experimental curves are similar to an Andrade law (power law with exponent $\approx 1/3$) and can be mastered by a shift factor a_{T_e}.

It was found (*Struik law*) that

$$\mu \equiv -\frac{\mathrm{d} \ln a_{T_e}}{\mathrm{d} \ln t_w} \approx 1\,, \quad T_e > T_\beta\,, \tag{4.156}$$

and that the Struik exponent μ decreases for temperatures below T_β where the aging effect is small. The $\mu = 1$ case is sometimes referred to as normal, and the $\mu < 1$ case [668] as anomalous aging. The $\mu > 1$ case is also discussed in the recent literature [669]. The Struik exponent depends on the activity. For example, it is different for shear γ, enthalpy H, and volume V recovery. In polycarbonate, for instance, $\mu_\gamma = 0.87$, $\mu_H = 0.49$, and $\mu_V = 0.56$ have been observed [670].

The crossover from short $t \ll t_w$ to long $t \gg t_w$ creep times is indicated in Fig. 4.6b [6,671]. At very long times t of order the equilibrium retardation time at $T = T_e$, the curves lead to the equilibrium compliance.

The Struik law (4.156) can be interpreted by using a scaling concept for the self-similarity of the Andrade $t^{1/3}$ (or t^a, a = 0.2–0.4) law with a corresponding dispersion law for a spatial picture. Two preliminary notes need mentioning before we continue.

(i) Consider the Arrhenius diagram of Fig. 4.6c. Let I and II be Andrade modes with length scales $\lambda_{\mathrm{I}} < \lambda_{\mathrm{II}}$. Choose T_e so that, for a given cooling rate, II is frozen in and I is at the borderline between frozen-in and working for reaching T_e at $t_w = 0$. Physical aging ($1 \to 2$) is the successive mobilization of the Andrade spectrum from I to II. Let II be reached at the aging time t_w, II = II(t_w). Then all modes in the β through I to II range are working, even if modified by the larger frozen-in modes between II and the glass transition (λ_{II} to ξ_α). The creep experiment (with creep time interval t starting from $t = 0$ at t_w) successively tests the mobilized modes with regard to their shear retardation activity, from β to II, for example.

(ii) Figure 4.6a tells us how the structural relaxation (II shifts with increasing waiting times t_w towards the glass transition, $\lambda_{\mathrm{II}} \to \xi_\alpha$) modifies the mobilized modes (β to II): the growing density (decreasing free volume) diminishes the shear compliance.

Equation (4.156), including the statement that it is valid for the whole time scale t (but $t \ll t_w$), expresses the fact that both the waiting time t_w [shift of borderline II = II(t_w)] and the testing or probing time t proceed at similar rates. Both are real times, t_w for the shift factor a_{T_e}, and t for the successive testing of all mobilized Andrade modes. This follows from the ω identity (t identity here) of the FDT. Due to self-similarity, both processes do approximately 'the same'. It is the length scale (smaller for t than for t_w) that is different. Struik exponents $\mu \neq 1$ indicate the effect of 'getting' mobility (t_w) on 'having' mobility (t) or, in simpler words, the effect of short aging times on longer ones. Deviations from the Struik law can arise from the vicinity of a typical or characteristic length, e.g., ξ_α for the glass transition α process.

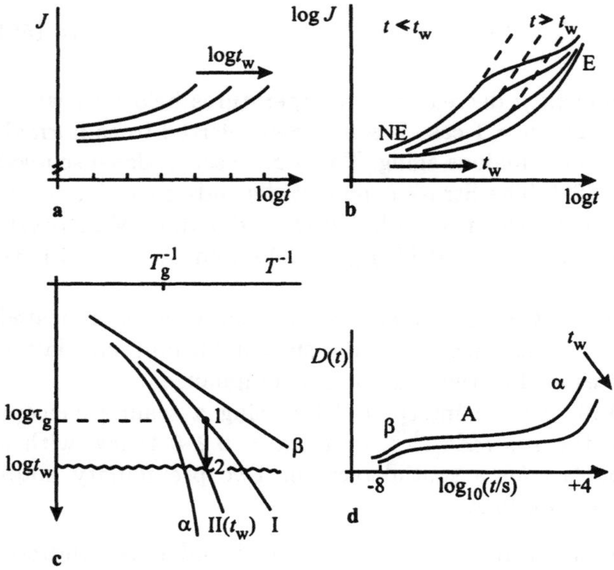

Fig. 4.6. Physical aging. (**a**) Shear compliance J at $T_e < T_g$ as a function of probing time interval t after waiting time t_w as a parameter for $t \ll t_w$. (**b**) The same experiment continued for $t \gg t_w$, up to equilibrium E (schematic). NE nonequilibrium. (**c**) The same experiment mapped into an Arrhenius diagram (*ordinate* log time). The annealing time t_w acts near 2 and shifts the Andrade mode $\mathrm{II} = \mathrm{II}(t_w)$ to the left. The probing time t acts near 1 and goes from 'below β' to 2. The case $t \gg t_w$ corresponds, in a way, to $t = t_w$ because then the experiment actually observes the shear response during the shift of borderline $\mathrm{II}(t_w)$. (**d**) Schematic illustration of a creep experiment in PVC near T_g (minute scale) from very short times (near β) to very long times. $D(t)$ is the creep function, $D(t) = (1/3)J(t) + (1/9)B(t)$. β is a local mode, A are Andrade modes, and α is the main transition

Let us repeat this complicated matter in other words. We refer to the general statement that dynamic t^a power laws are connected with the lack of an additional typical length. The frozen-in structure beyond the borderline $\mathrm{II}(t_w)$ defines an 'operating' length $\lambda_{\mathrm{II}}(t_w) < \xi_\alpha$ separating the frozen-in modes from the mobile modes (e.g., near the Glarum–Levy defects). According to the scaling conception of Andrade modes, there is no additional typical length between 0.5 nm and λ_{II}. Consequently, the decreasing free volume for larger waiting times t_w does not alter the power law character, and the process remains self-similar for $t \ll t_w$, as shown in Fig. 4.6a. It is only for $t \gtrsim t_w$ that the length $\lambda_{\mathrm{II}}(t_w)$ is additionally involved in the response. We thus observe a deviation from the power law (Fig. 4.6b). The driving force to equilibrium acts directly for $t \gtrsim t_w$, beyond the power law.

When the aging experiment, not too far below T_g, is continued to very large times, e.g., over 12 logarithmic decades, then the time scale comes up with the α process and the shear compliance increases once more (Fig. 4.6d) [672].

Physical aging damps not only the shorter Andrade modes but also the β relaxation and even the low-temperature processes [673]. The accompanying densification lowers the relaxation strength of the β relaxation and the heat capacity at $T < 1$ K. The possibilities for short modes are restricted because they are influenced by the loss of free volume in the environment.

The reference [669] also contains a review of aging in other substance classes, together with more theoretical concepts. An example of aging in computer simulations can be found in [674].

5. Epilogue

This chapter contains two sections. Section 5.1 tests our approach against general questions raised among the glass transition community over the past few years [46, 52–54]. These questions aim to focus the exploding number of experimental investigations on points that seem important, and to warn theorists, and people that like to make convincing models, about the complexity of the phenomena under investigation. I have chosen the 'Angell catechism', which consists of ten questions. It is a catechism in the sense that it is a set of formal questions put as a test. Angell wrote [53]: "We formulate ten *questions* which we believe must be properly answered by any successful theory of structural glassformers." Section 5.2 is a reminder of the large volume of practical knowledge about the glass transition which we have available for application, e.g., in materials science or technology. I chose the glass temperature T_g as an example to sketch trends that influence its numerical value. Today, of course, there is no problem in measuring the T_g of an interesting substance. For engineer's screening, however, it also seems important to know some general tendencies amongst the parameters that may shift T_g to higher or lower temperatures. In fact, this turns out to be much easier than estimating whether crystallization is likely to disturb the engineer's objective at the new T_g.

5.1 Ten Questions

Our approach, with its characteristic length ξ_α of cooperativity N_α for dynamic heterogeneity $\delta \ln \omega$, temperature fluctuation δT of the smallest representative subsystems functional to the α or a process (cooperatively rearranging regions CRRs), Glarum–Levy defects as a reason why we find one island of mobility in each CRR, and preference for dynamic arguments, will be tested against the ten questions. Since these questions are formulated from a different point of view – they are partly 'orthogonal' to our approach – they comprise a sure test for the consistency and robustness of our concepts.

Question One. Why is there a pending entropy crisis for structural glass formers and not for other glass types?

Answer One. The step in heat capacity Δc_p for these glassformers is often so large that the configurational entropy $S_c(T)$ reaches a large enough slope for the entropy crisis (Kauzmann paradox), with the interesting detail that the thermodynamic fragility $m_g(S_c)$ of (2.29) has a similar ranking to the dynamic fragility $m_g(\eta)$ of (2.3). Even for structural supercooled glass formers, however, the crisis is different, as can be seen from an approximate formula resulting from Fig. 2.5d, viz.,

$$\ln(T_m/T_0) \approx \frac{\Delta S_m}{\Delta c_p}\,, \tag{5.1}$$

where T_m is the melting temperature, T_0 the Vogel temperature, and ΔS_m the melting entropy [7]. There is no crisis for small Δc_p, and Δc_p does not correlate with fragility. For large Δc_p values and $c_p^{\text{glass}} \approx c_p^{\text{cryst}}$, the crisis is a thermodynamic phenomenon, whereas in our approach, the glass transition is believed to be a dynamic phenomenon (Sect. 2.5). The actual reason for the dynamic glass transition is lack of free volume. There is no crisis for the warm liquid (Table 2.8). The metastable entropy crisis in supercooled cold liquids is not a general presupposition for the glass transition because the latter can also be observed in stable phases (Table 2.5). A dynamic alternative to the entropy crisis is the exhaustion of the α process for too small a fluctuating free volume (Sect. 2.5.1).

Question Two. Why do structural glasses exhibit such a range of fragilities?

Answer Two. The fragility m_g of the α process depends on several parameters that vary with the substance but are not strongly correlated among themselves. This may be seen from the $\tilde{m} = 2.3m = T\delta\ln\omega/\delta T$ formula (4.81) combined with the formula defining the cooperativity N_α (2.77) and the Levy exponent from dynamic heterogeneity, $\alpha \approx 1/\delta\ln\omega$ (Table 2.3):

$$m(T) \approx \frac{1}{2.3}\,\frac{1}{\alpha(T)}\,\frac{\Delta c_p(T)}{\bar{c}_p(T)}\,\frac{R}{M_0\bar{c}_p(T)}\,\frac{1}{N_\alpha(T)}\,, \tag{5.2}$$

where $m_g = m(T_g)$. The range of fragility reflects the range of parameters in this equation, and the accidental values of the parameters (not strongly correlated with m_g) reflect individual molecular properties of the substances (Sect. 2.2.5). The slowing down of relaxation times is regulated [Fig. 2.23a (6)] in the warm liquid (above the crossover) with constant CRR size ($N_\alpha \approx 1$, Sect. 4.1), and in the cold liquid (below the crossover) with increasing CRR size ($N_\alpha \gg 1$, Sect. 4.3).

Question Three. Why do fragile glass formers often (but not always) show the α–β bifurcation at $T_{\alpha-\beta}$, and why does $T_{\alpha-\beta}$ correspond to T_c from the mode-coupling theory?

Answer Three. The crossover region, on the one hand, originates from the exhaustion of MCT possibilities in the warm liquid, i.e., from the arrest predicted by idealized MCT at the crossover temperature T_c (Sect. 4.1). On the other hand, it is overcome in the cold liquid due to the Glarum–Levy defects of cooperativity that concentrate free volume, resulting in possibilities for the local β process (Sect. 2.5.3). The non-existence of the β process seems to be a problem of missing possibilities or activities. The β process is not the origin for the crossover; there are crossover regions without detected β processes.

Question Four. Why does the mean-squared particle displacement MSD, measured on ps time scales, show a break at or near T_g, where the α relaxation time is 200 s?

Answer Four. The MSD has a steep increase at higher temperatures because the 'additional' cage-rattling amplitudes in the live α-process cages (susceptible to relaxation response) of the Glarum–Levy defects, and in the live a cages of the Williams–Götze process, increase when there is an ample supply of free volume. Conversely, the vault effect (Sect. 2.2.6) allows the cage-rattling relaxation to survive freezing several tens of kelvins below the glass temperature T_g. The steepness of the increase follows from the steepness of repulsion potentials making the mobility very sensitive to free volume. In dead cages (with only a vibrational response) below the break, the MSD is a constant response proportional to temperature (Sect. 2.3.1). Freezing-in time and cage-rattling time are not directly connected but the MSD break and freezing-in (at a glass time depending on the cooling rate) have a common cause: shortage of free volume.

Question Five. What is the origin of the boson peak, and the relation between the boson peak, the motions responsible for the MSD behavior discussed in Question Four, and the two-level systems responsible for the low-temperature specific heat anomalies?

Answer Five. The boson peak seems to be a general phenomenon of disordered molecular systems and not necessarily connected with a glass transition in the preparation. Many models with arbitrary disorder, not organized by a dynamic glass transition, also show a boson peak. A comprehensive review of the many models is still lacking. A direct relation between the boson peak length scale and the characteristic length of the glass transition does not exist since materials where the Williams–Götze process with small cooperativity is frozen also show an ordinary boson peak. The boson peak parameters are not correlated to the systematic part of tunneling-system density, since the vault-breakdown effect on the tunneling systems is not reflected in boson peak parameters (Sects. 2.2.6, 2.4.3) [590].

Question Six. Why is the relaxation function nonexponential at temperatures where the relaxation time is non-Arrhenius?

Answer Six. The relation between nonexponentiality ($1/\alpha > 1$) and non-Arrhenius behavior (fragility m_g) is indirect, and individual accidentalness is important. This follows from (5.2), where we see that $m_g \propto 1/\alpha$ is modified by individual proportionality constants. For both the α process and the a process, nonexponentiality is connected with the Levy exponent α, $\delta \ln \omega \approx 1/\alpha$. The cooperativity N_a for the a process is small and nearly constant, whilst for the α process, it increases steeply towards low temperatures. If non-Arrhenius behavior (fragility m_g) is connected with a WLF-type equation (Sect. 4.3), then nonexponentiality is expressed by the Vogel temperature T_0, $m_g = (1/2.3)[T_g/(T_g - T_0)] \ln(\Omega/\omega_g)$ (4.76). If T_0 is connected with landscape roughness (Fig. 2.24f), then the cooperative α process landscape has a smaller roughness than that of the a process (Figs. 2.1c and 2.6b).

Question Seven. What is the connection between microheterogeneous dynamics, seen in computer simulation studies below the onset temperature of two-step relaxation (a and c processes are meant here), the microheterogeneous dynamics seen in experimental studies near T_g, and the nonexponentiality of relaxation.

Answer Seven. Up to the time of writing (2000), molecular dynamics computer simulations have been restricted to the warm liquid (above the crossover) and are evaluated about spatial heterogeneities of molecular displacements. This defines a 'kinematic length' (Sect. 2.3.4). The calorimetrically determined 'characteristic length' is believed to come from a pattern of entropy fluctuations disengaged from molecular structure (Sect. 2.2.5). As a kinematic length also counts regions with, for example, parallel molecular displacements or long strings, both having zero or small entropy variation, the entropic pattern is more finely granulated. The characteristic length is the smaller one. All experimental studies near T_g, in the cold liquid (below the crossover), see lengths of order 1 to 5 nm, i.e., of order the characteristic length. The lengths obtained recently [292] from spin diffusion in multidimensional NMR experiments show the same temperature dependence as the characteristic length. The NMR lengths are smaller because they count only the immobile fraction of the pattern, whereas the characteristic length counts the whole. The connection between characteristic length ($\xi_\alpha^3 \propto N_\alpha$) and nonexponentiality ($1/\alpha$) is given by the Levy scaling relation for the fluctuating free volume, $v_f \propto 1/N_\alpha^{1/\alpha}$, where α in the exponent is the Levy exponent (Sect. 4.3).

Question Eight. Why does the Stokes–Einstein relation between viscosity and diffusivity in single component systems break down near and below the crossover temperature $T_X(T_c, T_B)$?

Answer Eight. Above the crossover region, the representative subsystem is small (N_a of order 1), and the difference between viscosity and diffusivity is only related to temporal aspects of transport-activity differences, e.g., between $\langle\tau\rangle$ and $\langle 1/\tau\rangle$ averages. Below the crossover, the size of representative subsystems (N_α) increases and spatial aspects of the Green–Kubo integral can amplify the differences (Sect. 2.2.3). Moreover, an increasing nonexponentiality will also amplify the temporal aspects, because larger $\delta \ln \omega$ permit larger differences between the averages.

Question Nine. Why are the kinetics of annealing (aging, equilibration) so nonlinear (structure dependent) for fragile liquids?

Answer Nine. If the temperature amplitude of the perturbation ΔT_0 is reduced by the temperature fluctuation δT of the smallest representative subsystems, the CRRs, then the nonlinearities in equilibrium ($T > T_g$) and nonequilibrium ($T < T_g$) are controlled by the same ratio $\Delta T_0/\delta T \gtrsim 1$ (Sect. 2.1.5). However, the nonlinearity below T_g is striking because

- the first-order terms are not quenched by symmetry as in the equilibrium response,
- the recovery tails are especially sensitive to nonlinearities (2.48).

The nonlinearity increases with fragility m because the reduction parameter δT is small for large m, $\delta T \approx 0.65\ T_g/m$ (2.39). This relation is a consequence of time–temperature equivalence applied to δT and $\delta \ln \omega$ ($mT \propto \mathrm{d}T/\mathrm{d}\ln\omega\,|_{\text{along}} = \delta T/\delta \ln \omega$, Sect. 2.4.5) together with the small variation in the mobility dispersion, $\delta \ln \omega \approx 1/\alpha$, when compared to the large variation in the fragility m. The details are complicated (Sect. 2.1.5).

Question Ten. Why does the excitation of the structural degrees of freedom in some overconstrained systems (e.g., Si, Ge) become first order in character, like a weak melting transition?

Answer Ten. I try to understand the glass transition by a dynamic approach (Sect. 2.5), so that thermodynamic arguments are not essential for the existence of the glass transition. Then it is a phase transition in exceptional glasses that generates two phases which can have different glass transitions, with different glass temperatures, different fragilities, different Δc_p values, and so on. The phase transition is not originated by the dynamic glass transition in one or both phases, but by their chemical potentials. It is only the kinetics of phase transition that can be influenced by the dynamic glass transition in one or both phases, i.e., by the liquid dynamics.

5.2 Control of T_g

As mentioned above, the measurement of glass temperature T_g for an interesting substance generally presents no problem. Moreover, in the current literature a large number of T_g values, T_g correlations, and T_g estimation methods is available for all substance classes that can freeze in, including food materials, for instance. A small selection is listed in our Tables 2.5, 2.9, 2.10, and 2.11. For instance, we refer to extensive lists for polymers in [675, 719], and for silicate and related glasses in [20, 676], whilst for small-molecule liquids and liquid mixtures, we refer to the tables in [677]. The trends for T_g control in polymers are summarized in the classical paper von Shen and Eisenberg [30].

From a physicochemical point of view, we have a simple but effective rule of thumb, mentioned in the Introduction, namely that T_g will increase for larger cohesion energy and larger molecular order. This follows from

$$T_g = \frac{\Delta H}{\Delta S}, \tag{5.3}$$

where $\Delta X = \Delta H$ (enthalpy) and $\Delta X = \Delta S$ (entropy) is explained in the isobaric Fig. 5.1a. Increasing order diminishes the measure of molecular disorder ΔS. The same tendencies follow from our landscape-roughness picture (Fig.2.24f) for the Vogel temperature T_0, if the roughness is scaled by the cohesion energy. Higher order means that the system point in the energy landscape visits deeper valleys; this increases the roughness. In the following, we confine ourselves to a few examples.

Consider first the cohesion energy. The glass temperature T_g is scaled by the melting temperature T_m, or the liquidus temperature T_l for mixtures. A well known rule of thumb is

$$T_g/T_m \approx 2/3 \tag{5.4}$$

for organic liquids and molten oxides (Fig. 5.1b) [678]. As a function of mixture composition, the glass temperature T_g is less sensitive to the details than the liquidus temperature T_l resulting from equality of chemical potentials for liquid and solid (Fig. 5.1c) [679]. A further example of an energetic effect is

▶

Fig. 5.1. Control of glass temperature T_g. (**a**) Definition of $\Delta X = \{\Delta H, \Delta S\}$ in (5.3). (**b**) Correlation of glass temperature T_g with melting temperature T_m (*upper part*) and liquidus temperature T_l for mixtures (*lower part*) [678]. The lines are the 2/3 rules, e.g., corresponding to (5.4) for T_m. (**c**) A comparison for glass temperature T_g and liquidus temperature T_l for two inorganic glass-forming mixtures [679]. (**d**) Polyelectrolytes. Increasing T_g caused by the additional Coulomb energy q/a in phosphates P and silicates S [30]. q is the average cation charge in elementary units and d is the average distance in Å between O^{2-} and the cations as estimated from van der Waals radii. Phosphates XPO_3 with $X = H$, Na, Ca, K, Li inclusive mixtures; silicates are from the Na_2SiO_3+$CaSiO_3$ system. (**e**) Arrhenius diagram (or relaxation chart) for a hypothetical polyethylene PE sample having all relaxations at 'typical' positions. α_c the crystalline, $\beta(U)$ the upper, and $\beta(L)$ the lower dynamic glass transition. γ_I and γ_{II} are two more or less hindered (Sect. 2.4.2) glass transitions, γ_{III} the local mode. A common crossover region (except for α_c) is indicated in the 100 megahertz range. (**f**) A similar picture for polyethylene oxide PEO [680]. The flow transition is dropped in both relaxation charts

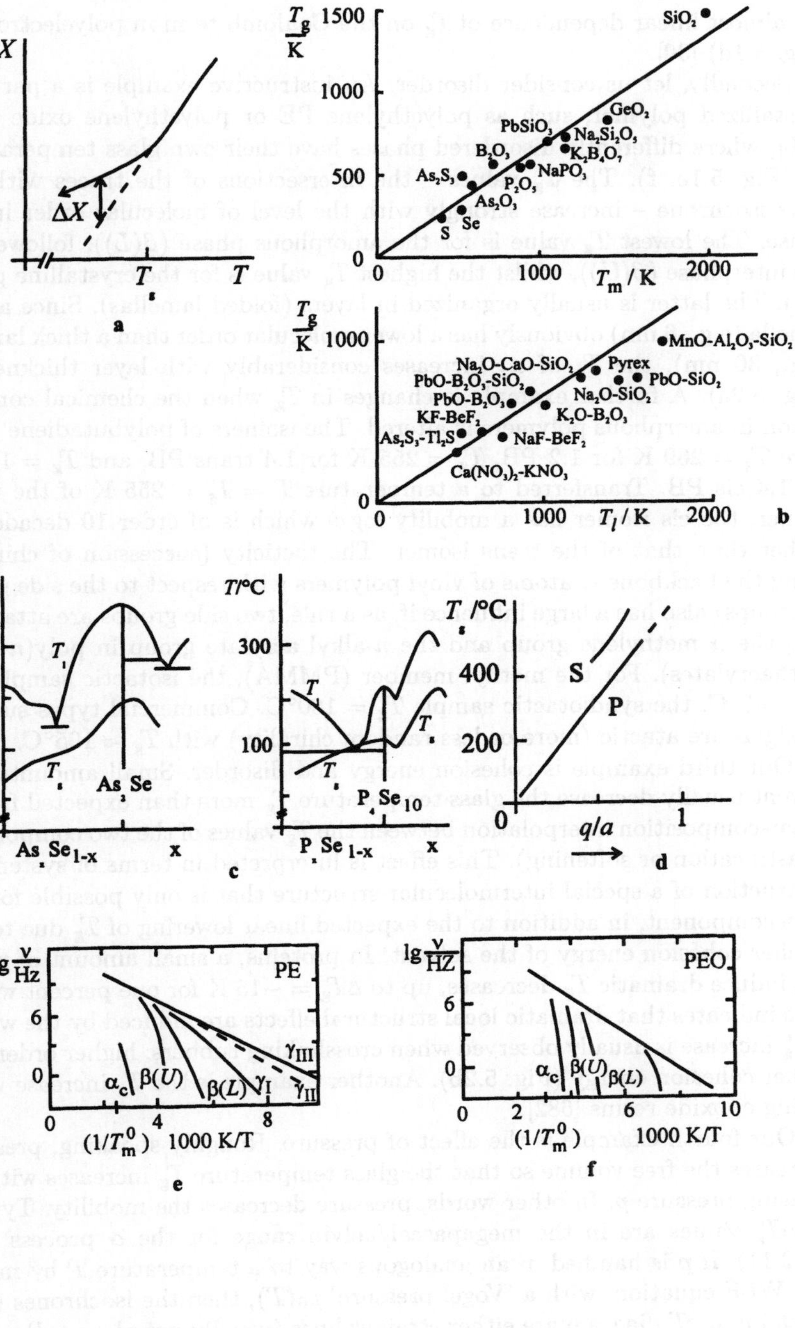
X
ΔX
T
a
Tg/K
1500
1000
500
0
SiO2
GeO2
PbSiO3
Na2Si2O5
K2B4O7
B2O3
NaPO3
P2O5
As2S3
As2O3
S
Se
1000
2000
Tm / K
MnO-Al2O3-SiO2
Na2O-CaO-SiO2
Pyrex
PbO-B2O3-SiO2
PbO-SiO2
PbO-B2O3
Na2O-SiO2
KF-BeF2
K2O-B2O3
As2S3-Tl2S
NaF-BeF2
Ca(NO3)2-KNO3
Tl / K
b
T/°C
300
100
As Se
AsxSe1-x
x
PxSe1-x
P4Se10
c
Ts /°C
400
200
S
P
0
1
q/a
d
lg ν/Hz
PE
6
0
αc
β(U)
β(L)
γI
γII
γIII
4
8
(1/T0m)
1000 K/T
e
lg ν/HZ
PEO
2
6
10
f

the almost linear dependence of T_g on the Coulomb term in polyelectrolytes (Fig. 5.1d) [30].

Secondly, let us consider disorder. An instructive example is a partially crystallized polymer, such as polyethylene PE or polyethylene oxide PEO [680], where differently disordered phases have their own glass temperatures T_g (Fig. 5.1e, f). The T_g values – the intersections of the traces with the mHz isochrone – increase strongly with the level of molecular order in the phase. The lowest T_g value is for the amorphous phase ($\beta(L)$), followed by the interphase ($\beta(U)$), whilst the highest T_g value is for the crystalline phase (α_c). The latter is usually organized in layers (folded lamellas). Since a thin lamella (e.g., 6 nm) obviously has a lower molecular order than a thick lamella (e.g., 30 nm), the T_g of α_c increases considerably with layer thickness l_f (Fig. 5.2a). A further example is changes in T_g when the chemical configuration in amorphous polymers is altered. The isomers of polybutadiene (PB) have $T_g = 269$ K for 1.2 PB, $T_g = 255$ K for 1.4 trans PB, and $T_g = 165$ K for 1.4 cis PB. Transferred to a temperature $T = T_g = 255$ K of the trans isomer, the cis isomer has a mobility $\log \omega$ which is of order 10 decades (!) higher than that of the trans isomer. The tacticity (succession of chirality along the backbone C-atoms of vinyl polymers with respect to the side group or groups) also has a large influence if, as a rule, two side groups are attached, e.g., the α methylene group and the n-alkyl acrylate group in poly(n-alkyl methacrylates). For the methyl member (PMMA), the isotactic sample has $T_g = 43$°C, the syndiotactic sample $T_g = 160$°C. Commercial types such as plexiglas are atactic (more or less random chirality) with $T_g \approx 105$°C.

Our third example is cohesion energy and disorder. Small amounts of a solvent usually decrease the glass temperature T_g more than expected from a linear-composition interpolation between the T_g values of the two components (plastification or softening). This effect is interpreted in terms of systematic destruction of a special intermolecular structure that is only possible for the pure component, in addition to the expected linear lowering of T_g due to the smaller cohesion energy of the solvent. In proteins, a small amount of water can induce dramatic T_g decreases, up to $\Delta T_g = -15$ K for one percent water. This indicates that dramatic local structural effects are induced by the water. A T_g increase is usually observed when crosslinking rubbers: higher order and higher cohesion energy (Fig. 5.2b). Another example is the T_g increase when curing epoxide resins [682].

Our fourth example is the effect of pressure. Roughly speaking, pressure decreases the free volume so that the glass temperature T_g increases with increasing pressure p. In other words, pressure decreases the mobility. Typical $\mathrm{d}p/\mathrm{d}T_g$ values are in the megapascal/kelvin range for the α process (Table 2.11). If p is handled in an analogous way to a temperature T by means of a WLF equation, with a 'Vogel pressure' $p_0(T)$, then the isochrones ($\omega =$ const.) in a pT diagram are either straight lines (usually not observed) or hyperbolas with common asymptotes for different frequencies (Fig. 5.2c) [683].

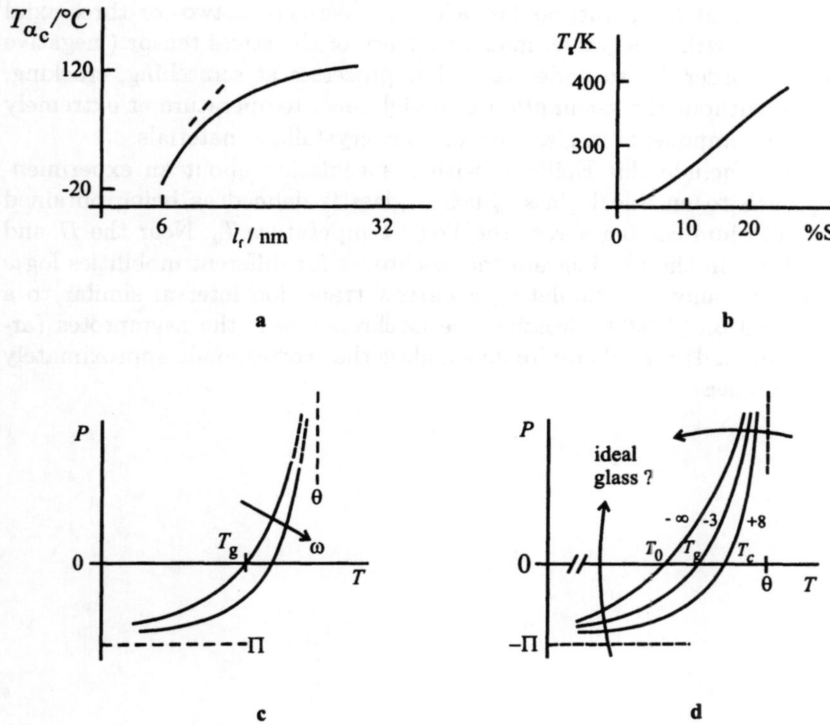

Fig. 5.2. Control of glass temperature T_g (continued). (**a**) Correlation of the α_c glass temperature in the crystalline layers with the lamella thickness l_f for PE [681]. The exceptions are LLDPE (linear low density polyethylene) types. (**b**) T_g increase in natural rubber due to vulcanization with sulfur (%S = bonded sulfur). (**c**) Dynamic glass transition isochrones (ω = const.) in a pressure–temperature PT diagram. $\Pi < 0$ the pressure asymptote, $\theta > T_g$ the temperature asymptote. (**d**) Crossing the dynamic glass transition near the temperature (θ) or pressure (Π) asymptotes, an ideal glass may be approached. The parameter is the mobility $\log \omega$ for the hyperbolic isochrones. $\log \omega = -\infty$ means the Vogel line $T_0(p)$ or $p_0(T)$

The *pressure asymptote* is denoted by Π, the temperature asymptote by θ. Since excessive negative pressures exclude any glass transition, due to surplus free volume, we actually find a curvature in the pT diagram, $\mathrm{d}^2p/\mathrm{d}T_g^2 < 0$, and no straight lines. The extrapolated Π values are usually in the 0.1 GPa range, and the extrapolated θ values are 100 K or more above T_g. Note that approaching Π and θ means more free volume, i.e., we expect a crossover from the α process to the a process by changing the WLF parameters. This also implies a change in the Π and θ asymptotes.

As mentioned in the Introduction (Fig. 1.4), the existence of a pressure asymptote Π implies that relatively small negative pressures (when compared to linear extrapolations of $\mathrm{d}T_g/\mathrm{d}p$ at normal pressure) are sufficient to induce flow processes at temperatures far below T_g. We need a two- or three-axial tension state with a negative, moderate trace of the stress tensor ('negative pressure' of order Π) to understand flow processes at scratching, cracking, or crazing without the assumption of a high local temperature or extremely large shear components in glass-forming noncrystalline materials.

Let us conclude this Epilogue with a speculation about an experimental approach to an 'ideal glass'. Such a glass is defined as being obtained from equilibrium cooling across the Vogel temperature T_0. Near the Π and θ asymptotes in the pT diagram the isochrones for different mobilities $\log \omega$ approach one another, simulating a narrow transition interval similar to a phase transition [7, 683]. Crossing the isochrones near the asymptotes (arrows in Fig. 5.2d) may therefore give a glass that corresponds approximately to an ideal glass.

6. Conclusion

In summary, the glass transition is the transformation of a disordered state with molecular mobility to an immobilized state of the same or similar structure by means of decreasing temperature, increasing pressure (densification), or extraction of a solvent. The standard example is freezing of a liquid to a glass. The transformation is caused by a continuous increase in the liquid relaxation time up to an experimental time, usually of order 100 seconds. For given pressure and composition, this glass transition is characterized by a glass temperature T_g, a transformation interval, and continuous steps in different susceptibilities. An amorphous solid is a wider concept than a glass, the former includes disordered materials without a glass transition in the preparation history, such as those produced by certain vapor deposition techniques on cold surfaces. Apart from liquids, glass transitions are also observed in other disordered materials such as liquid and plastic crystals, or for charge density waves. Undercooling the liquid is not a prerequisite for a glass transition, because the glass transition can also be observed in stable phases.

To describe the increasing relaxation time in the stable or metastable equilibrium state, the concept of dynamic glass transition is used. The immobilization (freezing-in at T_g) described above is then called a thermal glass transition. The dynamic glass transition is characterized by several traces in an Arrhenius diagram (Fig. 6.1). Such a 'relaxation chart' is typical for all glass formers of moderate complexity. It contains more traces but is still typical for more complex substances such as polymers or liquid crystals. If phase transitions in the relevant temperature range can be excluded, no alternative relaxation charts for molecular dynamics in disordered materials have so far been established. This also means that the relaxation dynamics of liquids is synonymous with the dynamic glass transition. The surprise and the mystery of the glass transition is that the multifarious molecular structures and intermolecular configurations of different glass formers, such as inorganic glasses, organic liquids, polymers, salt mixtures and metallic glasses, have such a general and common arrangement of traces in the Arrhenius diagram.

The change in mobility $\log \omega$ along the traces, and the mobility differences between the traces, may reach many orders of magnitude. There is one remarkable mobility and temperature region in the Arrhenius diagram, the crossover region C in Fig. 6.1, with a crossover frequency and a crossover temperature characteristic of each glass former. The crossover region separates two distinct processes of the main transition: a high-temperature a process in the warm liquid, and a cooperative α process in the cold liquid.

(i) The a process can be approximately described by the mode-coupling theory due to Götze (1984) with a spatial scale of order one nanometer or smaller for a cage formed by dynamically equivalent molecules in the neighborhood. The diffusion is organized by cooperative movements of the cage molecules to allow escaping the cage. The characteristic length is believed to be of the same order as the size of the cage door (about 0.5 nm).
(ii) The α process is believed to be organized by a cooperative motion of a larger number of molecules. The corresponding cooperatively rearranging regions according to Adam and Gibbs (1965) are believed to increase for falling temperatures. The characteristic length reached at T_g is estimated to be in the range between 1 and 4 nm.

The main problems discussed today are as follows.

- The disengagement of the general dynamics from the multifarious structures, a problem which obviously requires a high degree of abstraction.
- The curvature (non-Arrhenius process) of the main-transition traces in the Arrhenius diagram.
- The nonexponentiality of the relaxation – significantly stretched in comparison to an e^{-t} exponential (Debye) decay in time t – can often be approximated by a Kohlrausch function $\exp(-t^\beta)$, $0 < \beta < 1$. It seems important to establish a relation with the Levy distribution of probability theory.
- The interrelations between the various traces in the Arrhenius diagram of Fig. 6.1.
- Dynamic heterogeneity and the characteristic length for the α process.
- The reason for, and details of the crossover region.
- The question as to whether the glass transition can be understood by a thermodynamic approach on the basis of conventional a Gibbs distribution, or whether a nonconventional dynamic approach is needed, possibly along the lines of the von Laue treatment of thermodynamics.
- Construction of a spectral picture of partially frozen-in states near T_g in the frequency domain.

A comprehensive microscopic theory for the crossover region and the α process, a central aim of glass transition research, still remains to be invented. It is hardly possible to assess the value of the many physicochemical and spin-glass assisted models suggested for this aim. Molecular dynamics methods in computer simulations cannot yet reach such long times, several decades below the crossover region, which are needed to study the molecular cooperativity of the α process. Promising experimental methods with direct nanometer sensitivity, that may reach such long times (e.g., microseconds)

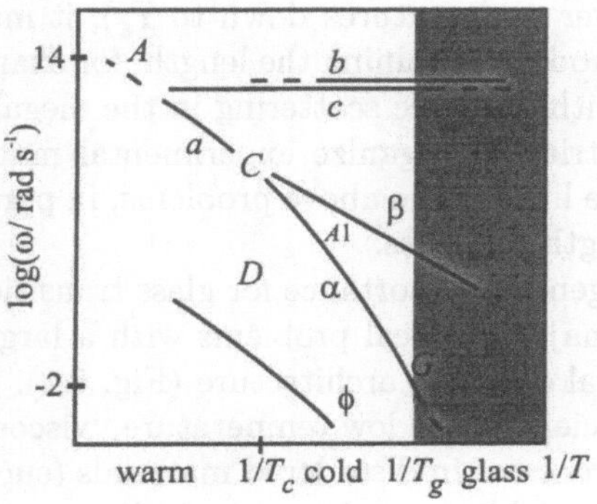

A	molecular transient
*A*1	α process precursor
a	high-temperature process, *a* process, Williams Götze process
b	boson peak
c	cage rattling
C	crossover region
D	diffusion modes
G	thermal glass transition
T_c	crossover temperature
gray	glass state
white	liquid state
	$T > T_c$ warm liquid
	$T < T_c$ cold liquid
ϕ	ultra slow modes, Fischer modes
α	cooperative process, α process
{*a*,α}	main transition
β	β process, Johari Goldstein process, local mode

Fig. 6.1. The glass transition is a problem with a large intrinsic complexity but a clear, general dynamic architecture

over the next five years, are dynamic neutron and X-ray scattering. In order to obtain sure information about characteristic lengths at much longer times (and correspondingly lower temperatures down to T_g), it may be crucial to test the calorimetric method for obtaining the length, for example, by overlap of periodic calorimetry with dynamic scattering in the megahertz region.

In this book, I have tried to organize experimental material and judge existing theories along the lines of the above problems, in particular, through the consequent use of length concepts.

I see three aspects of general importance for glass transition research. For physics, it is one of the major physical problems with a large intrinsic complexity but a clear, general dynamic architecture (Fig. 6.1). For science as a whole, including the life sciences, the low temperature, 'viscosity-determined' branch of all nanoscale processes in disordered materials (such as nucleation, or the physically understandable part of protein folding, or the crack tip for material fatigue) is affected by general glass transition dynamics. Other areas of interest include numerous applications of dynamic and thermal glass transitions in materials science, food science, and technology. Physics, physical chemistry, biology, medicine, geology, sociology, the stock market, and gamblers all stand to gain by progress made in this field.

A. Synonyms

This list contains also laboratory slang and journal idiom synonyms. In the present book, some attempt has been made to differentiate between synonyms. For example, WLF and VFT are attributed different meanings in Sect. 2.1.2, and it makes me feel uncomfortable to speak about melting a glass or about a glass melt.

a process, Williams–Götze process, high temperature (a) process, $\alpha\beta$ process, escaping-the-cage process

c process, cage rattling, cage process, fast β process, β_{fast} process

Crossover region, T_c of mode coupling theory, $\alpha\beta$ merging, $\alpha\beta$ splitting, $\alpha\beta$ coalescence, $\alpha\beta$ bifurcation

Glass temperature T_g, glass transition temperature, glass point, glass transformation temperature, freezing(-in) temperature

Aging, tempering, annealing, recovery, equilibration, stabilization, relaxation

ϕ modes, Fischer modes, ultraslow modes, usm

α process, cooperative (α) process, main transition (narrow sense), dynamic glass transition (narrow sense), glass-to-rubber transition (polymers)

β process, Johari–Goldstein process, local (β) mode, slow β process, β_{slow} process

Transformation interval, dispersion of thermal glass transition

Thermal glass transition, liquid–glass transition, liquid–glass transformation, liquid-to-glass transformation, freezing(-in), vitrification, softening, thawing, melting

Flow transition (polymers), terminal transition, terminal zone

Dispersion zone, relaxation zone, transition zone, process

Time–temperature equivalence, temperature–time equivalence, time–temperature superposition, time–temperature shift, time–temperature scaling, WLF scaling, VFT scaling, master curve construction, mastering

Arrhenius diagram, Arrhenius plot, relaxation chart, activation plot, activation diagram

Response, susceptibility, relaxation (wide sense)

Levy distribution, stable distribution (of probability theory)

Exponential decay, Debye relaxation

VFT (Vogel–Fulcher–Tammann), WLF (Williams–Landel–Ferry)

B. List of New Concepts

Nonconventional Phenomenological Concepts Used for the Description of the Cooperative α Process

The succession of concepts is intended to indicate a possible deductive representation of the cooperative α process.

1. Freely Fluctuating Subsystems. The fundamental concepts of thermodynamics are system and environment (as compared with particles and fields in classical mechanics). Freely fluctuating means that no thermodynamic variable is excluded from the thermal fluctuation. We have fluctuations of temperature, entropy, pressure, volume, etc., that are, for subsystems inside the total system and far from the total system walls, not influenced by boundary conditions of the total system (e.g., isochoric or isothermal, etc.). The environment of a freely fluctuating subsystem is the set of equivalent subsystems in the neighborhood, themselves freely fluctuating. Free fluctuations are less restricted than those of Gibbs ensembles. See Sect. 3.1, and Figs. 3.1a, 3.2b, 3.6a.

2. Von Laue Treatment of Thermodynamics. This approach starts from the definition of the thermodynamic fundamental form $\mathrm{d}U = T\mathrm{d}S - p\mathrm{d}V \pm \ldots$ for each freely fluctuating subsystem. The relationship with mechanics is later established via minimal work. This work is also used to generate the fluctuation. The relationship with statistics is mediated by the total entropy change of subsystem plus environment. See Sects. 3.2.2–3.2.3, the Stradola equation (3.31), equation (3.32) for the probability of fluctuation, and Figs. 3.2b–d. A microscopic von Laue distribution has not yet been invented.

3. Representative Subsystems. Defined by a gedankenexperiment for partition in subsystems. A freely fluctuating subsystem is representative if it has the same linear response as the larger systems or the total system. See Sect. 2.2.5 (gedankenexperiment), and Sect. 3.1 (representativeness theorem introducing the Levy distribution for extensive spectral densities in the dynamics of liquids).

4. Functionality. Relation of fluctuations or dynamic response to the frequency range of the dispersion zone for a given process of liquid dynamics (e.g., α process). See Sects. 2.2.5 and 3.5, and Figs. 2.2c, 2.3a, 2.9d, 2.15d, 2.18b, 2.24b, 2.28a.

5. Different Activities. For given temperature and given dispersion zone (e.g., α process), the various compliances (entropy, volume, shear angle, dielectric function, coherent and incoherent scattering, etc.) in the time domain, and correspondingly in the frequency domain, have different retardation times, maximum frequencies of loss parts, and shapes. The same is valid for moduli (temperature, pressure, shear stress, etc.). The various activities cannot be reduced to one origin, e.g., density fluctuation. Different activities indicate molecular individualities, dynamic heterogeneity, or cooperativity. An example is given in Fig. 2.19d (difference between coherent and incoherent scattering). See Sect. 2.1.3.

6. Cooperatively Rearranging Region CRR. This is the smallest functionally representative freely fluctuating subsystem of the α (or a) process. It is the only freely fluctuating subsystem in liquids whose size is labelled by nature itself via the response shape and may therefore be called a 'natural subsystem'. This definition is the application of the von Laue approach to the original idea of Adams and Gibbs. Representativeness leads to explicit formulas for its size from entropy compliance or volume (bulk) compliance, i.e., from dynamic calorimetry or compressibility. See Sects. 2.2.5, 3.1, 3.3, equations (2.76)–(2.78), Fig. 2.15g (characteristic length), Fig. 2.23a part 6, and the temperature dependence equation (4.111).

7. Cooperativity N_α. This is the number of particles (e.g., molecules, monomeric units for polymers, etc.) in an average CRR at a given temperature. The characteristic length ξ_α is the size of a cube representing the average CRR.

8. Temperature Fluctuation δT of CRR. In general, temperature fluctuation is possible according to the von Laue approach. The conceptual problem is that the temperature fluctuation depends on the size of the subsystem. For a CRR as the smallest representative subsystem, we must look to see whether the temperature fluctuation δT can be deduced from the entropy compliance J_S^* [i.e., from the dynamic heat capacity $C_p^*(\omega, T)$ and the corresponding temperature modulus $G_T^* = 1/J_S^*$]. Intuitively, the temperature fluctuation is given by the width of an isochronous section of the $C_p''(\log \omega, T)$ peak in the α (or a) dispersion zone. See Example 1 of Sect. 3.3, and Fig. 3.5a for a proof that δT can be obtained from dynamic calorimetry. Furthermore, δT is the thermodynamic partner of nonexponentiality (mobility dispersion $\delta \ln \omega$) in local time–temperature equivalence, $\mathrm{d} \ln \omega / \mathrm{d}T \mid_{\text{along}} = \delta \ln \omega / \delta T$ (Figs. 2.15e–f, 2.28g, h, 4.2e, f, and Sect. 3.5).

9. Disengagement of General Dynamics from Multifarious Structure. This expresses a most striking fact of liquid dynamics, namely that process traces in the Arrhenius diagram are general and possess a clear architecture even though the molecular structure is so varied for different classes of glass former (silicate glasses, salt melts, small molecules, metallic glasses, Lennard–Jones mixtures, etc.). The concept of 'structure–property relations' seems of limited value for glass transition research. Disengagement is the main motivation for seeking a dynamic approach to the dynamic glass transition, and this seems possible in the von Laue approach. See Fig. 2.32 (Arrhenius diagram), the introduction to Sect. 2.5, and Sect. 2.5.2 (dynamic approach).

10. Mobility Field $\log \omega(\mathbf{r}, t)$. This is the field of local frequencies fluctuating in space and time. Frequencies are defined by success frequencies (events) after many attempts for the slow process considered (e.g., α process), and attempts are defined by the events of the faster processes (e.g., β or c process). See Sects. 2.2.5 and 3.5. For a proof of large enough event density to guarantee a quasi-continuous treatment, see Table 3.2. This allows also the definition of small CRRs with cooperativities of order one particle (Figs. 2.21a, 2.33, and 2.34a for a cage door event).

11. Fluctuation–Dissipation Theorem FDT as Experimental Equation. The FDT is considered as an equation for the quantum mechanical measuring process. This combines ideas due to Nyquist (1928) and Szilard (1929) with the von Laue approach to thermodynamics (1917). The quantum mechanical subject and the apparatus for the experiment are now identified with representative subsystems. See Sect. 3.4, and Figs. 3.6a–b. Application of the FDT to CRRs is necessary in order to get the characteristic lengths of α and a processes.

12. ω Identity of the FDT. The many locally or functionally distributed frequency variables are mapped to one frequency variable of a susceptibility measured by the FDT. This includes dynamic scattering on bulk samples. (A counterexample is provided by the laser speckles of Fischer modes). A further aspect is the identity of frequencies in dynamic response and spectral density of molecular fluctuation. Frequencies are mediated by quanta $\hbar\omega$ in Nyquist's transmission lines between the quantum mechanical subject and the measuring apparatus. See Sect. 3.4.5, and Figs. 3.6a, b, h.

13. Dynamic Heterogeneity. This is the pattern of the mobility field as related to a definite activity, e.g., to entropy fluctuations measured by the FDT as dynamic heat capacity. The length scale of this pattern for the α process is the characteristic length $\xi_\alpha(T)$ as determined from heat capacity spectroscopy. The same length would be obtained from density fluctuations as measured by the FDT for bulk compliance. For a given temperature, the parameters of the pattern are $\delta \ln \omega$ (mobility amplitude or mobility dispersion, equivalent to the isothermal width of the dispersion zone), characteristic length ξ_α, and retardation time τ_α. A kinematic pattern from computer simulations based on molecular displacements, e.g., in molecular dynamic strings or expressed by kinematic correlations in mobile clusters, can be different if not related to entropy fluctuations. This is an example of different activities. Since the FDT for susceptibilities has no spatial resolution, a theoretical input is at present needed to construct a pattern for dynamic heterogeneity. A combination of dynamic scattering and periodic calorimetry in the crossover would improve the situation (Fig. 2.19b). See Sects. 2.2.5, 2.3.4, and Fig. 2.23a (comparison with computer simulation pattern).

14. Minimal Coupling. This provides a phenomenological model for cooperativity suited to disengagement. Irrespective of molecular structure, local control of mobility by local free volume is assumed inside each CRR. The redistribution of free volume is balanced in each CRR. See Sect. 4.3, and Figs. 4.2a, b.

15. Fluctuating Free Volume. The minimal coupling model uses a partition of each CRR into partial systems. This is allowed by the high spatio-temporal event density of the mobility field. The fluctuating free volume v_f is the free volume parameter invariant under partition. See Sect. 4.3, and equations (4.52)–(4.54).

16. Glarum–Levy Defect. The minimal coupling model is a Levy situation with the spectral density of free volume fluctuation as Levy distribution (probability density with frequency measure $\mathrm{d}\omega$) and with a Levy exponent $\alpha \leq 1$ for relaxation. The preponderating Levy component of this distribution is identified with one Glarum defect inside each CRR. The defect is an island of high mobility where free volume is concentrated, whilst the remainder of the CRR is a cooperativity shell with low mobility. See Sects. 2.2.5, 4.3, Figs. 2.12d, 2.15b, 2.16e, 2.17a, 2.27e, 2.30b, 2.34a (for a pattern of dynamic heterogeneity with such defects), and equation (4.124) for the Glarum–Levy defect diffusion. The basic relation between free volume, cooperativity, and Levy exponent is $v_\mathrm{f} \propto N_\alpha^{-1/\alpha}$ [see (4.63)–(4.69)].

17. Vault Effect. Cooling a CRR below the glass temperature T_g, the cooperativity shell freezes first and manifests lower thermal contraction than the defect inside. If the cooperativity shell embraces at least the first coordination shell ($N_\alpha > 15$), the shell forms a vault that helps the defect mobility to survive several tens of kelvins below T_g. See Sect. 2.2.6, Fig. 2.16d for a model, and Figs. 2.16e, 2.26f, g (vault effect on tunneling systems at 1 K temperatures).

C. Acronyms

BIBE	Benzoin iso butyl ether
BMMPC	Bis methyl methoxy phenyl cyclohexane
CKN	$(KNO_3)_{0.6}$ $(Ca(NO_3)_2)_{0.4}$ salt mixture, KCN is also used
CN	CN^- ion (in KBrCN)
CPU	Central Processing Unit
CRR	Cooperatively Rearranging Region
DGEBA	Diglycidyl ether of bisphenyl-A
DSC	Dynamic Scanning Calorimetry
DZ	Dispersion zone
EPON 828	Commerical epoxy resin, Epon 828 is also used
EPR	Einstein–Podolski–Rosen (quantum effect)
EXAFS	Extended X-ray Absorption Fine Structure
FDT	Fluctuation–dissipation theorem
FWHM	Full width at half maximum
FZ	Flow zone
GZ	Glass zone
HCS	Heat Capacity Spectroscopy
HF	High frequency
HN	Havriliak–Negami
KBr	Potassium bromide
KBrCN	Mixed $(KBr)_{1-x}(KCN)_x$ crystal
KWW	Kohlrausch–Williams–Watts
LLDPE	Linear Low Density Polyethylene
MC	Monte Carlo
MCT	Mode Coupling Theory
MD	Molecular Dynamics
MSD	Mean squared particle displacement
NE	Nonequilibrium
NMR	Nuclear Magnetic Resonance
OTP	Ortho terphenyl
PB	Polybutadiene
PC	Polycarbonate
PCS	Photon Correlation Spectroscopy
PE	Polyethylene

PEMA	Poly(ethyl methacrylate)
PEO	Polyethylene oxide
PET	Polyethylene terephthalate
PIB	Polyisobutylene
PMMA	Polymethylmethacrylate
PMPS	Poly(methyl phenyl siloxane)
PS	Polystyrene
PT	Pressure–temperature
PVAC	Polyvinylacetate
PVC	Polyvinylchloride
QM	Quantum mechanics
SAXS	Small Angle X-ray Scattering
SED	Stokes–Einstein–Debye
SFM	Spin-facilitated kinetic Ising model
TCP	Tricresylphosphate
TIP	Thermodynamics of irreversible processes
TL	Transmission lines
TLS	Two-level states
TMDSC	Temperature Modulated Dynamic Scanning Calorimetry
TPCP-BO	6-(4-benzyloxyphenyl)-1,2,3,4 tetraphenylfulvene
TS	Tunneling systems
VFT	Vogel–Fulcher–Tammann
WKB	Wentzel–Kramers–Brillouin (quantum approximation)
WLF	Williams–Landel–Ferry

References

1. G. Tamman: *Der Glaszustand* (Voss, Leipzig 1933) (in German)
2. N.G. McCrum, B.E. Read, G. Williams: *Anelastic and Dielectric Effects in Polymeric Solids* (Dover, New York 1967)
3. J.D. Ferry: *Viscoelastic Properties of Polymers*, 3rd ed. (Wiley, New York 1980)
4. G. Harrison: *The Dynamic Properties of Supercooled Liquids* (Academic Press, London 1976)
5. S. Brawer: *Relaxation in Viscous Liquids and Glasses* (American Ceramic Society, Columbus 1985)
6. L.C.E. Struik: *Physical Ageing in Amorphous Polymers and Other Materials* (Elsevier, Amsterdam 1978)
7. E. Donth: *Glasübergang* (Akademie-Verlag, Berlin 1981) (in German)
8. V.G. Rostiashvili, V.I. Irshak, B.A. Rozenberg: *Steklovanie polimerov* (Khimiya, Leningrad 1987) (in Russian)
9. G.W. Scherer: *Relaxation in Glass and Composites* (Wiley, New York 1986)
10. P.G. Debenedetti: *Metastable Liquids: Concepts and Principles* (Princeton University Press, Princeton 1996)
11. G.O. Jones: *Glass* (Methuen, London 1956)
12. J. Wong, C.A. Angell: *Glass: Structure by Spectroscopy* (Dekker, New York 1976)
13. O.V. Mazurin: *Steklovanie i stabilizatsiya neorganicheskikh stekol* (Nauka, Leningrad 1978) (in Russian)
14. G.M. Bartenev, Yu.V. Zelenev: *Kurs fiziki polimerov* (Khimiya, Leningrad 1976) (in Russian)
15. D.S. Sanditov, G.M. Bartenev: *Fizicheskie svoistva neuporyadochennykh struktur* (Nauka, Novosibirsk 1982) (in Russian)
16. R. Zallen: *The Physics of Amorphous Solids* (Wiley, New York 1983)
17. H. Scholze: *Glass: Nature, Structure, and Properties* (Springer, New York 1991)
18. E. Donth: *Relaxation and Thermodynamics in Polymers. Glass Transition* (Akademie-Verlag, Berlin 1992)
19. A.K. Varshneya: *Fundamentals of Inorganic Glasses* (Academic Press, Boston 1994)
20. R.H. Doremus: *Glass Science*, 2nd ed. (Wiley, New York 1994)
21. I. Gutzow, J. Schmelzer: *The Vitreous State* (Springer, Berlin 1995)
22. J. Zarzycki: *Glasses and the vitreous state* (Cambridge University Press, Cambridge 1991)
23. F. Simon: Ergebn. exakt. Naturwiss. **9**, 222–274 (1930)
24. W. Kauzmann: Chem. Rev. **43**, 219–256 (1948)
25. R.O. Davies, G.O. Jones: Adv. Phys. **2**, 370–410 (1953)
26. D. Turnbull: Contempor. Phys. **10**, 473–488 (1969)

27. A.J. Kovacs: J. Fortschr. Hochpolym.-Forsch. **3**, 394–507 (1963)
28. R.F. Boyer: Rubber Chem. Technol. **36**, 1303–1421 (1963)
29. F.R. Schwarzl, L.C.E. Struik: Advan. Mol. Relax. Proc. **1**, 201–255 (1967–1968)
30. M.C. Shen, A. Eisenberg: Rubber Chem. Technol. **43**, 95–155 (1970); A. Eisenberg, M.C. Shen: Rubber Chem. Technol. **43**, 156–170 (1970)
31. G.B. McKenna: Glass Formation and Glassy Behavior. In: *Comprehensive Polymer Science, Vol.2, Polymer Properties*, ed. by C. Booth, C. Price (Pergamon, Oxford 1989) pp. 311–362
32. J. Jäckle: Rep. Prog. Phys. **49**, 171–232 (1986)
33. W. Götze, L. Sjögren: Rep. Prog. Phys. **55**, 241–376 (1992)
34. C.A. Angell: J. Phys. Chem. Solids **49**, 863–871 (1988)
35. C.A. Angell: Proc. Natl. Acad. Sci. **92**, 6675–6682 (1995)
36. C.A. Angell: Comput. Mat. Sci. **4**, 285–291 (1995)
37. K. Binder: Theory of glass transition in spin glasses, orientational glasses and structural glasses. In: *25 Years of Non-Equilibrium Statistical Mechanics: Proceedings of the XIII Sitges Conference* (Springer, Berlin 1995)
38. V.P. Privalko: J. Materials Educ. **20**, 57–74 (1999)
39. C.A. Angell: Science **267**, 1924–1935 (1995)
40. I.M. Hodge: Science **267**, 1945–1947 (1995)
41. F.H. Stillinger: Science **267**, 1935–1939 (1995)
42. A.L. Greer: Science **267**, 1947–1953 (1995)
43. B. Frick, D. Richter: Science **267**, 1939–1945 (1995)
44. K.L. Ngai, D.J. Plazek: Rubber Chem. Technol. **68**, 376–434 (1995)
45. H. Sillescu: J. Non-Cryst. Solids **243**, 81–108 (1999)
46. M.D. Ediger, C.A. Angell, S.R. Nagel: J. Phys. Chem. **100**, 13200–13212 (1996)
47. H.Z. Cummins, G. Li, Y.H. Hwang, G.Q. Shen, W.M. Du, J. Hernández, N.J. Tao: Z. Phys. B **103**, 501–519 (1997)
48. K. Funke: Prog. Solid State Chem. **22**, 111–195 (1993)
49. P. Lunkenheimer, U. Schneider, R. Brand, A. Loidl: Contemp. Phys. **41**, 15–36 (2000)
50. M.D. Ediger: Annu. Rev. Phys. Chem. **51**, 99–128 (2000)
51. J. Rault: J. Non-Cryst. Solids **271**, 177–217 (2000)
52. C.A. Angell, K.L. Ngai, G.B. McKenna, P.F. McMillan, S.W. Martin: J. Appl. Phys. **88**, 3113–3157 (2000)
53. C.A. Angell: J. Phys. Condens. Matter **12**, 6463–6475 (2000)
54. K.L. Ngai: J. Non-Cryst. Solids **275**, 7–51 (2000)
55. K.L. Ngai: J. Phys. Condens. Matter **12**, 6437–6451 (2000)
56. *The Glass Transition and the Nature of the Glassy State*, ed. by M. Goldstein, R. Simha, Ann. New York Acad. Sci. **279** (1976); *Structure and Mobility in Molecular and Atomic Glasses*, ed. by J.M. O'Reilly, M. Goldstein, Ann. New York Acad. Sci. **371** (1981); *Dynamic Aspects of Structural Change in Liquids and Glasses*, ed. by C.A. Angell, M. Goldstein, Ann. New York Acad. Sci. **484** (1986)
57. *Relaxations in Complex Systems*, ed. by K.L. Ngai, G.B. Wright (National Technical Information Service, Springfield 1984)
58. *1st International Discussion Meeting on Relaxations in Complex Systems* (Herakleon, Greece 1991), J. Non-Cryst. Solids **131–133** (1991); *2nd International Discussion Meeting on Relaxations in Complex Systems* (Alicante, Spain 1994), J. Non-Cryst. Solids **172–174** (1994); *3rd International Discussion Meeting on Relaxations in Complex Systems* (Vigo, Spain 1997), J. Non-Cryst. Solids **235–237** (1998)

59. *40 Years of Entropy and the Glass Transition*, J. Res. Natl. Inst. Stand. Technol. **102**, 135–248 (1997)
60. *Liquids, Freezing and Glass Transition* (Les Houches, France 1989), ed. by J.P. Hansen, D. Lévesque, J. Zinn-Justin (North-Holland, Amsterdam 1991)
61. *Dynamics of Disordered Materials I* (Grenoble, France 26–28 September 1988), ed. by D. Richter et al. (Springer, Berlin 1989); *Dynamics of Disordered Materials II* (Grenoble, France 22–24 March 1993), ed. by A.J. Dianoux, W. Petry, D. Richter (North-Holland, Amsterdam 1993)
62. *Non-equilibrium phenomena in supercooled fluids, glasses and amorphous materials* (Pisa, Italy, 25–29 September 1995), ed. by M. Giordano, D. Leporini, M.P. Tosi (World Scientific, Singapore 1996); *Second Workshop on Non-Equilibrium Phenomena in Supercooled Fluids, Glasses and Amorphous Materials* (Pisa, Italy, 27 September–2 October 1998), J. Phys. Condens. Matter **11**, A1–A377 (1999)
63. ICTP-NIS Conference on *Unifying concepts in glass physics* (Trieste, Italy 15–18 September 1999), Special issue of J. Phys. Condens. Matter **12**, 6295–6682 (2000)
64. R. Hosemann, S.N. Bagchi: *Direct Analysis of Diffraction by Matter* (North-Holland, Amsterdam 1962)
65. P.W. Anderson: In: *Les Houches, Session 31* (France 1978), ed. by R. Balian et al. (North-Holland, Amsterdam 1979) pp. 159–261
66. J.F. Sadoc, R. Mosseri: Amorphous Structure Description Starting from Ordered Structures in Curved Space. In: *Amorphous Materials: Modeling of Structure and Properties*, ed. by V. Vitek, (Metallurgical Soc. of AIME, Warrendale 1983); J.F. Sadoc, R. Mosseri: *Geometrical Frustration* (Cambridge University Press, Cambridge 1999)
67. N. Rivier: Structure of glasses from a topological viewpoint. In: *Structure of Non-Crystalline Materials*, ed. by P.H. Gaskell, E.A. Davies, J.M. Parker (Taylor and Francis, Cambridge 1982)
68. U. Buchenau, Yu.M. Halperin, V.L. Gurevich, H.R. Schober: Phys. Rev. B **43**, 5039–5045 (1991)
69. C.A. Angell, W. Sichina: Ann. New York Acad. Sci. **279**, 53–67 (1976)
70. C.A. Angell: In: *Relaxations in Complex Systems*, ed. by K.L. Ngai, G.B. Wright (National Technical Information Service, Springfield 1985) p. 3
71. I.M. Hodge: J. Non-Cryst. Solids **202**, 164–172 (1996)
72. R. Richert, C.A. Angell: J. Chem. Phys. **108**, 9016–9026 (1998)
73. C.A. Angell, B.E. Richards, V. Velikov: J. Phys. Condens. Matter **11**, A75–A94 (1999)
74. H. Vogel: Phys. Z. **22**, 645 (1921)
75. G.S. Fulcher: J. Amer. Ceram. Soc. **8**, 339–355 (1925)
76. G. Tammann, G. Hesse: Z. Anorg. Allg. Chem. **156**, 245–257 (1926)
77. R. Böhmer, K.L. Ngai, C.A. Angell, D.J. Plazek: J. Chem. Phys. **99**, 4201–4209 (1993)
78. A.J. Barlow, J. Lamb, A.J. Matheson: Proc. R. Soc. Lond. A **292**, 322–342 (1966)
79. W.T. Laughlin, D.R. Uhlmann: J. Phys. Chem. **76**, 2317–2325 (1972)
80. M. Cukierman, J.W. Lane, D.R. Uhlmann: J. Chem. Phys. **59**, 3639–3644 (1973)
81. D.J. Plazek, J.H. Magill: J. Chem. Phys. **49**, 3678–3682 (1968)
82. P.A. O'Connell, G.B. McKenna: J. Chem. Phys. **110**, 11054–11060 (1999)
83. A. Schönhals, F. Kremer, A. Hofmann, E.W. Fischer, E. Schlosser: Phys. Rev. Lett. **70**, 3459–3462 (1993)
84. F. Stickel: PhD Thesis (Shaker, Universität Mainz 1995)

85. F. Stickel, E.W. Fischer, R. Richert: J. Chem. Phys. **102**, 6251–6257 (1995)
86. F. Stickel, E.W. Fischer, R. Richert: J. Chem. Phys. **104**, 2043–2055 (1996)
87. C. Hansen, F. Stickel, R. Richert, E.W. Fischer: J. Chem. Phys. **108**, 6408–6415 (1998)
88. W. Oldekop: Glastechn. Ber. **30**, 8–14 (1957)
89. G.P. Johari: J. Chem. Phys. **112**, 8958–8969 (2000)
90. E. Rössler, A.P. Sokolov: Chem. Geol. **128**, 143–153 (1996)
91. N.O. Birge, S.R. Nagel: Phys. Rev. Lett. **54**, 2674–2677 (1985)
92. T. Christensen: J. Phys. France (Colloq. 8) **12**, 635–637 (1985)
93. Y.-H. Jeong: Thermochim. Acta **304**, 67–98 (1997)
94. H. Nyquist: Phys. Rev. **32**, 110–113 (1928)
95. H.B. Callen, R.F. Greene: Phys. Rev. **86**, 702–710 (1952)
96. N.W. Tschoegl: *The Phenomenological Theory of Linear Viscoelastic Behavior* (Springer, Berlin 1989)
97. R.F. Greene, H.B. Callen: Phys. Rev. **88**, 1387–1391 (1952)
98. C. Bauer, R. Böhmer, S. Moreno-Flores, R. Richert, H. Sillescu, D. Neher: Phys. Rev. E **61**, 1755–1764 (2000)
99. G. Diezemann, R. Böhmer, G. Hinze, H. Sillescu: J. Non-Cryst. Solids **235–237**, 121–127 (1998); G. Diezemann, H. Sillescu, G. Hinze, R. Böhmer: Phys. Rev. E **57**, 4398–4410 (1998); G. Diezemann, K. Nelson: J. Phys. Chem. B **103**, 4089–4096 (1999)
100. K. Schröter: unpublished work (1999)
101. K. Schröter, G. Wilde, R. Willnecker, M. Weiss, K. Samwer, E. Donth: Eur. Phys. J. B **5**, 1–5 (1998)
102. L. Duffrene, R. Gy: J. Non-Cryst. Solids **211**, 30–38 (1997)
103. D.J. Plazek, V.M. O'Rourke: J. Polym. Sci. A2 **9**, 209–243 (1971)
104. K.L. Ngai, D.J. Plazek, I. Echeverria: Macromolecules **29**, 7937–7942 (1996)
105. S. Reissig, M. Beiner, S. Vieweg, K. Schröter, E. Donth: Macromolecules **29**, 3996–3999 (1996)
106. D.J. Plazek, C.A. Bero, I.C. Chay: J. Non-Cryst. Solids **172**, 181–190 (1994)
107. D.J. Plazek, C.A. Bero, S. Neumeister, G. Floudas, G. Fytas, K.L. Ngai: Colloid & Polym. Sci. **272**, 1430–1438 (1994)
108. P. Debye: *Polare Molekeln* (Hirzel, Leipzig 1929)
109. A.K. Jonscher: *Dielectric Relaxation in Solids* (Chelsea Dielectrics Press, London 1983)
110. S. Havriliak, S. Negami: J. Polym. Sci. C **14**, 99–117 (1966)
111. A. Schönhals, F. Kremer, E. Schlosser: Phys. Rev. Lett. **67**, 999–1002 (1991)
112. R. Kohlrausch: Ann. Phys. Chem. **72**, 393–398 (1847)
113. G. Williams, D.C. Watts: Trans. Faraday Soc. **66**, 80–85 (1970); G. Williams, D.C. Watts, S.B. Dev, A.M. North: Trans. Faraday Soc. II **67**, 1323–1335 (1971)
114. G.C. Berry, D.J. Plazek: Rheol. Acta **36**, 320–329 (1997)
115. D. Lellinger, E. Donth: unpublished work (1994)
116. S. Kahle: unpublished work (1998)
117. S. Reissig, M. Beiner, S. Zeeb, S. Höring, E. Donth: Macromolecules **32**, 5701–5703 (1999)
118. M.L. Williams, R.F. Landel, J.D. Ferry: J. Amer. Chem. Soc. **77**, 3701–3707 (1955)
119. M.S. Green: J. Chem. Phys. **22**, 398–413 (1954) ; R. Kubo: J. Phys. Soc. Jpn. **12**, 570–586 (1957)
120. F. Fujara, B. Geil, H. Sillescu, G. Fleischer: Z. Phys. B Condens. Matter **88**, 195–204 (1992)

121. M. Beiner, J. Korus, H. Lockwenz, K. Schröter, E. Donth: Macromolecules **29**, 5183–5189 (1996)
122. G. Adam, J.H. Gibbs: J. Chem. Phys. **43**, 139–146 (1965)
123. S.H. Glarum: J. Chem. Phys. **33**, 639–643 (1960)
124. J. Koppelmann: In: *Physics of Non-Crystalline Solids*, ed. by J.A. Prins, (North-Holland, Amsterdam 1965) p. 229
125. G.B. McKenna: *Lecture at the Europhysics Conference on Macromolecular Physics* (Merseburg, Germany September 27–October 1, 1998)
126. M. von Laue: Phys. Z. **18**, 542–544 (1917)
127. C.A. Angell: J. Phys. Chem. **70**, 2793–2803 (1966)
128. C.T. Moynihan, C.A. Angell: J. Non-Cryst. Solids **274**, 131–138
129. S. Matsuoka: *Relaxation Phenomena in Polymers* (Hanser, München 1992) Chap. 2
130. O. Yamamuro, I. Tsukushi, A. Lindqvist, S. Takahara, M. Ishikawa, T. Matsuo: J. Phys. Chem. B **102**, 1605–1609 (1998)
131. S. Kästner: private communication (1977)
132. G.P. Johari: Ann. New York Acad. Sci. **279**, 117–140 (1976)
133. J.H. Gibbs: J. Chem. Phys. **25**, 185–186 (1956); J.H. Gibbs, E.A. DiMarzio: J. Chem. Phys. **28**, 373–383 (1958); E.A. DiMarzio, J.H. Gibbs, P.D. Fleming, I.C. Sanchez: Macromolecules **9**, 763–771 (1976)
134. J.K. Krüger, P. Mesquida, J. Baller: Phys. Rev. B **60**, 10037–10041 (1999)
135. J.K. Krüger, K.-P. Bohn, R. Jimenez, J. Schreiber: Colloid & Polymer Sci. **274**, 490–495 (1996)
136. G.B. McKenna: private communication (1996)
137. e.g., DIN 52324 of Feb. 1984 (Deutsche Industrie Norm, German Industrial Standard)
138. E. Donth: Plaste & Kautschuk **25**, 617–626 (1978)
139. A. Arbe, J. Colmenero, M. Monkenbusch, D. Richter: Phys. Rev. Lett. **82**, 1336 (1999); A. Heuer, H.W. Spiess: Phys. Rev. Lett. **82**, 1335 (1999)
140. S. Kahle: private communication (1999)
141. J. Qian, R. Hentschke, A. Heuer: J. Chem. Phys. **111**, 10177–10182 (1999)
142. E. Flikkema, G. van Ekenstein, G. ten Brinke: Macromolecules **31**, 892–898 (1998)
143. C.T. Moynihan, J. Schroeder: J. Non-Cryst. Solids **160**, 52–59 (1993)
144. S.L. Simon, G.B. McKenna: J. Chem. Phys. **107**, 8678–8685 (1997)
145. C. Schick: private communication (1998 and 1999); S. Weyer, M. Merzlyakov, C. Schick: Thermochim. Acta (2001) submitted
146. M.W. Zemansky: *Heat and Thermodynamics*, 5th edn. (McGraw-Hill, New York 1968)
147. R.-J. Roe: *Lecture at the 1st International Discussion Meeting on Relaxations in Complex Systems* (Herakleon, Greece 1991)
148. J.E. McKinney, M. Goldstein: J. Res. Nat. Bur. Stand. **78**, 331–353 (1974)
149. G.P. Johari: J. Chem. Phys. **112**, 7518–7523 (2000)
150. G.P. Johari: Phil. Mag. B **41**, 41–47 (1980)
151. W. Kauzmann: Chem. Rev. **43**, 219–256 (1948)
152. C.A. Angell, D.L. Smith: J. Phys. Chem. **86**, 3845–3852 (1982)
153. H. Suga: Thermal Anal. Calor. **60**, 957–974 (2000)
154. K. Kishimoto, H. Suga, S. Seki: Bull. Chem. Soc. Jpn. **46**, 3020–3031 (1973)
155. G.P. Johari: J. Chem. Phys. **113**, 751–761 (2000)
156. M.M. Santore, R.S. Duran, G.B. McKenna: Polymer **32**, 2377–2381 (1991); L.C.E. Struik: Polymer **38**, 4053–4057 (1997)
157. F.H. Stillinger, E.A. DiMarzio, R.L. Kornegay: J. Chem. Phys. **40**, 1564–1576 (1964)

158. M. Goldstein: J. Chem. Phys. **51**, 3728–3739 (1969)
159. C.A. Angell: Nature **393**, 521–522 (1998)
160. S. Sastry, P.G. Debenedetti, F.H. Stillinger: Nature **393**, 554–557 (1998)
161. J.N. Onuchic, P.G. Wolynes, Z. Luthey-Schulten, N.D. Socci: Proc. Natl. Acad. Sci. **92**, 3626–3630 (1995)
162. F.A. Stillinger, T.A. Weber: Science **225**, 983–989 (1984)
163. F. Sciortino, W. Kob, P. Tartaglia: J. Phys. Condens. Matter **12**, 6525–6534 (2000)
164. A. Crisanti, F. Ritort: J. Phys. Condens. Matter **12**, 6413–6422 (2000)
165. F.H. Stillinger: Phys. Rev. A **28**, 2408–2416 (1983)
166. S. Büchner, A. Heuer: Phys. Rev. Lett. **84**, 2168–2171 (2000)
167. E. Donth: Wiss. Z. TH Leuna-Merseburg **24**, 475–484 (1982)
168. K. Schmidt-Rohr, H. Spiess: Phys. Rev. Lett. **66**, 3020–3023 (1991); U. Tracht, M. Wilhelm, A. Heuer, H. Feng, K. Schmidt-Rohr, H.W. Spiess: Phys. Rev. Lett. **81**, 2727–2730 (1998)
169. K. Schmidt-Rohr, H.W. Spiess: *Multidimensional Solid-State NMR and Polymers* (Academic, London 1994)
170. H.N. Ritland: J. Am. Ceram. Soc. **37**, 370–378 (1954)
171. L.C.E. Struik: Polymer **38**, 4677–4685 (1997)
172. W. Retting, H.M. Laun: *Kunststoff-Physik* (Hanser, München 1991)
173. C.W. Macosko: *Rheology. Principles, Measurements, and Applications* (VCH, New York 1994)
174. W.P. Cox, E.H. Merz: J. Polym. Sci. **28**, 619–622 (1958)
175. C. Schick, M. Merzlyakov, A. Hensel: J. Chem. Phys. **111**, 2695–2700 (1999)
176. A.Q. Tool: J. Res. Nat. Bur. Stand. **37**, 73–90 (1946); J. Am. Ceram. Soc. **29**, 240 (1946)
177. I.L. Hopkins: J. Polym. Sci. **28**, 631–633 (1958)
178. L.W. Morland, E.H. Lee: Trans. Soc. Rheol. **4**, 233–263 (1960)
179. O.S. Narayanaswamy: J. Am. Ceram. Soc. **54**, 491–498 (1971)
180. C.T. Moynihan, P.B. Macedo, C.J. Montrose: In: *The Glass Transition and the Nature of the Glassy State*, ed. by M. Goldstein, R. Simha, Ann. New York Acad. Sci. **279**, 15–35 (1976)
181. O.V. Mazurin, S.M. Rekhson, Yu.K. Startsev: Sov. J. Glass Phys. Chem. **1**, 412–416 (1975) (English translation)
182. C. Schick, A. Hensel, S. Weyer, E. Hempel, H. Huth: private communication (1999 and 2000)
183. I.M. Hodge, J.M. O'Reilly: J. Phys. Chem. B **103**, 4171–4176 (1999)
184. J.L. Gomez Ribelles, M. Monleon Pradas: Macromolecules **28**, 5867–5877 (1995)
185. J.L. Gomez Ribelles, A. Vidaurre Garayo, J.M.G. Cowie, R. Ferguson, S. Harris, I.J. McEwen: Polymer **40**, 183–192 (1999)
186. J.M. Hutchinson, S. Montserrat, Y. Calventus, P. Cortes: Macromolecules **33**, 5252–5262 (2000)
187. E. Hempel, E. Donth: to be published
188. M.V. Vol'kenshtein, Yu.A. Sharonov: Vysokomolek. Soed. **3**, 1739–1745 (1961) (in Russian)
189. M.V. Vol'kenshtein, O.B. Ptitsyn: Zhurnal Tekh. Fiz. **26**, 2204–2222 (1956) (in Russian)
190. K.-H. Illers: Makromol. Chem. **127**, 1–33 (1969)
191. H. Wilski: Kolloid-Z. Z. Polymere **210**, 37–45 (1966)
192. M. Reading: Trends Polym. Sci. **1**, 248–253 (1993)

193. J. Heijboer: Mechanical Properties and Molecular Structure of Organic Polymers. In: *Physics of non-crystalline solids*, ed. by J.A. Prins (North-Holland, Amsterdam 1965) pp. 231–253
194. G.P. Johari: Faraday Symp. Chem. Soc. **6**, 42–44 (1972)
195. A. Kudlik, C. Tschirwitz, T. Blochowitz, S. Benkhof, E. Rössler: J. Non-Cryst. Solids **235–237**, 406–411 (1998)
196. S. Yagihara, M. Yamada, M. Asano, Y. Kanai, N. Shinyashiki, S. Mashimo, K.L. Ngai: J. Non-Cryst. Solids **235–237**, 412–415 (1998)
197. A. Kudlik, S. Benkhof, T. Blochowicz, C. Tschirwitz, E. Rössler: J. Mol. Struct. **479**, 201–218 (1999)
198. S.C. Kuebler, D.J. Schaefer, C. Boeffel, U. Pawelzik, H.W. Spiess: Macromolecules **30**, 6597–6609 (1997)
199. E. Helfand: J. Chem. Phys. **54**, 4651–4661 (1971)
200. A. Abe, R.L. Jernigan, P.J. Flory: J. Am. Chem. Soc. **88**, 631–639 (1966)
201. G. Diezemann, U. Mohanty, I. Oppenheim: Phys. Rev. E **59**, 2067–2083 (1999)
202. M. Schulz, E. Donth: J. Non-Cryst. Solids **168**, 186–194 (1994); S. Kahle, M. Schulz, E. Donth: J. Non-Cryst. Solids **238**, 234–243 (1998)
203. J.Y. Cavaille, J. Perez, G.P. Johari: Phys. Rev. B **39**, 2411–2422 (1989)
204. M. Vogel: PhD Thesis (Bayreuth 2000)
205. J. Koppelmann: Koll. Z. **189**, 1–6 (1963)
206. G. Williams: Trans. Faraday Soc. **62**, 2091–2102 (1966)
207. S. Reissig, M. Beiner, S. Zeeb, S. Höring, E. Donth: Macromolecules **32**, 5701–5703 (1999)
208. F. Garwe, A. Schönhals, H. Lockwenz, M. Beiner, K. Schröter, E. Donth: Macromolecules **29**, 247–253 (1996)
209. M. Beiner, S. Kahle, E. Hempel, K. Schröter, E. Donth: Europhys. Lett. **44**, 321–327 (1998)
210. G.P. Johari, M. Goldstein: J. Chem. Phys. **53**, 2372–2388 (1970); **55**, 4245–4252 (1971); J. Phys. Chem. **74**, 2034–2035 (1970)
211. H. Wagner, R. Richert: J. Phys. Chem. B **103**, 4071–4077 (1999)
212. J. Fan, E.I. Cooper, C.A. Angell: J. Phys. Chem. **98**, 9345–9349 (1994)
213. J.-N. Roux, J.-L. Barrat, J.-P. Hansen: J. Phys. Cond. Matter **1**, 7171–7186 (1989)
214. G. Wahnström: Phys. Rev. A **44**, 3752–3764 (1991)
215. J. Korus, E. Hempel, M. Beiner, S. Kahle, E. Donth: Acta Polymer. **48**, 369–378 (1997)
216. H. Huth, M. Beiner, E. Donth: Phys. Rev. B **61**, 15092–15101 (2000)
217. S. Kahle, K. Schröter, E. Hempel, E. Donth: J. Chem. Phys. **111**, 6462–6470 (1999)
218. K.L. Ngai: Comments Solid State Phys. **9**, 127–140 (1979); K.L. Ngai: Phil. Mag. B **79**, 1783–1797 (1999)
219. E. Schlosser: Polym. Bull. **8**, 461–467 (1982)
220. C.H. Wang, E.W. Fischer: J. Chem. Phys. **82**, 632–638 (1985)
221. P.K. Dixon, S.R. Nagel: Phys. Rev. Lett. **61**, 341–344 (1988)
222. C. Schick, E. Donth: Physica Scripta **43**, 423–429 (1991)
223. C. Leon, K.L. Ngai: J. Phys. Chem. B **103**, 4045–4051 (1999)
224. M.F. Shlesinger: Ann. Rev. Phys. Chem. **39**, 269–290 (1988)
225. K.L. Ngai: J. Chem. Phys. **109**, 6982–6994 (1998)
226. E. Donth: J. Non-Cryst. Solids **53**, 325–330 (1982)
227. E. Rössler: Phys. Rev. Lett. **65**, 1595–1598 (1990)
228. W. Steffen: private communication (1994)
229. E.W. Fischer, E. Donth, W. Steffen: Phys. Rev. Lett. **68**, 2344–2346 (1992)

230. R. Zorn, A. Arbe, J. Colmenero, B. Frick, D. Richter, U. Buchenau: Phys. Rev. E **52**, 781–795 (1995); V.N. Novikov: Phys. Rev. B **58**, 8367–8378 (1998)
231. P.K. Dixon, L. Wu, S.R. Nagel, B.D. Williams, J.P. Carini: Phys. Rev. Lett. **65**, 1108–1111 (1990)
232. E.N. Andrade: Phil. Mag. **7**, 2003–2014 (1962); D.J. Plazek, E. Riande, H. Markovitz, N. Raghlinpathi: J. Polym. Sci., Polym. Phys. Ed. **17**, 2189–2213 (1979)
233. A. Schönhals, E. Donth: Acta Polymer. **37**, 475–480 (1986)
234. A. Arbe, D. Richter, J. Colmenero, B. Farago: Phys. Rev. E **54**, 3853–3869 (1996)
235. E. Donth, K. Schröter, S. Kahle: Phys. Rev. E **60**, 1099–1102 (1999)
236. D.J. Plazek: J. Rheol. **36**, 1671–1690 (1992)
237. K. Schröter, R. Unger, S. Reissig, F. Garwe, S. Kahle, M. Beiner, E. Donth: Macromolecules **31**, 8966–8972 (1998)
238. G. Williams, D.C. Watts: In: *NMR Basic Principles and Progress*, ed. by P. Diehl, E. Fluck, R. Kosfeld (Springer, Berlin 1971) Vol. 4, p. 271
239. S.S.N. Murthy: J. Chem. Soc. Faraday Trans. II **85**, 581–596 (1989); J. Mol. Liquids **44**, 51–61 (1989)
240. A. Schönhals: In: *Non-Equilibrium Phenomena in Supercooled Fluids, Glasses and Amorphous Materials*, ed. by M. Giordano et al. (World Scientific, Singapore 1997) p. 210
241. T. Fujima, H. Frusawa, K. Ito, R. Hayakawa: Jpn. J. Appl. Phys. **39**, L744–L764 (2000)
242. A. Schönhals, F. Kremer: J. Non-Cryst. Solids **172–174**, 336–343 (1994)
243. C. Hansen, F. Stickel, T. Berger, R. Richert, E.W. Fischer: J. Chem. Phys. **107**, 1086–1093 (1997); C. Hansen, F. Stickel, R. Richert, E.W. Fischer: J. Chem. Phys. **108**, 6408–6415 (1998)
244. M. Beiner, S. Kahle, E. Hempel, K. Schröter, E. Donth: Macromolecules **31**, 8973–8980 (1998)
245. S. Kahle, J. Korus, E. Hempel, R. Unger, S. Höring, K. Schröter, E. Donth: Macromolecules **30**, 7214–7223 (1997)
246. E. Casalini, P. Fioretto, A. Livi, M. Lucchesi, P.A. Rolla: Phys. Rev. B **56**, 3016–3021 (1997)
247. S. Corezzi, E. Campani, P.A. Rolla, S. Capaccioli, D. Fioretto: J. Chem. Phys. **111**, 9343–9351 (1999)
248. S. Corezzi, K. Schröter, H. Huth, M. Beiner et al.: to be published
249. A. Kudlik, C. Tschirwitz, S. Benkhof, T. Blochowicz, E. Rössler: Europhys. Lett. **40**, 649–654 (1997)
250. A.G. Oblad, R.F. Newton: J. Amer. Chem. Soc. **59**, 2495–2499 (1937)
251. S. Corezzi, P.A. Rolla, M. Paluch, J. Ziolo, D. Fioretto: Phys. Rev. E **60**, 4444–4452 (1999)
252. M. Beiner, K. Schröter, E. Hempel, S. Reissig, E. Donth: Macromolecules **32**, 6278–6282 (1999)
253. G.P. Johari: private communication (1998)
254. S. Sen, J.F. Stebbins: Phys. Rev. Lett. **78**, 3495–3498 (1997)
255. K. Kawasaki: Bussei Kenkyu **69**, 810–825 (1998)
256. S.C. Glotzer, V.N. Novikov, T.B. Schröder: J. Chem. Phys. **112**, 509–512 (2000)
257. K. Kawasaki: Jpn. J. Appl. Phys. (Part 1) **37** Suppl. 1, 36–40 (1998)
258. M.T. Cicerone, P.A. Wagner, M.D. Ediger: J. Phys. Chem. B **101**, 8727–8734 (1997)
259. I. Chang, F. Fujara, B. Geil, G. Heuberger, T. Mangel, H. Sillescu: J. Non-Cryst. Solids **172**, 248–255 (1994)

260. M.T. Cicerone, M.D. Ediger: J. Chem. Phys. **104**, 7210–7218 (1996)
261. C.T. Moynihan, N. Balitactac, L. Boone, T.A. Litovitz: J. Chem. Phys. **55**, 3013–3019 (1971)
262. C.A. Angell: Solid State Ionics **9** & **10**, 3–16 (1983)
263. M. Paluch, J. Ziolo: Europhys. Lett. **44**, 315–320 (1998)
264. M.D. Ingram: J. Non-Cryst. Solids **131–133**, 955–960 (1991)
265. H. Jain: J. Non-Cryst. Solids **131–133**, 961–968 (1991)
266. K.L. Ngai, G.N. Greaves, C.T. Moynihan: Phys. Rev. Lett. **80**, 1018–1021 (1998)
267. K. Funke, C. Cramer: Curr. Opin. Solid State Mater. Sci. **2**, 483–490 (1997)
268. M.D. Ingram, K. Funke: Ion Dynamics and Structural Relaxation in Glassy Materials. In: *Non-equilibrium phenomena in supercooled fluids, glasses and amorphous materials* (Pisa, Italy, 25–29 September 1995) ed. by M. Giordano, D. Leporini, M.P. Tosi (World Scientific, Singapore 1996) pp. 61–70
269. K. Funke, D. Wilmer: Mater. Res. Soc. Symp. Proc. **548**, 403–414 (1999)
270. D.L. Sidebottom: Phys. Rev. Lett. **82**, 3653–3656 (1999)
271. T.B. Schröder, J.C. Dyre: Phys. Rev. Lett. **84**, 310–313 (2000)
272. B. Roling, A. Happe, K. Funke, M.D. Ingram: Phys. Rev. Lett. **78**, 2160–2163 (1997)
273. B. Roling: private communication (1999)
274. S.R. de Groot, P. Mazur: *Non-Equilibrium Thermodynamics* (Dover, New York 1984)
275. I. Prigogine: *Introduction to Thermodynamics of Irreversible Processes*, 3rd edn. (Interscience Publ., New York 1967) pp. 144ff
276. J.P. Boon, S. Yip: *Molecular Hydrodynamics* (Dover, New York 1991)
277. H.B. Callen, T.A. Welton: Phys. Rev. **83**, 34–40 (1951)
278. J. Honerkamp, J. Weese: Continuum Mech. Thermodyn. **2**, 17–30 (1990); Macromolecules **22**, 4372–4377 (1989)
279. W. Feller: *An Introduction to Probability Theory and Its Applications*, Vol. II, 2nd edn. (Wiley, New York 1971) pp. 120, 134, 169–176, 569–570, 574–581, and Chap. XIX
280. F. Reif: *Statistische Physik und Theorie der Wärme* (de Gruyter, Berlin 1987)
281. A. Arbe, J. Colmenero, M. Monkenbusch, D. Richter: Phys. Rev. Lett. **81**, 590–593 (1998)
282. M. Fuchs: J. Non-Cryst. Solids **172**, 241–247 (1994)
283. S. Havriliak, Jr., S.J. Havriliak: J. Non-Cryst. Solids **172**, 297–310 (1994)
284. A. Schönhals: In: *Dielectric Properties of Amorphous Polymers*, ed. by J.J. Fitzgerald, J.P. Runt (ACS-Books, Washington 1997) pp. 81–106
285. R.G. Palmer, D.L. Stein, E. Abrahams, P.W. Anderson: Phys. Rev. Lett. **53**, 958–961 (1984)
286. M.F. Shears, G. Williams: J. Chem. Soc. Faraday Trans. II **69**, 608–621 (1973)
287. M.C. Phillips, A.J. Barlow, J. Lamb: Proc. R. Soc. Lond. A **329**, 193–218 (1972)
288. C.A. Angell, K.J. Rao: J. Chem. Phys. **57**, 470–481 (1972)
289. R. Böhmer, R.V. Chamberlin, G. Diezemann, B. Geil, A. Heuer, G. Hinze, S.C. Kuebler, R. Richert, B. Schiener, H. Sillescu, H.W. Spiess, U. Tracht, M. Wilhelm: J. Non-Cryst. Solids **235–237**, 1–9 (1998)
290. R. Richert: Chem. Phys. Lett. **216**, 223–227 (1993)
291. S.C. Kuebler, A. Heuer, H.W. Spiess: Phys. Rev. E **56**, 741–749 (1997)
292. S.A. Reinsberg, X.H. Qiu, M. Wilhelm, H.W. Spiess, M.D. Ediger: Preprint (2000)

293. C. Donati, J.F. Douglas, W. Kob, S.J. Plimpton, P.H. Poole, S.C. Glotzer: Phys. Rev. Lett. **80**, 2338–2341 (1998); W. Kob, C. Donati, S.J. Plimpton, P.H. Poole, S.C. Glotzer: Phys. Rev. Lett. **79**, 2827–2830 (1997)
294. S.C. Glotzer: J. Non-Cryst. Solids **274**, 342–355 (2000)
295. B. Doliwa, A. Heuer: Phys. Rev. E **61**, 6898–6908 (2000)
296. E. Donth: Acta Polymer. **50**, 240–251 (1999)
297. H.W. Spiess: *Lecture at the DPG-Spring Conference* (Potsdam 2000)
298. E. Donth: Acta Polymer. **30**, 481–485 (1979)
299. E. Hempel, G. Hempel, A. Hensel, C. Schick, E. Donth: J. Phys. Chem. B **104**, 2460–2466 (2000)
300. B. Schiener, R. Böhmer, A. Loidl, R. Chamberlin: Science **274**, 752–754 (1996)
301. T. Inoue, M. Cicerone, M.D. Ediger: Macromolecules **28**, 3425–3433 (1995)
302. R. Richert: J. Phys. Chem. B **101**, 6323–6326 (1997)
303. G. Floudas, S. Paraskeva, N. Hadjichristidis, G. Fytas, B. Chu, A. Semenov: J. Chem. Phys. **107**, 5502–5509 (1997)
304. M. Arndt, R. Stannarius, H. Groothues, E. Hempel, F. Kremer: Phys. Rev. Lett. **79**, 2077–2088 (1997)
305. E. Hempel, A. Huwe, K. Otto, F. Janowski, K. Schröter, E. Donth: Thermochim. Acta **337**, 163–168 (1999)
306. E.W. Fischer: Physica A **201**, 183–206 (1993)
307. H.-W. Hu, G.A. Carson, S. Granick: Phys. Rev. Lett. **66**, 2758–2761 (1991)
308. B.D. Fitz, J. Mijovic: Macromolecules **32**, 3518–3527 (1999)
309. A.K. Rizos, K.L. Ngai: Phys. Rev. E **59**, 612–617 (1999)
310. R.L. Leheny, N. Menon, S.R. Nagel, D.L. Price, K. Suzuya, P. Thiyagarajan: J. Chem. Phys. **105**, 7783–7794 (1996)
311. R.E. Robertson: J. Polym. Sci., Polym. Symp. **63**, 173–183 (1978)
312. J.M. O'Reilly: J. Polym. Sci., Polym. Symp. **63**, 165–172 (1978)
313. L.D. Landau, E.M. Lifshitz: *Course of Theoretical Physics. Vol.5 Statistical Physics*, 3rd edn. (Pergamon Press, Oxford 1980)
314. E. Donth: unpublished work (1999)
315. E. Eckstein, J. Qian, R. Hentschke, T. Thurn-Albrecht, W. Steffen, E.W. Fischer: J. Chem. Phys. **113**, 4751–4762 (2000)
316. W. Chen, B. Wunderlich: Macromol. Chem. Phys. **200**, 283–311 (1999)
317. A. Hensel, C. Schick: J. Non-Cryst. Solids **235–237**, 510–516 (1998); E. Donth, J. Korus, E. Hempel, M. Beiner: Thermochim. Acta **304/305**, 239–249 (1997)
318. E. Donth: to be published
319. M. Beiner, S. Kahle, S. Abens, E. Hempel, M. Meissner, E. Donth: to be published
320. H. Suga, M. Oguni: Amorphous Materials and Their Elucidation by Adiabatic Calorimetry. In: *Chemical Thermodynamics*, ed. by T.M. Letcher (Blackwell Science, Oxford 1999) pp. 227–237
321. G. Rehage, W. Borchard: In: *The Physics of Glassy Polymers*, ed. by E.N. Haward (Appl. Sci. Publ., London 1973) p. 54
322. G. Rehage: Ber. Bunsenges. physik. Chem. **81**, 969 (1977)
323. S.R. Elliott: Europhys. Lett. **19**, 201–206 (1992); U. Buchenau: J. Mol. Struct. **296**, 275–283 (1993)
324. A.P. Sokolov, E. Rössler, A. Kisliuk, D. Quitmann: Phys. Rev. Lett. **71**, 2062–2065 (1993)
325. H.R. Schober, C. Gaukel, C. Oligschleger: Prog. Theor. Phys. Suppl. **126**, 67–74 (1997)
326. A.J. Martin, W. Brenig: Phys. Stat. Sol. (b) **64**, 163–172 (1974)
327. I. Tsukushi; T. Kanaya, K. Kaji: J. Non-Cryst. Solids **235–237**, 250–253 (1998)

328. T. Uchino, Y. Toshinobu: J. Chem. Phys. **108**, 8130–8138 (1998)
329. Y. Inamura, M. Arai, T. Otomo, N. Kitamura, U. Buchenau: Physica B **284–288**, 1157–1158 (2000)
330. W. Schirmacher, G. Diezemann, C. Ganter: Phys. Rev. Lett. **81**, 136–139 (1998)
331. P. Debye, A.M. Bueche: J. Appl. Phys. **20**, 518–525 (1949)
332. A. Patkowski, Th. Thurn-Albrecht, E. Banachowicz, W. Steffen, P.T. Bösecke, T. Narayanan, E.W. Fischer: Phys. Rev. E **61**, 6909–6913 (2000)
333. A. Guinier: *X-Ray Diffraction In Crystals, Imperfect Crystals, and Amorphous Bodies* (Dover, New York 1994)
334. P.A. Egelstaff: *An Introduction to the Liquid State*, 2nd edn. (Clarendon Press, Oxford 1994) Sect. 10.3
335. G. Meier, F. Kremer, G. Fytas, A. Rizos: J. Polym. Sci. B: Polym. Phys. **34**, 1391–1401 (1996)
336. E.W. Fischer: private communication (1992)
337. G. Floudas, T. Pakula, E.W. Fischer: Macromolecules **27**, 917–922 (1994)
338. E.W. Fischer: In: *The Physics of Non-Crystalline Solids*, ed. by G.H. Frischat (Trans. Tech. Publ. 1977) pp. 34–51; J. Rathje, W. Ruland: Colloid Polym. Sci. **254**, 358–370 (1976)
339. L. Comez, D. Fioretto, L. Palmieri, L. Verdini, P.A. Rolla, J. Gapinski, T. Pakula, A. Patkowski, W. Steffen, E.W. Fischer: Phys. Rev. E **60**, 3086–3096 (1999)
340. J.K. Krüger: private communication (1999)
341. S.A. Kivelson, X. Zhao, D. Kivelson, T.M. Fischer, C.M. Knobler: J. Chem. Phys. **101**, 2391–2397 (1994)
342. E.W. Fischer, A.S. Bakai: In: *Slow Dynamics in Complex Systems*, ed. by M. Tokuyama, I. Oppenheim, AIP Conf. Proc. No. 469 (AIP, New York 1999) p. 325
343. J.T. Bendler, M.F. Shlesinger: J. Phys. Chem. **96**, 3970–3973 (1992)
344. E.W. Fischer: private communication (1996)
345. See [313], Sect. 110
346. J. Fan, E.I. Cooper, C.A. Angell: J. Phys. Chem. **98**, 9345–9349 (1994)
347. I.V. Sochava: Biophys. Chem. **69**, 31–41 (1997)
348. G. Sartor, E. Mayer, G.P. Johari: Biophys. J. **66**, 249–258 (1994)
349. H. Suga, S. Seki: Faraday Discuss. Roy. Soc. Chem. **69**, 221–240 (1980)
350. G.P. Johari, J.W. Goodby, G.E. Johnson: Nature **297**, 315–317 (1982)
351. T.R. Kirkpatrick, P.G. Wolynes: Phys. Rev. A **35**, 3072–3080 (1987); T.R. Kirkpatrick, D. Thirumalai: Phys. Rev. Lett. **58**, 2091–2094 (1987); S. Franz, G. Parisi: J. Phys. I France **5**, 1401–1415 (1995); T.M. Nieuwenhuizen: Phys. Rev. Lett. **74**, 4289–4292 (1995)
352. K. Biljaković, J.C. Lasjaunias, P. Monceau: J. Phys. IV France **3**, 335–342 (1993); J. Odin, J.C. Lasjaunias, K. Biljaković, P. Monceau: Solid State Commun. **91**, 523–527 (1994); K. Biljaković, D. Staresinic, K. Hosseini, W. Brütting: Synth. Met. **103**, 2616–2619 (1999)
353. C.A. Angell: J. Solid State Chem. (2000) preprint
354. F. Mezei, W. Knaak, B. Farago: Phys. Rev. Lett. **58**, 571–574 (1987)
355. D. Richter, J.B. Hayter, F. Mezei, B. Ewen: Phys. Rev. Lett. **41**, 1484–1487 (1978)
356. S.W. Lovesey: *Theory of Neutron Scattering from Condensed Matter*, Vols. 1 and 2 (Clarendon Press, Oxford 1984)
357. B. Ewen, D. Richter: Adv. Polym. Sci. **134**, 1–129 (1997)
358. A. Arbe, J. Colmenero, M. Monkenbusch, D. Richter: Phys. Rev. Lett. **81**, 590–593 (1998)

359. J.P. Hansen, I.R. MacDonald: *Theory of Simple Liquids* (Academic Press, London 1986)
360. J. Wuttke, W. Petry, S. Pouget: J. Chem. Phys. **105**, 5177–5182 (1996); J. Wuttke, I. Chang, O.G. Randl, F. Fujara, W. Petry: Phys. Rev. E **54**, 5364–5369 (1996)
361. G. Floudas, J.S. Higgins, F. Kremer, E.W. Fischer: Macromolecules **25**, 4955–4961 (1992)
362. E. Kartini, M.F. Collins, B. Collier, F. Mezei, E.C. Svensson: Phys. Rev. B **54**, 6292–6300 (1996)
363. F. Mezei, M. Russina: J. Phys. Condens. Matter **11**, A341–A354 (1999)
364. M. Russina, F. Mezei, R. Lechner, S. Longeville, B. Urban: Phys. Rev. Lett. **84**, 3630–3633 (2000)
365. G. Li, W.M. Du, A. Sakai, H.Z. Cummins: Phys. Rev. A **46**, 3343–3356 (1992)
366. P. Lunkenheimer, A. Pimenov, M. Dressel, Yu.G. Goncharov, R. Böhmer, A. Loidl: Phys. Rev. Lett. **77**, 318–321 (1996)
367. J. Wuttke, J. Hernandez, G. Li, G. Coddens, H.Z. Cummins, F. Fujara, W. Petry, H. Sillescu: Phys. Rev. Lett. **72**, 3052–3055 (1994)
368. E. Jenckel: Z. Phys. Chem. **184**, 309–319 (1939)
369. J.L. Barrat, W. Götze, A. Latz: J. Phys. Condens. Matter **1**, 7163–7170 (1989)
370. H.Z. Cummins, G. Li, Y.H. Hwang, G.Q. Shen, W.M. Du, J. Hernandez, N.J. Tao: Z. Phys. B **103**, 501–519 (1997)
371. K. Kawasaki: Physica A **281**, 348–360 (2000)
372. E. Leutheusser: J. Phys. C Solid State Phys. **15**, 2801–2843 (1982)
373. W. Götze: Z. Phys. B **56**, 139–154 (1984); **60**, 195–203 (1985)
374. U. Bengtzelius, W. Götze, A. Sjölander: J. Phys. C Solid State Phys. **17**, 5915–5934 (1984)
375. W. Götze: Cond. Matt. Phys. **1**, 873–904 (1998)
376. W. Götze: Aspects of Structural Glass Transition. In: *Liquids, Freezing and the Glass Transition*, ed. by J.P. Hansen, D. Lévesque, J. Zinn-Justin (North Holland, Amsterdam 1989) pp. 1–187
377. H. Eliasson, B.-E. Mellander, L. Sjögren: J. Non-Cryst. Solids **235–237**, 101–105 (1998)
378. J. Wuttke, M. Seidl, G. Hinze, A. Tölle, W. Petry, G. Coddens: Eur. Phys. J. B **1**, 169–172 (1998)
379. V.I. Arnold: *Catastrophe Theory*, 2nd edn. (Springer, Berlin 1986)
380. R. Gilmore: *Catastrophe Theory for Scientists and Engineers* (Dover, New York 1981)
381. C. Bennemann, J. Baschnagel, W. Paul, K. Binder: Comput. Theor. Polym. Sci. **9**, 217–226 (1999)
382. J. Horbach, W. Kob, K. Binder: Phil. Mag. B **77**, 297–303 (1998)
383. G. Johnson, A.I. Mel'cuk, H. Gould, W. Klein, R.D. Mountain: Phys. Rev. E **57**, 5707–5718 (1998)
384. R. Yamamoto, A. Okuni: Phys. Rev. E **58**, 3515–3529 (1998)
385. P.N. Pusey, W. van Megen: Phys. Rev. Lett. **59**, 2083–2086 (1987)
386. W. van Megen, S.M. Underwood: Nature **362**, 616–618 (1993)
387. W. van Megen, S.M. Underwood: Phys. Rev. E **49**, 4206–4220 (1994)
388. A.H. Marcus, J. Schofield, S.A. Rice: Phys. Rev. E **60**, 5725–5736 (1999)
389. W.K. Kegel, A. van Blaaderen: Science **287**, 290–293 (2000)
390. E.R. Weeks, J.C. Crocker, A.C. Levitt, A. Schofield, D.A. Weitz: Science **287**, 827–631 (2000)
391. A. Kasper, E. Bartsch, H. Sillescu: Langumir **14**, 5004–5010 (1998)
392. K. Binder: Colloid Polym. Sci. **266**, 871–885 (1988); K. Kremer, K. Binder: Comp. Phys. Rep. **7**, 261–310 (1988)

393. C.A. Angell, J.H.R. Clarke, L.V. Woodcock: Adv. Chem. Phys. **48**, 397–453 (1981)
394. F.H. Stillinger, T.A. Weber: J. Chem. Phys. **70**, 4879–4883 (1979)
395. D. Rigby, R.-J. Roe: J. Chem. Phys. **87**, 7285–7292 (1987)
396. W. Kob, H.C. Andersen: Phys. Rev. E **51**, 4626–4641 (1995)
397. M. Nauroth, W. Kob: Phys. Rev. E **55**, 657–667 (1997)
398. P. Allegrini, J.F. Douglas, S.C. Glotzer: Phys. Rev. E **60**, 5714–5724 (1999)
399. E. Donth, H. Huth, M. Beiner: to be published
400. S. Sastry, P.G. Debenedetti, F.H. Stillinger, T.B. Schröder, J.C. Dyre, S.C. Glotzer: Physica A **270**, 301–308 (1999)
401. D.N. Perera: J. Phys. Condens. Matter **10**, 10115–10134 (1998); D.N. Perera, P. Harrowell: J. Non.-Cryst. Solids **235–237**, 314–319 (1998)
402. B. Doliwa, A. Heuer: Phys. Rev. Lett. **80**, 4915–4918 (1998)
403. Y. Hiwatari, T. Muranaka: J. Non-Cryst. Solids **235–237**, 19–26 (1998)
404. W. Kob: J. Phys. Condens. Matter **11**, R85–R115 (1999)
405. P.H. Poole, C. Donati, S.C. Glotzer: Physica A **261**, 51–59 (1998)
406. C. Bennemann, C. Donati, J. Baschnagel, S. Glotzer: Nature **399**, 246–249 (1999)
407. C. Donati, S.C. Glotzer, P.H. Poole, W. Kob, S.J. Plimpton: Phys. Rev. E **60**, 3107–3119 (1999)
408. C. Donati, S.C. Glotzer, P.H. Poole: Phys. Rev. Lett. **82**, 5064–5067 (1999)
409. T.B. Schroeder, S. Sastry, J.C. Dyre, S.C. Glotzer: J. Chem. Phys. **112**, 9834–9840 (2000)
410. K. Binder: Comp. Phys. Comm. **122**, 168–175 (1999)
411. N.G. van Kampen: *Stochastic Processes in Physics and Chemistry* (Elsevier, Amsterdam 1992) p. 57
412. K. Binder (Ed.): *Monte Carlo and Molecular Dynamics Simulations in Polymer Science* (Oxford University Press, New York 1995)
413. J. Baschnagel: J. Phys. Condens. Matter **8**, 9599–9603 (1996); K. Binder, W. Paul: J. Polym. Sci. B Polym. Phys. **35**, 1–31 (1997); D. Morineau, G. Dosseh, R.J.-M. Pellenq, M.C. Bellissent-Funel, C. Alba-Simionesco: Mol. Simul. **20**, 95–113 (1997); M. Wittkop, Th. Hölzl, S. Kreitmeier, D. Göritz: J. Non-Cryst. Solids **201**, 199–210 (1996)
414. I. Carmesin, K. Kremer: Macromolecules **21**, 2819–2823 (1988)
415. W. Paul, K. Binder, K. Kremer, D.W. Heermann: Macromolecules **24**, 6332–6334 (1991)
416. G. Bhanot: Rep. Prog. Phys. **51**, 429–457 (1988)
417. P.J. Flory: Proc. Natl. Acad. Sci. **79**, 4510–4514 (1982)
418. B. Wunderlich, D.M. Bodily, M.H. Kaplan: J. Appl. Phys. **35**, 95–101 (1964)
419. M. Wolfgardt, J. Baschnagel, W. Paul, K. Binder: Phys. Rev. E **54**, 1535–1543 (1996)
420. A. Milchev, J. Gutzow: J. Macromol. Sci.-Phys. **21**, 583–615 (1982)
421. L. Santen, W. Krauth: Nature **405**, 550–551 (2000)
422. K. Binder, J. Baschnagel, C. Bennemann, W. Paul: J. Phys. Condens. Matter **11**, A47–A55 (1999)
423. K. Binder, C. Bennemann, J. Baschnagel, W. Paul: AIP Conf. Proc. **469**, 193–204 (1999)
424. J. Jäckle: Prog. Theor. Phys. Suppl. **126**, 53–60 (1997)
425. G.H. Fredrickson, H.C. Anderson: J. Chem. Phys. **83**, 5822–5831 (1985)
426. B. Schulz, M. Schulz, S. Trimper: Phys. Rev. E **58**, 3368–3371 (1998)
427. S. Butler, P. Harrowell: J. Chem. Phys. **95**, 4454–4465 (1991); P. Harrowell: Phys. Rev. E **48**, 4359–4363 (1993)
428. R.F. Boyer: J. Polym. Sci. C **14**, 3–14 (1966) and other papers in this volume

429. J.M. O'Reilly, J.S. Sedita: J. Non-Cryst. Solids **131–133**, 1140–1144 (1991); *Lecture at the Creta Conference* (1990), see [58]
430. E. Donth, M. Beiner, S. Reissig, J. Korus, F. Garwe, S. Vieweg, S. Kahle, E. Hempel, K. Schröter: Macromolecules **29**, 6589–6600 (1996)
431. S. Reissig: PhD Thesis (University Halle 1999)
432. J. Heijboer: Molecular Origin of Relaxations in Polymers. In: *The Glass Transition and the Nature of the Glassy State*, ed. by M. Goldstein, R. Simha: Ann. New York Acad. Sci. **279**, 104–116 (1976)
433. T.F. Schatzki: J. Polym. Sci. **57**, 496 (1962); R.H. Boyd, S.M. Breitling: Macromolecules **7**, 855–862 (1974)
434. P.E. Rouse: J. Chem. Phys. **21**, 1272–1280 (1953)
435. J.S. Higgins, L.K. Nicholson, J.B. Hayter: Polymer **22**, 183–167 (1981)
436. D. Richter: Physica B **180**, 7–14 (1992)
437. W. Paul, G.D. Smith, D.Y. Yoon, B. Farago, S. Rathgeber, A. Zirkel, L. Willner, D. Richter: Phys. Rev. Lett. **80**, 2346–2349 (1998)
438. G. Floudas, C. Gravalides, T. Reisinger, G. Wegner: J. Chem. Phys. **111**, 9847–9852 (1999)
439. D.J. Plazek: Polym. J. **12**, 43–53 (1980); D.L. Plazek, D.J. Plazek: Macromolecules **16**, 1469–1475 (1983)
440. W.W. Graessley: Adv. Polym. Sci. **16**, 1–179 (1974); J. Polym. Sci., Polym. Phys. Ed. **18**, 27–34 (1980); D. Richter, L. Willner, A. Zirkel, B. Farago, L.J. Fetters, J.S. Huang: Phys. Rev. Lett. **71**, 4158–4161 (1993)
441. L.J. Fetters, D.J. Lohse, W. Graessley: J. Polym. Sci. B Polym. Phys. **37**, 1023–1033 (1999)
442. P.G. de Gennes: *Scaling Concepts in Polymer Physics* (Cornell University Press, Ithaca 1979)
443. M. Doi, S.F. Edwards: *The Theory of Polymer Dynamics* (Clarendon Press, Oxford 1986)
444. W. Pfandl, G. Link, F.R. Schwarzl: Rheol. Acta **23**, 277–290 (1984)
445. M. Beiner, S. Reissig, K. Schröter, E. Donth: Rheol. Acta **36**, 187–196 (1997)
446. H. Kresse, S. Ernst, W. Wedler, D. Demus, F. Kremer: Ber. Bunsenges. Phys. Chem. **94**, 1478–1483 (1990)
447. H. Kresse: private communication (1998)
448. M. Antonietti, T. Pakula, W. Bremser: Macromolecules **28**, 4227–4233 (1995)
449. R.F. Boyer: Eur. Polym. J. **17**, 661–673 (1981)
450. D.J. Plazek: J. Polym. Sci., Polym. Phys. Ed. **20**, 1533–1550 (1982); D.J. Plazek, G.-F. Gu: J. Polym. Sci., Polym. Phys. Ed. **20**, 1551–1564 (1982); J. Chen, C. Kow, L.J. Fetters, D.J. Plazek: J. Polym. Sci., Polym. Phys. Ed. **20**, 1565–1574 (1982); S.J. Orbon, D.J. Plazek: J. Polym. Sci., Polym. Phys. Ed. **20**, 1575–1583 (1982)
451. R.F. Boyer: Multiple Transitions and Relaxation in Synthetic Organic Amorphous Polymers and Copolymers: An Overview. In: *Computational Modeling of Polymers*, ed. by J. Bicerano (Dekker, New York 1992) pp. 1–52
452. J. Baschnagel, C. Bennemann, W. Paul, K. Binder: J. Phys. Condens. Matter **12**, 6365–6374 (2000)
453. R. Conrad, W.-D. Hergeth, E. Donth: unpublished work (1982)
454. G. Groeninckx, H. Reynaers, H. Berghmans, G. Smets: J. Polym. Sci., Polym. Phys. Ed. **18**, 1311–1324 (1980); G. Groeninckx, H. Reynaers: J. Polym. Sci., Polym. Phys. Ed. **18**, 1325–1341 (1980)
455. C. Schick: *Untersuchungen zum Einfluß der Morphologie auf die molekulare Beweglichkeit in den amorphen Bereichen teilkristalliner Polymere,* Habilitationsschrift (Pädagogische Hochschule, Güstrow 1988)
456. H. Sillescu: Acta Polymer. **45**, 2 (1994)

457. E. Donth, E. Hempel, C. Schick: J. Phys. Condens. Matter **12**, L281–L286 (2000)
458. F. Kremer, A. Huwe, M. Arndt, P. Behrens, W. Schwieger: J. Phys. Condens. Matter **11**, A175–A188 (1999)
459. J.K. Krüger, R. Hotwick, A. le Coutre, J. Baller: Nano Struct. Mater. **12**, 519–522 (1999)
460. J.-Y. Park, G.B. McKenna: Phys. Rev. B **61**, 6667–6676 (2000)
461. J. Baschnagel, C. Mischler, K. Binder: J. Phys. IV France, Proc. Grenoble Pr7 **10**, 9–14 (2000)
462. Conference report: J. Phys. IV France Grenoble **10**, Pr7 (2000); G.B. McKenna: J. Phys. IV France **10**, 343–346 (2000)
463. J. Schüller, R. Richert, E.W. Fischer: Phys. Rev. B **52**, 15232–15238 (1995)
464. M. Arndt, R. Stannarius, H. Groothues, E. Hempel, F. Kremer: Phys. Rev. Lett. **79**, 2077–2080 (1997)
465. C.L. Jackson, G.B. McKenna: J. Non-Cryst. Solids **131–133**, 221–224 (1991)
466. J. Zhang, G. Liu, J. Jonas: J. Phys. Chem. **96**, 3478–3480 (1992)
467. S. Vieweg, R. Unger, E. Hempel, E. Donth: J. Non-Cryst. Solids **235–237**, 470–475 (1998)
468. P. Pissis, D. Daoukaki-Diamanti, L. Apekis, C. Christodoulides: J. Phys. Condens. Matter **6**, L325–L328 (1994)
469. J.L. Keddie, R.A.L. Jones, R.A. Cory: Europhys. Lett. **27**, 59–64 (1994)
470. J.A. Forrest, J. Mattsson: Phys. Rev. E **61**, R53–R56 (2000)
471. J.A. Forrest, K. Dalnoki-Veress: preprint (2000)
472. P. Pissis, A. Kyritsis, G. Barut, R. Pelster, G. Nimtz: J. Non-Cryst. Solids **235–237**, 444–449 (1998)
473. J. Schüller, Yu.B. Mel'nichenko, R. Richert, E.W. Fischer: Phys. Rev. Lett. **73**, 2224–2227 (1994)
474. M. Arndt, R. Stannarius, W. Gorbatschow, F. Kremer: Phys. Rev. E **54**, 5377–5390 (1996)
475. A. Huwe, F. Kremer, P. Behrens, W. Schwieger: Phys. Rev. Lett. **82**, 2338–2341 (1999)
476. R.C. Zeller, R.O. Pohl: Phys. Rev. B **4**, 2029–2041 (1971)
477. W.A. Phillips: Rep. Prog. Phys. **50**, 1657–1708 (1987)
478. P.W. Anderson, B.I. Halperin, C.M. Varma: Phil. Mag. **25**, 1–9 (1972)
479. W.A. Phillips: J. Low Temp. Phys. **7**, 351–360 (1972)
480. S. Hunklinger, A.K. Raychaudhuri: Thermal and Elastic Anomalies in Glasses at Low Temperatures. In: *Prog. Low Temp. Phys*, Vol. 9, ed. by D.F. Brewer (Elsevier, Amsterdam 1986) pp. 267–344
481. J.L. Black: Phys. Rev. B **17**, 2740–2761 (1978)
482. J.F. Berret, M. Meißner: Z. Phys. B **70**, 65–72 (1988)
483. J. Classen, C. Enss, C. Bechinger, G. Weiss, S Hunklinger: Ann. Physik **3**, 315–335 (1994)
484. P. Esquinazi (Ed.): *Tunneling Systems in Amorphous and Crystalline Solids* (Springer, Berlin 1998)
485. P. Strehlow, C. Enss, S. Hunklinger: Phys. Rev. Lett. **80**, 5361–5364 (1998)
486. A.L. Burin, Yu. Kagan: Physica B **194–196**, 393–394 (1993)
487. M. Deye, P. Esquinazi: Z. Phys. B **76**, 283–288 (1989)
488. R.M. Ernst, L. Wu, S.R. Nagel: Phys. Rev. B **38**, 6246–6256 (1988)
489. J.J. De Yoreo, W. Knaak, M. Meissner, R.O. Pohl: Phys. Rev. B **34**, 8828–8842 (1986)
490. W. Knaak, M. Meißner: Thermal Relaxation Phenomena in $(KBR)_{1-x}(CKN)_x$ Single Crystals at Low Temperatures. In: *Disordered Systems and New Materials*, ed. by M. Borissov, N. Kirov, A. Vavrek (World Scientific, Varna 1989)

491. S. Bhattacharya, S.R. Nagel, L. Fleishman, S. Susman: Phys. Rev. Lett. **48**, 1267–1270 (1982); N.O. Birge, Y.H. Jeong, S.R. Nagel, S. Bhattacharya, S. Susman: Phys. Rev. B **30**, 2306–2308 (1984)
492. J.P. Sethna: Ann. New York Acad. Sci. **484**, 130–149 (1986)
493. G.P. Johari: Phys. Rev. B **33**, 7201–7204 (1986)
494. D.E. Day: J. Non-Cryst. Solids **21**, 343–372 (1976)
495. M.D. Ingram: Glastechn. Ber. Glass Sci. Technol. **67**, 151–155 (1994)
496. J.E. Shelby: J. Non-Cryst. Solids **263** & **264**, 271–276 (2000)
497. B. Roling, M.D. Ingram: J. Non-Cryst. Solids **265**, 113–119 (2000)
498. G. Berg, A. Ludwig: J. Non-Cryst. Solids **170**, 109–111 (1994)
499. P. Maass, A. Bunde, M.D. Ingram: Phys. Rev. Lett. **68**, 3064–3067 (1992); P. Maass, M. Meyer, A. Bunde, W. Dietrich: Phys. Rev. Lett. **77**, 1528–1531 (1996)
500. P. Maass: J. Non-Cryst. Solids **255**, 35–46 (1999)
501. W.C. LaCourse: J. Non-Cryst. Solids **95–96**, 905–912 (1987)
502. C.T. Moynihan, A.J. Easteal, D.C. Tran, J.A. Wilder, E.P. Donovan: J. Am. Ceram. Soc. **59**, 137–140 (1976)
503. Y. Hiwatari, J. Habasaki: J. Phys. Condens. Matter **12**, 6405–6412 (2000)
504. B. Roling, A. Happe, M.D. Ingram: J. Phys. Chem. B **103**, 4122–4127 (1999)
505. J. Habasaki, Y. Hiwatari: Phys. Rev. E **58**, 5111–5114 (1998)
506. G.N. Greaves, K.L. Ngai: Phys. Rev. B **52**, 6358–6380 (1995)
507. C.A. Bero: *Universal creep behavior of amorphous materials near their glass temperature*, PhD Thesis (University of Pittsburgh 1994)
508. E.N. da C. Andrade: Proc. R. Soc. Lond. A **84**, 1–12 (1910)
509. E.N. da C. Andrade, K.H. Jolliffe: Proc. R. Soc. Lond. A **254**, 291–316 (1960)
510. G.S. Cohen, M.H. Grest: Liquids, Glasses, and the Transition: A Free-Volume Approach. In: *Advances in Chemical Physics*, ed. by I. Prigogine, S.A. Rice (Wiley, New York 1981) pp. 455–525
511. A. Schönhals, E. Donth: Phys. Stat. Sol. (b) **124**, 515–524 (1984)
512. L. Boehm, M.D. Ingram, C.A. Angell: J. Non-Cryst. Solids **44**, 305–313 (1981)
513. S. Kamath, R.H. Colby, S.K. Kumar, K. Karatasos, G. Floudas, G. Fytas, J.E.L. Roovers: J. Chem. Phys. **11**, 6121–2126 (1999)
514. J. Souletie: J. Phys. I France **51**, 883–898 (1990); J. Souletie, D. Bertrand: J. Phys. I France **1**, 1627–1637 (1991)
515. R.H. Colby: Phys. Rev. E **61**, 1783–1792 (2000)
516. R.L. Leheny, S.R. Nagel: Europhys. Lett. **39**, 447–452 (1997)
517. P. Lunkenheimer, A. Pimenov, B. Schiener, R. Böhmer, A. Loidl: Europhys. Lett. **33**, 611–616 (1996)
518. K. Schröter, E. Donth: J. Chem. Phys. **113**, 9101–9108 (2000)
519. E. Rössler: private communication (2000)
520. U. Schneider, R. Brand, P. Lunkenheimer, A. Loidl: Phys. Rev. Lett. **84**, 5560–5563 (2000)
521. A. Kudlik, T. Blochowicz, S. Benkhof, E. Rössler: Europhys. Lett. **36**, 475–476 (1996)
522. J. Korus: PhD. Thesis (Halle University 1997)
523. K. Schneider, A. Schönhals, E. Donth: Acta Polymer. **32**, 471–475 (1981); A. Hensel: PhD. Thesis (University of Rostock 1998)
524. W. Vogel: *Glaschemie* (VEB Deutscher Verlag für Grundstoffindustrie, Leipzig 1979)
525. D. Turnbull, J.C. Fisher: J. Chem. Phys. **17**, 71–73 (1949)
526. D.R. Uhlmann, H. Yinnou: Glass: *Science and Technology*, Vol. 1 (Academic Press, San Diego 1983) p. 1

527. P.F. James: *Nucleation and Crystallization in Glasses*, Vol. 4 of *Advances in Ceramics*, ed. by J.H. Simmons, D.R. Uhlmann, G.H. Beall (Columbus, Ohio 1982) pp. 1–48
528. T.L. Hill: *Thermodynamics of Small Systems* (W.A. Benjamin Inc., New York 1963)
529. K.L. Ngai, J.H. Magill, D.J. Plazek: J. Chem. Phys. **112**, 1887–1892 (2000)
530. See [313], Sect. 112
531. J. Thoen: *Lecture at the Lähnwitz seminar* (Kühlungsborn 2000); J. Caerels, C. Glorieux, J. Thoen: Rev. Sci. Instr. **69**, 2452–2458 (1998)
532. O. Yamamuro, S. Takahara, A. Inaba, T. Matsuo, H. Suga: J. Phys. Condens. Matter **6**, L169–L172 (1994)
533. E. Hempel et al.: to be published
534. M. Mezard, G. Parisi: J. Phys. Condens. Matter **12**, 6655–6673 (2000)
535. K. Kawasaki: J. Phys. Condens. Matter **12**, 6343–6351 (2000)
536. E. Hempel, M. Beiner, T. Renner, E. Donth: Acta Polymer. **47**, 525–529 (1996)
537. T.R. Kirkpatrick, D. Thirumalai: Phys. Rev. Lett. **58**, 2091–2094 (1987); T.R. Kirkpatrick, P.G. Wolynes: Phys. Rev. A **35**, 3072–3080 (1987)
538. J. Souletie: J. Phys. I France **51**, 883–898 (1990)
539. P.K. Gupta, C.T. Moynihan: J. Chem. Phys. **65**, 4136–4140 (1976)
540. S. Franz, G. Parisi: J. Phys. I France **5**, 1401–1415 (1995)
541. A.J. Kovacs, J.M. Hutchinson, J.J. Aklonis: In: *The Structure of Non-Crystalline Materials*, ed. by P.H. Gaskell (Taylor and Francis, London 1977) pp. 153–163
542. A.J. Kovacs, J.J. Aklonis, J.M. Hutchinson, A.R. Ramos: J. Polym. Sci., Polym. Phys. Ed. **17**, 1097–1162 (1979)
543. S.F. Edwards, P.W. Anderson: J. Phys. F: Met. Phys. **5**, 965–974 (1975)
544. M. Mézard, G. Parisi: J. Phys. A Math. Gen. **29**, 6515–6524 (1996)
545. D. Kivelson, S.A. Kivelson, X. Zhao, Z. Nussinov, G. Tarjus: Physica A **219**, 27–38 (1995)
546. M. Mezard, G. Parisi: J. Phys. I France **1**, 809–836 (1991)
547. M. Mezard, G. Parisi: Phys. Rev. Lett. **82**, 747–750 (1999)
548. X. Xia, P.G. Wolynes: Proc. Natl. Acad. Sci. **97**, 2990–2994 (2000)
549. T.R. Kirkpatrick, D. Thirumalai: Phys. Rev. Lett. **58**, 2091–2094 (1987); Phys. Rev. B **36**, 5388–5397 (1987); T.R. Kirkpatrick, D. Thirumalai, P.G. Wolynes: Phys. Rev. A **40**, 1045–1054 (1989)
550. D.J. Gross, M. Mezard: Nucl. Phys. B **240**, 431–452 (1984)
551. J.-P. Bouchaud, M. Mezard: J. Phys. I (France) **4**, 1109–1114 (1994); E. Marinari, G. Parisi, F. Ritort: J. Phys. A **27**, 7615–7645 (1994); ibid. 7647–7668
552. B. Coluzzi, G. Parisi, P. Verrocchio: Phys. Rev. Lett. **84**, 306–309 (2000)
553. D.J. Gross, I. Kanter, H. Sompolinsky: Phys. Rev. Lett. **55**, 304–307 (1985)
554. M. Mezard, G. Parisi, M.A. Virasoro: *Spin Glass Theory and Beyond* (World Scientific, Singapore 1987)
555. E. Gardner: Nucl. Phys. B **257**, 747–765 (1985)
556. A. Crisanti, F. Ritort: Physica A **280**, 155–160 (2000)
557. T.R. Kirkpatrick, P.G. Wolynes: Phys. Rev. B **36**, 8552–8564 (1987)
558. A. Coniglio, A. de Candia, A. Fierro, M. Nicodemi: J. Phys. Condens. Matter **11**, A167–A174 (1999)
559. S. Franz, G. Parisi: Physica A **261**, 317–339 (1998)
560. H. Tanaka: Phys. Rev. Lett. **80**, 5750–5753 (1998)
561. W. Kob, H.C. Andersen: Phys. Rev. Lett. **73**, 1376–1379 (1994)
562. F. Sciortino, W. Kob, P. Tartaglia: Phys. Rev. Lett. **83**, 3214–3217 (1999)
563. K. Binder, A.P. Young: Rev. Mod. Phys. **58**, 801–976 (1986)
564. E. Donth: Z. Physik. Chem. **234**, 235–257 (1967)

565. K.S. Pitzer: J. Chem. Phys. **7**, 583–590 (1939)
566. M. Kleman, J.F. Sadoc: J. Phys. (France) Lett. **40**, L569–L574 (1979)
567. N. Rivier: Structure of Glasses from a Topological Viewpoint. In: *Structure of Non-Crystalline Materials* (Taylor and Francis, Cambridge 1983)
568. T. Tomida, T. Egami: Phys. Rev. B **52**, 3290–3308 (1995)
569. E.W. Fischer, A.S. Bakai: In: *Slow Dynamics in Complex Systems*, ed. by M. Tokuyama, I. Oppenheim, AIP Conf. Proc. **469**, 325–338 (1999)
570. E. McLanghlin, A.R. Ubbelohde: Trans. Faraday Soc. **54**, 1804–1810 (1958)
571. D. Kivelson, G. Tarjus: J. Non-Cryst. Solids **235–237**, 86–100 (1998)
572. D. Kivelson, G. Tarjus: Phil. Mag. B **77**, 245–256 (1998)
573. G. Tarjus, D. Kivelson, P. Viot: J. Phys. Condens. Matter **12**, 6497–6508 (2000)
574. P. Viot, G. Tarjus, D. Kivelson: J. Chem. Phys. **112**, 10368–10378 (2000)
575. S.S. Chang, A.B. Bestul: J. Chem. Phys. **56**, 503–516 (1972)
576. E. Donth: J. Phys. I France **6**, 1189–1202 (1996)
577. I. Prigogine, R. Defay: *Chemical Thermodynamics* (Longmans, Green and Co., New York 1954) Chap. 19
578. R.O. Davies, G.O. Jones: Adv. Phys. **2**, 370–410 (1953)
579. A.V. Lesikar, C.T. Moynihan: J. Chem. Phys. **73**, 1932–1939 (1980)
580. J. Jäckle: J. Chem. Phys. **79**, 4463–4467 (1983)
581. J.M. O'Reilly: J. Polym. Sci. **57**, 429–444 (1962); CRC Crit. Rev. Macromol. Sci. **13**, 259–277 (1987)
582. T.M. Nieuwenhuizen: Phys. Rev. Lett. **79**, 1317–1320 (1997)
583. R. Brand, P. Lunkenheimer, U. Schneider, A. Loidl: Phys. Rev, Lett. **82**, 1951–1954 (1999)
584. E. Rössler: private communication (1999)
585. C.A. Angell: J. Phys. Chem. **70**, 2793–2803 (1966)
586. O. Mishima, H.E. Stanley: Nature **396**, 329–335 (1998)
587. K. Ito, C.T. Moynihan, C.A. Angell: Nature **398**, 492–495 (1999)
588. L. Angelani, G. Parisi, G. Ruocco, G. Viliani: Phys. Rev. E **61**, 1681–1691 (2000)
589. H. Murakami, T. Kushida, H. Tashiro: J. Chem. Phys. **108**, 10309–10318 (1998)
590. M. Beiner, S. Kahle, S. Abens, E. Hempel, S. Höring, M. Meissner, E. Donth: to be published
591. K. Schmidt-Rohr, A.S. Kudlik, H.W. Beckham, A. Ohlemacher, U. Pawelzik, C. Boeffel, H.W. Spiess: Macromolecules **27**, 4733–4745 (1994)
592. M. Vogel, E. Rössler: private communication (2000); M. Vogel: PhD Thesis (University of Bayreuth 2000); M. Vogel, E. Rössler: J. Phys. Chem. B **104**, 4285–4287 (2000)
593. G.P. Johari, E. Whalley: Faraday Symp. Chem. Soc. **6**, 23–47 (1972)
594. C. Hansen, R. Richert: Acta Polymer. **48**, 484–489 (1997)
595. H. Suga, S. Seki: J. Non-Cryst. Solids **16**, 171–194 (1974)
596. M. Beiner, H. Huth, K. Schröter: J. Non-Cryst. Solids **279**, 126–135 (2001)
597. A.N. Kolmogorov: *Osnovnye ponyatiya teorii veroyatnostei* (Nauka, Moskau 1974) p. 16
598. B.W. Gnedenko: *Lehrbuch der Wahrscheinlichkeitstheorie*, 10th edn. (Deutsch, Frankfurt, 2nd edn. 1997)
599. G. Katana, E.W. Fischer, Th. Hack, V. Abetz, F. Kremer: Macromolecules **28**, 2714–2722 (1995)
600. T. Christensen, N.B. Olsen: J. Non-Cryst. Solids **235–237**, 296–301 (1998)
601. T. Christensen: private communication (2000)
602. H.B. Callen: *Thermodynamics* (Wiley, New York 1960)

603. J.K. Nielsen: Phys. Rev. E **60**, 471–481 (1999)
604. J.K. Nielsen, J.C. Dyre: Phys. Rev. B **54**, 15754–15761 (1996)
605. A.C. Pipkin: *Lectures on Viscoelasticity Theory. Applied Mathematical Sciences*, Vol. 7 (Springer, New York 1972)
606. D. Forster: *Hydrodynamic Fluctuations, Broken Symmetry, and Correlation Functions* (W.A. Benjamin Inc., London 1975)
607. K. Bennewitz, H. Rötger: Phys. Z. **40**, 416–428 (1939); H. Rötger: Silikattechnik **10**, 57–62 (1959)
608. H. Bateman, A. Erdelyi et al.: *Tables of Integral Transforms*, Vol. II. (McCraw-Hill, New York 1954) Chap. XV
609. J. Jäckle, H.L. Frisch: J. Polym. Sci. Polym. Phys. Ed. **23**, 675–682 (1985); J. Jäckle: Z. Physik B **64**, 41–54 (1986)
610. G. Kluge, S. Bark-Zollmann: Wiss. Z. Univ. Jena, Naturwissenschaftliche Reihe **39**, 81–84 (1990); G. Kluge: private communication (1990)
611. S.W. Provencher: Comput. Phys. Comm. **27**, 213–227 (1982); J. Honerkamp, J. Weese: Macromolecules **22**, 4372–4377 (1989); S.-L. Nyeo, B. Chu: Macromolecules **22**, 3998–4009 (1989)
612. K. Schröter: private communication (2000)
613. J.E. McKinney, H.V. Belcher: J. Res. Nat. Bur. Stand. **A67**, 43–53 (1963); C.H. Wang, B.Y. Li, R.W. Rendell, K.L. Ngai: J. Non-Cryst. Solids **131–133**, 870–876 (1991)
614. E.H. Lee: Quart. Appl. Math. **13**, 183–190 (1955)
615. E. Donth: J. Phys. Condens. Matter **12**, 10371–10388 (2000)
616. L. van Hove: Phys. Rev. **95**, 249–262 (1954)
617. E. Donth, H.E. Hempel, C. Schick: J. Phys. Condens. Matter **12**, L281–L286 (2000)
618. E. Donth, H. Huth, M. Beiner: to be published
619. B.V. Gnedenko: *Kurs Teorii Veroyatnostej*, 3rd edn. (Gosud. Izd., Moscow 1961)
620. See [598], p. 125
621. G. Williams: Chem. Rev. **72**, 55–69 (1972)
622. E. Burkel: Rep. Prog. Phys. **63**, 171–232 (2000)
623. H.-J. Rossberg: Addendum to B.W. Gnedenko. In: *Lehrbuch der Wahrscheinlichkeitstheorie* 10th edn. (Harry Deutsch, Frankfurt/M. 1997), pp. 365–368
624. L. Szilard: Z. Phys. **32**, 753–788 (1925); **53**, 840–856 (1929)
625. J. Bagott: In: *The Meaning of Quantum Theory* (University Press, Oxford 1992)
626. L. Onsager: Phys. Rev. **37**, 405–426 (1931)
627. Yu.L. Klimontovich: Uspechi. Fiz. Nauk. **151**, 309–332 (1987) (in Russian)
628. V.L. Ginzburg, L.P. Pitaevskij: ibid **151**, 333–339 (1987) (in Russian)
629. E. Hempel, S. Kahle, R. Unger, E. Donth: Thermochim. Acta **329**, 97–108 (1999)
630. E.W. Montroll, J.T. Bendler: J. Stat. Phys. **34**, 129–162 (1984)
631. M.F. Shlesinger: Ann. Rev. Phys. Chem. **39**, 269–290 (1988)
632. D. Holt, E.L. Crow: J. Res. Nat. Bur. Stand. B. Math. Sci. **77B**, 143–198 (1973)
633. F.I. Mopsik: Rev. Sci. Instrum. **55**, 79–87 (1984); J. Res. Natl. Inst. Stand. Technol. **104**, 180–192 (1999)
634. D. Lellinger: private communication (2000)
635. C. Burger: PhD Thesis (University of Marburg 1994)
636. H.-J. Rossberg, B. Jesiak, G. Siegel: *Analytic Methods of Probability Theory* (Akademie-Verlag, Berlin 1985)
637. M. Fuchs: J. Non-Cryst. Solids **172–174**, 241–247 (1994)

638. F. Alvarez, A. Alegria, J. Colmenero: Phys. Rev. B **47**, 125–130 (1993)
639. M. Fuchs: private communication (1997)
640. W. Götze Z. Phys. B **60**, 195–203 (1985)
641. W. Götze, L. Sjögren: The Mode Coupling Theory of Structural Relaxations. In: *Transport theory and statistical physics.* Special issue devoted to relaxation kinetics in supercooled liquid-mode coupling theory and its experimental tests, ed. by P. Nelson (Dekker, New York 1995)
642. W. Götze, L. Sjögren: Chem. Phys. **212**, 47–59 (1996)
643. *Encyclopedia of Mathematics*, ed. by M. Hazewinkel (Reidel, Dordrecht 1988) Vol. 7, p. 132
644. H.Z. Cummins, G. Li, W.M. Du, J. Hernandez: Physica A **204**, 169–201 (1994)
645. J. Heijboer: In: *Static and Dynamic Properties of the Polymeric Solid State*, ed. by R.A. Pethrick, R.W. Richards (Reidel, London 1982) p. 197. See also Ann. New York Acad. Sci. **279**, 104 (1976)
646. H. Eyring: J. Chem. Phys. **3**, 107–115 (1935); H. Eyring, E.M. Eyring: *Modern Chemical Kinetics* (Van Nostrand Reinhold, New York 1965)
647. H.A. Kramers: Physica **7**, 284–304 (1940)
648. J. Koppelmann: Progr. Colloid Polym. Sci. **66**, 235–251 (1979)
649. E. Donth, S. Kahle, J. Korus, M. Beiner: J. Phys. I France **7**, 581–598 (1997)
650. L. von Ranke: *Meisterwerke*, Vol. 2 (Duncker & Humblot, Munich 1994) p. 213
651. E. Kamke: *Differentialgleichungen, Lösungsmethoden und Lösungen*, Bd.II. (Akad. Verlagsgesellschaft, Leipzig 1959) Sect. 10.3
652. S. Vieweg: private communication (1997)
653. J.M. Hutchinson: private communication (2000)
654. J.M. Hutchinson: to be published
655. I.M. Hodge: Macromolecules **20**, 2897–2908 (1987)
656. A. Brunacci, J.M.G. Cowie, R. Ferguson, J.L. Gomez Ribelles, A.V. Garayo: Macromolecules **29**, 7976–7988 (1996)
657. S. Kahle, E. Hempel, M. Beiner, R. Unger, K. Schröter, E. Donth: J. Mol. Struct. **479**, 149–162 (1999)
658. E. Hempel, S. Kahle, R. Unger, E. Donth: Thermochim. Acta **329**, 97–108 (1999)
659. S. Montserrat, J.L. Gomez Ribelles, J.M. Meseguer: Polymer **39**, 3801–3807 (1998)
660. R. Di Leonardo, L. Angelani, G. Parisi, G. Ruocco: Phys. Rev. Lett. **84**, 6054–6057 (2000)
661. G. Parisi: Phys. Rev. Lett. **79**, 3660–3663 (1997)
662. L.F. Cugliandolo, J. Kurchan: Phys. Rev. Lett. **71**, 173–176 (1993)
663. J.-L. Barrat,W. Kob: Europhys. Lett. **46**, 637–642 (1999)
664. T.S. Grigera, N.E. Israeloff: Phys. Rev. Lett. **83**, 5038–5041 (1999)
665. R. Exartier, L. Peliti: Eur. Phys. J. B **16**, 119–126 (2000)
666. G.B. McKenna, M.G. Vangel, A.L. Rukhin, S.D. Leigh, B. Lotz, C. Straupe: Polymer **40**, 5183–5205 (1999)
667. L.C.E. Struik: Polymer **38**, 5243–5245 (1997)
668. W. Zippold, R. Kühn, H. Horner: Eur. Phys. J. B **13**, 531–537 (2000)
669. J.P. Bouchaud: In: *Soft and Fragile Matter: Nonequilibrium Dynamics, Metastability, and Flow*, ed. by M.E. Cates, M.R. Evans (IOP Publ., Bristol 2000) pp. 285–304
670. J.M. Hutchinson, S. Smith, B. Horne, G.M. Gourlay: Macromolecules **32**, 5046–5061 (1999)
671. F.R. Schwarzl, G. Link, F. Greiner, F. Zahradnik: Progr. Coll. Polym. Sci. **71**, 180–190 (1985)
672. B.E. Read: Polymer **22**, 1580–1586 (1981)

673. G.P. Johari: J. Chem. Phys. **77**, 4619–4626 (1982); K. Pathmanathan, J.Y. Cavaille, G.P. Johari: J. Polym. Sci. **B27**, 1519–1527 (1989)
674. W. Kob, J.-L. Barrat, F. Sciortino, P. Tartaglia: J. Phys. Condens. Matter **12**, 6385–6394 (2000)
675. *Polymer Handbook*, ed. by J. Brandrup, E.H. Immergut (Wiley, New York 1989)
676. O.V. Mazurin, M.V. Strelsina, J.P. Shvaiko-Shvaikovskaya (Eds.): *Handbook of Glass Data. Elsevier, Amsterdam. Part A: Vitreous Silica and Binary Silicate Glasses* (1983); *Part B: Single-Component and Binary Non-Silicate Oxide Glasses* (1985); *Part C: Ternary Silicate Glasses* (1987)
677. C.A. Angell, J.M. Sare, E.J. Sare: J. Phys. Chem. **82**, 2622–2629 (1978)
678. S. Sakka, J.D. Mackenzie: J. Non-Cryst. Solids **6**, 145–162 (1971)
679. J.C. Phillips: J. Non-Cryst. Solids **41**, 179–187 (1980)
680. J.B. Enns, R. Simha: J. Macromol. Sci. Phys. **13**, 25–47 (1977)
681. R. Popli, M. Glotin, L. Mandelkern, R.S. Benson: J. Polym. Sci., Polym. Phys. Ed. **22**, 407–448 (1984)
682. B.D. Fitz, J. Mijovic: Macromolecules **32**, 3518–3527 (1999)
683. E. Donth, R. Conrad: Acta Polymer. **31**, 47–51 (1980)
684. F. Stickel: PhD Thesis, Universität Mainz, Shaker (1995)
685. R. Richert, C.A. Angell: J. Chem. Phys. **108**, 9016–9026 (1998)
686. F. Stickel, E.W. Fischer, R. Richert: J. Chem. Phys. **104**, 2043–2055 (1996)
687. F. Garwe, A. Schönhals, H. Lockwenz, M. Beiner, K. Schröter, E. Donth: Macromolecules **29**, 247–253 (1996)
688. S. Kahle, J. Korus, E. Hempel, R. Unger, S. Höring, K. Schröter, E. Donth: Macromolecules **30**, 7214–7223 (1997)
689. M. Beiner, J. Korus, H. Lockwenz, K. Schröter, E. Donth: Macromolecules **29**, 5183–5189 (1996)
690. R. Nozaki, D. Suzuki, S. Ozawa, Y. Shiozaki: J. Non-Cryst. Solids **235–237**, 393–398 (1998)
691. Y.H. Jeong, I.K. Moon: Phys. Rev. B **52**, 6381–6385 (1995)
692. S. Kahle, K. Schröter, E. Hempel, E. Donth: J. Chem. Phys. **111**, 6462–6470 (1999)
693. A. Hensel: PhD Thesis, University of Rostock (1998)
694. P.K. Gupta, C.T. Moynihan: J. Chem. Phys. **65**, 4136–4140 (1976)
695. J.M. O'Reilly: J. Polym. Sci., Polym. Symp. 165–172 (1978)
696. J.M. O'Reilly: J. Appl. Phys. **48**, 4043–4048 (1977)
697. K. Samwer, R. Busch, W.L. Johnson: Phys. Rev. Lett. **82**, 580–583 (1999)
698. O. Yamamuro, Y. Oishi, M. Nishizawa, T. Matsuo: J. Non-Cryst. Solids **235–237**, 517–521 (1998)
699. S. Takahara, O. Yamamuro, H. Suga: J. Non-Cryst. Solids **171**, 259–270 (1994)
700. S.S. Chang, A.B. Bestul: J. Chem. Thermodyn. **6**, 325–344 (1974)
701. *Polymer Handbook*, ed. by J. Brandrup, E.H. Immergut (Wiley, New York 1989)
702. K. Schneider, A. Schönhals, E. Donth: Acta Polymer. **32**, 471–475 (1981)
703. J.M. O'Reilly: J. Polym. Sci. **57**, 429–444 (1962)
704. *Disorder Effects on Relaxational Processes. Glasses, Polymers, Proteins* (Bad Honnef, Germany, April 1992), ed. by R. Richert, A. Blumen (Springer, Berlin 1994)
705. H. Huth, B. Beiner et al.: to be published
706. E. Donth, M. Beiner: unpublished (2000)
707. M. Descamps, V. Legrand, Y. Guinet, A. Amazzal, C. Alba, J. Dore: Progr. Theoret. Phys. Jpn. Suppl. **126**, 207–212 (1999)

708. W. Petry, E. Bartsch, F. Fujara, M. Kiebel, H. Sillescu, B. Farago: Z. Phys. B **83**, 175–184 (1991)
709. J. Wong, C.A. Angell: *Glass: Structure by Spectroscopy* (Dekker, New York 1976)
710. H.C. Andersen, G.H. Fredickson: Ann. Rev. Phys. Chem. **39**, 149–180 (1988)
711. E. Williams, C.A. Angell: J. Phys. Chem. **81**, 232–237 (1977)
712. R.D. Corsaro: Phys. Chem. Glasses **17**, 13–22 (1976)
713. D. Reichert, H. Schneider: private communication (2001)
714. S. Takahara, M. Ishikawa, O. Yamamuro, T. Matsuo: J. Phys. Chem. **103**, 792–796 (1999)
715. T.L. Hill: *Statistical Mechanics* (McGraw–Hill, New York 1956; Dover, New York 1987)
716. L.D. Landau, E.M. Lifshitz: Course of Theoretical Physics, Vol. 3, *Quantum Mechanics* 3rd edn. (Butterworth–Heinemann, Oxford 1997)
717. H. Frauenfelder, G.U. Nienhaus, R.D. Young: 'Relaxation and disorder in proteins', in [704], 591–614
718. N.G. McCrum: Polym. Comm. **25**, 2–4 (1984)
719. V.R. Privalko: 'Svojstva Polymerov v Blochnom Sostoyanii', in *Spravochnik po Fizicheskoj Chimii Polymerov* Vol. 2, ed. by Yu.S. Lipatov et al. (Naukova Dumka, Kiev 1984, 1985) (in Russian)

Index

Springer Series in
MATERIALS SCIENCE

Editors: R. Hull   R. M. Osgood, Jr.   H. Sakaki   A. Zunger

1 **Chemical Processing with Lasers***
By D. Bäuerle

2 **Laser-Beam Interactions with Materials**
Physical Principles and Applications
By M. von Allmen and A. Blatter
2nd Edition

3 **Laser Processing of Thin Films and Microstructures**
Oxidation, Deposition and Etching of Insulators
By. I. W. Boyd

4 **Microclusters**
Editors: S. Sugano, Y. Nishina, and S. Ohnishi

5 **Graphite Fibers and Filaments**
By M. S. Dresselhaus, G. Dresselhaus, K. Sugihara, I. L. Spain, and H. A. Goldberg

6 **Elemental and Molecular Clusters**
Editors: G. Benedek, T. P. Martin, and G. Pacchioni

7 **Molecular Beam Epitaxy**
Fundamentals and Current Status
By M. A. Herman and H. Sitter 2nd Edition

8 **Physical Chemistry of, in and on Silicon**
By G. F. Cerofolini and L. Meda

9 **Tritium and Helium-3 in Metals**
By R. Lässer

10 **Computer Simulation of Ion-Solid Interactions**
By W. Eckstein

11 **Mechanisms of High Temperature Superconductivity**
Editors: H. Kamimura and A. Oshiyama

12 **Dislocation Dynamics and Plasticity**
By T. Suzuki, S. Takeuchi, and H. Yoshinaga

13 **Semiconductor Silicon**
Materials Science and Technology
Editors: G. Harbeke and M. J. Schulz

14 **Graphite Intercalation Compounds I**
Structure and Dynamics
Editors: H. Zabel and S. A. Solin

15 **Crystal Chemistry of High-T_c Superconducting Copper Oxides**
By B. Raveau, C. Michel, M. Hervieu, and D. Groult

16 **Hydrogen in Semiconductors**
By S. J. Pearton, M. Stavola, and J. W. Corbett

17 **Ordering at Surfaces and Interfaces**
Editors: A. Yoshimori, T. Shinjo, and H. Watanabe

18 **Graphite Intercalation Compounds II**
Editors: S. A. Solin and H. Zabel

19 **Laser-Assisted Microtechnology**
By S. M. Metev and V. P. Veiko
2nd Edition

20 **Microcluster Physics**
By S. Sugano and H. Koizumi
2nd Edition

21 **The Metal-Hydrogen System**
By Y. Fukai

22 **Ion Implantation in Diamond, Graphite and Related Materials**
By M. S. Dresselhaus and R. Kalish

23 **The Real Structure of High-T_c Superconductors**
Editor: V. Sh. Shekhtman

24 **Metal Impurities in Silicon-Device Fabrication**
By K. Graff 2nd Edition

25 **Optical Properties of Metal Clusters**
By U. Kreibig and M. Vollmer

26 **Gas Source Molecular Beam Epitaxy**
Growth and Properties of Phosphorus Containing III–V Heterostructures
By M. B. Panish and H. Temkin

MIX
Papier aus verantwortungsvollen Quellen
Paper from responsible sources
FSC® C105338

If you have any concerns about our products,
you can contact us on
ProductSafety@springernature.com

In case Publisher is established outside the EU,
the EU authorized representative is:
Springer Nature Customer Service Center GmbH
Europaplatz 3, 69115 Heidelberg, Germany

Printed by Libri Plureos GmbH
in Hamburg, Germany